BIHAR AGRICULTURE

ABOUT THE AUTHOR

P.C. Bansil heads the Techno Economic Research Institute, New Delhi, founded by him in 1984. He belongs to the first batch of Indian Economic Service and has worked with the FAO for about 15 years in Sri Lanka, Thailand, Rome, Zambia and Libya. He has authored 30 books, besides a large number of reports and papers prepared for various Governments. He has published over 400 papers in specialized technical journals. His work experience relates to the formulation, implementation, evaluation and formulation of regional plans for agriculture, statistics, marketing, livestock, irrigation, fisheries, forestry, wildlife, national parks, tourism, agro-industrial and social sectors like demography, health, education and housing, as well as project identification, formulation and evaluation in these fields.

Dr. Bansil was appointed by the Planning Commission as Chairman of the Tenth and Eleventh Plans Working Group on Animal Husbandry Economics and Statistics. He completed the study "Livestock in India 2000-2030" sponsored by the Food and Agriculture Organisation of the United Nations. During his FAO assignments in Sri Lanka, Zambia and Libya as Senior Planning Advisor, he undertook micro level studies on issues relating to water.

BIHAR AGRICULTURE
A Perspective

P.C. Bansil

Director, Techno Economic Research
Institute, New Delhi

CONCEPT PUBLISHING COMPANY PVT. LTD.,
NEW DELHI-110059

ISBN-13: 978-81-8069-743-2

First Published 2011

@ P.C. Bansil (*b.* 1919)

Published and Printed by
Ashok Kumar Mittal

Concept Publishing Company Pvt. Ltd.
Regd. Office:
A/15-16, Commercial Block, Mohan Garden,
New Delhi-110059 (INDIA)
Phones : 25351460, 25351794, *Fax* : 091-11-25357109
Email : publishing@conceptpub.com,
Website: www.conceptpub.com

Editorial Office:
H-13, Bali Nagar, New Delhi-110 015, India.

Cataloging in Publication Data--*Courtesy:* D.K. Agencies (P) Ltd. <docinfo@dkagencies.com>

Bansil, P. C.
Bihar agriculture : a perspective / P.C. Bansil.
p. cm.
Includes bibliographical references (p.) and index.
ISBN 9788180697432

1. Agriculture--India--Bihar. 2. Agriculture and state--India--Bihar. 3. Agriculture -Economic aspects--India--Bihar. I. Title.

DDC 338.10954123 22

R.L. BHATIA
Governor Bihar

RAJ BHAVAN
PATNA-800022
3-11-2008

FOREWORD

The present book entitled "Bihar Agriculture: A Perspective" written by Dr. P.C. Bansil, Director, Techno Economic Research Institute depicts the agricultural development issues of Bihar.

Our country is talking of the Second Green Revolution. This book has focused on Bihar and other eastern States which have the potential to emerge as the fertile crescent. It may be mentioned here that Bihar can really be the leader of the Second Green Revolution.

The author has recommended interlinking and desilting of rivers and hydro projects as a solution to ameliorate the damages from floods and rainwater.

I do hope, this book will be highly useful in introducing new technology among small/marginal farmers and also solve their marketing problems.

I appreciate the endeavour of Dr. P.C. Bansil for this valuable work and congratulate him for it.

R.L. Bhatia
(Governor, Bihar)

PREFACE

The food economy of India has gone through a number of phases since the late sixties when seeds of Green Revolution were sown. Having achieved self-sufficiency, the country has unmanageable stocks for which India ran short of storage capacity. The position today is that the Green Revolution States, Punjab and Haryana in particular, are suffering from technology fatigue and are unable to show any significant growth. There is a growing imbalance between the demand and the supply of cereals, pulses and edible oilseeds. The country is seriously talking about Second Green Revolution. This concept, however, has not taken any concrete shape.

It is in this context that the Institute (TERI) approached the Planning Commission for an in-depth study of the potentials of agriculture development in Bihar. This state which has immense agricultural resources, to be the leader in the Second Green Revolution, has so far failed to exploit those resources to the desired levels. Fifteen years into India's economic reforms, Bihar, country's third largest state, is finally gearing itself to catch up with the rest of India, according to a World Bank report. From being the first state to abolish the Agriculture Marketing Control (APMC) Act providing additional avenues to farmers from sale produce, repealing the Urban Land Ceiling Act and adopting the Model Municipal Act expected to attract organized construction and real estate firms to Bihar, to providing a Single Window Clearance Act 2006 and State Investment Promotion Board (SIPB) to facilitate Industrial Investments, Bihar has indeed come a long way.

Perspective Plan in the study is based on the theory of 'Look Within, Look Around and Look Beyond'. This means that we first study the potential within the state. We find that there are wide gaps in the productivity levels as between different districts under similar agro-climatic conditions and those between State level and at the research on the farmer's field in the State. The most important lesson for Bihar is to look around. We have quoted in the text specific success stories from some states like Rajasthan, Karnataka, Punjab, Haryana and

Maharashtra. Let the State Government start with an open mind, learn from the experience of these States, and try to adapt those found feasible to achieve higher goals in agriculture. There is also the need to avoid the pitfalls which have led to the decline of Green Revolution leaders — Punjab and Haryana. Bihar poised for ushering the Second Green Revolution should take all precautionary steps right now and not fritter away its valuable soil and water resources as has happened in Punjab and Haryana. It is interesting to observe that Bihar was more or less at the same level as Punjab in agricultural development till the late sixties and has today gone far behind. Factors responsible for development of Punjab agriculture and those of China and India (Under Looking Beyond) are carefully examined. Based on all this information, supply projections are made for agriculture and livestock products. This is accompanied by implementable recommendations. By looking at the national and international data, a case has also been made for this poorest state that agricultural development is directly related with Poverty Alleviation. In fact Bihar is one of the few states which is endowed with the best agro-climatic conditions plenty of water and good soil, trained labour and a forward looking Government. There is then every reason to believe that in the coming years, Bihar will have a place of pride and shine in its prestine glory.

Recent unprecedented Kosi floods have no doubt wrought havoc in Bihar and pessimists would certainly depict a gloomy picture for the future. It may, however, turn out to be a blessing in disguise as the problems of floods and their solutions are known. Perhaps, it needed an intensity of this nature to give a shock-therapy to all those who have been sleeping over the matter. It should provide both sides (Nepal and India) an opportunity for a much-delayed review of the Indo-Nepal water-sharing treaty. The quantum of destruction, loss of life and displacement now make it imperative that both sides work out a more permanent solution to the unpredictability of Kosi that has earned it the sobriquet 'river of sorrow'. The crisis will also affect labour migration patterns from Nepal to Bihar, bringing with it greater economic hardship for a country trying to find its feet after the regime change. Once remedial measures are taken to control the floods as discussed here, Bihar can certainly take its place among the developing States in the country.

New Delhi
May, 2009

P.C. Bansil

ACKNOWLEDGEMENTS

The Institute is grateful to the Planning Commission for giving it an opportunity to initiate the study on agricultural perspectives in Bihar. Since the study is based on secondary sources of data, all the possible data sources–various journals, books and reports–have been listed in the Bibliography. Wherever possible, appropriate acknowledgement has also been made. Thanks are due to all such eminent authors.

Mr. B.K. Taimini (Ex Food Secretary), Dr. M.L. Srivastav and Dr P.N. Radha Krishnan—Institute Faculty—took upon themselves the arduous job of going through the first draft—a very painful exercise. Their detailed comments and suggestions changed the shape of the book drastically. Thanks to them.

We also benefited a lot by having very fruitful discussions with Dr. T. Haque, the then Chairman, Commission for Agricultural Costs and Prices; Dr. S.C. Jha, Chairman, Bihar Task Force, Planning Commission and his Team: Dr. V.V. Sadamate, Advisor, Agriculture, Planning Commission; Dr. H.P. Singh. DDG (Horticulture), ICAR, Government of India; and Mr. Avinash Misra, Director, Water Resources Division, Planning Commission.

The guidance and help received from Institute Faculty members—Dr. Praduman Kumar, Dr. Khatana, Dr. Padam Singh, Dr. S.R. Mohnot and Prof. S.K. Gupta, are gratefully acknowledged. Their precious contributions helped in improving the quality of the study. The contribution of Institute's team – Mr. K.R. Dasgupta, Dr. S.S. Sirohi and Mr. C.P. Sharma, is highly appreciated and acknowledged. Last but certainly not the least, the total responsibility of providing office support fell on our two diligent staff—Mrs. Ajitha Gangatharan and Mrs. Vasantha R. Pillai who cheerfully helped in printing/reprinting of various draft reports from time to time.

The study presented here is the revised, updated and expanded version of the original report which was hardly one-fifth the present size. The report is based on several published/unpublished sources. Although every effort has been made to acknowledge the source wherever possible, yet it might have been left out unintentionally at a few places for which I would beg to be excused. The views expressed and conclusions arrived at are, however, the total responsibility of the author alone.

New Delhi
May, 2009

P.C. Bansil

CONTENTS

LIST OF GRAPHS

LIST OF TABLES

1

INTRODUCTION

The state of Bihar is bounded by Nepal in the North, Jharkhand in the South, West Bengal in the East and Uttar Pradesh in the West. Prior to November 15, 2000, Jharkhand also formed a part of Bihar. The reorganized State has an area of 94163 sq. kms. and a population of 83.0 million[1], growing at the rate of 2 per cent per annum. Following the creation of Jharkhand, Bihar was virtually bereft of an industrial sector. It was left with the smallest industrial base in the country, contributing only about 12 per cent toward the state's Gross State Domestic Product (GSDP), and employing less than 10 per cent of its workforce. The state shares about 2.9 per cent of India's total geographical area as against 8.07 per cent of the country's total population. The state has a density of 881 persons per sq. kilometre, one of the highest in India.[2]

It is a land-locked state in the eastern part of India and is situated between 83°-30' to 88°-00 longitude. The state is roughly quadrilateral in shape. Agriculture is the backbone of Bihar's economy, employing 77 per cent of the workforce and generating nearly 35 per cent of the State Domestic Product. The State has the natural endowment of fertile soil, good rainfall, plenty of water resources, and agro-climatic conditions suitable for growing three crops a year and almost all types of crops. Agriculture is not only the source of livelihood but it also generates raw material for the agro based industries which have immense potential in the state.

River Ganges divides the state into two parts—North Bihar (53.3 thousand sqr. kms.) and South Bihar (40.9 thousand sq. kms.). Seven river systems of Ghagra, Burhi Gandak, Bagmati, Kamla, Kosi and Mahananda originate in Tibet and Nepal and flow through North Bihar. Any rainfall occurring in Tibet and Nepal directly affects the flow in this river system. Flooding of North Bihar is thus a recurrent phenomenon. The six river system of Karmanasa, Sone, Punpun, Kiulharhar, Badue and Chandan originate in the Indian territories in the states of Bihar, Jharkhand and Chhatisgarh and flow through southern part of Bihar. All these rivers drain into the main Ganga system. Basic demographic, geographical and administrative data for Bihar is as in Table 1.1.

Table 1.1: Basic Demographic, Geographical and Administrative Data

Population in Millions—as per 2001 census

	Male	*Female*	*Total*
(i) Total population (Millions)	43.15	39.72	82.87
(ii) Rural population (Millions)	38.51	35.68	74.19
(iii) Urban population (Millions)	4.64	4.03	8.68
(iv) Density of population	Rural-803, Urban-4809, Total-880 persons per sq. km.		
(v) SC population (Millions)	9.98 (15.5% of the total population)		
(vi) ST population (Millions)	0.58 (0.9% of the total population)		
(vii) Literacy Rate:%	47.53		
Male%	60.32		
Female%	33.10		
(viii) BPL population%	42.56		
(ix) Per capita income (2004-05)	Rs. 5772/- (Current Prices)		
	Rural	*Urban*	*Total*
(x) Sex Ratio (F/000' M)	927	869	921
(xi) Literacy%	44.42%	72.91%	47.53%

Source: Census of India, Series 11, Bihar, Director of census operation, July 2001

Geographical Features

(i) Latitude	21°58'10" to 27°31'15"N
(ii) Longitude	83°15'15 to 88° 17'49"E
(iii) Height above Sea level	53 mtrs.
(iv) Average Rain fall	1271.90 mm
(v) Average no. of rainy days	53.9
(vi) Geographical Area	94163 sq.km.
(vii) Main rivers flowing through the State	Ganges, Gandak, Kosi, Bagmati,

Source: Bihar at a Glance, Directorate of Statistics and Evaluation, Bihar, Patna.

Administrative Units

(i) No. of Divisions	9
(ii) No. of Districts	38
(iii) No. of sub-divisions	101

(*Contd.*)

Map: Bihar before Nov., 2000

(*Contd.*)

Administrative Units	
(iv) No. of blocks	533
(v) No. of villages	45337
(vi) No. of villages electrified	20006
(vii) Villages connected by roads	14510
(viii) National Highways	2057 km.
(ix) State Highways	10442 km.
(x) Health Centers	1643
(xi) Panchayats	8471
(xii) No. of Urban Agglomeration	9
(xiii) No. of Towns	130
(xiv) Schools	56825
(xv) Universities	12

Source: Bihar at a Glance, 2004, Directorate of Statistics and Evaluation, Bihar, Patna.

Bihar Administrative Divisions

The modern state of Bihar, came into existence in 1956. It is bounded by Nepal on the North, West Bengal on the East, Uttar Pradesh in the West and Jharkhand on the South. The state has two distinct physiographic regions, viz. Sub-Himalayan area (North Bihar Plains) and Gangetic plains (South Bihar Plains), marked by the river Ganges which flows from west to east to join the Bay of Bengal. In November 2000 the state of Jharkhand was carved out of the state of Bihar by transfer of 13 districts to the new state (Map). The remaining 29 districts have been reorganized into 38 districts nine of them divided into 2 each (Table 1-2).

Table 1.2: Present details of the reorganization of districts.

Table 1.2 Reorganization of Districts in Bihar

Original District	New Districts after Reorganization
Bhojpur	Buxar
Gaya	Jehanabad
Jehanabad	Arwal
Munger	Lakhisarai, Sheikpura, Khagaria
Purnia	Araria, Kishanganij
Rohtas	Kaimur (Bhabua)
Saharsa	Supaul, Madhepura
Sitamarhi	Sheohar

Administratively, Bihar now has 9 divisions, 101 sub-divisions and 533 community development blocks. There are 130 towns, including 125 statutory towns and 5 census towns, and 45,098 villages, out of which 39,015 are inhabited. There are 9032 gram panchayats, 7 municipal corporations, 42 municipalities, 3 nagar panchayats, and 853 police stations. The households number 13,744,130, of which 12,407,132 are rural and 1,336,998 are urban. Bihar has a land area of 94,163 sq km, which is 2.86 per cent of the land area of India (3,287,240 sq km). The state's rural land area is 92,358.40 sq km (98.08%) and urban land area is 1804.60 sq km (1.92%).

With the bifurcation of Bihar in 2000, the newly carved state was deprived of its mineral deposits. It is significant because more than 98 per cent of GDP which was earlier originating from mining and quarrying, went to Jharkhand. As against this, the share of population which Bihar has to support is two and a half times of its share of land resources. After bifurcation, except for the Barauni Complex, entire industrial base—Bokaro and Jamshedpur Steel Plants and mineral wealth were lost. However, it is endowed with abundant natural resources—rich water and soil with a high potential for agricultural growth. In fact agriculture is now the dominant sector with over 40 per cent share in the SGDP as against around 18 per cent at the all India level. At the same time, it has also inherited drought prone area. Bihar with the present per capita income of Rs. 5,772 as against all India level of Rs. 22,946 (2004-05) is certainly at the bottom of the ladder.

History of Bihar

Bihar, the oldest citadel of Indian glory, has a glorious past. In fact, a large part of India's history is enshrined in that of Bihar. It is a matter of pride for the State that many great people lived and worked there. It is the same land where the seeds of first Indian republic were sown. The ancient kingdoms of Magadh and of Lichavis during 7th and 8th century B.C. produced rulers who devised a system of administration that truly is the progenitor of the modern art of statecraft. Vishnugupta Kautilya, the author of 'Arthashastra', the first treatise of the modern science of economics lived there. Known as Chanakya, he was the chief advisor to the Magadh king, Chandragupta Maurya. Another Mauryan king, Ashok, around 270 B.C. was the first to formulate certain basic principles for the governance of people. In fact, cultural, political and economic development of Bihar, was at the peak during this period which attracted the attention of lot of people from within and outside the country. It is also the land of Mahavir, and a citadel of education—Nalanda, Vikramshilla. Nalanda housed more than 6,000 scholars. Its alumini encompasses a large canvas of luminaries, including grammarian Panini and Chinese scholar—travellers Hiuen-Tsang, Fa-Hien and Itsing. The contributions of Bihar to India's rich heritage are many. It has gifted the *Gayatri mantra* to the country and scholars such as Yagyavalkya, author of *Shukla Yajur Veda* and the Yagyavalkyasmriti, Kautilya (Arthashastra), mathematician-astronomer Aryabhatta, as well as women scholars such as Gargi and Bharati, wife of Mandan Mishra, among others.

This is not surprising as this area had been a cradle of arts, besides philosophy, mathematics and science. Bihar boasts of several legendary dancers such as Amrapali, Salvati, Kosha and Padmavati, whose ancestry can be traced to the sixth century B.C. Nalanda University's curriculum included courses on fine arts inclusive of performing arts and handicrafts along with courses on social, religious and political texts. The treatise on hand gestures—*Shri Hastamuktavali*—written by the Mithila ruler Maharaja Shubhankar (1516-1607) of the Kharaoire dynasty, Vaidyapati padavali by legendary poet Vidyapati and the Varna Ratnakar by Jyotishwar Thakur, are some of the gems that form a part of the vast cultural canvas of the state. All these indicated a high degree of interest and high level of proficiency in literature and performing arts. However, the glorious history of Bihar lasted till the middle of 8th Century i.e., the Gupta period. In medieval times, the state lost its prestige as the political and cultural center of India. Till now it has not regained its past glory. Rather the state is known today as the most underdeveloped region in the country—BIMARU (a sick state).

Historically, Champaran holds the centrestage as a turning point in Indian politics as it was here that Gandhi's Satyagrah (non-violence) came to be associated for the first time with the fight for freedom. It was on April 10, 1917, when Gandhi reportedly set foot upon Champaran's soil in response to the call of submerged humanity. The European planters had rallied to obstruct and oppose Gandhi in his struggle to release the poor farmers of Indigo plantations there. It was left to Bihar to produce the First President of India—the only one having two terms. Recently, the second freedom struggle—the Jaiprakash Narain movement—flourished there, without much of an impact on the political economy of the State.

Nalanda, the oldest University in the country which had gone to bad days is now being given a new life as an International University in collaboration with countries like Japan and Korea. In a visionary gesture, Government has decided to revive the university that was perhaps the biggest international seat of learning. Nalanda, which drew scholars from all over the world for nearly 800 years, starting from the 5th century, lost its grandeur at the end of the 12th century, just when Oxford University was getting established. It was the first residential academic center that attracted scholars from as far as China, Tibet, Korea, Japan, Indonesia, Persia and Turkey. The proposed university will be situated 16 km. from the ruins of Nalanda at the foot of the hills in Rajgir (earlier known as Rajgriha) and start functioning from 2009. It will be a unique institution being the sense that it will be owned jointly by several countries, especially from South and Southeast Asia. The University Act clearly talks about setting up a consortium of international partners and friendly countries. The project has already attracted the attention of some of the most dynamic economies of East and Southeast Asia, ringing hope in a state lagging far behind in all indices of development. The Nalanda project hopes to revive the historical ties, this region enjoyed in areas like trade, science, mathematics, astronomy, religion and philosophy.

A teenage girl from Bihar has made it to the textbook published by the National Council of Educational Research and Training (NCERT) of her own merit. The 17 year old Anita Kushwaha from a remote village in Muzaffarpur truly vindicated the belief that real India, as often argued, lived in mostly unexplored small towns and villages. Her success story in educating herself against all odds while rearing honeybees for a livelihood, drew her admiration from the UNICEF too. Till UNICEF spotted her by early 2006, Anita was an unknown face in Pattiayasam, a nondescript village, mainly populated by Dalits, in Muzaffarpur. In appreciation of the qualities of Anita, NCERT recently decided to include a new chapter on her success story, titled "Anita and the honey bees", in a Class IV book on environmental studies called "Looking Around".[3]

There is again the shining example of 'Poster Girls of a changing Bihar'. The *12 Bihar ki betiyan* as they are called—all from the minority community, scheduled castes or tribes—feature in a UNICEF Bihar Education Project (BEP) calendar for their unusual achievements and were chosen from among 2,500 girls who took part in the Kishori Mahotsava recently. They are the new role models of the state. Two of these girls, Gudiya Khatoon and Lalita, also figure in a calendar prepared by the Government of India. All of them are achievers in their own way. Take the case of Abhilasha Kumari, 10 from Nalanda, who has played a major role in convincing parents to send their daughters to school. She persuaded 50 children to attend school—none of whom have dropped out as yet.

Lalita of Sitamarhi district comes from the Musahar community, the most backward among Dalits. A jodo instructor, she is a source of inspiration not only to her community but to other castes as well. Gudiya Khatoon of Gaya and Angoori Khatoon of Muzaffarpur had to overcome religious dogmas and their community's opposition to education. Gudiya was also selected for UNICEF's World's Children report in London. Paual, Savitir and Baby from Begusarai have made a mark in swimming; Sheela Kumari (Chhapra) and Chandani (Vaishali) are both afflicted with polio but are burning example of not being a handicap. The calendars with their photographs have been distributed to all the schools free of cost. They can become motivating factors for others. Bihar, under no circumstances, can now be classified as a basket case and is moving forward to get out of so called BIMARU stigma to an economic powerhouse.

Many State Governments can now replicate the RTI model of Bihar, in which, instead of writing an application for seeking information, one has to dial a helpline number. The Rs. 10 RTI application fee would be added to the telephone call charge. There is plenty of water, fertile land, and people are hard working. These are good things. But Bihar has been associated with bad things as well—its feudal set-up, the lack of work culture and caste-ridden society. At one point Buddhism flourished in Bihar; now one hardly finds any Buddhists there, but it should not be forgotten that the state is a sleeping giant. When it wakes up, it won't lag behind any other state in the country. History will repeat itself and Bihar will attain the glory it used to have. Its people are hard-working and talented and have proved their mettle elsewhere. They will contribute to its upward surge. Many of the farm labour which used to migrate to Punjab/ Haryana, is having work at home now, under the National Rural Employment Guarantee Scheme (NREGS).

Bihar is the first State in India to introduce a number of innovative projects during the last few years.

(a) To implement 50 per cent reservations of Panchayat posts for women in Panchayat elections throughout the state.
(b) To develop a call centre dedicated to RTI queries, where any citizen can file an RTI application using the telephone, thereby making the system transparent, affordable and accessible.
(c) To enact and implement a new Police Act. The Bihar Police Act, 2007 takes into account the respect of human rights of citizens and redefines the role of the police, its duties and responsibilities by taking into account the emerging challenges of policing and security in the state.
(d) To incorporate the empowerment of girl children and the

participation of children in school administration at elementary school level through the mandatory constitution of 'Bal Sansad' (Children's Parliament) and 'Meena Manch' (a group for gender education and sensitization).

(e) To launch the 'Hunar programme for poor Muslim girls so that they may be empowered and given skills to become productive citizens in the emerging economic order.

Human Capital

A brief overview of the socio-economic aspect of the growth theory shows that Agriculture and human development are the two sides of the same coin and the development of one is directly related with the other. Farmers are human being first and have their families. If so, agricultural development would really mean looking after the basic needs of the farmer and his family—their health and education, in particular and environ as well as infrastructure in general. Farmers are inseparably attached to the lands since their birth, and agriculture, as is well known, is a 'way of life' for them. Unlike other professions, a farmer can hardly offer to leave agriculture and switch over to any other avocation, even if his earning in monetary term may be comparatively lower.

About a quarter of Indians getting hospitalized every year slip below poverty line (BPL)—hit by medical expenses for which they rarely are insured; more than a third of our population hold their savings in cash at home; and close to 70 per cent families depend on single member's earnings. This in other words would mean that health of the farmer family is much more important than all the subsidies. These are the findings of a wide ranging survey on how India earns, spends and saves. They tell the story of rising India, as also the saga of those who have been left out of the economic boom of the past one and half decades. According to a survey conducted by insurance company Max New York Life and National Council for Applied Economic Research, a New Delhi-based research body, hospitalisation of one earning member of the household unshakes the myth and pushes a middle-class household below the poverty line. Twenty-four per cent of people hospitalized in India in a year fall below poverty line. The position in Bihar may be worse. Moreover, an astounding 96 per cent of the 63,016 respondents across 250 districts, participating in the survey, said they would not survive for more than a year in case of a loss of their major source of income. And about 68 per cent of them said their families are dependent on a single member's income.

A farmer is the key behind any agricultural development and it is, therefore, imperative to make every effort for his all round development. In order to bring a turn around in the state's near stagnant agricultural economy, the focus has to be on the farmer—the engine of agricultural growth—and his improvement. For this, he needs food, health, shelter, education to enhance his knowledge and skill, and proper environment as well as infrastructure. If the farmer himself remains on empty stomach or suffers from disease, it will be impossible on his part to put hard labour needed in this profession. Likewise, he must have the minimum basic education so essential to use the technologies, required for modern day agriculture. Education helps him to manage production risks, to take decisions with regard to what to produce, when to produce, how much to produce and when to dispose off the produce?

The Way we are

- One out of four Indians falls below the poverty line.
- Seven out of TEN families are dependent on a single member's earnings.
- No one will survive for more than a year, if the earning member loses his job.

Evidence all over the world has established the fact, that with improved opportunities of education and health, the farmers have been able to efficiently manage production and market risks and bring prosperity in agriculture. Besides providing education and health facilities, the farmers require infrastructure for bringing the desired growth momentum in agriculture. All this is more important in the case of a state like Bihar, which has otherwise immense unexploited potentialities in the form of plenty of water resources, good soil, sufficient rainfall and sunshine as well as religious harmony. This so called BIMARU State has shown a sustained agricultural growth at the rate of around 3 per cent during one of the last decade—highest among all the major states. It has highest productivity regime in maize, gram and sunflower, rapeseed and mustard. The State has thus all the built in qualities to be the leader of Second Green Revolution.

High concentration of population in agriculture and low yields of the major cereal crops are considered to be the main cause of high poverty ratio in Bihar. Hence, the goal of agricultural production system has to be to maximize income of land owning and landless ruralite for improvement of their livelihood. It could be tackled by enhancing productivity and intensity of farming and by generating more employment in non-farm activities along with better education and health facilities. While doing so, care has to be taken to preserve the health of land, water and other natural resources, so that the agricultural production system remains ever productive and sustainable. With additional irrigation facilities, along with an increase in the double cropped area, it may be possible to reduce area under fallow land bringing the same under irrigated cultivation. Special attention has to be given to Tal, Diara, Chaur and Saline lands. These areas have to be properly utilized for crop production, horticulture and fisheries, etc. If all this is to be done, human capital development of Bihar farmers has to be the first priority.

Growth Pattern

The decadal growth rate for the state averaged 4.66 per cent during 1980-91, and 4.89 per cent during the 12 years from 1993-94 to 2004-05. Bihar is usually classified as a relatively poor performer. Yet 4.89 per cent growth can hardly be called zero development. It is much higher than the 3.5 per cent that India averaged during the early decades. Indeed, Bihar has been growing faster than Punjab, which used to be India's richest state. Between 1993-94 and 2004-05, Punjab grew at the rate of only 4.39 per cent annually—slower than some other backward states such as Orissa (4.86) per cent and Madhya Pradesh (4.55) per cent. (Table 1.3)

Bihar has the lowest literacy rate among major Indian states, yet while India's overall literacy improved by roughly 23 per cent in the decade 1991-2001, that of Bihar improved by 27 per cent. That cannot be called a decade of lost development. By some yardsticks, Bihar exceeds the national average. Bihar's infant mortality rate of 61 per 1,000 in 2002 was better than the national average 63 per 1,000 and better than the 62 per 1,000 recorded by relatively developed states like Andhra Pradesh and Haryana. Bihar's life expectancy for males 65.66 years is actually higher than the national average of 63.87 years. The data suggests that it is not correct to state that Bihar had little or no development.

Table 1.3: Annual Average Growth of Gross State Domestic Product for 14 Major States

(*Percentage*)

State	1993-94 to 2004-05	1980-81 to 1990-91
Gujarat	7.46	5.08
West Bengal	7.06	4.71
Rajasthan	6.39	6.60
Haryana	6.33	6.43
Karnataka	7.26	5.29
Kerala	6.18	3.57
Andhra Pradesh	5.95	5.65
Maharashtra	5.59	6.02
Tamil Nadu	5.46	5.38
Madhya Pradesh	4.55	4.56
Orissa	4.86	4.29
Punjab	4.39	5.34
Uttar Pradesh	4.28	4.95
Bihar	4.89	4.66
All India	6.32	5.47

Source: Calculated from CSO website and Ahluwalia (2000). Excludes minor and special category states (e.g. Delhi, Jammu and Kashmir and new states (Uttarakhand, Jharkhand, Chhatisgarh) carved out of old states, *Economic Times*.

Given the poor investment climate, had Bihar's migrants invested in Bihar itself, their savings would have fetched little return. Since they invested in the rest of India, they enjoyed substantial returns. Education collapsed in the state, but its people went to Delhi and elsewhere for higher education. Bihar which has a serious law and order problem, poor roads and electricity and serious education and telecom problems, but has grown at close to 5 per cent per year for over a decade mainly because Biharies could migrate to other states for jobs, not only send remittances home but also bring back new skills. This is how *Economic Times* has summarized the Bihar scenario.

Present Study

Indian agriculture is at cross roads today. Green Revolution forerunner states like Punjab and Haryana are suffering from technology fatigue and are faced with the second generation problems. Bihar is the first state which had a prosperous sugar industry and first Agriculture Research Institute. The State has rich resources for a dynamic agricultural growth and untapped as well as unexploited potential, not only in major crops, but also in horticulture, fisheries and forestry as well. Even a modest increase of one to one and a half tonnes per hect. of rice and wheat will go a long way to meet the increasing demands at the national level. Bihar is a 'Sleeping Lion' and can feed the whole of India. First priority may be given to Flood control measures which will serve the dual purpose i.e., to save the huge devastation and provide added irrigation facilities. Recent devastation created by Kosi, is a clarion call for all concerned to take immediate remedial measures.

In fact, agriculture in Bihar, is presently being included in the backward category though there exists immense scope for an accelerated growth with proper crop intensification/diversification. The short study prepared for the Planning Commission, kindled our curiosity to undertake an indepth study on the roadblocks responsible for this dismal situation in the state and propose a new road map for future development; hence this comprehensive book which has no resemblance to our earlier study—both in content and coverage.

The Study is based both, on empirical and descriptive data base. The empirical aspect is a combination of both published and unpublished material. The sources of data are various Reports on Bihar agriculture, publications of the Census of India, NSSO, Planning Commission, Directorate of Economics and Statistics under the Ministry of Agriculture, Government of India, publications of the relevant departments of the State Government and writings of experts-published/unpublished. All of them have been carefully studied/ analysed and researched to arrive at some meaningful results.

While collecting/compiling data, the study has taken care that it presents a composite picture, covering not only crops, but also livestock, fisheries, horticulture and agro processing etc. As the study aims at identifying the potential and constraints of agricultural growth in the state and suggesting measures for following an alternative growth path, it may be immensely helpful for the administrators, policy makers, researchers and investors both in the public and private sector, particularly when Bihar has been declared as one of the thrust area in the context of Second Green Revolution.

The Study has been divided into 14 chapters. After the Introductory Chapter-1, Chapter two deals with Socio Economic Profile of the State. A critical examination of the Agronomic Situation is made in Chapter 3. With the importance of water, Chapter 4 deals with irrigation and Flood Control. These four chapters thus provide an indepth reach into the socio-economic and agro-climatic situation in the State. Chapter 5—Agricultural Scenario—then addresses itself with individual crops—rice, wheat, maize, pulses, sugarcane, oilseeds and Jute etc., Growth Patterns, constraints and present Government policies in respect of each of them have thoroughly been gone through. Bihar has done well in Horticulture and Livestock. Chapters 6 and 7 are devoted to each of them. With this ground situation before us, Chapter 8 deals with Input Management, Chapter 9 Research and Extension, Chapter 10 Infrastructure, and Chapter 11 with Policy Frameweork.

There is a lot for Bihar to learn from the experience of other parts within the country. Some to the them have been discussed in Chapter 12 under Replicable Models. Being fully armed with the ground situation, Chapter 13 deals with Development Strategy. Appreciating that Bihar has the potential to be the forerunner state under our Second Green Revolution Campaign, Chapter 14 presents a crisp picture of Agricultural Perspectives. This is followed by a detailed Bibliography, a reference for the benefit of all those interested in the study of Bihar agriculture. Quite a valuable material and basic tables which could not be a accommodated in the text for its smooth readability, have been given in Appendices. This has, of course, been duly referred to in the text as cross references.

Notes

1. Estimated at 90.0 million in 2007.
2. Govt. of Bihar, Economic Survey, 2006-07.
3. *The Tribune*, 19-1-08.

2
SOCIO-ECONOMIC PROFILE

Human Capital

The population of Bihar (83 million) represents 8 per cent of the total Indian population. It has grown at an average rate of 2.8 per cent during 1991-2001, compared to 1.7 per cent for West Bengal, 1.6 per cent for Orissa, and 2.1 per cent for all India. Like other neighbouring states, Bihar's population features a low level of urbanization: only 10 per cent of its population lives in urban areas, compared to 28 per cent for the entire country. Other socio-economic indicators show that although literacy rate has gone up from 13.5 per cent in 1951 to only 47 per cent in 2001, (Graph 2.1), it is still much below the all India level at 67 per cent. The state ranks particularly low for literacy rates among rural women—just 33.1 per cent against the national figure of 54 per cent (Table 2.1). This position is still worse with regard to rural females where it is only 29.6 per cent. This is certainly one of the main reason for its backwardness. Along with this, the school drop out rate is one of the highest. This is basically due to inadequate number of schools as well as teachers.

Table 2.1: Literacy Rates of Population Aged 7 years and above, 1991 and 2001

States/Union Territories	*1991*			*2001*		
	Persons	*Males*	*Females*	*Persons*	*Males*	*Females*
1. Andhra Pradesh	44.09	55.13	32.72	61.11	70.85	51.17
2. Assam	52.89	61.87	43.03	64.28	71.93	56.03
3. Bihar	38.48	52.49	22.89	47.53	60.32	33.10
4. Gujarat	61.29	73.13	48.64	69.97	80.50	58.60
5. Haryana	55.85	69.10	40.47	68.59	79.25	56.31

(*Contd.*)

Table 2.1 (*Contd.*)

States/Union Territories	1991			2001		
	Persons	*Males*	*Females*	*Persons*	*Males*	*Females*
6. Himachal Pradesh	63.86	75.36	52.13	77.13	86.02	68.08
7. Jammu and Kashmir	0.00	0.00	0.00	54.46	65.75	41.82
8. Karnataka	56.04	67.26	44.34	67.04	76.29	57.45
9. Kerala	89.81	93.62	86.17	90.92	94.20	87.86
10. Madhya Pradesh	44.20	58.42	28.85	64.11	76.80	50.28
11. Maharashtra	64.87	76.56	52.32	77.27	86.27	67.51
12. Orissa	49.09	63.09	34.68	63.61	75.95	50.97
13. Punjab	58.51	65.66	50.41	69.95	75.63	63.55
14. Rajasthan	38.55	54.99	20.44	61.03	76.46	44.34
15. Tamil Nadu	62.66	73.75	51.33	73.47	82.33	64.55
16. Uttar Pradesh	41.60	55.73	25.31	57.36	70.23	42.98
17. Uttaranchal	57.75	72.79	41.63	72.28	84.01	60.26
18. West Bengal	57.70	67.81	46.56	69.22	77.58	60.22

* Census was not conducted in J&K in 1991.

Source: Census of India, Ministry of Home Affairs

Graph 2.1: Literacy Rate in Bihar, 1951-2001

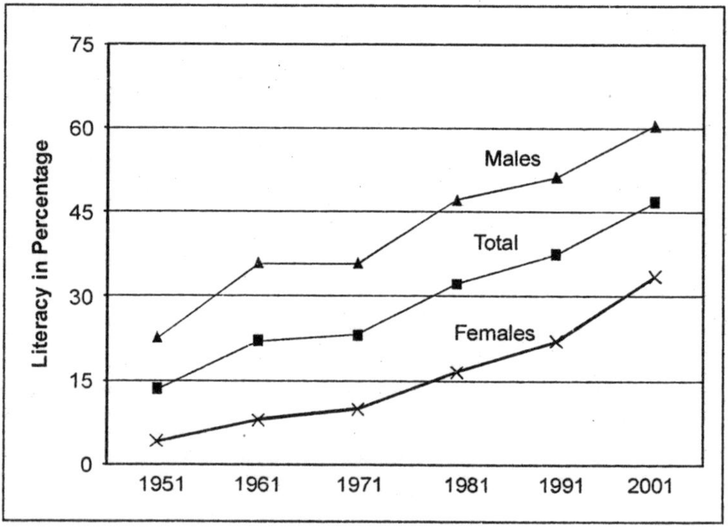

As many as 80 per cent of the bottom quintile household heads have no education, health care or well paid jobs. (Bihar XI Plan) Female literacy is as low as 7.6 per cent in a District like Kishanganj (Table 2.3).

Table 2.2: Literacy Rates for Scheduled Castes and Scheduled Tribes, Bihar Census 2001

(*Numbers*)

	Total			*Male*			*Female*		
	Total	*Rural*	*Urban*	*Total*	*Rural*	*Urban*	*Total*	*Rural*	*Urban*
Overall	31109577 (47.0)	25876919 (43.9)	5232658 (71.9)	20644376 (59.7)	17513010 (57.1)	3131366 (79.9)	10465201 (33.1)	8363909 (29.6)	2101292 (62.6)
Scheduled Caste	2880895	2536205	344690	2128948	1900047	228901	751947	636158	115789
Scheduled Tribe	169895	147420	22475	125048	111086	13962	44847	36334	8513

Note: Figures in bracket are percentage.

Table 2.3: District-wise Rural Female Literacy Rate (2001)

State/District		*State/District*		*State/District*	
Patna	26.90	Gopalganj	16.70	Samastipur	NA
Nalanda	26.30	Muzzaferpur	18.10	Sahrsa	12.80
Bhojpur	24.10	Sitamarhi	13.70	Araria	11.90
Rohtas	24.30	Sheohar	NA	Purnia	13.50
Gaya	19.00	Begusarai	21.10	Madhepura	12.90
Jehanabad	25.70	West Champaran	11.10	Kishanganj	7.60
Aurangabad	28.40	East Champaran	11.70	Katihar	13.00
Nawada	19.70	Vaishali	22.70	Bhagalpur	20.00
Saran	20.70	Darbhanga	16.90	Munger	20.20
Siwan	19.90	Madhubani	15.80	Khagaria	17.90

Source: Census of India, Ministry of Home Affairs

EDUCATIONAL LEVELS

Out-of-School Students

Bihar has a total of 70,523 recognised primary schools of which 49,598 are lower primary schools (grades 1 to 5). There are 17,307 middle schools (grades 1 to 8) and 3,618 upper primary schools (grades 6 to 8). The State has an estimated 19.5 million children between the ages of 6 to 14. According to the education department, there are about 15,40,813 out of school children (about 8 per cent of the child population). About 50 per cent of these children are girls.

School education in Bihar calls for immediate attention. There are as many as 11 districts in the State where out of school children in the age group 6-14 are more than 50,000 each while neighbouring West Bengal has nine such districts. Overall, there are 24 districts in India having more than 50,000 out of school children (OOSC). These 24 district contribute 19.33 lakh OOSC to the total OOSC figure of 70.18 lakh in the country of—roughly 27 per cent as of March 31, 2007. In Bihar, the 11 districts are West Champaran (89.113), East Champaran (74.343), Madhubani (85,820), Katihar (70,100), Madhepura (61,073), Samastipur (55,503), Begusarai (55,183), Patna (Rural 52,584), Khagaria (52,079), Nawada (50,481) and Gaya (50,236). Drop out rate of children at the primary level is the highest—Bihar (Table 2.4).

Table 2.4: Dropout Rates at Primary Level (2004-05)

State	Primary			Upper Primary	Secondary
	Total	Girls	Boys		
Bihar	51.59	48.62	53.37	74.69	83.06
Kerala	0.00	0.00	0.00	0.00	7.15
Punjab	23.96	19.91	27.42	33.67	44.06
All India	29.00	25.42	31.81	50.84	61.92

Source: Selected Educational Statistics, 2004-05; Ministry of Human Resource Development.

Inadequate Schools and Teachers

Mahatma Gandhi introduced the idea of basic education in Bihar during the freedom struggle. Contrary to Gandhi's love for Bihar, about 43 basic schools in the state are now struggling for survival. They reportedly not only suffer from a shortage of teachers, but even land meant for them is in the illegal possession of powerful people. Like the other dreams of Mahatma Gandhi, his dream of basic schools too has been left in the lurch. Successive state governments here in the past, allowed the schools to languish. At present there are 143 teachers as against the sanctioned posts of 461 in 43 basic schools.

Bihar Government schools lead the country in student and teacher absenteeism. Less than 40 per cent students and 75 per cent teachers attend classes on any given day according to a Human Resource Development ministry survey. But more surprisingly, educationally sound Delhi scores just 51 per cent student attendance and ranks lower than educationally backward states like Orissa, Uttar Pradesh, Rajasthan and Chattisgarh. While teacher absenteeism is not too high-just 5 per cent—their accountability in ensuring student attendance is very low. Student attendance at the primary level in Bihar was only 42 per cent, while in the upper-primary classes just 37 per cent. Teacher attendance for these two categories was 76 and 75 per cent respectively.

According to the Report of Bihar Common School System Commission (CSSC) total estimated child population for primary, middle, secondary and higher secondary age groups in 2006-07 was a little over 30 million, but only 15 million children are in school today. There were only 34,800 primary and 15,500 middle schools. This means that 25,900 primary and 15,500 middle schools were required for the

5 year period 2008-09 to 2012-13 — a 75 per cent increase—for the projected population. During this 5 year period, the number of primary school teachers required will have to be increased by 205 per cent and middle school teachers by 440 per cent. This shows the magnitude of the problem with regard to basic education.

There is also the allied problem of quality of teachers. A study by the National University of Educational Planning and Administration (NUEPA) has found that about half of the 47 lakh elementary school teachers in India have themselves not studied beyond the higher secondary level (Class XII). Just about one-third, or 35 per cent to be precise, of all those who teach Classes I to VIII in the country are graduates and another 17 per cent are post-graduates and above. But that leaves 45 per cent who have never been beyond school level. Against this all India picture, in Bihar only about 21 per cent of teachers have qualification of having a certificate of secondary level or below. Although the problem is mind boggling, the Government of Bihar has accepted the recommendations of CSSC, and one can hope that some good results will follow in the near future.

Despite all the problems, Bihar and West Bengal, always the subject of ridicule among policy planners, have created a record of sorts by contributing the maximum in bringing down the national figure of out-of school children in the age-group 6-14 from 75.97 lakh in March 2007 to 45.05 lakh in March 2008. What lends further credence to the efforts of these states is that the maximum number of children who have started going back to school are Muslims, SCs and STs take Bihar. The state has paid an immense tribute to the late Madan Mohan Jha, jokingly called a part—time bureaucrat and full-fledged educationist, who died as principal secretary, education in 2007 and made tireless efforts to drag Bihar from the bottom of the educational heap. The state has done the unthinkable by bringing down the out-of school children of Muslims, SC and ST in the age group of 6-14[1] (Table 2.5)

We have no hesitation in highlighting that crux of the farming problem is the lack of adequate education in enterprise and entrepreneurship. This has made the rural population prone to exploitation as requisite professionalism adopted in urban management enterprises could not percolate to farm management till today. Food processing plants and export potentials are being dangled as proverbial carrots and touted as the panacea. However, meeting the tough global competition and fast track changes is a herculean task. The level of higher education in Bihar still continues to be very low. (Table 2.6)

Table 2.5: Out-of-School Children (6-14 years) in Bihar 2008

(lakhs)

Muslim Children	March 2007	4.24
	March 2008	2.83
SCs children		
	March 2007	4.99
	March 2008	2.71
STs children		
	March, 2007	0.45
	March, 2008	0.26

Source: *Times of India*, 5th June 2008.

Table 2.6: Percentage of Persons in 15+ Age Group by General Education Level (2004-05)

States	*Not Literate*	*Literate up to Primary*	*Middle*	*Secon-dary*	*Higher Secondary*	*Diploma/ certificate*	*Graduates and above*
			Rural				
Bihar	54.5	20.1	11.3	8.4	3.3	0.2	2.1
Kerala	10.3	27.9	31.2	15.2	6.0	5.3	4.1
All India	45.2	43.8	15.2	8.2	4.1	0.7	2.5
			Urban				
Bihar	25.8	17.3	15.6	16.5	10.8	0.4	13.5
Kerala	6.6	23.3	28.3	15.8	7.3	7.4	11.1
All India	19.6	20.0	18.2	15.3	10.4	2.6	14.0
			Combined				
Bihar	51.6	19.8	11.8	9.2	4.0	0.2	3.3
Kerala	9.4	26.8	30.5	15.4	6.3	5.8	5.8
All India	38.2	22.8	16.0	10.2	5.8	1.2	5.7

Source: Estimated from NSSO Report No. 517 on "State of Education and Vocational Training in India."

The best way out of this mess is not economic doles, but due emphasis on education and skills. Rural people are not so conversant with the intricacies of modern-day marketing backed by management research and science as compared to their urban counter parts. Thus, the Centre and states should develop adequate educational infrastructure for entrepreneurship so that they can develop strategies to match their consumption with incomes. We need to teach them to establish village brands and promote agriculturism for farm—gate sales to market their produce innovatively and reaching the consumer directly. This will be the fast track help to the self-respecting rural population, bridge the rural urban divide, and keep food, vegetable and fruit prices in affordable range. There is then the need to equip them to decide and not stress them to get lost. Percolation of knowledge will facilitate better decision-making abilities and reduce the stress induced decisions in their households.

An Encouraging Development

It is encouraging to note that the Government of Bihar has launched two projects—'Humur' for providing free skills training to 12,000 chosen Muslim girls enrolled with the National Institute of Open Schooling (NIOS) centres in Bihar; and a scheme for free education to SC and ST students entering NIOS system Bihar Government has set aside Rs. 1.82 crore for skills training of 12,000 Muslim girls in the State; next year (2009) the number would be 50,000 to girls who acquire skills under "Humur" will also be provide free tools like sewing machines. All 14,000 madrassas and maktabs in Bihar will be brought under Sarva Shiksha Abhiyan (SSA). All high schools in Bihar will be made NIOS centres so that students have better access.

DEMOGRAPHIC PROFILE

Demographic Scenario

In 1991, with a population of 64.5 million the state ranked fifth in the country in terms of population. In the 2001 census it ranked third, (83.0 million) after Uttar Pradesh and Maharashtra. The state along with Uttar Pradesh, Maharashtra, West Bengal, Andhra Pradesh, Tamil Nadu, Madhya Pradesh, Rajasthan and Karnataka constitutes almost 50 per cent of the population of India. Table 2.7 and Graphs 2.2 and 2.3 present details.

Crude Birth Rate (CBR) for Bihar (Graph 2.4) is higher than India and CDR (Graph 2.5) is lower (Table 2.7A). The State had a decadal growth rate of 28.62 per cent, the highest among all the States in India, the national growth rate being 21.34 per cent. Sex ratio in Bihar at 919 females per thousand males is slightly lower

Table 2.7: Population Growth in Bihar Compared to that of India, Census, 1901-2001

Year	Bihar		India		Population Proportion of Bihar in India
	Population	Decadal growth rate	Population	Decadal growth rate	
1901	27,311,865	—	238,396,327	—	11.46
1911	28,314,281	1.52	252,093,390	5.75	11.23
1921	28,126,675	-0.97	251,321,213	-0.31	11.19
1931	31,347,108	9.74	278,977,238	11	11.24
1941	35,170,840	12.22	318,660,580	14.22	11.04
1951	38,782,271	10.58	361,088,090	13.31	10.74
1961	46,447,457	19.79	436,234,771	21.64	10.65
1971	56,353,369	20.91	548,159,652	24.8	10.28
1981	69,914,734	24.16	683,329,097	24.66	10.23
1991	86,374,465	23.38	846,421,039	23.86	10.20
2001	82,998,509	28.62	1,028,737,436	21.54	8.07

Graph 2.2: Growth of Population in Bihar

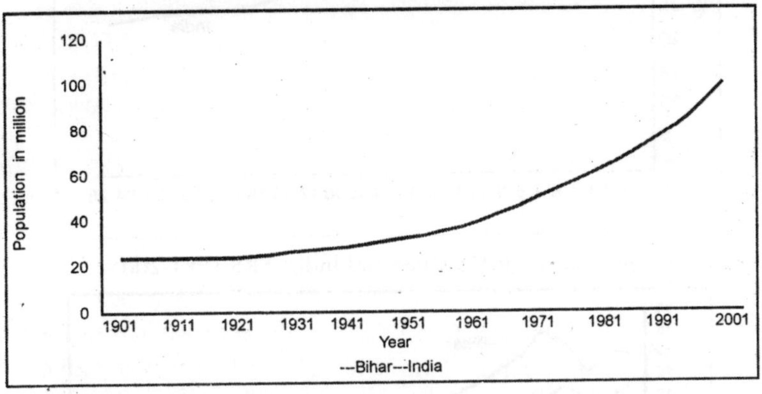

than the national ratio at 932, but the juvenile sex ratio in Bihar is higher than the national figure (Table 2.7 B). It is interesting to observe that improvement in sex ratio over India took place only over the last two decades.

Graph 2.3: Decadal Population Growth Rate in Bihar and India

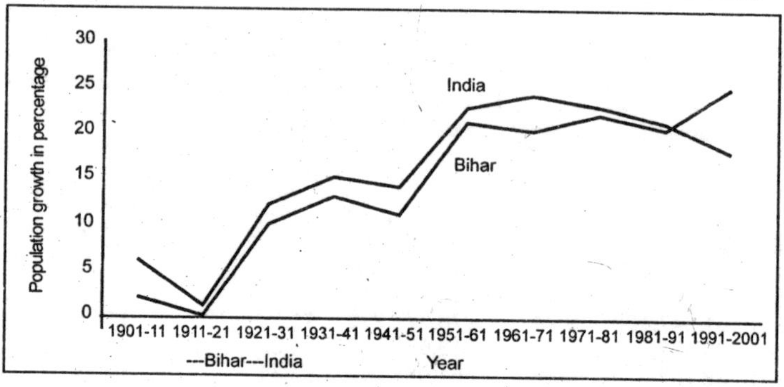

Graph 2.4: CBR for Bihar and India, SRS 1981-2005

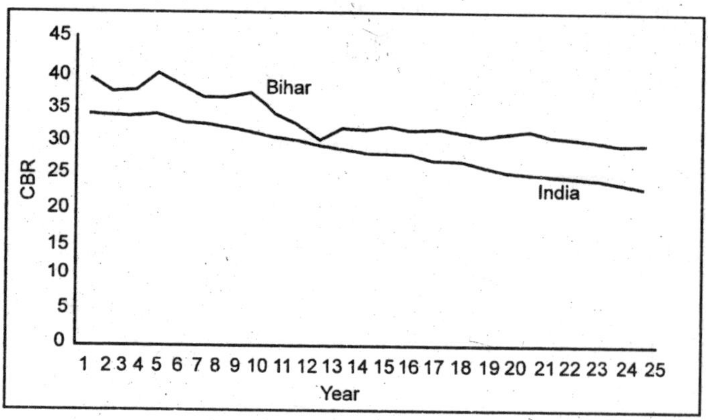

Graph 2.5: CDR for Bihar and India, SRS 1981-2005

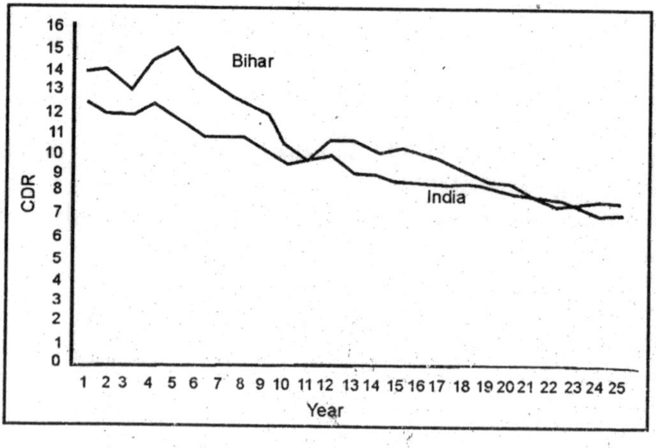

Table 2.7A: Socio-Economic Statistics in Bihar—A comparative position

	Parameters	Unit	India	Bihar	Orissa	West Bengal	Kerala
(i)	Birth rate	Per '000	25.4	31.2	23.4	20.5	17.2
(ii)	Death rate	Per '000	8.4	8.2	10.2	6.8	6.6
(iii)	Infant mortality rate	Per '000	66	61	90	51	11
(iv)	Literacy rate	Per cent	65.49	47.53	63.61	69.22	90.92
(v)	BPL Population	Per cent	26.10	42.10	47.15	27.02	12.72

Source: Eleventh Plan, Bihar

Table 2.7B: Change in Sex Ratio in Bihar and India, Census 1901-2001

Year	1901	1911	1921	1931	1941	1951	1961	1971	1981	1991	2001
India	972	964	955	950	945	946	941	930	934	927	933
Variance		–8	–9	–5	–5	+1	–5	–11	+4	–7	+6
Bihar	1061	1051	1020	995	1002	1000	1005	957	948	907	919
Variance		–10	–31	–25	+7	–2	+5	–48	–9	–41	+12

Age Distribution

As regards the age distribution of population, Bihar has larger percentage of population at lower age-groups of 0.6 and 7-14 years, primarily because of higher birth rates. The high fertility in the decades 1961-2001 is reflected in a young age structure for the country, with about 54 per cent of population being below the age of 24 years (Census 2001). Of this 35 per cent are in the age group of 10-14 years and 19 per cent in the age group of 15-24 years. In Bihar as well, the age pyramid reflects the growing base of youth population in the state. The population of the state is very young, with 42.0 per cent below the age of 15 (see Table 2.5). As per World Youth, 2006 Data Sheet, produced by the Population Research Bureau, there are about 33.1 crore young people (10 to 24 years) in India, representing a little less than one-third of the population.

The term 'youth' encompasses a diverse set of young people-urban, school going, rural youth, married adolescents and out-of-school youth. In addition, one has two consider youth from marginalized sections of society who have faced historical discrimininination. The United Nations Population Fund (UNFPA) defines adolescence as 10-19 years, with

10-14 years being early adolescence and 15-19 years as late adolescence. The National Youth Policy of India (2003) recognizes in the age group of 13 to 35 years as youth.

Table 2.8: Age-wise Population Distribution (%) in Bihar, Census 2001

Age Group	*Total*	*Male*	*Female*	*Rural*	*Urban*
0-4	13.3	13.0	13.5	13.6	10.6
5-9	15.4	15.5	15.3	15.7	13.2
10-14	13.3	13.8	12.8	13.3	13.6
15-19	8.7	9.3	7.9	8.4	10.7
20-24	7.6	7.3	7.9	7.5	8.8
25-29	7.1	6.7	7.6	7.1	7.5
30-34	6.7	6.3	7.1	6.7	6.8
35-39	6.1	6.0	6.3	6.1	6.5
40-44	5.0	5.1	4.9	5.0	5.4
45-49	4.2	4.1	4.2	4.1	4.6
50-54	3.3	3.6	2.9	3.2	3.3
55-59	2.5	2.3	2.7	2.5	2.5
60-64	2.5	2.6	2.5	2.6	2.3
65-69	1.6	1.6	1.7	1.6	1.6
70-74	1.2	1.3	1.2	1.3	1.1
75-79	0.5	0.5	0.5	0.5	0.5
80+	0.9	0.7	0.6	0.9	0.9

Source: Census of India 2001, Registrar General of India, New Delhi.

According to 2001 population Census, about 89.6 per cent of Bihar's total population of 83 million lived in rural areas, as against 72.2 per cent in the country as a whole. Thus the level of urbanization is low and along with a low sex ratio in the state, more than three quarters (22 million out of an estimated 28 million) of workforce are employed in agriculture. Nearly 51 per cent of the total population was in the age group of 15 to 59 years. Among main workers, cultivators accounted for 29.3 per cent and agricultural labourers as high as 48 per cent, while in the country as a whole, agricultural labourers constituted only 26.5 per cent of the main workers. (Table 2.9).

Scheduled Caste population of the newly constituted State of Bihar, is placed at 15.47 per cent whereas the tribal population component was a mere 0.91 per cent in 2001. It is worth noticing here that undivided Bihar had 7.66 per cent of tribal population in 1991. Most of the Scheduled Caste population live below poverty line.

Table 2.9: Demographic Profile of Bihar

	Bihar		India	
	1991	*2001*	*1991*	*2001*
Area (1000 sq. Km.)		94		3287
1. Population (million)	65.1	83.0	838	1029
2. Rural Population (%)	88.8	89.6	74.3	72.2
Population growth rate	—	1.8	—	1.6
Population Density sq.km.	—	313	—	881
3. Sex Ratio (Females/1000 males) (All)	911	921	927	932
4. Sex Ratio (Females/1000 males). (0-6 yrs)	959	942	945	927
Age Distribution of Population (%)				
0-6 Years	13.5	13.3	12.2	10.7
7.14 Years	27.6	28.7	25.1	24.6
15-59 Years	51.3	51.2	55.5	56.9
60-64 Years	2.8	2.5	2.7	2.7
65 years and above	3.8	4.3	4.5	5.1
All Years	100.0	100.0	100.0	100.0
6. Life expectancy at birth: male ***	—	61.4	—	61.6
Life expectancy at birth: female ***	—	59.5	—	63.3
Infant mortality rate **	—	61	—	63
Underweight children (below 3 years) (%)	—	54.4	—	47.0
Gross enrollment ratio (6-14 years, %)	—	47	—	65
Complete immunization (all) before 24 months) (%)	—	12.8	—	56.6
Maternal mortality rate (MMR)	—	371	—	301
7. Work Participation Ratio (%)				
Male (Rural)	48.5	40.8	36.2	44.3
Male (Urban)	41.7	37.6	48.6	47.2
Male (All)	47.6	40.5	40.9	45.1
Female (Rural)	10.8	9.4	18.6	16.7
Female (Urban)	4.3	4.7	8.1	9.4
Female (All)	11.8	18.8	15.9	14.7
8. Scheduled Caste Population (%)	15.47	15.7	16.5	16.2
9. Sectoral Distribution of Main Workers %				
Cultivators	43.6	29.3	38.7	31.7
Agricultural Labourers	37.1	48.0	26.1	26.5
Household Industry Workers	1.7	3.9	2.4	4.2
Other Workers	17.6	18.8	32.8	37.6
All Workers	100.0	100.0	100.0	100.0
10. Power Consumption (Kwh)	54.9*	74.5****	334.3*	606.0****

Note: *—1998, ** 2003-05, SRS
*** 1998-02 (CSO), ****—2005

Vital Socio-Economic Statistics: A few vital statistics given in Tables 2.10, 2.10A and 2.11 indicate the extent of backwardness of the state.

Table 2.10: Infant Mortality Rate (2005) and Child Mortality Rate (2000)

States	*Infant Mortality Rate 2002 (as 1000 live births)*			*Infant Mortality Rate 2005 (as 1000 live births)*			*Child Mortality 2000 (as 1000 live births)*	
	Total	*Male*	*Female*	*Total*	*Rural*	*Urban*	*Child Mortality*	*Under Five Mortality*
Bihar	61	56	66	61	62	47	34.7	105.1
Jharkhand	41	44	37	50	53	33	—	—
Kerala	10	9	12	14	15	12	2.6	18.8
Punjab	51	38	66	44	49	37	15.9	72.1
All India	63	62	65	58	64	40	29.3	94.9

Source: National Institute of Public **Cooperation** and Child Devclopment (NIPCCD), Statistics on Women in India, 2007.

Table 2.10(A): Infant Mortality Rate for Bihar and India by Residence, SRS 1981-2005

Year	*Bihar*			*India*		
	Total	*Rural*	*Urban*	*Total*	*Rural*	*Urban*
1981	118	124	60	110	119	62
1982	112	116	60	105	114	65
1983	99	102	65	105	114	66
1984	95	97	79	104	113	66
1985	106	109	62	97	107	59
1986	101	104	68	96	105	62
1987	101	104	72	95	104	61
1988	97	99	70	94	102	62
1989	91	93	63	91	98	58
1990	75	77	46	80	86	50
1991	69	71	46	80	87	53
1992	73	74	49	79	85	53
1993	70	73	41	74	82	45
1994	67	68	61	74	80	52
1995	73	74	57	74	80	48
1996	71	73	54	72	77	46
1997	71	73	53	71	77	45
1998	67	68	51	72	77	45
1999	63	64	55	70	75	44
2000	62	63	53	68	74	44
2001	62	63	52	66	72	42
2002	61	62	50	63	69	40
2003	60	62	49	60	66	38
2004	61	63	47	58	64	40
2005	61	62	47	58	64	40

Table 2.11: Sex Ratio, Expectation of Life at Birth, Maternal Mortality Rate, 2001-2006.

States	*Sex Ratio 2001*	*Expectation of Life at Birth 2001-2006*		*Material Mortality Rate (per 1,00,000 Live Births) 2003*
	Total	*Male*	*Female*	
Bihar	919	65.66	64.79	371
Jharkhand	941	—	—	371
Kerala	1058	71.67	75.00	110
Punjab	876	69.78	72.00	178
All India	933	63.87	66.91	301

Source: National Institute of Public Cooperation and Child Development (NIPCCD), Statistics on Women in India, 2007.

POVERTY

The State Specific Poverty Line for rural Bihar in 2004-05 was Rs. 354.36 per capita per month; its urban equivalent was Rs. 435.00 per capita per month, compared to all-India figures of Rs. 356.30 and Rs. 538.60, respectively. As per the latest NSSO 61st round based on Maximum Retail Price (MRP) consumption, 32.5 per cent of the population in Bihar was below the poverty line in 2004-05 as compared to 21.87 per cent of the population for India. The only state worse than Bihar in this respect is Orissa, with a 39.9 per cent figure. People living below the poverty line in rural Bihar comprise 32.9 per cent of the population as compared to 21.8 per cent of rural India; the corresponding figures for urban areas are 28.9 per cent and 21.7 per cent, respectively. Per capita income in Bihar in 2005 was Rs. 6719. Bihar has the lowest per capita Net State Domestic Product (Rs. 5772) among the states.

Table 2.12: Poverty Indices (% change) for Bihar and India, 1993-94 to 1999-2000

	Head count Index			*Poverty Gap Index*		
	Urban	*Rural*	*Overall*	*Urban*	*Rural*	*Overall*
Bihar	–7.5	–15.4	–15.0	–10.7	–20.6	–20.8
India	–32.6	–20.3	–22.3	–37.8	–25.7	–27.4

Source: India Development Report, 2004-05.

The head count indices declined in Bihar by 6.9 percentage points between 1993-94 and 1999-2000. The decline was 7.5 percentage points for rural Bihar, which again points to the continuing rural urban gap. While the reduction in the head count levels was similar to the national average of 6.5 percentage points, both rural head count at 41 per cent and urban head count at 24.7 per cent were significantly higher than the national average (26.3 per cent for rural and 21 per cent for urban: see Table 2.12.

The status of women in Bihar is low in terms of income and poverty. Not surprisingly, the state is also far below the national average on key health indicators (Table 2.13). For most dimensions of human development, the performance of the state during the 1990s fell well short of what is needed to achieve the Millennium Development Goals by 2015 (Table 2.14).

Table 2.13: Selected Socio-Economic Indicators of Bihar

		India	*Bihar*	*Ranking among 16 Major States*	*Ranking Status*
Socio-Economic Indicators	Per capita income (Rs.) 2005	12,416	6719	16	Lowest
	Persons Below Poverty Line (%) (NSSO, 2004-05)	21.8	32.5	15	2nd lowest
	Female Literacy Rate (%)	53.7	33.1	16	Lowest
	(%) of Girls in School (11-14yrs), NFHS-2 (1998-99)	67.0	53.0	16	Lowest
	Human Development Index (HDI)	57	47	16	Lowest
	Human Poverty Index (HPI)	39	52	16	Lowest
Health Indicators	TFR NFHS-3 (2005-06)	2.9	4.3	15	2nd lowest
	CPR*	56.3	34.1	15	2nd lowest
	IMR	58	61	11	6th lowest
	Safe Delivery (%)	47.6	29.5	14	3rd lowest
	Anaemic Children 6-35 months NFHS-3 (2005-06) (%)	79.2	87.6	14	3rd lowest
	Children Fully Immunized (%)	20.18	32.8	16	Lowest
	Maternal Mortality Rate	301	371	12	5th highest

*Contraceptive Prevalence Rate.

Source: Bihar; Towards a Development Strategy, a World Bank Report, 2007.

Today, by almost every known development index, Bihar, if not at the bottom of the state's list, is very near to the bottom. The composite development index, formulated by the Centre for Monitoring Indian Economy (CMIE), places Bihar in 13th place in its list of 15 states (Table 2.15). Although this index has not been updated since 2000, it is generally agreed that, in all likelihood, Bihar has fallen to last place among the current list of 28 states. The United

Nations and the Indian Government utilize the Human Development Index (HDI), as its gauge for evaluating human well-being and it serves as a fairly sensitive measure of the impact of socio-economic policies on quality of life. Table 2.16 provides human development index for Indian States in rank order for the years 1981, 1991 and 2001.[2]

Table 2.14: Selected Millennium Development Goals-Indicator for Bihar

	MDG	*Bihar 1993*	*Bihar 1999*	*India 1999*
Eradicate extreme poverty and hunger	Poverty head count (%)	45.9	39.0	28.6
	Poverty gap (%)	0.10	0.08	—
	Prevalence of child (%) malnutrition/underweight children Below 5 years (%)	62.6	54.4	47.0
Achieve universal primary education	Net primary enrolment ratio (%)	54	52	77
Promote gender equality	Literacy gap (%)	0.44	0.56	0.71
Reduce child mortality	Infant mortality rate (IMR) (per 1000 live births)	89.2	72.9	67.6
		127.5	105.1	94.9
	Child mortality rate (per 1000 live births)	10.7	11.0	42.0
	Immunization (measles) (% children below 12 months of age)			
Improve maternal health	Maternal mortality ratio (per 1 lakh live births)		451	408
	Delivery by skilled birth attendant (%)	19	23.4	42.3
Combat HIV/AIDS, malaria and other diseases	Incidence of tuberculosis 9 per lakh)	595	989	544
	Contraceptive prevalence rate (%)	23.1	24.5	48.2
Access of safe drinking water and sanitation	Access to improved water sources (%)	63	75.4	77.9
	Access to improved sanitations (%)	16.5	16.8	36.0

Literacy gap figures for 2001 census

Source: Bihar; Towards a Development Strategy, A World Bank Report, 2007.

Irrespective of the index used, Bihar has consistently, occupied the lowest rung of the development ladder, year after year. Overall, while there has been some improvement in the state-wise indices between 1981 and 2001, the disparities have not been reduced significantly and the rank order of the states has not changed. Kerala is consistently at the top of the list and Bihar at the bottom. The HDI figures seem to indicate that the relative position of Bihar has, in fact, worsened. (Table 2.16)

Table 2.15: Composite Development Index

State	*CDI Value*	*Rank*
Punjab	171.92	1
Kerala	162.42	2
Tamil Nadu	145.62	3
Haryana	133.12	4
Uttar Pradesh	112.04	5
Maharashtra	106.77	6
Karnataka	106.12	7
Gujarat	105.33	8
Assam	104.39	9
Andhra Pradesh	104.01	10
West Bengal	102.09	11
Orissa	101.45	12
Bihar	91.31	13
Rajasthan	87.27	14
Madhya Pradesh	86.66	15
All India	**100**	

Source: Planning Commission, Government of India, 2002.

Table 2.16: Human Development Index for India, 1981 and 2001

Rank State	*HDI 1981*	*HDI 1991*	*HDI 2001*
Kerala	0.50	0.59	0.64
Punjab	0.41	0.48	0.54
Tamil Nadu	0.34	0.47	0.53
Maharashtra	0.36	0.45	0.52
Haryana	0.36	0.44	0.51
Gujarat	0.36	0.43	0.48
Karnataka	0.35	0.41	0.48
West Bengal	0.31	0.40	0.47
Andhra Pradesh	0.30	0.38	0.42
Rajasthan	0.26	0.35	0.42
Orissa	0.27	0.35	0.40
Assam	0.27	0.35	0.39
Madhya Pradesh	0.25	0.33	0.39
Uttar Pradesh	0.26	0.31	0.39
Bihar	0.24	0.31	0.37
All-India	0.30	0.38	0.47

Source: Planning Commission, Government of India, 2002.

A recent on-line newspaper article reported that, "The state [of Bihar] ranks at the bottom with respect to Human Development Indicator (HDI) with the HDI for Bihar being about 20 per cent lower than the national HDI" (*The Times of India* 2007)[3]. This would mean that, using the last available HDI for India, 2006, with a score of 0.611, Bihar has a score of roughly 0.488. Most important, however, is the percentage of the population below the poverty line (BPL). The national average is 27.5 per cent, but for Bihar the figure is 43.2 per cent. As a corollary to the BPL, Bihar has the highest landless agricultural labourer rate in India—Bihar = 48 per cent, All India = 26.5 per cent [Census of India 2001].[4]

While there is little debate about Bihar's poverty, there is much debate concerning the causes of that poverty. The World Bank rightly characterized Bihar's poverty as complex.[5] There is no doubt that much of the state's poverty stems from its complex social stratification, insufficient infrastructure, lack of investment strategies and incentives, low rates of productivity, especially in agriculture, poor urban-rural differentiation, and weak governance. Agriculture is the foundation of Bihar's economy, and employs roughly 80 per cent of the state's workforce, but generates only 40 per cent of its Gross Domestic Product (GDP). Much of the state's agricultural production is largely subsistence due to productivity levels that are well below the national average. Basically, there is little surplus available

Table 2.17: Poverty Ratios in Bihar and India

Sector	*Year*	*Bihar*	*India*
	1983-84	64.4	45.7
	1987-88	52.6	39.1
Rural	1993-94	58.2	37.3
	1999-00	44.3	27.1
	2004-05	42.1	28.3
	1983-84	47.3	40.8
	1987-88	48.7	38.2
Urban	1993-94	34.5	32.4
	1999-00	32.9	23.6
	2004-05	34.6	25.7
	1983-84	62.2	44.5
	1987-88	52.1	38.9
Combined	1993-94	55.0	36.0
	1999-00	42.6	26.1
	2004-05	41.4	27.5

Source: CSO, NSSO Survey Reports.

for the people's participation in the cash market. Present poverty ratio in Bihar is 44.3 (Rural) against all India 28.3 and 34.6 urban against all India 25.7 (Table 2.17).

Without question, the future of Bihar and its economic resurrection has to build on its existing foundation—agriculture. It makes no sense to perpetuate or foster the illusion that investors outside the state, or the country for that matter, will arrive as knights in shining armour to save the day by making massive, economy-changing investments.

Rural Poverty

Rural poverty is severe and tends to be concentrated among landless agricultural labourers small and marginal households and schedule castes/tribes. (Table 2.18). After Orissa, Bihar is the second poorest state in India. In 1999-2000, some 30-38 million (40-48) per cent of the 76 million rural inhabitants were classified as poor according to various estimates (Tables 2.19 and 2.20).

Table 2.18: Rural Poverty Incidence by Land Ownership

Land owned (ha.)	*% of Rural Population*	*Poverty Inci-dence*	*% Share of the Poor*	*% of Rural Population*	*Poverty Inci-dence*	*% Share of the Poor*
No land	9	51	12	10	56	14
0<*<= 0.4ha.	43	51	55	53	46	61
0.4 <*<= ha.	24	34	20	20	29	15
1 <*<= 2ha	15	28	10	10	30	7
2 <*<= ha	7	18	3	4	16	2
> 4ha	3	6	0	2	18	1
Overall	100	40	100	100	40	100

Source: NSSO 50th and 55th Round.

The Planning Commission estimates that one in every 6 poor people in the country lives in Bihar: this rises to 1 in 5 when considering rural poverty only. More than half (54) per cent of the landless agricultural labourers are in the lower 40 per cent of Bihar's income scale. Agriculture-dependent households, account for more than 75 per cent—all of them are categorized as poor. Households of scheduled castes and scheduled tribes, representing a little over one fifth of the rural population, are overwhelmingly poor with average monthly per capita expenditure of Rs. 390 compared to

Rs. 451 for all rural households in the State (NSS 58th round, 2003).

Table 2.19: Poverty Estimates for Bihar and India (1999-2000)

Poverty estimates for Bihar	*Head count % Bihar*			*No. of people in million*		
	Rural	*Urban*	*Overall*	*Rural*	*Urban*	*Overall*
Planning Commission estimates (2000)	44	33	43	38	5	43
Deaton and Dreze (2002)	41	25	39	31	2	33
Kijima and Lanjouw (2003)	48	31	47	36	3	39
Poverty estimates for all India	*Head count % India*			*No. of people in million*		
	Rural	*Urban*	*Overall*	*Rural*	*Urban*	*Overall*
Planning Commission estimates (2000)	27	24	26	193	62	255
Deaton and Dreze (2002)	26	12	22	188	32	220

Source: Planning Commission (1999-2000); Deaton and Dreze (2002); Kijma and Lanjouw (2003).

Table 2.20: Demographic and the Rural Economy

	India	*Bihar*
BPL (% 1999-2000)	26.1	42.6
Rural BPL (%)	27.1	44.3
Rural BPL Population (Crore)	20.12	3.29
Projected rural BPL Population (for 2006 in crore)	21.78	3.60
MDG target BPL families (lakh)	12.26	2.02
Target no. of BPL families (lakh)	282	50.50
Poverty growth elasticity	–0.73	–0.43
Shelterless families (lakh)		42.1
Rural Population with access to safe drinking water (2000) (%)	72.3	68.5

Source: Bihar Approach to Eleventh Five Year Plan.

Rural Non-farm Sector

One of the long standing problem in Bihar's economy has been the absence of economic rural diversification. Rural non-farm sector is the only bridge, which could lead to provide sustainable livelihood to the landless rural population of the state. (Table 2.21) Bihar has a low percentage of rural workers engaged in non-farm activities. In India, rural non-farm employment has significantly expanded during the last two decades. However, Bihar is one of the few states where the expansion has been extremely low-between 1983 and 1987-88 when there was hardly any growth in non-farm employment. Thereafter Between 1987-88 and 1999-2000, the percentage of rural workers in non-farm sector has actually fallen from 19 per cent to 16 per cent.

Table 2.21: Proportion of Households that Do Not Own Land (2003-04)

(*percentage*)

State	*(Households that do not own any land including homestead)*	*Households that do not own. any land other than homestead)*	*Households that neither own. any land other than homestead. not cultivate on owned homestead*
J&K	3.29	10.97	7.29
Himachal Pradesh	15.0	22.68	21.87
Punjab	4.57	56.89	29.51
Uttarakhand	10.64	26.4	21.15
Haryana	9.21	49.49	25.96
Uttar Pradesh	3.82	26.2	16.31
Rajasthan	5.65	19.95	12.73
Chhattisgarh	12.09	27.31	20.8
Madhya Pradesh	12.05	31.81	22.76
Gujarat	13.6	44.11	35.37
Maharashtra	17.66	44.78	38.27
Andhra Pradesh	14.33	53.19	48.75
Karnataka	14.09	40.47	30.76
Kerala	4.8	68.36	36.74
Tamil Nadu	16.55	64.52	55.43
Bihar	7.6	38.8	31.01
West Bengal	6.15	46.52	34.69
Jharkhand	4.8	39.25	18.43
Orissa	9.56	38.48	31.07
Sikkim	30.67	44.4	37.96
Arunachal Pradesh	21.59	23.5	22.59
Nagaland	8.02	15.45	10.85
Manipur	2.68	30.3	13.53
Mizoram	2.34	14.1	6.67
Tripura	8.69	59.52	33.22
Meghalaya	6.7	29.01	19.93
Assam	8.05	40.3	23.0
India	10.04	41.63	31.12

Source: EPW, Vol. XLIII No. 10, March 14, 2008. NSSO.

Owing to poverty and high fluctuation in the availability of employment during the course of the year a large number of rural labourers and marginal farmers migrated to other States to seek employment. After introduction of new economic policies in 1991, rural areas, especially that of Bihar are reported to have borne the brunt of the workforce restructuring process, with agriculture in particular reverting to its role as the residual sector. With almost 42 per cent of the total inter-state migrants in search of work originating in Bihar and Eastern U.P., the decline in secondary sector employment at an all India level has inevitably hit the workers of these States. In these emerging circumstances Non-Farm Sector (NFS) has assumed greater importance for providing sustainable livelihood in rural areas of Bihar. Despite enormous potential in the form of agro-based raw materials, skill and demand for setting up of cottage, tiny and village industries, credit flow under NFS has stagnated at around 30 per cent during the preceding three years period.

Rural Diversification in Bihar

At the All India level, employment in sectors other than crop production and plantation grew faster in rural areas in comparison with urban areas between 1998 and 2005. Millions of Indians shifted to jobs in the retail trade, manufacturing and allied agricultural activities such as livestock farming and fishing, according to the Fifth Economic Census 2005. It is, however, not clear if this shift was triggered by distress in the agricultural sector or Indicates enhanced entrepreneurship among the country's rural folk. It may be because people got pushed out of agriculture or were pulled into business because of higher economic returns. Data clearly shows higher economic activity in rural areas compared to urban India. In 2005, 100 million people were in areas establishments as owners or employees. And more than half of them were in rural areas. The higher employment rate in rural areas was primarily because villagers ventured into self-owned retail and manufacturing business—the two sectors with the maximum growth between 1998-2005. Growth of self owned establishments run by family members doubled in rural areas during the census period as compared to 1990-98, when it was 2.25 per cent. Similarly, establishments with up to six workers witnessed a three fold jump in rural areas.

Along with agriculture the rural non-farm sector provides an important source of income for rural households in Bihar. It accounted for about 39 per cent of total household income in 1993-94 (NCAER,

2000). However, agriculture is still Bihar's major source of rural income: 53 per cent of income is derived from cultivation and 6.7 from agricultural labour (Table 2.22). The share of non-farm income sources to total income is significant across all income quintiles, with poor households receiving much of their non-farm income from self-employment, while wealthier households rely on regular non-farm employment.

The proportion of own-farm income to total income tends to rise as per capita income increases, with the exception of the highest income quintile (i.e. the riches 20 per cent of households), where it falls steeply. Similarly, the income share generated from agricultural wages falls as income increases, with a sharp decline for the top 40 per cent of rural households. Agricultural wage labour accounts for 24 per cent of income for the poorest 20 per cent of households, but just 0.4 per cent income for the richest 20 per cent of households.

Agriculture cannot sustain a growing population, especially when land-holdings are getting fragmented and input costs go on rising. New technology, farming techniques, farm machinery and latest seeds and chemicals require heavy investment, which only large farmers can afford. With land prices rising, it makes economic sense to the small land owner to sell the unviable piece of land to take up a new occupation. The transformation is no doubt quite painful for the first generation of migrants.

Table 2.22: Non-Farm Income in Bihar

(*Percentage*)

Quartile	Cultivation	Agriculture Wage Labour	Non-farm Wage Labour	Non-farm Self-employment	Non-farm Regular Employment	Total non-farm Sources	Other Sources	Real per capita Income
Lowest	29.8	24	23.5	18.6	3.1	45.2	1	1064
Q2	30.9	19.1	16.6	26.1	5.4	48.1	1.9	1951
Q3	43.8	14	11.4	21.1	7.4	39.9	2.3	2870
Q4	51.4	4	4.7	21.7	16	42.4	2.2	4137
Highest	4.5	0.4	1	10	23	34	1.1	8418
Total	52.8	6.7	6.4	16.5	16	38.9	1.6	3690

Source: Lanjoured and Shariff, 2000.

There are pull and push factors: those who get pulled by better opportunities and incomes in other occupations and in other states and those who are pushed out of land by poverty. It is the second category that needs immediate help. The Government should provide them training and resources to make a new beginning and ensure that the transition is less painful. For lack of awareness, resources and motivation, many farmers and farm workers are driven to the labour market, crime, and drugs.

A cultivator has an emotional relationship with his land and the parting is not easy. It is the small farmers who should get the government subsidies, but in reality large farmers corner a major part of the government largess given in the name of the poor farmer. The displaced farmer and farm worker need jobs, which can be created by encouraging industry. Agriculture needs to be made more lucrative through higher productivity and higher prices of farm produce. It is in this context that NABARD has taken a number of initiatives in the Rural NFS.

District Rural Industries Project (DRIP)

DRIP is an integrated area based project for development of Rural Non-Farm Sector (RNFS) through credit intensification process in collaboration with project partners to provide sustainable employment opportunities to rural populations by facilitating setting up of cottage village, tiny and small industries in the rural areas. DRIP has been operationalized in Muzaffarpur, Vaishali and Gaya districts. NABARD has conducted various programmes for sensitisation of the functionaries of Project Partners (Govt./Banks/NGOs/VAs). The progress in implementation of the project and issues relating to development of non-farm sector are discussed and reviewed on quarterly basis by the Project Coordination and Guidance Committee at the district level and on half yearly basis at the State Level. It has been extended to three more districts viz. Nalanda, Bhojpur and Purnea districts during the year 2005-06. Based on the Goal Oriented project planning workshops for these districts the achievement is as follows:

SUDHA (Scheme for Skill-upgradation and Design Development for Handloom Weavers)

To improve the quality of Handloom products through skill upgradation and design development and to tap the market at domestic and international level, NABARD has evolved SUDHA scheme. Under this scheme promotional assistance is available to

Apex Handloom Cooperative Marketing Federations, Weavers Cooperative Societies, NGOs/VAs etc.

Cluster Development under the National Programme for Rural Industrialisation (NPRI)

Government of India has focused attention on development of 'artisans based' small enterprises in clusters with a target of development of 100 clusters over a period of 5 years i.e. 1999-2000 to 2003-04.

Credit Linked Capital Subsidy Scheme (CLCSS) for Technology Upgradation of SSI

The Credit Linked Capital Subsidy Scheme (CLCSS) for technology upgradation of SSI units was launched by Government of India during 2000-01, with an aim to facilitate technology upgradation by SSI Units in the specified products/sub-sectors by providing 12 per cent subsidy for induction of proven technologies. NABARD and SIDBI have been identified as nodal agencies for implementation of the scheme. However no response has been received from any entrepreneur/Bank for the scheme in Bihar till date.

Rural Haat

Considering the importance of rural '*haats*' (markets) in the socio-economic development of rural areas, NABARD had decided to extend the grant assistance to two Gram Panchayats in each district for setting up of rural '*haats*'.

Position Today

Farmers in Punjab are having to fork out more for their farmhands this sowing season. The whole labour market scene has changed today (2008). Their regular supply of cheap labour from Bihar has dried up. The Bihar worker is finding employment at home as the state government has ramped up its spendings on development activities. The increased economic activity in the state is a story that is beginning to attract attention as the first visible real evidence becomes Bihar's economic growth has become a moment of crisis for other states, who have for long fuelled their growth on the backs of the low-paid Bihari workers.

The Bihar construction workers from Nasik must be having the last laugh. In February, Raj Thackeray's Maharashtra Navnirman Sena attacked Bihari workers and drove them out of Nasik. Little did Mr. Thackeray know that these workers, many employed in the construction sector, weren't going to return. With the conditions in their home state improving, the workers from Bihar no longer need to risk life. The end result is that Nasik's construction business landed in a tight spot on account of scarcity of workers.

Bihar's economy is turning around, and it is no longer merely in the realm of Government documents. The changes are now becoming visible on the ground. And the turnaround is impacting other states as well. In Punjab, small farmers are finding that the traditional Bihari farm labourer is no longer easily available. It would seem that Bihar is no longer the obvious choice for cheap labour. Last year, farmers in Punjab and Haryana paid their migrant farmhand Rs. 500 to Rs. 600 per acre of the season. This year, the farmers had to hike the rates by as much as 50 per cent, in a bid to attract farm labour.

The Bihari labourers, who would have travelled beyond the state for low paid work, are finding work within the state and will venture out only if the pay is worth it. Road works in the state could explain the story. In the first two years in office, the JD(U)-BJP Government has spent Rs. 4,722 crore on roads, both rural and PWD. Contrast this figure to Rs. 366 crores spent by the RJD Government in its last year in office on PWD roads and none on rural roads. This year 2008, the State Government proposes to spend Rs. 3,000 crore on roads alone. An increase in investment by as much as 800 per cent. More roads, means more jobs besides better connectivity and greater economic activity.

It is not just roadworks that are receiving attention. The state has also been ramping up its spending—Plan expenditure for the current year has been set at Rs. 13,500 crore, while last year (2007-08) it was at Rs. 10,400 crore. Compare this to the average plan layout in the RJD regime, which was at Rs. 2,000 crore with the spending never exceeding Rs. 1,500 crore. Pumping in thousands of crores of funds into the system imply a flurry of construction work—roads, schools, hospitals. Exact figures of how many Bihari workers have chosen to remain in the state is difficult to procure, as much of the information is anecdotal. It would seem that local block level officers find that many of those who habitually migrated for work are staying on even if the earnings are marginally lower.

Where are Men

- Farmers in Punjab struggle with paucity of labour. Have even increased wages by almost 50 per cent to retain migrant workers.
- Raj Thackeray taken heart, Nasik's construction business has landed in a tight spot due to lack of workers.
- Surat millowners recently asked Shushil Kumar Modi to appeal to Bihari workers to return after Chhat and Durga Puja.

The rise of employment opportunities is the most important reason which explains why Bihar is no longer exporting cheap labour. Other factors that have helped include investment in education and health. There has been a marked improvement in the utilization of funds in social sector efforts like universalising elementary education—the state's ability to use funds increased from 48 per cent to 78 per cent. The state's primary healthcare centres are now functional. From an earlier average of 39 patients a month, these centres now cater to as many as 150 patients a day. The Bihari labourer is no longer treated with disdain; he is a valued and sought after worker. No longer does a trip back to Bihar for Chhat means loss of job. Which would explain why Surat millowners recently requested the Deputy Chief Minister, Sushil Kumar Modi to appeal to Bihari workers to return to Surat after Chhat and Durga Puja.

The World Bank has now started lending support to Bihar with a US$ 63 million. Bihar Livelihoods Project, named Jeevika to improve the livelihoods of the rural poor in 4,000 villages in the poorest districts of Nalanda, Gaya, Muzaffarpur, Madhubani, Khagaria, and Purnea. It is expected to directly benefit about 2.9 million people belonging to 590,000 households. The state is also gearing up to harness its comparative advantage in agriculture. It has already taken a step towards this end by abolishing the Agriculture Product Marketing Control (APMC) Act, and thus providing additional avenue to farmers. The state would now like to move into value added and processed agriculture.

> "While Bihar has a lot of catching up to do, there are signs of hope. If the current momentum for progress can be

sustained, the state can hope to bring prosperity to its people. The road to development will now need the combined efforts of the state and the central government, the private sector, civil society, Bihar migrant population, as well as the international community". (Ms. Isabel Guerrero, the World Bank Country Director for India)

STRUCTURE OF BIHAR'S ECONOMY

The structure of Bihar economy is substantially different from the national economy, mainly because of the larger share of rural agricultural sector in Bihar. Consequently, the livelihood patterns are also very different in Bihar than those of other states of India as revealed by the Work Participation Ratio (WPR) of their respective population. Both for males and females, the WPR is lower in Bihar as its traditional economy offers less opportunities for economic work. It is also significant to note that, between 1991 and 2001, the WPR for male decreased in Bihar, but increased in India as a whole. The distribution of the main workers among the major occupation groups is again very different in Bihar, compared to the national pattern. The proportion of cultivators is nearly the same in Bihar and India, 29.3 and 31.7 per cent respectively; but because of a much smaller urban industrial economy in Bihar, not less than 48.0 per cent of its main workers are agricultural labourers. For Indian as a whole, agricultural labourers constitute 26.5 per cent of main workers, a substantial 37.6 per cent workers being engaged as other workers, mostly in the non-agricultural occupations (Table 2.23).

Table 2.23: Labour Force in Agriculture in Bihar and Other States, 2001

Labour Force in Agriculture	*Bihar*	*All India*
As percentage of total	77.4	59.2
Cultivators (% of total)	29.2	31.6
Agric labourers (% of total)	48.2	27.6
GSDP/capita (2000) (Rs.)	3,650	11,752
Agricultural GSDP/worker (Rs.)	4,650	7,603
Agricultural wage (1999-00) (Rs. per man day)	42.85	
(Kharif, Rs./man-day)	40,04	50.77
Rural Poverty rates.	44%	27%

Source: Population Census, 2001; and Planning Commission.

The bifurcation of the State in 2000 into Jharkhand and Bihar, had an impact on the structure of the state economy. As 75 per cent of population remained with Bihar, it lead to an increase in population density. The contribution of the secondary sector, which was 15.39 per cent in 1980-81, rose to 20.98 per cent in 1990-91 and fell to 10.55 per cent in 2000-01, less than half the share of the secondary sector of the national level which is around 25 per cent. The tertiary sector has grown disproportionately and without concomitant growth in the secondary sector. Its contribution was 25.84 per cent in 1980-81, but its high growth in the 1990s raised its contribution to 49.77 per cent of SDP in 2000-01, which is at par with the country at large. In the year 2004-05, the contribution of Agriculture, Industry and Services sectors in GSDP at 1993-94 prices was 40, 12 and 48 per cent respectively as against 20, 27 and 53 per cent respectively at 1999-2000 prices for the country as a whole.

STATE GROSS DOMESTIC PRODUCT (SGDP)

On the basis of per capita income of various states in India, Bihar is just at the bottom (Table 2.24) and Graph 2.6. The Gross State Domestic Product (GSDP) in Bihar at constant 1993-94 prices was estimated at Rs. 39332.10 crore in 2004-06. The annual growth rate of Gross State Domestic Product during 1993-94 to 2004-05 was 5.08 per cent (Table 2.25). Due to slow growth of income and high growth rate of population, the average per capita income in Bihar continued to be the lowest in the country. As of 2003-04, the per capita net state domestic product at current prices was only Rs. 5,780, as against the national average of Rs. 21,142. Although the tertiary sector accounted for about 49 per cent of the gross state domestic product in the state, the share of agriculture and animal husbandry was as high as 38.2 per cent, as against the national average of 18 per cent. Table 2.26 shows the sectoral composition of GSDP at constant prices. Annexures 2.1 to 2.3 provide detailed information regarding Net and Gross State Domestic Product, it is interesting to observe that during the 8 years since 1992-2000 while the share of primary sector has shown a marginal decline, that of tertiary sector has gone up.

Table 2.24: Per Capita Net State Domestic Production at Current Prices

(Rupees)

State	*2001-02*	*Rank*	*2002-03*	*Rank*	*2003-04*	*Rank*	*2004-05*	*Rank*
Andhra Pradesh	17932	(9)	19087	(9)	21372	(8)	23153	(9)
Assam	11423	(14)	12247	(13)	12821	(14)	13633	(14)
Bihar	5004	(18)	5606	(18)	5362	(18)	5772	(18)
Chhattisgarh	12032	(13)	12369	(12)	14963	(12)	15073	(12)
Gujarat	19713	(7)	22624	(6)	16672	(4)	28355	(4)
Haryana	24883	(2)	26818	(2)	29504	(1)	32712	(1)
Himachal Pradesh	21570	(4)	22902	(4)	25059	(5)	27486	(5)
Jharkhand	10129	(15)	11139	(15)	11999	(16)	13013	(16)
Karnataka	18091	(8)	19576	(8)	21238	(9)	23945	(8)
Kerala	20287	(6)	22776	(5)	24492	(6)	27048	(6)
Madhya Pradesh	12209	(12)	11500	(14)	13722	(13)	14069	(13)
Maharashtra	24044	(3)	26858	(1)	28848	(2)	32170	(2)
Orissa	9879	(16)	10164	(16)	12645	(15)	13601	(15)
Punjab	25868	(1)	26395	(3)	28607	(3)	30701	(3)
Rajasthan	13621	(11)	12641	(11)	15738	(11)	16212	(11)
Tamil Nadu	20326	(5)	21740	(7)	23358	(7)	25965	(7)
Uttar Pradesh	9320	(17)	9963	(17)	10637	(17)	11477	(17)
West Bengal	17499	(10)	18494	(10)	20548	(10)	22497	(10)
India	17800		18899		20936		22946	

Source: Economic Survey 2006-07, Government of India.
Note: Figures in brackets indicate the ranking of the districts

Graph 2.6: GSDP at Constant (1999-00 Prices)—Bihar

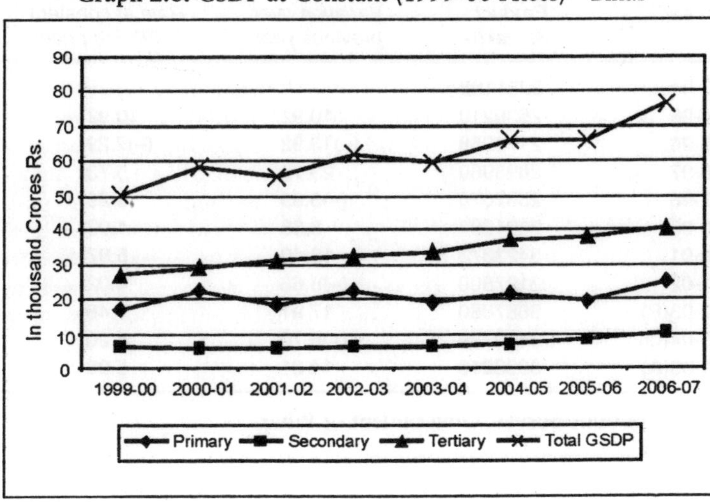

Graph 2.6a: Per Capita Income : Inter-State Comparison

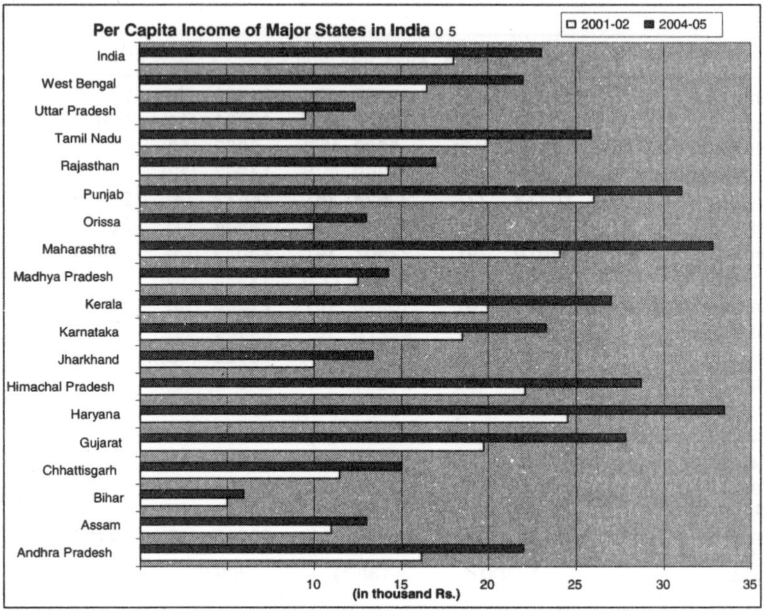

Table 2.25: Gross State Domestic Product (GSDP) of Bihar

Year	*State Domestic Product (Rs. lakh)*	*Percentage Variation over previous year*	*Annual average growth Rate at constant (1993-94) prices*
1993-94	2281198		
1994-95	2530210	10.92	10.92
1995-96	2178068	(–)13.92	(–)2.29
1996-97	2695960	23.78	5.73
1997-98	2592076	(–)3.85	3.25
1999-00	2891397	3.68	4.03
2000-01	3423373	18.40	5.97
2001-02	3127800	(–)8.63	4.02
2002-03(P)	3687880	17.91	5.48
2003-04(Q)	3440738	(–)6.70	4.20
2004-05(A)	3933210	14.31	5.08

Source: Economic Survey, Government of Bihar.

Graph 2.7: Sectoral Composition (%) of GSDP at Constant (1999-00) Prices—Bihar

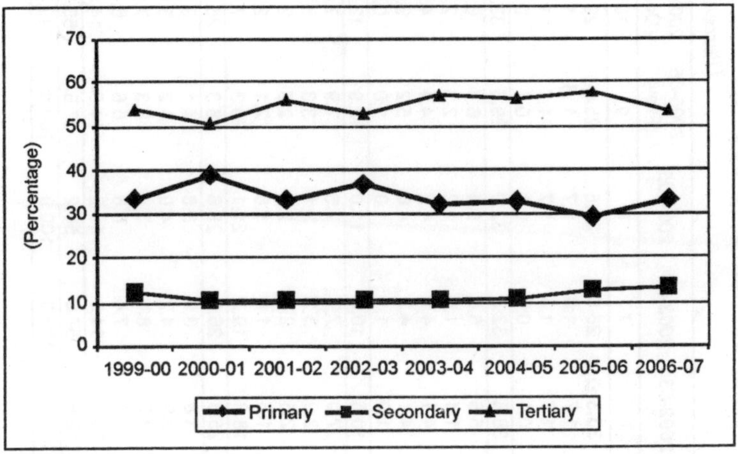

Growth Rate

Growth pattern of Bihar's economy during the Ninth Plan was actually negative even though it picked up somewhat during the Tenth Plan. But the recent 3 years (2001-02 to 2005) have shown a complete change. While Agriculture GDP growth is just near the all India level, Industry is higher (10.58%) as against all India (6.96%) Table (2.27).

Agriculture is the mainstay of the people of the state. The share of agriculture in gross state domestic product at constant prices (1993-94 prices) was as high as 38.3 per cent in 2002-03 and ranked first in the country with Punjab—a close second with 37.2 per cent. The latest Economic Survey of Bihar (Govt. of Bihar, 2007) however, highlights that there is large inequality in average per capita income in various districts of the state. This is highest in Patna (Rs. 6,958) followed by Rohtas (Rs. 4,615), Munger (Rs. 4,321), Buxar (Rs. 4,285), Nalanda (Rs. 3,879), Bhabhua (Rs. 3,894) and Bhojpur (Rs. 3,728). The lowest in the ladder is Sheohar (Rs. 2,219), followed by Gopalganj (Rs. 2,800), Araria (Rs. 2,879), Madhubani (Rs. 2,880), Sitamarhi (Rs. 2,955), Saran (Rs. 2,966), Darbhanga (Rs. 2,970) and Jehanabad (Rs. 2,975) and Siwan (Rs. 3,067). It may be seen that districts which are on the top, —Rohtas, Bhabhua, Buxar and Bhojpur barring Patna, the capital—fall in the rice belt, with assured irrigation from

Table 2.26: **Sectoral Composition (Percentage) of GSDP, Factor Cost at Constant 1999-00 Prices**

(Percentage)

Sl. No.	Sector	1999-00	2000-01	2001-02	2002-03	2003-04	2004-05	2005-06 (Prov)	2006-07 (Quick)
1	2	3	4	5	6	7	8	9	10
1.	Agriculture and Animal Husbandry	30.3	35.8	29.4	33.3	28.5	29.5	25.9	29.8
2.	Forestry and Logging	1.8	1.7	1.8	1.7	1.8	1.7	1 .8	1.6
3.	Fishing	1.4	1.3	1.8	1.7	1.8	1.7	1.7	1.6
4.	Mining and quarrying	0.2	0.2	0.4	0.1	0.1	0.1	0.1	0.1
	Sub-Total (Primary)	33.7	39.0	33.4	36.8	32.2	32.9	29.5	33.1
5.	Manufacturing	7.2	5.8	5.7	5.5	5.6	5.4	6.7	6.0
5.1	Registered	2.3	1.3	1.2	1.4	1.0	1.2	2.3	2.0
5.2	Un-registered	4.9	4.5	4.4	4.1	4.5	4.2	4.4	4.0
6.	Construction	3.8	3.4	3.8	4.1	4.2	4.6	5.2	6.0
7.	Electricity, Water, Supply and Gas	1.4	1.3	1.1	1.0	1.1	1.0	1,0	0.9
	Sub-Total (Secondary)	12.5	10.5	10.6	10.5	10.8	11.0	12.9	13.6
8.	Transport Storage and Communication	7.4	7.0	7.3	7.0	7.7	7.5	7.9	7.4
8.1	Railways	3.1	3.0	3.2	2.9	3.3	3.1	3.3	3.1
8.2	Transport by other means and Storage	2.8	2.6	2.7	2.5	2.7	2.5	2.5	2.3
8.3	Communication	1.5	1.4	1.3	1.5	1.8	1.8	2.1	2.0
9.	Trade, Hotel and Restaurants	15.0	14.9	17.2	18.3	19.2	21.1	20.3	17.6
	Sub-Total (Transport and Hotel)	22.4	21.9	24.5	25.3	26.9	28.6	28.3	25.0
10.	Banking and Insurance	3.6	3.5	4.5	3.8	3.9	3.8	4.1	3.8
11.	Real Estate, Ownership of Dwellings	4.2	3.8	4.2	3.9	4.2	4.0	4.2	3.8
	Sub-Total (Finance and Real Estate)	7.8	7.3	8.6	7.7	8.1	7.8	8.3	7.6
12.	Public Administration	7.6	7.1	8.1	6.2	7.4	6.6	6.6	6.6
13.	Other Services	16.1	14.2	14.9	13.4	14.6	13.2	14.5	14.0
	Sub-Total (Tertiary)	53.6	50.5	56.0	52.6	57.0	56.2	57.6	53.3
		100.0	100.0	100.0	100.0	100.0	100.0	100.0	100.0

Source: CSO, Government of India and Government of Bihar.
Note: Detailed data are given in Annex 2.1

Table 2.27: Income Growth and GSDP: Bihar *vs.* India (Ninth and Tenth Plans)

(*Percentage*)

	Ninth Plan		*Tenth Plan*		*According to Planning Commission CAGR 3 Years (2001-02 to 2004-05) Growth*	
	India	*Bihar*	*India*	*Bihar*	*India*	*Bihar*
Per capita GDP growth rate	4.00	1.00	5.50	2.00	N.A	N.A
GDP Growth in:	5.50	2.90	7.00	4.01	6.45	5.67
(a) Agriculture	2.00	(-)1.14	1.80	0.96	1.03	5.01
(b) Industry	4.60	7.53	8.00	9.80	6.96	10.58
(c) Services	8.10	6.37	8.90	5.08	8.64	5.12

Source: Approach to Eleventh Five Year Plan, Government of Bihar.

Sone Canal System. Outside this, Munger is second whereas Nalanda is placed at fourth also due to agricultural development. The backward districts of Sheohar, Siwan, Gopalganj and Sitmarhi fall in adverse climatic areas. Apart from Muzaffarpur, all other districts of North Bihar follow the same pattern. Muzaffarpur has a higher rate (Rs. 4,058) because of its position as the biggest mercantile centre of North Bihar (Table 2.28). There is, however, regional disparity in development. After five decades of planning, wide intra-State disparity in income levels still exist.

This can be easily understood if we look at the figures of per capita GSDP at constant prices (1993-94) and 1999-2000 for all the 38 districts of Bihar. (Table 2.28)

Table 2.28: Gross District Domestic Product and Per Capita GDDP at Constant (1993-94) Prices

Sl. No	*Name of the Districts*	*GDDP at constant (1993-94) prices (Rs. in lakh)*		*Per capita GDDP at constant (1993-94) prices (Rs.)*	
		1998-99	*1999-2000*	*1998-99*	*1999-2000*
1	*2*	*3*	*4*	*5*	*6*
1	Patna	291482	314274	6638	6958
2	Nalanda	85609	89249	3791	3879
3	Bhojpur	87863	80340	4176	3728
4	Buxar	54965	57758	4193	4285
5	Rohtas	107516	108693	4689	4615

(*Contd.*)

(*Contd.*)

Sl. No	Name of the Districts	GDDP at constant (1993-94) prices (Rs. in lakh)		Per capita GDDP at constant (1993-94) prices (Rs.)	
		1998-99	1999-2000	1998-99	1999-2000
1	2	3	4	5	6
6	Bhaphua	46220	47934	3861	3894
7	Gaya	115172	121354	3565	3651
8&9	Jehanabad & Arwal	43336	43197	3067	2975
10	Nawada	53474	55962	3187	3235
11	Aurangabad	65043	67585	3480	3515
12	Saran	96813	92857	3171	2966
13	Siwan	81737	80142	3203	3067
14	Gopalganj	60317	57968	2987	2800
15	Muzaffarpur	157831	146241	4493	4058
16	E. Champaran	103848	113650	2827	3010
17	W. Champaran	109320	105718	3855	3623
18	Sitamarhi	63550	75464	2565	2955
19	Sheohar	9562	10874	2017	2219
20	Vaishali	83536	89163	3280	3414
21	Darbahanga	91270	93530	2984	2970
22	Madhubani	100928	99057	3009	2880
23	Samastipur	104247	108714	3249	3304
24	Begusarai	86532	99306	3953	4414
25	Munger	45019	47527	4172	4321
26	Shekhpura	20109	19163	4129	3817
27	Jamui	27421	29070	3594	3736
28	Lakhisarai	40679	42781	3139	3202
29	Khagaria	38796	38653	3255	3153
30	Bhagalpur	84563	86239	3712	3689
31	Namla	49544	54161	3268	3488
32	Saharsa	41561	45511	2975	3160
33	Supaul	51480	58891	3164	3518
34	Madhepura	46923	48958	3297	3346
35	Purnia	74041	80125	3155	3305
36	Kishanganj	37193	41277	3092	3331
37	Araria	53898	58566	2729	2879
38	Katihar	77394	81446	3478	3557
	Total	2788795	2891402	3598	3631

Source: Directorate of Statistics, Government of Bihar.

Relative Development Index (RDI)

Using all India average index of development as 100, the RDI of Bihar was just 43. Excepting Patna, all the districts of Bihar had very low RDI. (Table 2.29) More particularly, the districts of Sitamarhi,

Darbhanga, Madhubani, Saharsa, Madhepura and Khagaria are on the bottom of relative development index.

Table 2.29: Relative Development Index (CMIE)

State/District	*RDI*	*State/District*	*RDI*
Patna	84.00	West Champaran	45.00
Nalanda	46.00	East Champaran	36.00
Bhojpur	46.00	Vaishali	31.00
Rohtas	51.00	Darbhanga	29.00
Gaya	39.00	Madhubani	28.00
Jehanabad	39.00	Samastipur	33.00
Aurangabad	41.00	Sahrsa	24.00
Nawada	34.00	Araria	NA
Saran	39.00	Purnia	35.00
Siwan	37.00	Madhepura	28.00
Gopalganj	36.00	Kishanganj	NA
Muzzaferpur	39.00	Katihar	31.00
Sitamarhi	25.00	Bhagalpur	39.00
Sheohar	NA	Munger	38.00
West Champaran	45.00	Begusarai	39.00
East Champaran	36.00	Khagaria	27.00
Bihar	43.00	India	100.00

Source: CMIE

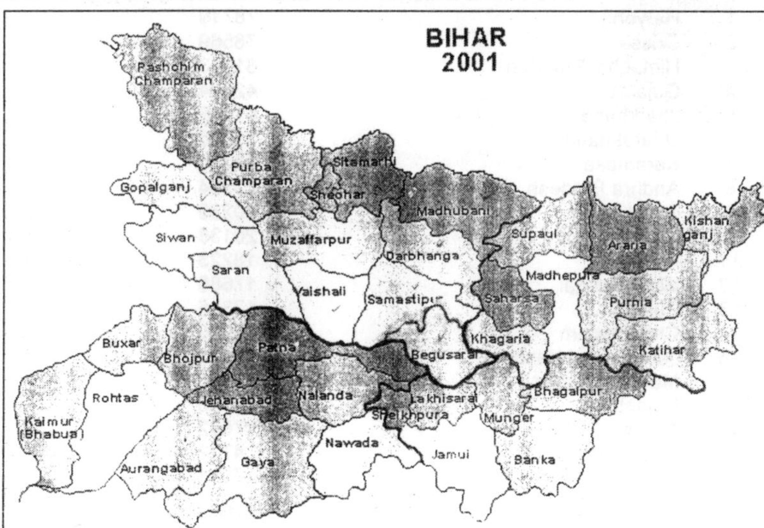

State Level Investment

Measured in nominal rupees, the per capita investment under implementation in India rose from Rs. 8,450 in December 1997 to Rs. 22,669 in December 2007. This reflected a rise of 112 per cent over the last five years and 168 per cent over the last 10 years. (Table 2.30). The top five states by investment outstanding (per capita) in December 2007 were: Haryana, Orissa, Himachal Pradesh, Gujarat and Jharkhand. The bottom five states were: Madhya Pradesh, Rajasthan, Assam, Uttar Pradesh and Bihar. These are the states that are likely to get the slowest GDP growth in the coming two to three years. The unfortunate story is Bihar. Despite high hopes of an improved Bihar under the new Government, the state did not see much improvement in 2007 compared to 2002. Per capita investment rose to Rs. 3,145 from Rs. 3,080. The surprise is that it had done much better in the previous five years under an RJD Government that seemed to be floundering on all fronts. Per capita investment had risen from Rs. 959 in 1997 to Rs. 3,145 in 2002. The Government clearly needs to do a lot more in terms of actual improvement in the state to attract the investment necessary to propel growth.

Table 2.30: Per Capita Investment State-wise

Rank	*States*	*Per capita Investment, 2007 (Rs.)*
1.	Haryana	78710
2.	Orissa	78569
3.	Himachal Pradesh	61643
4.	Gujarat	42594
5.	Jharkhand	41586
6.	Uttarakhand	35035
7.	Karnataka	34253
8.	Andhra Pradesh	21964
9.	Tamil Nadu	20239
10.	Maharashtra	20138
11.	Kerala	18223
12.	West Bengal	17685
13.	Punjab	17345
14.	Chhattisgarh	16960
15.	Jammu and Kashmir	15178
16.	Madhya Pradesh	13035
17.	Rajasthan	8743
18.	Assam	7502
19.	Uttar Pradesh	4722
20.	Bihar	3081
	All India Average	22669

Source: Ila Patnaik, Indian Express, 23-3-2008.

In the backdrop of this gloomy picture, this could be the first signs of an economic revival. The latest corporate tax collection figures for Bihar show that the state recorded up the highest growth in the first half of 2007 at 265 per cent, although this comes on a smaller base. The state shows high tax collections in absolute numbers as well. The collections, which includes collections from both Bihar and Jharkhand, grew from Rs. 123 crore in April—September 2006 to Rs. 451 crore in April—September 2007. Analyst say a better political system and efficient administration could be finally attracting investments in the state, particularly in construction sector. In fact, individuals too seem to be reaping the benefits of changing economic scenario. An indication of this is the rise in tax deducted at source collections. Patna witnessed a growth of 43.09 per cent in personal income-tax collections at Rs. 559 crore. The centre collected 40 per cent more in direct taxes at Rs. 1,11,055 crore in the first half of this fiscal 2007 compared to Rs. 79,208 crore a year ago. Corporate tax recorded a growth of 41 per cent to touch Rs. 72,240 crore, up from Rs. 49,813 crore an year-ago. Personal income-tax, including Fringe Benefit Tax (FBT), Securities Transactions Tax (STT) and Banking Cash Transaction Tax (BCTT), rose by 39 per cent to Rs. 40,744 crore in the period under review.

Economic Reforms Initiatives in the State

The first of these initiatives have been in the institutional arrangements in the formation of the Land Reforms Commission, the Farmer's Commission, the Administrative Reforms Commission and the Common Schooling System Commission. The mandates of these bodies are quite wide in terms of policy perspectives. The repeal of two statutes namely—the Agriculture Produce Market Act with the hope that this will ensure better prices and the Urban Land Ceilings Act to facilitate land acquisition for commercial purposes are measures informed by the logic of the market. Many of the government departments have prepared their own vision documents at a departmental level feeding into a thought-induced policy making process.

The second set of initiatives has been in the form of fiscal management policies that are premised on lowering the overall fiscal deficit to less than 3 per cent according to the Fiscal Responsibility and Budget Management (FRBM) Act that was promulgated in

February 2006. With the passing of the FRBM Act, the focus of the government has been on finding ways to finance the existing deficit through appropriate debt-management, expenditure rationalization and tax mobilization measures. The tight financial situation has made the government increasingly reliant on market loans as a form of deficit financing. The FRBM is also a tight rope walk for the State Government as the growth enhancing policies require public investment that have effect on the overall fiscal deficit.

Notwithstanding this dagger of the deficit hanging on the government, the new growth enhancing policies have been directed towards selective incentive policies for existing industries and agricultural resources. The third range of initiatives vary in extent and content, but on the whole, they are directed often towards skill enhancing training programmes in industry that are contingent on an increase and effective use of public investment. In some areas, like sugarcane, the responses to such initiatives have been recorded in the first year itself. The State Investment Promotion Board (SIPB) has approved 72 proposals for private investment in various sectors like sugar, steel, solvent extraction, cement, ethanol and export and fruit and vegetable packaging. Similar policy packages for agriculture have been announced, along with the emphasis on low input technology. However, the institutional impediments of translating these packages into growth is a more formidable exercise.

The fourth form of intervention has been human-centric, with a clear intent on making the service delivery mechanisms in health, crime prevention and education more efficient and accountable. This has been found to be part of the 'good governance' agenda and has an ingredient of growth enhancement and employment in the form of large scale appointment of teachers to bring the pupil teacher ratio close to the national average of 40 : 1 as compared to the existing 90 : 1, expansion of the police force and making the primary health care system work through mandatory attendance and provision of medicines. The timely payment of salaries of the teaching staff has been a long awaited step.

The universalizing of National Rural Employment Guarantee Programme (NREGP) and proper monitoring of BPL and Indira Awas Yojana beneficiaries are a significant policy step. The appointments under the ICDS (Integrated Child Development Scheme) project and increase in their number have been undertaken with a holistic approach towards the development of targeted group of children under the programme.

The so called economic and social backwardness of Bihar has always been an enigmatic phenomenon. In this background, the government is now determined to pay full attention to the basic challenge of development in the state, identify its weaknesses and strengths and then plan for its growth and development in a systematic manner. During the last decade, the overall Indian economy has experienced very high growth rates, making the relative economic position of Bihar even more disadvantaged. Thus, the agenda for Bihar development is now even more pressing for the State Government.

Notes

1. *Times of India*, 5-6-08.
2. J. Albert Rorabacher, Gerymandering, Poverty and Flooding: A Perenuial Story of Bihar. *EPW*, 16th February, 2008.
3. *The Times of India* (2007): Bihar's "First" Economic Survey Report Tables, *India Times NEWS, India Times*, March 7, http://timesofindia.indiatimes.com/NEWS/Cities/Patna/Bihars_first_Economic_Survey_Report_tables/articleshow/172960.cms.
4. Census of India (2007): "Educational Level", Census Data Finder—B Series Tables, Economic Tables: Educational Level, http://www.censusindia.gov.in/Census_Data_2001/Census_data_finder/B_Series/Educational_Level. htm (accessed August 28, 2007).
5. Sundberg, Mark and Kaul, Mandakini *et al.* (2005): Bihar: Towards a Development Strategy, A World Bank Report, World Bank.

ANNEXURE 2.1

Gross District Domestic Product (GDDP) and Per Caita GDDP for Different Districts of Bihar

Division/District	GDDP (1999-00 Prices) (in Rs. Crore)		Per Capita GDDP (1999-00 Prices) (In Rs.)	
	2003-04	**2004-05**	**2003-04**	**2004-50**
Patna	13721.3	15923.9	27605 (1)	31441 (1)
Nalanda	1425.3	1408.7	5849 (14)	5727 (17)
Bhojpur	1395.2	1372.6	5968 (11)	5786 (16)
Buxar	793.2	795.0	5385 (19)	5303 (27)
Rohtash	1780.3	1864.2	6932 (5)	7138 (6)
Kaimur	806.5	755.8	5931 (12)	5452 (12)
Gaya	2259.0	2245.1	6175 (7)	6023 (13)
Jehanabad	531.6	521.5	5465 (18)	5267 (29)
Arwal	306.4	297.0	4962 (28)	4726 (34)
Nawada	942.4	911.3	4918 (30)	4657 (35)
Aurangabad	1166.9	1143.4	5499 (16)	5287 (28)
Bhagalpu	2001.9	2129.3	8799 (3)	8268 (4)
Banka	804.5	902.7	4805 (32)	5316 (25)
Munger	1140.6	1197.7	9711 (2)	10087 (2)
Sekhpura	284.6	273.5	5109 (21)	4806 (33)
Jamui	695.1	742.9	4693 (33)	4913 (32)
Lakhisarai	495.2	516.3	5994 (10)	6188 (11)
Khagaria	758.0	849.5	5628 (15)	6194 (10)
Begusarai	1870.7	2342.0	7573 (4)	9312 (3)
Muzaffarpur	2674.3	3030.5	6824 (6)	7611 (5)
Sitamarhi	1210.4	1262.8	4260 (37)	4352 (37)
Sheohar	225.6	204.7	4101 (38)	3636 (38)
Vaishali	1401.9	1737.7	4933 (29)	6018 (14)
East Champaran	2066.[illegible]	2287.2	4989 (26)	5423 (23)
West Champaran	1901.3	1997.0	5930 (13)	6113 (12)
Darbhanga	1782.7	1976.4	5126 (20)	5574 (19)
Madhubani	1648.4	2598.6	4414 (35)	6851 (7)
Samastipur	1756.8	2075.0	4967 (27)	5783 (15)
Saran	1716.1	1830.9	5057 (23)	5312 (26)
Siwan	1370.6	1440.3	4846 (31)	5019 (31)
Gopalganj	1122.7	1166.5	4992 (25)	5107 (30)
Saharsa	971.5	1063.6	6082 (8)	6521 (9)
Supaul	853.9	1032.2	4692 (34)	5572 (20)
Madhepura	804.1	885.7	5008 (24)	5417 (24)
Purnia	1485.5	1549.9	5490 (17)	5600 (18)
Araiya	996.6	1070.8	4353 (36)	4578 (36)
Katihar	1519.0	1744.1	6019 (9)	6779 (8)
Kishanganj	698.9	762.6	5104 (22)	5461 (21)
Total	**59385.7**	**65908.6**	**6816**	**7434**

Source: Directorate of Statistics and Evaluation, Government of India.
Note: Figures in brackets indicate the ranking of the districts

ANNEXURE 2.2

Gross State Domestic Product (GSDP) at Factor Cost at Constant (1999-00) Prices

(Figures in Rs. Crore)

Sl. No.	Sector	1999-00	2000-01	2001-02	2002-03	2003-04	2004-05	2005-06 (Prov)	2006-07 (Quick)	Growth Rate
1.	Agriculture & Animal Husbandry	15202.5	20861.4	16287.3	20665.7	16899.2	19417.2	17069.6	22821.0	2.61
2.	Forestry & Logging	910.6	962.9	987.7	1023.2	1066.1	1116.8	1165.3	1215.4	4.11
3.	Fishing	696.7	769.2	984.8	1069.3	1091.5	1095.3	1144.7	1243.3	7.89
4.	Mining & Quarrying	94.4	133.4	242.5	68.0	55.6	43.9	54.7	68.5	-13.35
	Sub-Total Primary	16904.2	22727.0	18502.3	22826.1	19112.3	21673.2	19434.3	25348.2	2.84
5.	Manufacturing	3614.0	3389.9	3144.4	3386.1	3317.2	3526.5	4404.5	4593.2	4.02
	5.1 Registered	1150.7	785.8	678.1	841.6	621.8	790.4	1487.0	1553.8	6.70
	5.2 Un-registered	2463.4	2604.1	2466.3	2544.5	2695.4	2736.1	2917.6	3039.4	2.91
6.	Construction	1929.2	1952.6	2121.2	2522.2	2472.9	3037.1	3440.0	5082.9	13.54
7.	Electricity, Water Supply & Gas	718.5	778.0	602.1	615.7	631.7	654.4	672.5	717.0	-0.56
	Sub-Total (Secondary)	6261.7	6120.5	5867.7	6524.0	6421.8	7218.1	8517.0	10393.2	7.15
8.	Transport, Storage & Communication	3724.0	4054.0	4030.7	4319.4	4576.8	4938.0	5236.1	5646.2	5.96
	8.1 Railways	1563.8	1755.2	1789.1	1827.7	1940.6	2047.0	2204.6	2374.4	5.54
	8.2 Other Transport & Storage	1392.6	1503.8	1497.8	1575.4	1582.0	1672.9	1675.4	1755.9	3.02
	8.3 Communi-cation	767.6	795.0	743.8	916.3	1054.2	1218.1	1356.1	1515.9	11.38
9.	Trade, Hotel & Restaurant	7540.9	8700.9	9529.8	11357.2	11419.5	13880.5	13397.9	13496.3	9.17
	Sub-Total (Transport, Communication & Trade)	11264.8	12755.0	13560.5	15676.7	15996.2	18818.4	18634.0	19142.5	8.19
10.	Banking & Insurance	1819.1	2014.3	2473.6	2366.7	2311.2	2494.1	2704.3	2932.1	5.90
11.	Real Estate, ownership of Dwelling & Business	2097.1	2209.8	2304.4	2399.7	2508.6	2637.6	2768.9	2920.5	4.75
	Sub-Total (Finance & Real Estate)	3916.2	4224.1	4778.0	4766.4	4819.7	5131.7	5473.2	5852.6	5.29
12.	Public Administration	3793.6	4138.9	4471.5	3854.5	4372.3	4378.1	4348.5	5067.6	2.82
13.	Other Services	8059.4	8295.1	8267.4	8328.6	8663.4	8689.1	9549.1	10718.4	3.50
	Sub-Total (Tertiary)	27034.1	29413.0	31077.3	32626.1	33851.6	37017.3	38004.8	40781.1	5.78
	Total GSDP	**50199.9**	**58260.4**	**55447.4**	**61976.2**	**59385.7**	**56908.6**	**65956.2**	**76522.5**	**4.94**
	Per Capita GSDP (Rs.)	**6304**	**7116**	**6571**	**7243**	**6816**	**7434**	**7315**	**8351**	**2.92**

Source: Directorate of Statistics and Evaluation, Government of Bihar

ANNEXURE 2.3

Net State Domestic Product (NSDP) at Factor Cost at Constant (1999-00) Prices

(Figures in Rs. Crore)

Sl. No.	Sector	1999-00	2000-01	2001-02	2002-03	2003-04	2004-05	2005-06 (Prov)	2006-07 (Quick)	Growth Rate
1.	Agriculture & Animal Husbandry	14308.4	19861.7	15202.2	19520.4	15685.0	18138.0	15695.4	21344.7	-0.44
2.	Forestry & Logging	875.7	926.0	949.8	982.2	1022.6	1071.3	1117.8	1165.8	4.06
3.	Fishing	627.1	683.4	866.0	930.7	934.5	933.8	978.6	1072.5	7.13
4.	Mining & Quarrying	74.8	112.3	211.4	60.5	48.9	38.9	49.8	63.5	-11.75
	Sub-Total Primary	15885.9	21583.4	17229.3	21493.8	17690.9	20182.1	17841.6	23646.5	0.61
5.	Manufacturing	3144.7	2882.0	2611.6	2815.7	2694.3	2839.3	3662.8	3792.6	3.29
	5.1 Registered	997.1	620.4	501.4	653.6	419.6	569.8	1249.8	1294.8	3.29
	5.2 Un-registered	2147.5	2261.6	2110.2	2162.1	2274.7	2269.5	2413.0	2493.9	1.97
6.	Construction	1889.4	1899.1	2059.9	2460.3	2396.7	2945.2	3330.9	4953.5	15.31
7.	Electricity, Water Supply & Gas	407.7	353.1	233.2	300.2	258.3	245.5	240.5	260.7	-5.83
	Sub-Total (Secondary)	5441.8	5134.2	4904.7	5576.2	5349.3	6030.0	7234.2	9006.8	8.25
8.	Transport, Storage & Communication	2902.3	3275.3	3275.8	3508.7	3747.6	4096.4	4335.3	4699.5	6.77
	8.1 Railways	1072.2	1269.8	1324.9	1355.6	1459.5	1557.1	1704.0	1864.9	7.28
	8.2 Other Transport & Storage	1239.8	1354.5	1334.8	1386.8	1381.9	1477.4	1472.4	1542.8	2.71
	8.3 Communication	590.3	651.0	616.1	766.3	906.2	1062.0	1148.8	1291.9	12.81
9.	Trade, Hotel & Restaurant	7426.7	8575.4	9412.4	11236.5	11287.8	13759.7	13275.8	13372.8	9.27
	Sub-Total (Transport, Communication & Trade)	10329.0	11850.7	12688.1	14745.1	15035.3	17856.2	17601.1	18072.3	8.61
10.	Banking & Insurance	1766.4	1950.0	2395.7	2297.1	2240.1	2419.9	2626.0	2799.4	5.7
11.	Real Estate, ownership of Dwelling & Business	1341.4	1361.0	1325.9	1296.2	1274.3	1294.3	1261.8	1229.6	-1.2
	Sub-Total (Finance & Real Estate)	3107.8	3311.0	3721.6	3593.3	3514.4	3714.1	3887.7	4029.0	3.1
12.	Public Administration	3228.4	3478.4	3754.2	3218.0	3618.0	3848.4	3818.8	4537.9	3.6
13.	Other Services	7924.4	8146.6	8094.9	8138.2	8446.5	8401.1	9214.0	10328.5	3.1
	Sub-Total (Tertiary)	24589.6	26786.7	28258.8	29694.6	30614.3	33819.8	34521.7	36967.7	5.74
	Total NSDP	**45917.3**	**53504.2**	**50392.8**	**56764.6**	**53654.5**	**60031.8**	**59597.5**	**69621.0**	**4.60**
	Per Capita NSDP (Rs.)	**5766**	**6535**	**5972**	**6634**	**6158**	**6771**	**6610**	**7598**	**2.7**

Source: Directorate of Statistics and Evaluation, Government of Bihar

3

AGRONOMIC SITUATION

Agriculture sector in India in general, and Bihar in particular, continues to occupy the center-stage in overall economic progress and development. The role of agriculture including animal husbandry, dairy, and fisheries in generating broad based economic growth in India is evident from the fact that the sector has been contributing nearly 18 per cent of GDP, and giving employment to 65 per cent of the population in the country. As against this, population employed in agriculture production system in Bihar is still as high as 77 per cent, much higher than the national average of 65 per cent. Share of agriculture in the SGDP is also quite high being around 38 per cent. Bihar is lagging behind the national average on all socio-economic indicators like per capita income, average size of operational holding, per capita cultivated land, percentage of villages electrified, road length per thousand square km, per capita deposit, per capita bank credit, credit-deposit ratio, male and female literacy rate, and life expectancy, particularly females. Consequently about 42 per cent of the state population is below poverty line as against national average of 26 per cent.

High concentration of population in agriculture and low yields of major cereal crops are considered to be the main cause of high poverty ratio in Bihar. Hence, the goal of agricultural production system is to maximize income of land owning and landless ruralite for improvement of their livelihood. It could be tackled by enhancing productivity and intensity of farming and by generating more employment in non-farm activities. While doing so, care should be taken to preserve the health of land, water and other natural resources, so that the agricultural production system remains ever productive and sustainable. With additional irrigation facilities, along with an increase in the double cropping, it may be possible to reduce area under fallow lands and bring under irrigated cultivation. Special attention has to be given to Tal, Diara, Chaur and Saline areas. These areas have to be gainfully utilized for crop production, horticulture, fisheries, etc.

Graph 3.1: Rural Poverty Rates (%) in Bihar and Other Selected States (1999/00)

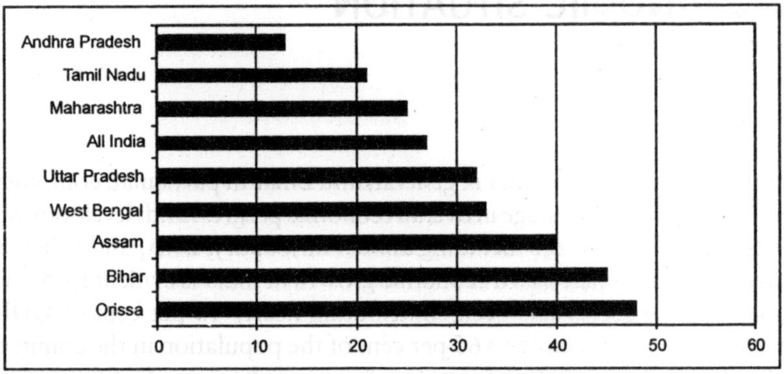

Source: Planning Commission Estimates (1999-2000)

AGRO-CLIMATIC ZONES

An agricultural zone or region is an extensive area of land with broadly similar conditions in respect of various parameters such as topography, soil, climate, temperature, rainfall, irrigation, etc. It is a homogeneous area having certain characteristics which enables it to be recognized as a unit distinguishing it from the surrounding area. Apart from geographers, biologists, sociologists, agronomists, and anthropologists, agricultural economists have also felt interested in the delineation of regions for a system study of constraints and production potentials of different regions. A region generally has a transitional frontier zone in which a change in the regional pattern occurs. The demarcation of a region, especially the agricultural region is no doubt a difficult task, but is extremely important.

An attempt was first made by the Census Commissioner 1951 to divide the country into three natural regions, seven sub regions and several divisions. The Indian Council of Agricultural Research then divided the country into five major zones/regions on the basis of consideration of maximum possible similarity in respect of latitude, altitude, temperature, rainfall, natural vegetation, soils, crops and animals, etc. Within the major agricultural regions, there were tracts showing variation from the general pattern. These were designated as special sub-regions. The designation of the Temperate Himalayan Regions as fruit Region and Eastern Region as rice Region or the Coastal Region as spices and plantation region implied that the predominance of one crop of a set of few crops (crop association) was regarded as a defining feature of a special sub-region.

The Planning Commission in 1964 divided the country into 15 regions and 61 divisions based on such criteria as topography, soil, climate, geological conditions, land utilization, irrigation and cropping patterns. The objective in demarcating the resource development regions was to have a unit which is internally homogeneous in physical and agricultural conditions and located entirely within the State. In 1972, ICAR in collaboration with the Directorate of Economics and Statistics, Ministry of Agriculture, Government of India proposed a new classification of the agricultural landscape of India where the concept of agricultural regions appears to be quite different from the earlier versions. Here, the country was divided into four broad zones for each of the most common crops of India on the basis of relative yield and spread of the crops. Reference William Easter in 1972, classified regions for agricultural planning and management into different categories. His agro-climatic regions were based on climate, soils, geology, and cropping patterns. He divided the country into three divisions, 10 sub-divisions and 52 regions. Similarly, Sengupta and Sdasyuk divided the country into 4 macro-agricultural regions, 25 macro regions and 60 micro regions.

Recently, the Planning Commission has tried to delineate 15 agro-climatic regions on the basis of population, density, share of work force in population, value productivity, cropping intensity, gross irrigated areas as percentage of gross cropped area and per capita net sown area. These include Western and Eastern Himalayan Regions, Lower Gangetic, Middle Gangetic, Upper Gangetic and Trans Gangetic Regions, Eastern, Western and Central Plateau and Hill Regions, East Coast and West Coast Plains and Ghats, Gujarat Plains and Hill Division, Western Dry Region and Island Region. While geographers have generally attempted to delineate the regions on the basis of simple but crucial parameters like cropping patterns, crop concentration and crop diversification, the economists and statisticians are making use of sophiscated quantitative techniques, such as, factor analysis, discriminant analysis or taxonomy for delineation of regions as a guide to regional planning.

According to this classification, Bihar with plains of Uttar Pradesh falls in the Sub-Humid Sutlej-Ganga alluvial plains. This region has dry sub-humid climate with medium to deep black red and yellow soil. Recommended crops for the region are rice, wheat, gram, linseed, jowar and groundnut. Composite Bihar according to the Agro-climatic Zones was divided into 3 Regions—Southern, Northern and Central. The Southern region has gone to Jharkhand and the other two have now been divided into three main agro-climatic zones, as follows:

(i) North Bihar plains/North West Alluvial Plains, covering the districts of Darbhanga, East Champaran, Gopalganj, Madhubani, Muzaffarpur, Samastipur, Saran, Sitamarhi, Sheohar, Siwan, Vaishali and West Champaran.
(ii) North Eastern Bihar Plains/North East Alluvial Plains, covering the districts of Araria, Begusarai, Katihar, Khagaria, Kishanganj, Madhepura, Purnea, Saharsa and Supaul.
(iii) South Bihar Alluvial plains covering Aurangabad, Bhabua, Rohtas, Buxar, Bhojpur, Jahanabad, Gaya, Nalanda, Patna, Nawada, Sheikhpura, Lakhisarai Jamui, Bhabua, Munger and Bhagalpur.

Quite often minimum temperature in the state is below 5°C in January. In Zone-I, the average rainfall ranges between 1040 and 1450 mm and the temperature varies from 7.7 °C to 36.6 °C with an average of 25.3 °C. Important crops grown include rice, wheat, maize, rapeseed, sweet potato and lentil. In Zone-II, the average rainfall is between 1200 mm and 1700 mm, while the average temperature varies between 8.8 and 33.8°C (average 24.5 °C). Important crops grown in this zone are rice, jute, wheat, potato, mustard, moong, toria and urad. In Zone-III, the average rainfall ranges between 990 mm and 1240 mm and the temperature varies from 7.8 °C and 37.1 °C (average 26.4 °C). Major crops grown in Zone-III are rice, wheat, gram lentil and rapeseed in Table 3.1.

Further, the seasonal distribution of rainfall, average, maximum and minimum temperature, relative humidity, soil texture, pH value, are also shown zone-wise in Table 3.2.

Table 3.1: Agro-Climatic Zones of Bihar and their Salient Features

Zone	Districts	(Area) Million ha.	Share in the State (%)	Population Density (no. per sq.km)	Animal Density (no. per sq km.)	Cropping Intensity	Literacy (%)	Poverty (%)	Flood Area (% CA)	Net Area Sown (2001) %	Net Irrigated Area %	Area under HYV %	Important Cropping Sequence
(1)	(2)	(3)	(4)	(5)	(6)	(7)	(8)	(9)	(10)	(11)	(12)	(13)	(14)
North West	West Champaran, East Champaran, Gopalganj, Siwan, Saran, Sitamarhi, Madhubani, Darbhanga, Muzaffarpur Vaishali, Samastipur, Sheohar	3.5	36.0	1073	2.73	142.17	45.20	50.83	36.3	40.2	29.7	37.6	Rice-Wheat, Rice-Rai, Rice Sweet Potato, Rice Maize (Rabi) Maize-wheat Maize-Sweet Potato, Maize Rai, Rice-lentil
North East	Saharsa, Madhepura, Purniea, Araria, Kishanganj, Katihar, Khagaria, Begusarai, Supaul	2.0	21.1	518	1.52	158.43	37.98	53.35	48.0	18.3	8.8	8.7	Jute-Rice, Jute, Wheat, Jute Potato, Jute Kalai, Jute Mustard, Rice Wheat-Moong, Rice-Toria

(*Contd.*)

Table 3.1 (*Contd.*)

Zone	Districts	(Area) Million ha.	Share in the State (%)	Population Density (no. per sq.km)	Animal Density (no. per Sq km.)	Cropping Intensity	Literacy (%)	Poverty (%)	Flood Area (% CA)	Net Area sown (2001) %	Net Irrigated area %	Area under HYV %	Important Cropping Sequence
(1)	(2)	(3)	(4)	(5)	(6)	(7)	(8)	(9)	(10)	(11)	(12)	(13)	(14)
South	Bhojpur, Rohtas, Patna, Jahanabad, Nalanda, Gaya, Nawada, Aurangabad, Munger, Bhagalpur, Banka, Buxar, Jamui, Lakhisarai, Sheikhpura, Bhabua (Kumarai)	4.1	42.9	880	2.83	115.11	47.42	48.05	21.3	41.5	61.5	53.8	Rice (41%) Gram (76%) Wheat Guava Mango Potato Rice-Wheat Rai-Gram Rice-Lentil, Rice-Rai

Table 3.2: Rainfall, Humidity and Soil Texture etc., in the Agro-climatic Zones of Bihar

Particulars	Zone I NW Alluvial Plain	Zone II NE Alluvial Plain	Zone III South Alluvial Plain
Climate			
(a) Rainfall (mm)	1344.35	1381.90	1165.45
(i) June to Sept.	1077.30	1105.90	935.55
(ii) Oct. to Nov.	121.51	112.30	111.95
(iii) Dec. to Feb.	35.15	28.70	31.95
(iv) March to May	110.75	135.00	86.00
(b) Temperature (degree C)			
(i) Average (Max and Min)	25.3	24.5	26.4
(ii) Maximum	41.8	40.1	43.8
(ii) Minimum	7.1	4.4	5.7
(c) Relative Humidity (%) 75			
(i) 8.30 A.M.	67.3	75.0	66.4
(ii) 5.30 P.M	58.8	57.	50.9
		Soil	
(i) Texture	Sandy loam to loam, clayed, medium acidic and heavy structured	Loamy sand to clay loam, height to medium structured	Sandy loam to loam with clay in some area
(ii) pH	6.5-8 Except in recent Calcareous alluvium		6.8-8.0 (7.9 in NE Areas)
(iii) Organic carbon %	0.5-1	0.2-1.0 (0.6-in NE Portion)	0.7-1
(iv) Nitrogen (%)	0.04-0.2	0.02-0.15	0.04-0.17
(v) Available P_2O_5 (kg./ha.)	20.50	10-20 (50-70 in NE Portion)	10-100
(vi) Available P_2O (kg./ha.)	200-300	200 (150-300 in NE Portion)	150-300

Source: NABARD Report, 2007.

North Eastern Region of the state is the most vulnerable to frequent floods. Nearly half of the total cropped area in this zone is damaged during floods. In some districts (e.g. Araria), floods are apt to wipe out the entire cropped area. Major crops grown in this region include rice, wheat, and maize (which comprise 50 per cent of the cropped area in district like Begusarai). The zone is also known for the production of pineapple.

Soil Conditions

Bihar is endowed with natural resources like fertile land, plenty of surface and sub-surface water, with predominantly an agricultural economy. Along with these natural advantages major problems affecting land and soils in the State of Bihar are as indicated in Table 3.3. In the south of the natural levee of the Ganga, there is a vast stretch of backwaters known as "Tal, (low order monsoon stream)" comprising lands extending from Buxar to Pakur where most of the rivers and rivulets coming from the south get lost. Estimated area under "Tals" is about 1 lakh ha. In addition to the Tal lands, the state has about 5 lakh ha of 'Diara lands (saucer shaped flood plain of a river) under the flood plains of rivers Ganga and Ghaghara. The state also has about 4 lakh ha of 'Chaur lands (remnants of river course), which are highly suitable for fish farming. The Tal, and Diara lands are inundated by water for varying periods, and are difficult to manage.

Table 3.3: Major Problems Affecting Land and Soil in Bihar

Sl. No.	*Particulars*	*Area in lakh ha.*	*% total problematic land*
1.	Saline and alkaline soil (chaur lands)	4.14	30.87
2.	Diara land	5.16	38.48
3.	Tal land	1.00	7.46
4.	Bhoodan land	0.92	6.86
5.	Culturable waste land	0.58	4.33
6.	Other fallow land	1.61	12.01
	TOTAL	13.41	100

Source: Rajendra Agricultural University (2006).

Soil and water conservation techniques in the individual farmers' field such as contour bunding, contour trenches, retaining wall, check dam etc. have been found to be of immense benefit in not only arresting soil erosion but also improving the productivity of crops.

A total area of 253 thousand ha. has been identified as the one requiring soil conservation treatment. Of this, works are underway to treat about 60,000 ha. under NWDPRA/DPAP, leaving a huge potential to be exploited through credit support. These areas are mainly located in the districts of Patna (24140 ha.), Nalanda (17800 ha.), Bhojpur (18060 ha.), Buxar (12710 ha.), Rohtas (23794 ha.), Bhabhua (17082 ha.), Gaya (32044 ha.), Jahanabad (11925 ha.), Nawada (5572 ha.), Aurangabad (19450 ha.), Munger (2615 ha.), Sheikhpura (4635 ha.), Lakhisarai (9785 ha.), Jamui (16212 ha.), Bhagalpur (19350 ha.) and Banka (17770 ha.). An area of about 5 lakh ha. could be retrieved by desalting of Ahars and Pynes at a cost of Rs. 15,000 ha. or so. The NREG Programme has also sufficient provision for recharge related schemes.

Most of the soils in Bihar are deficient in zinc and magnesium. Calcium and Bromine fall in the next category. Some soils also need Iron and Sulphur (Table 3.4). The rampant use of subsidized urea is undermining soil fertility in farmlands. Despite the fertilizer subsidy, farm productivity has been in decline and lags behind other states. Fertiliser use in India is dominated by sulphur-free fertilizers like urea, Dia-Ammonium Phosphate (DAP) and Muriate of Potash (MoP), reducing yields, depleting soil nutrients and demand for greater applications of these inputs. A study of Punjab Agriculture University reveals that each rupee invested in replenishing sulphur depleted soils gains a return of as much as Rs. 8-20 by improving crop quality and yields. The Agriculture Ministry has gazetted its Fertiliser Control (Third Amendment) Order, according 'straight-sulphur fertilizers' the status of a full fertilizer.

Table 3.4: Micronutrient Status of Soil in Bihar

Soil Types	*Percent soil sample deficient*					
	Zn	*Cu*	*Fe*	*Mn*	*B*	*S*
Sub-Himalayan hill and forest soil	45	16	2	Nil	—	—
Young alluvium calcareous soil	73	4	14	5	48	20
Young alluvium non-calcareous soil	66	2	5	3	—	—
Recent alluvium non-calcareous soil	50	7	5	1	41	—
Old alluvium soil of south Bihar	48	2	3	1	32	28
Tal land soil	49	1	4	1	35	15

Source: Rajendra Agricultural University (2006).

Zn–zinc, Cu–calcium, Fe–ferrous, Mn–magnesium, B–bromine and S–sulphur.

Sulphur, a major plant nutrient, has been placed alongside nitrogen (N), phosphorus (P) and potassium, or potash (K, for potassium). Applied through such sulphurous fertilizers as ammonium sulphate, single super phosphate (SSP), ammonium phosphate sulphate, and gypsum, sulphur are essential for the growth and development of all crops. Sulphur Mills Mumbai credits itself with producing the world's first granular water dispersible sulphur fertilizer under the brand name of Cosavet Fertis WG, a dosage of 3 kg of Cosavet per acre for short duration crops, translating into a cost less than Rs. 330 per acre, with a repeat application later for long duration crops gives good returns.

Soil Erosion

Added to this is the problem of soil erosion all over the country, including Bihar. Though soil erosion is fast eating into the country's agricultural productive lands, little appears has been done to stem its destructive impact. In fact, the Dehradun-based Central Soil and Water Conservation Research and Training Institute (CSWRTI) has warned the Union Agriculture Ministry about the dangers of rapid loss of prime agricultural lands due to soil erosion. According to a study conducted by the CSWRTI and the National Bureau of Soil Survey and Land Use Planning, Nagpur, on the potential erosion rates for different states, about 38.7 per cent area in the country has erosion rates of more than the permissible 10 tonnes per hectare per year, while about 11 per cent area falls in very severe category of 40 tonnes per hectare per year. An estimated 5,334 million tonnes of soil is lost annually, which is equivalent to 16.4 tonne per hectare per year, according to the CSWRTI. Of this, about 29 per cent is lost permanently to the sea, 10 per cent is deposited in the reservoirs and the remaining 51 per cent is displaced from one place to another.

As per CSWRT's studies, the annual erosion rates due to rains were found to vary from 5 tonnes per hectare per year for dense forests and cold desert regions to 80 tonne per hectare per year in the Shivalik region. Consequently, the decline in total agricultural output is estimated to vary from 5 to more than 50 per cent every year, which is equivalent to Rs. 5,250-8,400 crore.

"Besides informing the Union Government and various state governments, the potential erosion maps for different states are being printed by the CSWRTI. Such maps have already been brought for

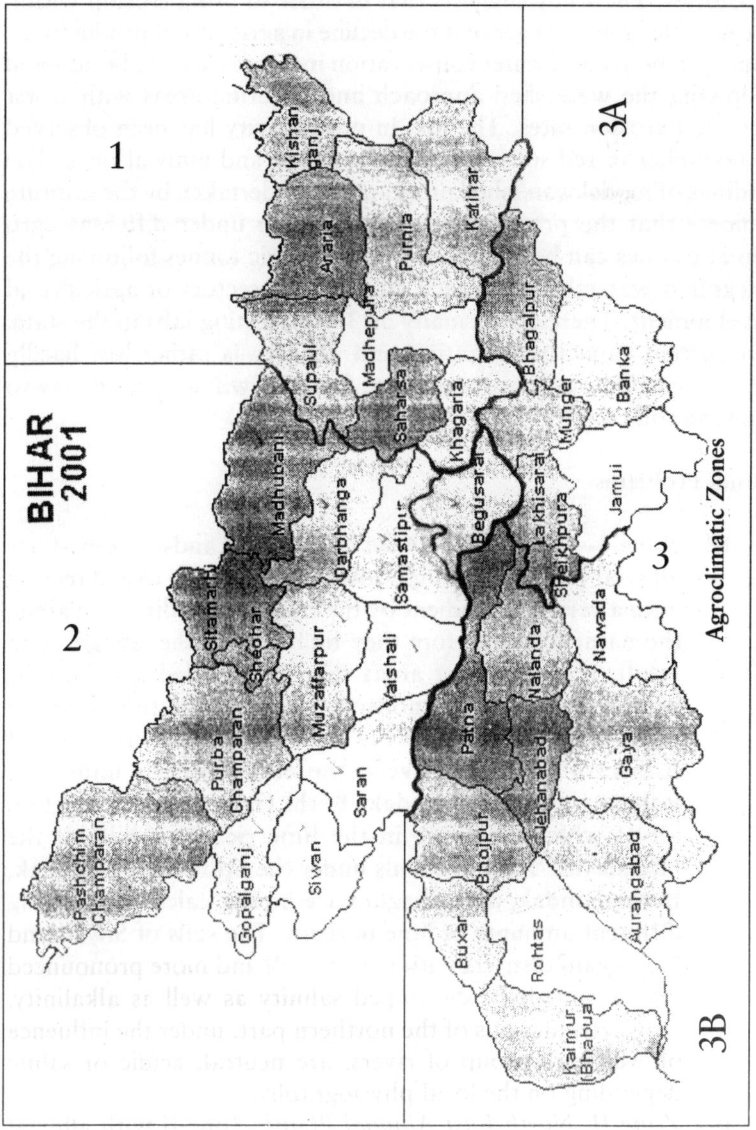

Map 3.1: Agroclimatic Zones

15 states including Bihar along with bulletins giving detailed description about the prevailing potential erosion rates under different categories. The study suggests that in order to bring erosion within permissible limits and prevent the decline in agricultural productivity, appropriate soil and water conservation measures need to be adopted following the watershed approach and covering areas with worst potential erosion rates. The loss in productivity has been observed to be higher in red soils, followed by black and alluvial ones. The findings of model watershed programmes undertaken by the institute indicate that the production in the country under different agro climatic zones can be enhanced by 150 metric tonnes following the integrated watershed approach covering all sectors of agricultural development. There are as many as 38 soil testing labs in the state. But current capacity utilization and delivery is rather low-hardly 10 per cent. Full utilization of these facilities will go a long way to improve soil conditions.

Zonal Problems

(i) *Zone 1—North West Alluvial Plains*: Lands in zone-I are alluvial plains that slope towards the south east direction with a very low gradient by the stream flow direction along the natural level before they finally meet the Ganga. As a result, there are vast areas that get flooded and become waterlogged during monsoon. Except for the northern portion in the west of the zone under the influence of Adhwara system of rivers, the entire zone is under the influence of rivers Gandak, Burhi Gandak, and Ghaghra, all of which originate in the lime rich foot hills of the Himalayas. Thus, the soils under the influence of Gandak, Burhi Gandak, and Ghaghra are mostly calcareous having different amounts of lime in them. The soils of Siwan and Gopalganj districts with less rainfall and more pronounced dry seasons have developed salinity as well as alkalinity. Similarly, the soils of the northern part, under the influence of Adhwara group of rivers, are neutral, acidic or saline depending on the local physiography.

(ii) *Zone-II, North East Alluvial Plains*: Zone-II with alluvial plains of Kosi, Mahananda and its tributaries and Ganga (narrow strip in the South) is slightly undulating to rolling landscape mixed with long stretches of nearly flat landscape with pockets of area having sub-normal relief. The area is full of streams with abandoned or dead channels of river Kosi. Its frequent and sudden change of course has left small lakes and shallow marshes. In the south, in between the natural levees of Ganga on the one hand and Kosi and Mahananda on the other, there are vast areas, which remain waterlogged over a considerable part of the year.

(iii) *Zone-III, South Bihar Alluvial Plains*: This Zone comprises of the alluvial plain of river Ganga on its southern side and the plains of all those rivers that flow into Ganga from the south having their origin in the Chotanagpur plateau. These rivers bring with them lot of fine sediments. The coarser sediments that they bring are either deposited in their beds or on their bunds. The soils are thus mostly medium to heavy textured throughout the depth of the profile. The general land slope moderate to low gradient is towards north and east. There are no marshy lands in the zone.

SIZE OF HOLDINGS

Distribution of Operational Holdings

As 77 per cent of the States' main workforce is employed in the agricultural sector, average size of operational holdings in Bihar is only 0.60 hectare as against the national level of 1.69 hectare (Government of India, 2005). The number of marginal holdings having less than 1 hectare is as much as 82.9 per cent. Marginal and small holders account for about 59.8 per cent of the total operated area. (Table 3.5.)

As per the Bihar Agricultural Census 1995-96, the average size of holdings in the state as a whole is very uneconomic being only 0.60 hectare. In zone-I it is smallest being only 0.49 hectare, followed by zone-II with 0.67 hectare and zone-III 0.70 hectare. Zone-I has as high as 88.45 per cent holdings less than one hectare of land. In zone II and III also, the concentration of holdings in this category was as high as 81.5 and 80.2 per cent respectively giving a state average of 84 per cent of land holdings operating 43 per cent of the land. On the other hand, share of holdings operating 5 hectares and above was 1.5 per cent in the State but operating almost 15 per cent of the land. The data indeed reveal how skewed is the distribution of agricultural land in the state. The zone-wise position is also the same, the condition in zone-I being worse (Table 3.6).

Table 3.5: Number and Area of Operational Holdings—Size-wise in Bihar State in 1985-86 to 2001

Year	*No. of Holdings ('000)*					*Area ('000 hectares)*				
	0-2	*2-4*	*4-10*	*Above 10*	*Total*	*0-2*	*2-4*	*4-10*	*Above 10*	*Total*
1985-86	10303	951	404	52	11710	5772	2593	2293	839	10897
1990-91	11631	945	351	39	12966	5545	2576	1982	640	10743
1995-96	12870	941	314	29	14155	5889	2566	1749	479	10682
2001	10812	589	164	9	11574	4203	1544	861	140	7648

Source: Agricultural Census Division, Ministry of Agriculture, New Delhi.

Note: Operational holding is defined as a techno-economic unit wholly or partly for agricultural production, and operated (directly/managed) by one person alone or with assistance of others, without regard to title size or location. A household is considered to have two operational holdings only when either: (1) two or more members of the household independently direct/manage two distinct economic units engaged in agricultural production, even if most of the technical resources used by them are the same or (2) there are two distinct technical units engaged in agricultural production, even if managed/ directed by the same set of persons.

Unproductive Agrarian Structure and Agrarian Relations

Small farms with fragmented landholdings along with insecure tenancies continue to affect agricultural performance in Bihar. While in districts Saran, Muzaffarpur, Sitamarhi and Vaishali, the average size of holding was even less than 0.50 hectare, there are Districts like Rohtas (1.3), Aurangabad (0.98), Khagaria (0.94) and Purnea (0.95) where the average size of the holding is very high (Table 3.7).

According to Bihar Agricultural Census, 1995-96, 93 per cent of the total operational holdings are below 2 hectares. Only about 1.4 per cent holdings are above 5 hectares. There are Districts like Aurangabad, Champaran East, Gaya, Katihar, Munger, Patna, Purnea and Vaishali where over 98 per cent of holders are small/marginal. (**Annexure 3.1**).

Table 3.6: Zone-wise Distribution of Number of Holdings and Operated Area in Bihar (1995-96)

(*No. in, '000 and area in, '000 ha.*)

Category	Zone I		Zone II		Zone-III		Bihar	
	No. of Holdings	*Operated Area*	*No. of Holdings*	*Operated Area*	*No. of Holdings*	*Operated Area*	*No. of Holdings*	*Operated Area*
< 1 ha.	4480 (88.45)	1241 (50.25)	1868 (81.48)	565 (36.83)	3230 (80.24)	1128 (40.21)	9578 (84.14)	2934 (43.09)
1-1-99 ha.	350 (6.91)	437 (17.68)	237 (10.33)	313 (20.41)	464 (11.53)	558 (19.89)	1051 (9.23)	1308 (19.20)
2-3-99 ha.	182 (3.59)	484 (19.62)	142 (6.21)	375 (24.46)	259 (6.43)	698 (24.89)	583 (5.12)	1558 (22.80)
4-9-99 ha.	50 (0.99)	262 (10.60)	43 (1.87)	239 (15.58)	69 (1.71)	368 (13.13)	162 (1.42)	869 (12.76)
10 ha. and above	3 (0.06)	46 (1.86)	3 (0.11)	42 (2.73)	4 (0.09)	53 (1.91)	9 (0.08)	141 (2.07)
All	5065	2469	2292	1534	4025	2806	11382	6810
Average size of holdings (ha.)	0.49		0.67		0.70		0.60	

Source: Bihar Agricultural Census, 1995-96 (through personal communication).

Note: Figures in parentheses indicate percentage to total. These figures for total do not tally exactly with those in Table 3.5

Table 3.7: Average Size of Land Holding by District 1995

(*Area in ha.*)

State/District	*Average Size of Holding*	*State/District*	*Average Size of Holding*
Patna	0.65	Madhubani	0.55
Nalanda	0.63	Samastipur	0.56
Bhojpur	0.95	Sahrsa	0.89
Rohtas	1.30	Araria	NA
Gaya	0.85	Purnia	0.95
Jehanabad	0.61	Madhepura	0.89
Aurangabad	0.98	Kishanganj	NA
Nawada	0.70	Katihar	0.85
Saran	0.45	Bhagalpur	0.82
Siwan	0.68	Munger	0.79
Gopalganj	0.63	Begusarai	0.64
Muzzafferpur	0.47	Khagaria	0.94
Sitamarhi	0.40	East Champaran	0.60
Sheohar	NA	Darbhanga	0.54
West Champaran	0.65	Bihar	0.87
Vaishali	0.36	India	1.69

The 48th round of NSSO further shows that about 1008 thousand rural households are absolutely landless. Due to poor implementation of land reforms, there has not been much change in the structure of land ownership and cultivation over time. Small and marginal farmers not only do not have any invisible surplus of their own, but also lack adequate access to institutional credit. Besides, this, other major problem is that land holdings are severally fragmented Table 3.8.

Table 3.8: Extent of Fragmentation of Holdings in Bihar (1990-91)

	No. of Parcels/Holding		*Area/Parcel (Ha.)*		*Area/Holding (Ha.)*	
	Bihar	*India*	*Bihar*	*India*	*Bihar*	*India*
<1 ha.	3.14	1.96	0.11	0.22	0.32	0.43
1-1.99 ha.	6.63	2.88	0.21	0.49	1.29	1.42
2-3.99 ha.	5.17	3.59	0.53	0.76	2.68	2.74
4-9.99 ha.	10.41	4.88	0.54	1.21	5.47	5.91
10 ha. and above	20.64	6.69	0.81	2.43	16.75	16.25

Source: All India Report on Agricultural Census, 1990-91, Bihar Agricultural Census, 1990-91.

The marginal farmers below 1 hectare have on an average 3 parcels per holding, while the small farmers have 6.63 parcels per holding. Even the large farms above 5 hectares are not consolidated to allow for proper crop planning and investment in tubewell etc. Those with 5-10 hect. have 10.4 parcels and the ones with over 10 hect. have an unacceptable 20.64 parcels. The implementation of consolidation of holding even on voluntary basis has been suspended in recent years. Inequality in land ownership and insecure land tenure have aggravated the rural poverty situation in Bihar and acted as a deterrent for investment. In addition, insecure tenancies and poor land administration have stood in the way of diversified growth. There is a legal ban on land leasing in Bihar, but informal leasing continues.

Various legal loopholes and implementation difficulties developed leasing of agricultural land, yet it is being done in many places in a concealed manner. This gives rise to a number of negative issues (including exclusion from formal credit channels), that can potentially exploit lessees, particularly the landless, marginal or small farmers who make up the majority of lessees (National Commission on Farmers, 2006). Though tenancy and land reforms policy has been discussed and has been under consideration for quite some time, there has been little progress. Reducing rural poverty in Bihar would thus require effective implementation of land reforms in order to secure the rights of tenants, who are mostly small and marginal farmers, and landless. (Table 3.9) Bargadari system of West Bengal could be a model in this respect. This will help the land holder to have not only right to land, but also ensure work on an economic holding.

Table 3.9: Percentage of Households Leasing in and Leased in Area in Bihar

Land size (Acres)	*Percentage of Households Leasing in*		*Percentage of Area Leased in to Total Area Cultivated*	
	1981-82	*1999-2000*	*1981-82*	*1999-2000*
Landless	34.53	22.63	100.00	100.00
0-1.0	53.57	36.09	62.63	58.15
1.0-2.5	36.99	28.57	37.52	27.62
2.5-5.0	34.48	5.19	18.87	3.88
5.0-10.0	10.00	0.00	4.21	0.00
10.0-20.0	3.85	11.11	0.97	5.04
20+	0.00	1.17	0.00	1.98

Source: A.N. Sinha, Institute of Social Sciences (1982) and Institute of Human Development (2002).

The 59th Round of National Sample Survey data reveal that about 12 per cent of the operated area is under tenancy, but some micro level studies indicate that as high as 25 to 35 per cent of the total cultivated area in the state is under informal tenancy which not only served as a disincentive to cultivate land efficiently, but also debars access to institutional credit for want of collateral securities. Share-cropping is the most dominant form of lease, but the leasing in on fixed rent basis is becoming popular, especially in irrigated areas. The Tenancy Act of 1885, as amended in 1986 prescribes 25 per cent of the gross produce as the maximum rent payable, but as a matter of general rule, 50 per cent of the gross produce is the prevailing rent in the state. Thus, there is the need for a new approach to land reform, involving legislation and formalisation of land leasing, which will help improve poor peoples' access to land through leasing and also encouraging the tenants to cultivate land efficiently.

Land Use Pattern

Undivided Bihar state was the largest in area among the states in the eastern region, having a geographical area of 17,388 thousand hectare, occupying around five per cent of the country's total. However, in the year 2000 A.D. the state was bifurcated and a new state of Jarkhand was created carving out the predominantly tribal inhabitated districts. As a result, the present geographical area of the state is only 9,416 thousand hectares—present Bihar state is only 54 per cent of the undivided state. Further, the reporting area for Land Utilization Statistics (LUS) is also marginally smaller being only 9360 thousand hectares.

Undivided Bihar had a considerable forest coverage of 2949 thousand hectares. But presently, the lion's share of this forest area has gone to Jharkhand. As per land utilization statistics of the state, area under forest is only 622 thousand hectares now. Thus, the forest coverage in the state is dismally low being less than 7 per cent of the geographical area necessitating implementation of a vigorous afforestation programme in the state. Area put to non-agricultural use is slowly increasing in the state. For example, in 2000-01 the estimate of such area was 1618 thousand hectares. But within a span of two years (i.e. by 2002-03) it increased to 1643 thousand hectares. However, the size of barren and unculturable land remained more or less the same at 436 thousand hectares during the above mentioned period. Land under permanent pastures and other grazing land also remained constant at 18 thousand hectares. Culturable waste land in the state was estimated to be nearly 46 thousand hectares. There was a marginal increase in area under miscellaneous tree crops and groves, from 231 thousand hectare in 2000-01 to 237 thousand hectares in 2003-04.

Table 3.10: Land Use Classification and Irrigated Area in Bihar State

('000 hectares)

Year	Geographical Area	Reporting area for land utilization statistics	Forest	Area put to non-agriculture use	Barren and un-cultural use	Total (Col. 5+6)	Permanent pastures and other grazi lands	Land under misc.tree crops and groves not included in area sown	Culturable waste lands nest	Total (Col. 8+9+10)	Fallow lands: Fallow lands other than current fallows	Fallow lands: Current fallows	Fallow lands: Total (col. 12+13)	Net Area sown	Gross Cropped area	Area sown more than once (col. 15-16)	Net Irrigated Area	Gross Irrigated Area
(1)	(2)	(3)	(4)	(5)	(6)	(7)	(8)	(9)	(10)	(11)	(12)	(13)	(14)	(15)	(16)	(17)	(18)	(19)
2000-01	9416	9360	616	1638	437	2075	18	231	46	295	135	576	711	5663	7992	2330	3429	4561
2001-02		9360	622	1642	436	2079	18	235	46	299	135	563	697	5664	7897	2233	3462	4539
2002-03		9360	622	1643	436	2079	18	237	46	301	133	499	632	5725	7957	2232	3462	4572
2003-04		9360	622	1645	436	2081	18	238	46	302	130	513	643	5712	7882	2170	3433	4567
2004-05		9360	622	1645	436	2081	18	238	46	302	130	513	643	5712	7802	2090	3433	4197
2005-06		9360	622	1645	436	2081	18	238	46	302	130	513	643	5712	7812	2100	3455	4325

Fallow lands are universally classified into two categories viz., fallow lands other than current fallows and current fallows. In this state while area fallow land remained more or less stagnant at around 135 thousand hectares, the size of current fallow land declined from 576 thousand in 2000-01 to 499 thousand hectares in 2003-04.

There were fluctuations in the net and gross area sown over the years due to variations in weather and rainfall conditions (Table 3.10). One of the worrying issue is that with more or less stagnant irrigated area and a fall in the double cropped area, both net and gross sown areas have declined over the years.

Over 61 per cent of the reported area is under cultivation. This is more or less the same as was 25 years ago in 1981. Land put to non-agricultural uses has, however, increased from about 12.5 to 17.6 per cent which is primarily at the cost of permanent fallows, cultivable waste and to a small extent permanent pastures and grazing lands. With urbanization and industrialization, and increasing road net work, net sown area may actually come down, even by a small margin. There is no possibility of any increase in the area in this respect. Gross sown area will, however, increase as new irrigated areas are added.

Zonal Land Use in Bihar (Table 3.11) indicates that Zone III (A and B) has the maximum land area. (49%) of the state which is followed by Zone I accounting for almost 37 per cent. Zone II, however, shares only 19 per cent of state's total area. While net sown area of Zone II and Zone I is 65.72 and 65.41 per cent respectively of their total land area, in Zone III it is only 56.29 per cent. Zone II tops with gross area sown at 103.8 per cent with a maximum double cropped area of 38 per cent. This is due to a little more than 50 per cent of its area being irrigated. (Table 3.11).

Table 3.11: Zonal Land Usage, and Percentage of Irrigated Land

'hectare'

Districts	*Land Area*	*Net Sown*	*Total Crop*	*Multi-Cropped Area*	*Total irrigated area*	*% irrigated area*
			Zone-I			
Saran	265,000	192938	233024	40086	117706	44.42
Siwan	224,000	162889	238635	75746	107262	47.88
Gopalganj	204,000	150524	232775	82251	107689	52.79
Muzaffarpur	315,000	207145	339364	132219	130164	41.32

(*Contd.*)

Table 3.11 (*Contd.*)

Districts	*Land Area*	*Net Sown*	*Total Crop*	*Multi Cropped Area*	*Total irrigated area*	*% irrigated area*
E. Champaran	432,000	288804	323302	34498	155655	36.03
W. Champaran	484,000	279758	366712	86954	176683	36.50
Sitamarhi	222,000	121048	188399	67351	71682	32.29
Sheohar	44,000	26483	43567	17084	12652	28.75
Vaishali	201,000	126650	190331	63671	77546	38.58
Darbhanga	254,000	172716	309949	37233	102087	40.19
Madhubani	354,000	225113	317083	91970	138551	39.14
Samastipur	262,000	184718	252196	67478	112387	42.90
Begusarai	188,000	117193	174313	57120	91089	48.45
Total	3,449,000	2255989	3109650	853661	1401153	40.62
% of Total Land Area	36,85	65,41	90.16	24.75	40.62	
			Zone-II			
Saharsa	165,000	109633	194022	84389	94983	57.57
Supaul	239,000	155251	268820	113569	142114	59.46
Madhepura	180,000	131531	205481	73950	133124	73.96
Purnia	314,000	221166	307003	85837	170504	54.30
Kishanganj	189,000	131105	195648	64543	170504	26.74
Araria	271,000	180983	282645	101662	50535	40.95
Katihar	291,000	167217	278835	111618	110962	44.47
Khagaria	149,000	84684	133902	49218	129408	55.97
Total	1,798,000	1,181,570	1,866,356	684,786	83393	50.89
% of Total Land Area	19	65.72	103.80	38.09	915023	
			Zone III A			
Munger	140,000	48029	69030	21001	37015	26.44
Sheikhpura	62,000	44217	62737	18520	53423	86.17
Lakhisarai	129,000	68044	78098	10054	41934	32.51
Jamui	305,000	81117	88723	7606	50780	16.65
Bhagalpur	254,000	145667	177576	31909	78364	30.85
Banka	306,000	153818	161325	7507	115698	37.81
Total	1,196,000	540,892	637,489	96,597	377214	31.54
% of Total Land Area	13	45.23	53.30	8.08	31.54	
			Zone III B			
Patna	317,000	206294	253848	47554	179595	56.65
Nalanda	232,000	180872	228353	47481	193288	83.31
Bhojpur	237,000	185364	227536	42172	188094	79.36
Bhuxar	167,000	138277	144944	6667	113910	68.21
Rohtas	391,000	254360	363159	108799	330834	84.61

(*Contd.*)

Table 3.11 (*Contd.*)

Districts	*Land Area*	*Net Sown*	*Total Crop*	*Multi-Cropped Area*	*Total irrigated area*	*% irrigated area*
Kaimur	342,000	154226	204719	50493	166537	48.70
Gaya	494,000	200333	277364	77031	235090	47.59
Jahanabad	94,000	63650	80024	16374	67206	71.50
Arwal	63,000	41784	58263	16479	48131	76.40
Nawada	249,000	110565	149046	38481	126558	50.83
Aurangabad	330,000	197912	281626	83714	224201	67.94
Total	2,916,000	1,733,637	2,268,882	535,245	1873444	64.25
% of Total Land Area	31	59.45	77.81	18.36	64.25	
Bihar	9,630,000	5,712,088	7,882,377	2,170,209	4,566,834	48.79

Source: Annual Report 2005-06, Department of Agriculture, Government of Bihar.

Note: The figures for Total Land Area in this report (9.36 million hectares) is different from the figures Census 2001 9.42 million hectares), but agrees with the Ministry of Agriculture Data (Table 3.12).

Graph 3.1: Land Use Classification and Irrigated Area in Bihar State

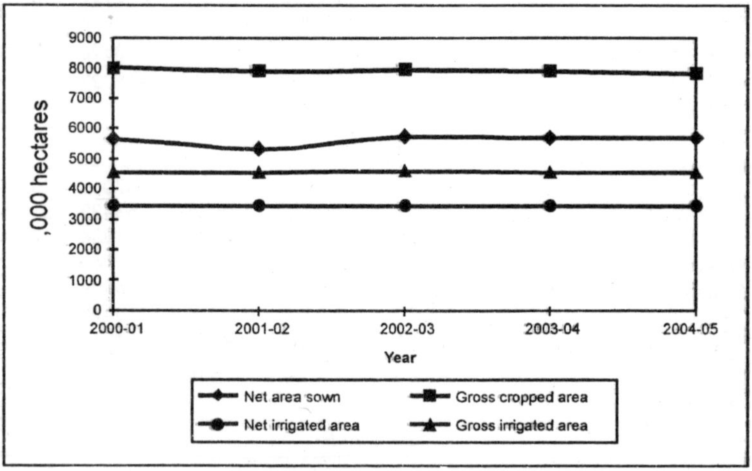

AGRICULTURAL SCENARIO

Agriculture is at the core of Bihar's economy. About 90 per cent of the state population of 83 million (2001) reside in rural areas and more than three quarters (62.2 million) of the workforce are employed in agriculture. Workforce participation rate in Bihar is 77.3 per cent as against 63.9 per cent in India. The contribution of agriculture and its allied services sector to the state domestic product was 40 per cent in 2003-04. Although there is a decline from 48.5 per cent in 1980, it remains amongst the highest at the all India level. (Table 3.12).

Table 3.12: Agriculture in Bihar and Other States

	Bihar	*Orissa*	*AP*	*UP*	*West Bengal*	*All India*
Real agriculture and allied sector growth						
1980-1992 (Per cent)	1.6	0.5	2.3	2.7	6.2	3.1
1993-2003 (Per cent)	2.7	0.1	2.8	2.2	5.4	2.2
GSDP/capita (2000) (Rs.)	3,650	6,278	11,154	6,638	10,757	11,752
Share of agri and allied (2003-04)	39.8	30.0	28.0	33.6	24.7	24.2
Agri and allied/capita (Rs.)	1,452	1,884	3,121	2,230	2,658	2,847

Source: World Bank, 2004. Bihar Agriculture Building on Emerging Models of Success.

During 1980-1992, Bihar's agriculture and allied sector grew at 1.6 per cent per annum, compared with the national average of 3.1 per cent. In fact it was the lowest among most of the states. However, since 1993, growth in Bihar's agricultural GSDP has accelerated. While the all-India agricultural GDP grew at 2.2 per cent per annum between 1993 and 2003, the corresponding figure for Bihar was 2.7 per cent which was next only to West Bengal and Andhra Pradesh among the major Indian states[1] (Table 3.12). A study for the period 1995-96 to 2004-05 not only confirms the above findings but also shows that, Bihar went ahead of both these states, though marginally. (Table 3.13).

Crop and livestock sector in Bihar account for over 90 per cent of the agriculture and allied sector. As stated earlier. It has grown at an annual rate of 2.7 per cent between 1993 and 2003, compared to the national average of 2.2 per cent. Of all components of the agriculture and allied sector, the most rapid growth occurred in the fisheries sub-sector (8%), albeit from a smaller base (Table 3.14).

Agricultural Growth by States: 1993-94 = 2003-04

(*Percentage*)

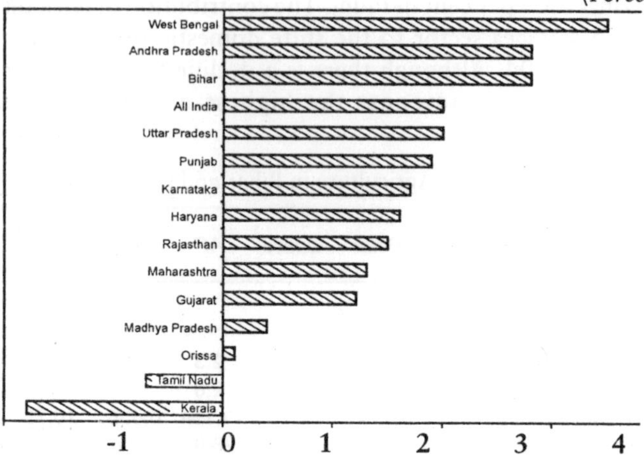

Note: Growth rates are for the Agriculture and Allied Sector GSDPs and are estimated using log-linear regressions.

Source: Estimated-Data from CSO, National Accounts Statistics

Table 3.13: Level of Aggregate Productivity and Growth in NSDP Agriculture in Various States

State	*NSDP Ag./Ha.NSA at Current Prices*			*Trend Growth Rate in NSDP Agri at 1993-94 Prices*	
	1984-85 and 1985-86	*1994-95 and 1995-96*	*2003-04 and 2004-05*	*1984-85 to 1995-96*	*1995-96 to 2004-05*
Rajasthan	2264	7912	13984	5.52	0.30
Chhattisgarh		8586	13013		-0.91
Madhya Pradesh (old)	2319	9406	16745	3.63	-0.23
Madhya Pradesh (new)		9667	17939		-0.05
Maharashtra	2863	12712	19375	6.66	0.10
Orissa	4168	13499	26115	-1.18	0.11
Gujarat	3748	14175	26404	5.09	0.48
Karnataka	3496	14481	19842	3.92	0.03
All India	4973	17763	34349	3.62	1.85
Bihar (new)		18622	28915		2.82
Bihar (old)	6654	18743	32654	-1.71	3.51
Jharkhand		19124	44727		5.39
Andhra Pradesh	4478	20204	40653	3.18	2.69
Uttar Pradesh (new)		20413	39811		1.91

(*Contd.*)

(Contd.)

State	NSDP Ag./Ha.NSA at Current Prices			Trend Growth Rate in NSDP Agri at 1993-94 Prices	
	1984-85 and 1985-86	1994-95 and 1995-96	2003-04 and 2004-05	1984-85 to 1995-96	1995-96 to 2004-05
Uttar Pradesh (old)	5863	20896	40426	2.82	1.87
Assam	6617	23283	40671	1.65	0.95
Tamil Nadu	5338	24558	36677	4.95	-1.36
Himachal Pradesh	6561	25270	71775	1.64	3.51
Haryana	6672	26604	48154	4.60	1.98
Jammu and Kashmir	8736	30554	69113	2.18	3.25
Uttaranchal		31296	53743		1.36
West Bengal	9499	34863	72295	4.63	2.67
Punjab	8467	35417	66864	4.00	2.16
Kerala	9564	38370	56341	3.60	-3.54

Source: (1) State Domestic Product (State Series), Central Statistical Organisation, GoI, New Delhi, various issues.

(2) Agricultural Statistics at a Glance, Ministry of Agriculture, GoI, New Delhi.

(TERI *Source*: *EPW*. Vol. XL II No. 26, Ramesh Chandra)

Table 3.14: Sub-Sectoral Agricultural Growth

	Bihar		India	
	Percentage Share	Growth 1993 to 2003 (% per annum)	Percentage share	Growth 1993 to 2003 (% per annum)
Crops (including Livestock)	91	2.5	91	2.1
Forestry	4	3.3	5	1.8
Fishing	5	8.0	4	4.3
Agriculture and Allied GSDP	100	2.7	100	2.2

Source: CSOI National Accounts Statistics.

It has, however, to be understood that Bihar's agricultural growth is highly unstable with significant year-to-year variations. Between 1993 and 2003 the variability of agricultural performance in Bihar was the third greatest among major Indian states. However, unlike states like Rajasthan or Gujarat where high variability arises because of frequent droughts, in Bihar the primary source of such variability is floods which frequently ravage large parts of northern Bihar during the monsoon. The rapid spread of HYVs and fertilizer use on rainfed lands have also contributed to the instability as output from these lands quite high in favourable years, but falls off significantly in unfavourable ones.

Crop Growth Pattern

Like all the states of eastern India, rice is the predominant crop in Bihar. During the triennium ending (TE) 2003-04, it occupied nearly 44 per cent of the total cropped area, though it was marginally lesser than that for TE 1980-81 when it was 45 per cent. On the other hand, the share of area under wheat increased to 26.6 per cent in TE 2003-04 as against 19.8 per cent in TE 1980-81. This was possible due to significant shift in area from maize, barley and other coarse cereals as gross return from wheat was higher than those of the other coarse cereals. Over the period, there was also considerable decline in the share of area under pulses from 12.9 to 9.8 per cent. Thus, but for wheat, a decline in the share of all other food crops was observed in the state. (Table 3.15).

Average yield of rice-wheat cropping system is 1.45-2.19 t/ha. as against the production potential (experimental yields at research farms) of 4.5-5.0 t/ha. Similarly, average maize yield in the State is about 2.38 t/ha. as against its yield potential of 6 t/ha. Even though the state is rich in soil and water resources, yield levels of rice, wheat and maize in the State are only about 32, 44 and 40 per cent respectively of potential yields. There is thus a considerable scope to increase the productivity of rice, wheat and maize in Bihar. The State has high potential for development of horticulture. Intervention for production, processing, post harvest management and marketing are expected to harness the full potential of this sector. Projects for development of mango in Darbhanga, litchi in Muzaffarpur, Makhana in Madhubani and spices in Samastipur are being already taken up.

Though share of area under non-food crops recorded an increase over the period, there was a marginal decline in the case of oilseeds by 0.10 per cent. Even in the case of fibre crop share in the gross cropped area recorded a decline from 2.4 to 2.0 per cent over the period. Spices and condiments, minor crops in the state, which occupied only 0.30 per cent of the gross cropped area in TE 1980-81, recorded further decline to 0.10 per cent in TE 2003-04. In the process

Table 3.15: Changes in Cropping Pattern Over Time in Bihar.

Crop	*Percentage of Total Cropped Area*	
	TE 1980-81	*TE 2003-04*
1. Rice	45.10	43.90
2. Wheat	19.80	26.60
3. Maize	8.30	3.20
4. Barley and other coarse cereals	2.90	0.70
5. Pulses	12.90	9.80
6. Total Foodgrains	89.00	84.20
7. Oils Seeds:	1.90	1.80
(a) Rapeseed and Mustard	0.80	1.10
(b) Linseed	1.00	0.60
(c) Sunflower	0.03	0.10
(d) Others	0.07	0.00
8. Fibers	2.40	2.00
9. Sugarcane	1.40	2.30
10. Fruits and Vegetables	4.30	9.20
11. Spies and Condiments	0.30	0.10
12. Others crops	0.70	0.10
13. Total	100.00	100.00

Source: Based on data collected from the Directorate of Economics and Statistics, Ministry of Agriculture, Govt. of India.

Changes in Cropping Pattern Over Time

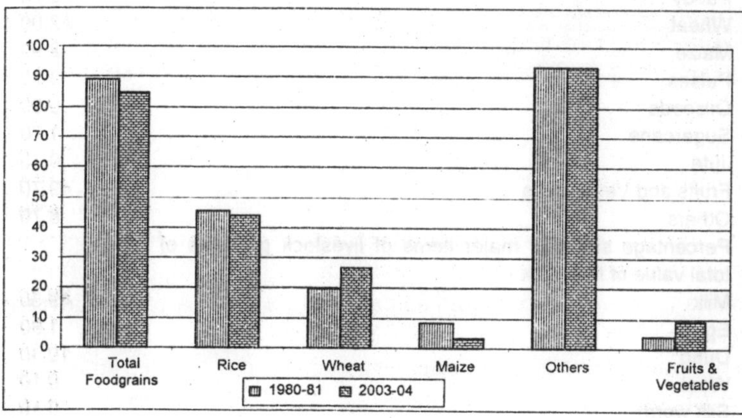

of shifting of land from foodgrain to non-foodgrain, maximum increase in area has taken place under fruits and vegetables where the share in the gross cropped area increased from per cent in TE

1980-81 to per cent in TE 2003-04. This change can be attributed to higher per hectare return. The share of sugarcane also went up from 1.40 to 2.30 per cent during the aforesaid period.

Structure of Agricultural Produce

It may be stated that with the growing commercialization of agriculture, cropping pattern changes according to the expectation of net return from the land possessed by the farmers. A bird's eye view of the data in Table 3.16 will reveal that in TE 2001, the total value of farm output including livestock was Rs. 31,70,930 lakh, (the share of livestock being Rs. 9,13,128 lakh or about 29%). In food crops, fruits and vegetables accounted for nearly 49.7 per cent of the total value, followed by rice 19.7 per cent, wheat 12.0 per cent

Table 3.16: Structure of Agricultural Production in Bihar (TE 2001)

Features	*Bihar*
1. Total Value of crop output at current prices in 2001 (Lakh Rupees)	2257802.00
2. Total value of livestock output at current prices in 2001 (Lakh Rupees)	913128.00
3. Percentage share of crops in the total value of crop output	
Paddy	19.70
Wheat	12.00
Maize	2.90
Pulses	3.70
Oilseeds	0.90
Sugarcane	1.70
Jute	0.70
Fruits and Vegetables	49.70
Others	8.70
4. Percentage share of major items of livestock products of total value of livestock	
Milk	49.30
Eggs	1.60
Dung	10.30
Wool	0.10
Silk worm	0.10
Increment in stock	14.20
Meat	24.40

Source: Estimates of Value of Agricultural output, Central Statistical Organization, Government of India.

pulses 3.7 per cent, maize 2.9 per cent and other crops 8.7 per cent. Among the livestock products which accounted for 29 per cent of the total value of agricultural output, milk constituted 49.3 per cent, followed by meat 24.4 per cent and other products 26.3 per cent.

Bihar also produces about 2.6 lakh tonnes of fish annually. There are about 2 lakh hectares of permanent water area and 1.2 lakh hectares of seasonal water area under fisheries in the state. There is also another 5 lakh hectares of paddy area under deep water conditions, which can be exploited for fisheries. The contribution of fish in the state GSDP in 2004-05 was 1.85 per cent.

It may be seen further that per hect. value of output of major crops in Bihar (Rs. 3,017) is a shade lower than all India (Rs. 3,576). But it is interesting to note that 7 districts namely, West Champaran (Rs. 4,170), Nalanda (Rs. 4,034), Siwan (Rs. 3,996), Gopalganj (Rs. 3,920), Darbhanga (Rs. 3,844), Saran (Rs. 3,633), Khagaria (Rs. 3,576) and Bhajpur (Rs. 3,521) have higher value than the all India average. One more observation will be important-districts Rohtas (Rs. 3,521), Nalanda (Rs. 3,529) (so called drought prone districts) are not far behind the all India average. Another drought prone district, Aurangabad (Rs. 3,074) is also doing well. (Table 3.17).

Table 3.17: District-wise Value of Output of Major Crops

(Rs. per ha.)

District	Value of output of major crops/ha	District	Value of output of major crops/ha
Patna	3215	Vaishali	2872
Nalanda	4034	Darbhanga	3844
Bhojpur	3521	Madhubani	2446
Rohtas	3531	Samastipur	NA
Gaya	2367	Sahrsa	2546
Jehanabad	NA	Araria	2218
Aurangabad	3074	Purnia	NA
Nawada	3529	Madhepura	2291
Saran	3633	Kishanganj	2861
Siwan	3996	Katihar	2946
Gopalganj	3920	Bhagalpur	3301
Muzaffarpur	2792	Munger	2298
Sitamarhi	2641	Begusarai	3017
Sheohar	NA	Khagaria	3576
West Champaran	4170	**Bihar**	**3017**
East Champaran	3560	**India**	**3576**

Crop Growth

Foodgrains comprising rice, wheat and maize are the major crops grown in the state having an area of around 7 million hectares producing around 11 million tonnes. Rice occupying nearly 51 per cent of the total area under foodgrains, shares around 59 per cent production. Wheat (2.12 million hectares in 2003-04), accounts for 30 per cent of area and around 36 per cent of the foodgrains

Table 3.18: Area, Production and Yield of Major Crops in Bihar During 2000-01 to 2006-07*

Area ('000 hectares)
Production ('000 tonnes)
Yield (kg./ha.)

	Bihar			*All India*		
	Area	*Production*	*Yield*	*Area*	*Production*	*Yield*
Rice						
2000-01	3656.3	5442.6	1489 (78.3)	44712.0	84976.6	1901
2003-04	3578.0	5447.8	1523 (73.3)	42592.0	71820.2	2078
2005-06	3222.0	3417.0	1061 (50.7)	43492.0	91039.0	2093
2006-07	3137.0	3488.0	1112 (55.9)	38079.0	75739.0	1989
Wheat						
2000-01	2067.8	4438.0	2146 (79.2)	25730.6	69680.9	2708
2003-04	2076.8	3688.9	1776 (65.5)	26594.7	72156.2	2713
2005-06	2038.0	3660.0	1796 (68.9)	26647.0	69481.0	2607
Maize						
2000-01	620.5	1497.2	2413 (132.4)	6611.3	12043.2	1822
2003-04	616.4	1473.5	2390 (117.1)	7343.4	14984.3	2041
2005-06	622.0	1424.0	2289 (117.2)	7727.0	15091.0	1953
Oilseeds						
2000-01	154.7	131.1	847 (104.6)	22770.0	18436.8	810
2003-04	140.6	123.8	881 (82.8)	23662.9	25186.1	1064
2005-06	136.0	119.0	875 (87.5)	27737.0	27731.0	1000
2006-07	6.0	5.0	833 (99.0)	15738.0	13239.0	841
Sugarcane						
2000-01	93.5	3987.6	42648 (62.2)	4315.7	295956.2	68577
2003-04	103.6	4285.9	41370 (69.7)	3938.4	233861.0	59380
2005-06	102.0	4346.0	42608 (65.0)	4245.0	278387.0	65580
2006-07	103.0	4249.0	41252 (63.8)	4386.0	283404.0	64615

*—First advance estimate

Note: Figures in brackets represent per cent of India.

Source: Agricultural Statistics Division, Ministry of Agriculture, New Delhi.

production. The rest is under maize crop. (Table 3.18). Latest data for 2006-07 are given in Appendix 3.1

Though rice is the main crop in the state contributing nearly 6 per cent towards the country's production, yield rate is one of the poorest (1112 kgs. per hectare) as against Punjab (4000 kgs.), Haryana (3000 kgs.) and all-India's average of 2000 kgs. This is despite the fact that around 54 per cent of the area is irrigated.

In so far as wheat is concerned the share of Bihar in area and production is 8 per cent and 5 per cent respectively. Even in this crop which is grown mostly. On irrigated area, yield rate is 1776 kg. as against 4207 kg. in Punjab, 3966 kgs. in Haryana and all-India average of 2713 kg. However, in maize, the position is better, ranking fourth both in area and production among the maize producing states in the country. The average productivity of the crop is 2371 kg. per hectare which is much higher than the all-India average of 1906 kg. per hectare. An IFAD-CIMMYT survey conducted in 2001 in the 3 districts of Bihar viz. Munger, Siwan and Begusarai showed that maize had the highest productivity of 5 to 6 mt. per hectare in the state.

Zone-wise Cropping Pattern

A study of the zonal growth pattern of agriculture in Bihar shows that rice is doing well in Zone-II, wheat in Zone I and II, Maize in Zone III and, Sugarcane in Zone I. Chickpea which has the highest productivity levels even at the international levels, has shown a declining trends in all the zones (Table 3.19). There are, of course, selected districts like Gopalganj where the growth rate is the highest (7.7 per cent) during 1990-04. Other districts like Bhojpur, Champaran East, Katihar, Muzaffarpur, Patna, Rohtas, Saran, Singbhua, Sitamarhi and Vaishali where growth rate in area is in the range of 2 to 4 per cent during the same period.

The state also grows a small quantity of oilseeds (0.51 per cent of the country's production), and jute and mesta. Though jute and mesta occupies only 0.17 million hectares, they rank second in the country, after West Bengal. The state shares nearly 12 per cent of India's production of raw jute.

Other crops which have some importance in the state are sugarcane (0.10 million hectares and 6.62 million tonnes), potato ranking third in the country (159 thousand hectares and 1539

Table 3.19: Distribution of Total Area between Different Crops-zone-wise

Sl. No.	Crop	Zone I		Zone II		Zone III		Bihar	
		TE 1981	TE 1999	TE 1981	TE 1999	TE 1981	TE 1999	TE 1981	TE 1999
1.	Rice	45.33	43.53	40.94	39.20	49.15	51.75	45.11	44.37
2.	Wheat	20.13	26.74	15.54	19.56	23.96	28.57	19.85	24.82
3.	Maize	9.31	7.52	13.22	12.66	2.08	1.28	8.35	7.51
	Coarse cereals	13.35	8.49	15.61	13.53	3.75	2.10	11.16	8.40
	Total cereals	78.81	78.76	72.09	72.29	76.85	82.42	76.13	77.59
4.	Chickpea	0.66	0.83	1.74	0.23	3.66	0.45	1.90	0.52
8.	Total Pulses	9.12	6.46	13.14	8.94	17.60	12.84	12.92	9.09
9.	Total foodgrains	87.92	85.22	85.23	81.23	94.45	95.26	89.05	86.68
10.	Jute, Mest and Sunhemp	0.38	0.15	7.02	5.88	0.07	0.01	1.17	2.05
11.	Sugarcane	2.85	2.96	0.36	0.24	0.53	0.26	1.38	1.28
12.	Potato	1.57	1.99	0.94	1.56	1.20	1.35	1.26	1.66
13.	S. Potato	0.93	0.14	0.86	0.14	0.04	0.01	0.64	0.10
14.	Onion	0.19	0.20	0.13	0.15	0.12	0.20	0.15	0.18
15.	Mango	2.21	1.85	0.83	0.64	0.51	0.28	1.27	1.00
16.	Banana	0.19	0.22	0.12	0.40	0.00	0.00	0.11	0.22
17.	Total fruits and vegetables	6.27	9.01	3.79	9.72	2.44	2.01	4.34	7.51
18.	Spices and condiments	0.39	0.18	0.41	0.23	0.12	0.05	0.33	0.16
19.	Miscellaneous	3.4	3.62	1.17	2.94	1.01	0.89	1.97	1.50
	Total	100.0	100.0	100.0	100.0	100.0	100.0	100.0	100.0

Source: Directorate of Statistics and Evaluation, Government of Bihar.

thousand tonnes) and tobacco (14.8 thousand hectares and 16.4 thousand tonnes). The state has 272 thousand hectares under different fruit, producing around 2877 thousand tonnes and shares nearly 7 per cent of the country's area and production.

One may conclude from all this that although lately agricultural performance of the State has been quite satisfactory, present productivity levels of major crops—rice and wheat are much lower than the desired levels. Various factors are responsible for this lower agricultural production and productivity. This is because the State is faced with the problem of recurrent floods and drought. Lack of rural power supply is another constraint. Power generation and utilization rates are among the lowest in India, with only 10 per cent of the rural households have electric connection, compared to 56 per cent for the country. Land holdings in the state are predominantly small with a high degree of fragmentation.

Rural credit delivery system is also awfully inadequate, with a weak cooperative sector and tepid credit flow from commercial banks. Bihar has one of the lowest CD ratio (28 to 31) in the country. Notwithstanding increasing pressure on land by a burgeoning population, challenge and opportunity of agricultural sector lies in the fact that even with present area under cultivation, there is tremendous scope for wealth creation through increased productivity.

Government Agriculture Policy for the State:

- Major focus on Rice, Maize and Pulses.
- Upland paddy to be replaced by Arhar and Urad. Wheat to be replaced by winter maize. Additional area to be brought under Pulses from current fallow with additional source of irrigation. Crop diversification—Fruits, Vegetables, Mushroom, Medicinal and Aromatic plants, Flowers, Honey Production and Spices to be encouraged.
- *Seed Policy*—Revival of Bihar Rajya Beej Nigam seed production on Government farms and encouragement to private seed companies.
- Organic farming, Integrated pest management and balanced use of fertiliser to be encouraged.
- Crop specific strategy to be firmed up or improving cultivation of rice, maize, wheat, pulses, oilseeds and fruits.
- *Objective*—Food security, Balanced nutrition, Raising Income level, Ecological Balance.

Key Concerns

Based on agro-climatic characteristics and other features mentioned above, one can say that the state of Bihar has tremendous potential for diversified and accelerated agricultural growth which in turn would help increase rural incomes and reduce poverty in rural areas. However, this would require formulating an agricultural and rural development strategy that would enable the state to respond more effectively to the new opportunities and challenges arising from an increasingly liberalizing agricultural environment domestically and internationally. For this, there is a need not only to analyse the major sources of agricultural growth in Bihar, but also to identify various physical, technological, infrastructural, institutional and policy constraints to growth as well as poverty reduction.

Note

1 The comparison of pre-and post 1993-94 growth performance is influenced mainly by data considerations. Prior to 1993-94 GSDP data are available only for the erstwhile undivided Bihar, but post 1993-94 GSDP series for divided Bihar and Jharkhand are available separately. The 1993-2003 growth rate of agricultural and allied DSDP in undivided Bihar was 3.5 per cent per annum, confirming that the acceleration from the 1980s is not simply because of difference in coverge between the pre-and post- 1993-94 GSDP series.

APPENDIX 3.1

Zonal Land Utilisation Pattern in Bihar

(Area in Hectares)

District	Geographical Area	Forests	Barren & Unculturable Land	Land Put to Non-Agricultural Use: Land Area	Water Area: Permanent	Water Area: Temporary	Total Col. 5-7	Culturable Waste Land
1	2	3	4	5	6	7	8	9
				Zone I				
Saran	264887	—	18077	26745	3464	2979	33188	165
Siwan	224410	0	8873	27409	1969	1599	30977	776
Gopalgni	203774	0	5571	29729	2073	438	32240	1427
E. Champaran	431715	118	8198	49215	10069	16070	75354	298
W. Champaran	484351	91745	3020	70012	15228	8482	93722	1323
Muzaffarpur	315351	0	5287	49401	7776	4866	62043	336
Sitamarhi	221891	0	1809	43761	2531	15847	62139	137
Sheohar	43475	0	416	9219	1269	1916	12404	28
Vaishali	201449	0	24257	28392	5827	2105	36324	151
Darbhanga	254077	0	1384	42949	9357	7789	60095	159
Madhubani	353498	0	2330	69735	13428	2316	85479	515
Samastipur	262390	0	3932	52929	8189	993	62111	0
Begusarai	187828	0	18125	29062	7668	4025	40755	41
Total Zone 1	3449096	91863	101279	528558	88848	69425	686831	5356
% Share in Total	**36.85**	**14.78**	**23.22**	**41.14**	**42.84**	**45.18**	**41.74**	**11.70**
				Zone 2				
Purnia	313883	113	12427	36438	6856	1487	44781	1162
Araria	271712	838	5088	39353	6438	4819	50610	567
Kishanganj	189080	354	11386	23862	7274	2574	33710	1240
Katihar	291349	1785	22279	39832	12128	4077	56037	792
Saharsa	164559	0	10883	21019	4780	2230	28029	479
Madhepura	179589	0	3949	25346	3859	1070	30275	0
Supaul	238603	0	20393	37859	9764	2912	50535	1558
Khagaria	149342	0	13686	18672	7715	4257	30644	641
Total Zone 2	1798117	3090	100091	242381	58814	23426	324621	6439
% Share in Total	**19.21**	**0.50**	**22.95**	**18.87**	**28.36**	**15.25**	**19.73**	**14.07**
				Zone 3A				
Munger	139793	28524	11501	19731	58.36	5427	30994	951
Shekhpura	62084	0	108	6748	875	2133	9756	242
Lakhisarai	128602	13445	7109	7709	1200	4822	13731	724
Jamui	305289	92855	28897	34271	2188	2873	39332	10467
Bhagalpur	254300	78	22610	51137	6564	9810	67511	2342
Banka	305621	46310	43081	35232	2905	3121	41258	7958
Total Zone 3A	1195689	181212	114278	154828	19568	28186	202582	22684
% Share in Total	**12.78**	**29.15**	**26.20**	**12.05**	**9.44**	**18.34**	**12.31**	**49.55**

(Contd.)

(*Contd.*)

District	Geographical Area	Forests	Barren & Uncultur-able Land	Land Put to Non-Agricultural Use				Culturable Waste Land
				Land Area	Water Area		Total Col. 5-7	
					Permanent	Temporary		
1	2	3	4	5	6	7	8	9
				Zone 3B				
Patna	317236	56	13119	62187	10332	2527	75046	767
Nalanda	232732	4640	1216	30031	2523	7874	40428	218
Bhojpur	237339	0	6883	28925	2823	1443	33191	628
Buxar	166999	0	2276	12187	3230	1338	16755	576
Rohtas	390722	66723	16948	37645	8961	325	46931	1118
Kaimur	342447	113039	19269	29769	2469	1341	33579	1395
Gaya	493774	77836	27646	61493	3759	6872	72124	3294
Jehanabad	94043	637	3290	12765	835	1066	14666	152
Arwal	62631	–	2191	8518	556	706	9780	101
Nawada	248732	63775	11296	24598	3031	7379	35008	1149
Auragabad	330011	18764	16430	50752	1645	1744	54141	1903
Total Zone 3B	2916666	345470	120564	358870	40164	32615	431649	11301
% Share in Total	**31.16**	**55.57**	**27.64**	**27.94**	**19.37**	**21.23**	**26.23**	**24.69**

4

IRRIGATION AND FLOOD CONTROL

As compared with most Indian states, Bihar is endowed with good irrigation infrastructure. The State is fortunate to have ample surface as well as ground water resources. Private investment is taking place in this sector mainly for exploitation of ground water through shallow tubewells, cavity wells, bamboo wells and also through dug wells and irrigation ponds in limited areas. All blocks in the State come under 'safe category'. The future availability of ground water for irrigation works out to 16.11 lakh ha.m. which may support approximately 8 lakh new tube-wells in the State. Out of which approximately 3 lakh units have been installed leaving a balance potential of approximately 5 lakh units.

Bihar the land of 14 rivers, is richly endowed with water resources and with a very healthy rainfall average of 1271.9 mm. Ultimate irrigation potential in the State is estimated to be around 10.9 million hectares which is far above the total cultivated area in the State. However, neither rainfall nor the distribution of the water resources is uniform causing uneven irrigation potential/coverage across the State.

The State has 45.72 thousand hectares of irrigated area against its total geographical area of 9416 thousand hectares, meaning thereby that around 46 per cent of total area is irrigated which has been stagnating around this figure for the last few years (Table 4.1). At the same time irrigation capacity is not evenly distributed across the State. Percentage of irrigated area varies greatly across different regions/districts. While Zone-2 receives earliest showers and highest rainfall (1382 mm) among all three agro-climate zones in the State, Zone-3 receives an average rainfall of only 1165 mm. A district level study shows a very wide variation in the irrigated area from District to District—from a low of 17 per cent in Jamui to around 86 per cent in Sheikhpura districts like East Champaran, Darbhanga,

Kishanganj, Madhubani, Muzaffarpur, Saharsa, Sitamarhi and Vaishali, have relatively lesser percentage of cropped area irrigated. (Table 4.2). In fact, both irrigation capacity and its source vary greatly across the 3 agro-climatic Zones of the State.

Table 4.1: Source-wise Gross Irrigated Area in Bihar

(*Lakh Hect.*)

Year	*Source*					
	Canal	*Tank*	*Tubewell*	*Other Well*	*Other Sources*	*Total Irrigated Area*
1948-51	3.62	2.56	N.A	2.61	9.03	17.82
1962-65	5.74	1.69	1.04	1.83	7.46	19.78
1978-81	12.87	0.89	10.09	1.77	7.83	33.48
1988-91	14.26	0.95	16.69	0.91	7.20	40.00
1998-99	13.28	1.63	26.64	0.21	3.92	45.68
2000-01	12.30	1.80	28.71	0.16	2.65	45.61
2001-02	12.37	1.40	29.65	0.16	1.82	45.39
2002-03	12.59	1.49	29.65	0.18	1.80	45.72
2003-04	11.43	1.50	31.03	0.13	1.57	45.67
2004-05	10.63	1.18	28.87	0.67	1.23	41.97
2005-06	10.61	1.83	28.95	0.12	1.74	43.25

Source: Department of Agriculture, Government of Bihar.

Table 4.2: Percentage of Gross Area Irrigated by District

Agro-climatic Zone	*District*	*% Area Irrigated*
Zone-1	Saran	44.4
	Sivan	47.9
	Gopalganj	52.8
	Muzzaffarpur	41.3
	E. Champaran	36.0
	W. Champaran	36.5
	Sitamarhi	32.3
	Sheohar	28.7
	Vaishali	38.6
	Darbhanga	40.2
	Madhubani	39.1
	Samastipur	42.9
	Begusarai	48.4
	Zone-1 Sub-Total	*40.6*
Zone-2	Saharsa	57.6
	Supaul	59.5

(*Contd.*)

Table 4.2: (*Contd.*)

Agro-climatic Zone	*District*	*% Area Irrigated*
	Madhepura	74.0
	Purnea	54.3
	Kishanganj	26.7
	Araria	40.9
	Katihar	44.5
	Khagaria	56.0
	Zone-2 Sub Total	*50.9*
Zone-3	Munger	26.4
	Sheikhpura	86.2
	Lakhi Sarai	32.5
	Jamui	16.6
	Bhagalpur	30.8
	Banke	37.8
	Patna	56.6
	Nalanda	83.3
	Bhojpur	79.4
	Buxar	68.2
	Rohtas	84.6
	Kaimur	48.7
	Gaya	47.6
	Jehanabad	71.5
	Arnol	76.4
	Navada	50.8
	Aurangabad	67.9
	Zone-3 Total	*58.8*

Zonal Distribution

While Zone-I accounts for nearly 37 per cent of total geographical area of the State, its share in total irrigation capacity is about 31 per cent. On the other hand, Zone-3B which also accounts for 31 per cent of geographical area has about 41 per cent of total irrigated land. Zone-3A having 13 per cent of geographical area of the State has only around 8 per cent of total irrigated area. Finally, while around 64 per cent of Zone-3B area is irrigated, Zone-3A has only around 32 per cent of its area irrigated (Table 4.3)

Table 4.3: Zonal Classification of Total Irrigated Area Through Different Sources

(Hectare)

	Name of District	*Canal*	*Tubewell*	*Other sources*	*Total irrigated*
			Zone-1		
	Saran	2414	11523	61	117706
	Siwan	6314	93353	7595	107262
	Gopalganj	50672	54027	2990	107689
	Muzaffarpur	—	130164	—	130164
	E.Champaran	794	154450	411	155655
	W. Champaran	128686	42971	5026	176683
	Sitamarhi	1491	63661	6530	71682
	Sheohar	—	12652	—	12652
	Vaishali	—	65655	11891	77546
	Darbhanga	—	95736	6351	102087
	Madhubani	—	41113	97438	138551
	Samastipur	—	112387	—	112387
	Begusarai	—	89416	1673	91089
Zone-1 Total		1,90,371 (13.59% of total irrigated area)	10,70,816 (76.42% of total irrigated area)	1,39,966 (9.99% of total irrigated area)	14,01,153 (30.68% of total irrigated area)
			Zone-2		
	Saharsa	5527	84562	4894	94983
	Supaul	67352	73678	1084	142114
	Madhepura	32795	92954	7375	133124
	Purnea	11729	158775	—	170504
	Kishanganj	—	50535	—	50535
	Araria	—	110962	—	110962
	Katihar	—	129408	—	129408
	Khagaria	—	76748	66445	83393
Zone-2 Total		117.403 (12.84% of total irrigated area)	777,622 (84.98% of total irrigated area)	19,998 (2.18% of total irrigated area)	9,15,023 (20.04% of total irrigated are)

(Contd.)

Table 4.3: (*Contd.*)

	Name of District	Canal	Tubewell	Other sources	Total irrigated
			Zone-3A		
	Jamui	1040	30439	5536	37015
	Sheikhpura	23951	525	28947	53423
	Munger	13314	24550	4070	41934
	Lakhisarai	2185	41107	7488	50780
	Bhagalpur	4297	57797	16270	78364
	Banka	90062	23893	1743	115698
Zone-3A Total		1,34,849 (35,75% of total zonal irrigated area)	1,78,311 (47.27% of total zonal irrigated area)	64,054 (16.98% of total zonal irrigated area)	3,77,214 (8.26% of total irrigated area)
			Zone-3B		
	Patna	51115	121049	7431	179595
	Nalanda	9442	164270	19576	193288
	Bhojpur	40781	135775	11538	188094
	Buxar	44421	66410	3079	113910
	Rohtas	262570	36037	32227	330834
	Kaimur	98493	54414	13630	166537
	Gaya	232303	—	2787	235090
	Jehanabad	596	57839	8771	67206
	Arwal	28712	18034	1385	48131
	Nawada	13944	108831	3783	126558
	Aurangabad	142664	65794	15743	224201
Zone-3B Total		9,25,041 (49.38% of total zonal irrigated area)	8,28,453 (44.22% of total zonal irrigated area)	1,19,950 (6.4% of total zonal irrigated area)	8,28,453 (44.22% of total zonal irrigated area)
Total		1367664 (29.95 of total irrigated area)	2855202 (62.52% of total zonal irrigated area)	343968 (7.53% of total irrigated area)	4566834 (Total irrigated area)

Source: Central Water Commission, Deptt. of Water Resources, Bihar and Central Ground Water Board

Variation in irrigation capacity across the regions also extends to the sources of irrigation. Over 64 per cent of total irrigation capacity in the State has tubewell as its source. Canal irrigation accounts for nearly 28 per cent of total irrigated area. Thus major and medium sources of irrigation together account for 93 per cent of total irrigation capacity in the State, with other sources (minor and micro-irrigation) accounting for only about 7 per cent of total irrigation. A very high dependence of irrigation on tubewells and operation of diesel tubewell due to lack of power in rural areas means a high cost and inefficient irrigation (Table 4.4).

Table 4.4: Development of Groundwater Extraction Structures in Bihar

(No. in '000)

As on end	*State Tubewell*	*Private Tubewell*	*Open Well*
1960-61	0.978	2	N.A
1968-69	1.468	40	68
1973-74	2.677	73	120
1976-77	3.511	190	164
1980-81	4.606	205	162
1985-86	4.953	365	150
1986-87	4.953	422	142
1993-94	5.289	571	104
2000-01	5.556	651	135

Source: For 1960-61 State Tubewells-Minor Irrigation Department, Government of Bihar. For 1968-69. Bihar Irrigation Commission Report, 1971. For 1976-77, 1980-81, 1985-86. Directorate of Agriculture Census Bihar. For 1986-87 and 1993-94 and 2000-01, Minor Irrigaiton Census. Ministry of Water Resources Government of India.

TERI *Source*: Bihar Development Report Institute of Human Development, 2006.

This variation in source wise irrigation has a very clear pattern across the agro-climatic zones of the State. While nearly 50 per cent of irrigation in Zone-3B has canal as its source, only 13 per cent of the irrigated area in Zone-2 receives canal irrigation. Reliance on tubewell is very high in Zones-1 and 2 (more than three-fourths of their irrigated area). On the other hand, canals have a greater role to play in Zones-3A and 3B and there is a greater balance between tubewell and canal irrigation in these regions. Incidentally, the latter two zones are also the zones with lowest rainfall among all the regions of the State. Thus, while importance of canal irrigation is much greater in southern parts of Bihar (Zone-3), there is an almost exclusive reliance on tubewell irrigation in the northern parts, especially in Zone-2. This is rather peculiar as most of the rivers in Bihar flow through its northern plains which should have had a higher canal irrigation capacity. As for minor and micro irrigation, here too, there is a great deal of variation in its use across the State. While only 2 per cent of irrigated area of Zone-2 has minor irrigation as its source, its contribution to irrigated area in Zone-3 is around 17 per cent.

Economics of Tubewell Irrigation

Since tubewells are the major source of irrigation in the state, it would be of interest to understand the economies of running these wells. Taking 2.31 litre diesel/ha.cu.m as the fuel need for pumping, a farmer is required to buy about 16.17 litre diesel to provide seven cu.m water for one irrigation at the farm head. For eight to 12 irrigations in Bihar, the farmer needs to purchase 129.36 to 194.04 litres diesel to supply water at the farm head. This value needs to be divided by the conveyance efficiency to account for the loss of water from the pumping station to the farm head. Even when the farmer carries water at the farm head through a plastic pipe, there is loss of water due to leakage in the pipe. The rate of leakage through the pipe will depend upon the quality of the material, and age and length of the pipe to account for loss of water in conveyance. Hence, the amount of diesel needed to pump water at the pump head for one-irrigation is computed as 20.2 litres, and for eight to 12 irrigations, it (the fuel requirement) varies from 162 to 24 3 litres.

Providing three irrigations to nursery will require six litres of diesel. Hence, the amount of diesel needed to grow one ha rice works out to be 168 to 249 litres. It amounts to an expenditure of Rs. 5,040 to Rs. 7,470 on the purchase of diesel for irrigating one ha rice in Bihar. In case water is carried to the farm head through unlined earthen channel, the fuel cost will increase further. Taking into consideration the real cost of pumping, the life of the diesel engine, pump, maintenance cost, and the wages of the pump operator will be much more than the cost of the fuel alone. Farmers usually sell their pumping hours at the rate that is approximately double the cost of fuel. Hence, the approximate cost of irrigation becomes Rs. 10,080 to Rs. 14,940/ha. Because of such a prohibitive cost, the farmers in Bihar are either forced to do away with irrigation or provide inadequate irrigation to their rice crop.

Of greater significance for the poor is the response of pump rental prices to the rise in diesel prices because the poorest strata of India's peasantry depend on water markets to secure their irrigation. Because water markets are natural oligopolies,[1] pump owners use diesel price increases to raise their pump rental rates in tandem with every major rise in diesel prices despite the fact that pumps themselves have become cheaper during 1990-2007. Figure below shows changes in the nominal prices of diesel *versus* the price of pump irrigation in Mirzapur, Uttar Pradesh. Between 1990 and 2007, diesel prices here rose from Rs. 4.6 to Rs. 34.8 but the rate buyers incur per hour of pump irrigation has increased from Rs. 23-25 to Rs. 90-95 per hour, far larger than needed to cover the increase in fuel cost. Another characteristic of this relationship has been the downward stickiness of pump irrigation prices. Every time there is a big increase in the diesel price, pump irrigation price tends to jump. However, the reverse is never the case.

Graph 4.1: Diesel Price Rise and Pump Irrigation Price in Mirzapur, U.P.

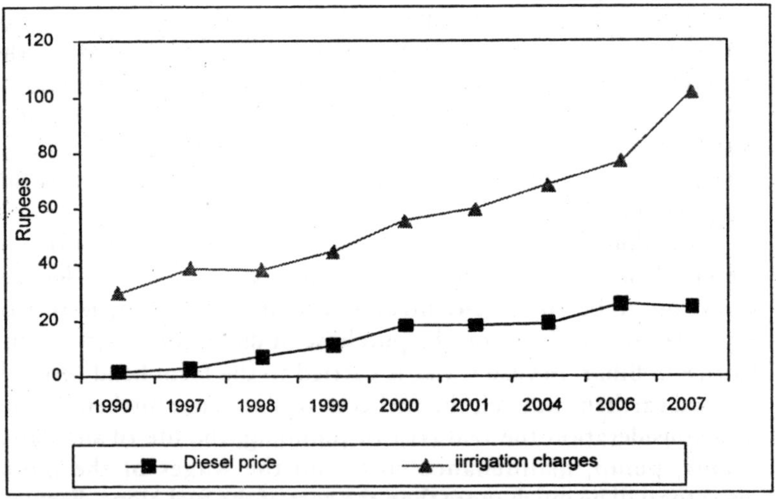

Source: Crop per Drop of Diesel? Energy Squeeze on India's Smallholder Irrigation, Tushaar Shah, *E.P.W.*, September 29, 2007. Shah, Tushaar, 1993, Groundwater Markets and Irrigation Development: Political Economy and Practical Policy, OUP, Bombay.

If the farmers of Bihar, like their counterparts in Punjab, have electric motor driven pump sets, their irrigation cost will reduce to (10/25), i.e., Rs. 4,032/ha. to Rs. 5,976/ha. At such a cost, there is likelihood that the farmers may maintain the desired level of moisture regime in their rice fields. It is pertinent to mention over here that the Government of Punjab is supplying free electricity to the tube-wells engaged in farming so that the production of rice and other crops remains remunerative to the farmers. The Bihar Government has also agreed to supply diesel to the farmers at Rs. 10* per litre for agriculture. On litre of diesel is equivalent to about 3.5 units of electricity. The scheme if properly implemented, will help the farmers to improve productivity of their crops and will certainly go a long way to improve the efficiency of irrigation. But the real solution lies in improving power availability.

How the energy squeeze is heralding wholesale cropping pattern changes from diesel—intensive to diesel saving crops was striking in the Simra village of Patna district. In 1990, this was a wholly rice-wheat village with little diversification away from this age-old rotation. Now, kharif paddy continues with or without irrigation but during rabi and summer, of its 300 ha., Simra has 150 ha. under rainfed masoor, 50 ha. under lightly irrigated gram, and 75 and 25 ha. respectively under intensively irrigated onion and coriander, the last fetching them the highest return per acre as well as per litre of diesel/kerosene. In the eastern UP village Akataha, farmers have switched from long duration to short duration paddy, and some irrigated paddy area has given way to diesel-saving groundnut and high-value potato crops.

Because of the excessive expense involved in irrigating rice, wheat, maize and other crops, the farmers in Bihar are unable to use optimum amount of irrigation water in crop production. Consequently, the productivity of most crops is quite low. In view of the conditions discussed over here, besides improving the supply situation, it is necessary to introduce water and energy saving irrigation technology for reducing the high expenditure on irrigation in Bihar. It will encourage the farmers to apply optimum amount of irrigation water which in conjunction with the fertilizers and other inputs will result in increased crop yield.

---

* One litre of diesel is equivalent to about 3.5 units of electricity. The scheme if properly implemented will help the farmers to improve crop productivity.

The ICAR research Complex for Eastern Region, Patna has developed a Low Energy Water Application (LEWA) device (Singh *et al.*, 2004; and Annual Report ICAR-RCER—2005-06), which sprinkles irrigation water satisfactorily at 0.4 to 0.6 kg/cm^2 pressure with wetter area diameter of 6 to 8 m. The standard hand moved sprinkler systems available in the country work satisfactorily at 1.5 to 2.5 kg/cm^2 pressure with the wetted area of 12 to 24 in diameter. As the LEWA irrigation system works at low pressure in comparison to standard hand moved sprinkler system, it requires less energy for its operation. In addition, it satisfactorily irrigates the small unconsolidated-fields of Bihar. The device has been designed to irrigate cereal as well as vegetable crops and is expected to effect a saving of water and diesel to the extent of 25 to 40 per cent. It, therefore, needs to be manufactured on a large-scale for adoption by the farmers.

Chinese Pumps—An Aid for Bengal's Agrarian Poor

In West Bengal, help has come to the "energy squeezed" farmer from unlikely quarters: Chinese kerosene-cum-diesel pumps. Boro rice is far more intensive in working capital, labour and irrigation but it is also land-saving—therefore appealing to marginal farmers and sharecroppers. It offers 7 mt/ha. of paddy yield against barely.1-1.5 mt/ha rainfed. Growing a small parcel of boro rice may liberate a farming family from subsistence worries for the whole year. It is, therefore, prized by the poor. For want of better alternatives, such as electric pumps, West Bengal's marginal farmers have been switching to Chinese pumps with gusto. They are cheap, costing Rs. 7,000 to Rs. 8,500 for 3.5 and 5 hp, against Rs. 16,000 for a 5 hp, Kirloskar. The Chinese 5 hp pump runs for two hours on a litre of diesel, which a Kirloskar 5 hp, pump burns up in an hour or less. Finally, while a Kirloskar needs a bullock cart to more around, the Chinese pump can be easily carried by a farmer on his shoulders.

In just five years, Chinese pumps have captured the irrigation pump market in West Bengal. How these pumps made in-roads into irrigation scene of that state remains something of a mystey. Apparently, some second-hand Chinese pump smuggled across the Bangladesh border, caught the farmers' fancy, and soon enough, there followed a deluge of Chinese pump smuggles across the

Bangladesh border. It was only a matter of time before official imports began in 1998. Now, out of every 100 new diesel pump assemblies purchased in these parts of West Bengal, over 90 have Chinese engines. Kolkata has emerged as the epicenter of Chinese pump diffusion. Several brands of Chinese and Chinese—Indian pump assemblies are on offer here (Greeves China, Tricircle China, GK200, Chhanta China, Zenith China, Changfa China ZL 175 China etc.) All these are selling at 35-40 per cent of the price of Kirloskar 4-5 hp engines, which remained market leaders for decades. Interviews with Chinese 4 hp pump dealers in Kolkata confirmed that farmers preferred these for their low price, their much higher fuel efficiency (0.35-0.4 1/hour), ability to work on kerosene, and easy portability. Chinese pump mechanics have come up in every village, and their spare parts are cheaper and easily available.

POOR MANAGEMENT OF WATER USER ASSOCIATIONS

Despite the involvement of Water User Associations (WUAs) under Participatory Irrigation Management (PIM) for canal irrigation, the progress made under the Million Shallow Tubewell (MST) Scheme as discussed here and progress of some key items under CADP (Table 4.5), gross irrigated area in Bihar has not increased over the past 5 years or so. It was 4539 thousand hect. during 2001-02 and according to Economic Survey, Government of Bihar, 2008, it is reported as 4325 thousand hect. in 2005-06, despite an investment of Rs. 11,890 million over the five year. (Table 4.6).

Table 4.5: Physical Progress of Some Key Items Under CADP in Bihar

('000 hectares)

Items	*1974-80*	*1980-90*	*1990-92*	*1992-02*	*Total*
Construction of Field Channels	13.76	824.32	550.58	50.10	1438.76
Construction of Field Drains	0.59	1.42	—	—	2.01
Land Levelling and Shaping	0.092	0.478	—	—	0.57
Reclamation of Water Logged area	—	—	—	6.58	6.58
Enforcement of Warabandi	—	71.45	65.54	8.76	145.75

Source: Agriculture (Special Programme) Department, Bihar, Patna.

Table 4.6: Development of Major/Medium Irrigation in Bihar (1951-2005)

Period	Plan Investment (in Rs. Billion)		Irrigation Potential Created Lakh Hectares		Per hectare cost of potential created (in Rs. 1000 at constant prices)
	Current Prices	1994-95 Prices	Cumulative	Annual Average	
Bihar					
Pre-Plan	N.A	N.A	4.03	N.A	N.A
1951-61*	0.38	6.50	11.97	0.79	8.19
1961-69*	1.22	13.09	15.98	0.50	32.65
1969-80*	5.10	24.19	22.00	0.55	40.19
1980-90*	20.44	43.78	26.95	0.50	88.63
1990-00	13.75	13.13	24.47	0.09	142.39
2000-05	11.89	7.90	26.19	0.34	46.47
India					
Pre-Plan	N.A	N.A	86.2	N.A.	N.A
1951-61	7.56	123.26	132.51	4.63	26.63
1961-69	10.06	108.48	170.12	4.70	28.85
1969-80	58.37	275.49	255.69	7.78	32.19
1980-90	184.77	402.42	298.63	4.29	93.80
1990-00	540.70	511.93	347.87	4.86	105.29
2000-02	214.46	152.32	370.76	11.45	66.52

* Data till 1990 relate to undivided Bihar.

Source: Central Water Commission, New Delhi, Department of Water Resources, Government of Bihar, Patna. Bihar Development Report, Institute for Human Development, 2006.

Surface irrigation system remains highly mismanaged in terms of water allocation, water use efficiency, cost recovery etc. Ground water resources have also not been utilized satisfactorily. In this respect, problems like energization, investment credit, public and private investment, etc. would have to be addressed on priority basis. Most public tubewells become defunct after a few years of operation due to poor maintenance. Therefore, private investment in tubewells would have to be encouraged. All this will have no meaning if power supply in the state is not improved. One of the reason for this sorry state of affairs is that during the 1990s public expenditures on Irrigation in Bihar declined significantly. Capital as well as revenue (current) expenditures on irrigation and flood control as a per cent of agriculture GSDP fell consistently for about a decade from around the late-1980s (Graph 4.2). While the decline in capital spending was steeper and more noticeable, the reduction in revenue expenditure had a more immediate effect on the efficiency of existing surface irrigation systems.

Graph 4.2: Expenditures on Irrigation and Flood Control as Percentage of Agriculture GSDP

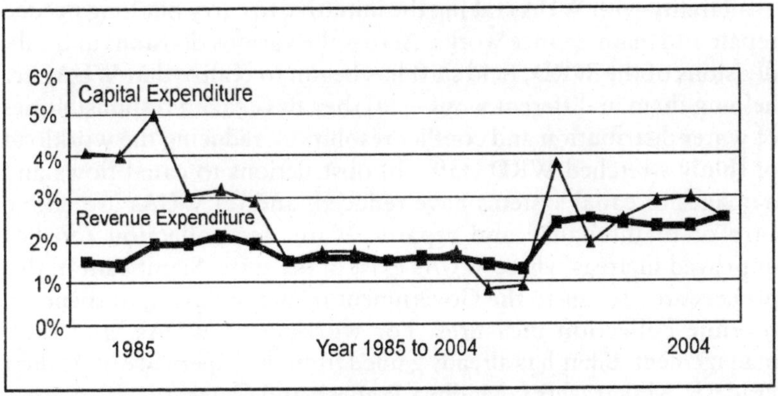

Participatory Irrigation Management (PIM)

Bihar is one of the earliest state to experiment with transferring management of irrigation schemes to Water Users Associations (WUAs). These Associations are intended to take care of the problems under watershed such as—fragmented planning and implementation, non-involvement of beneficiary communities, lack of local institutions for deciding and enforcing rules regarding continuing maintenance of facilities as well as sharing of costs and benefits within the community. A pilot project under PIM of Paliganj Distributary was started in late eighties but not much progress could be made even up to mid eighties. The issue was examined by the Second Bihar Irrigation and Drainage Commission, 1994. But not much progress was visible as is clear from 2005 Water Resources Department (WRD) Paper which says "the existing irrigation system was breaking down and it called for urgent remedial measures. It recommended the participation of the stake holders (farmers) in the management of irrigation system[2]. No wonder then that gross irrigated area which was 4534.5 thousand hect. during TE 1998 hovered around the same figure in the next 7-8 years. It actually declined during 2005-06 (Table 4.1).

The Participatory Irrigation Management (PIM) system has now been taken up on a priority basis. By February 2006 nearly 150,000 hectares of irrigated command area had been transferred to WUAs[3]. The Water Resources Department (WRD) has an ambitious plan of further expanding the PIM programme to cover an estimated 800,000 hectares by the end of 2007 and 1.6 million hectares by 2012[4]. Main gains from the implementation of the system include improved water availability with WUAs taking the initiative to carry out long needed repair and maintenance works. Across the various divisions and sub-divisions of the WRD, field staff has begun to realize that WUAs are helping them in different ways— (a) they have taken responsibilities of water distribution and conflict resolution, reducing the workload of thinly stretched WRD staff; (b) obstructions to canal flows and damages to canal systems have reduced; and (c) WUAs are taking care of maintenance and repairs. Water fee collection too has improved in areas where WUAs exist in the state. Significant in this process are savings to the Government from not having to spend on revenue collection measures. For working of WUAs, and their management, Bihar has already gained from the experience of Andhra Pradesh, Maharashtra, Madhya Pradesh and Gujarat.[5]

Million Shallow Tubewell (MST)

Groundwater resources have remained largely unutilized in the State. To tap the huge potential of ground water irrigation, Million Shallow Tubewell Programme has been prepared which would imply additional irrigation capacity of 20 lakh hectares. This is a central by sponsored scheme exclusively for Bihar. About 6.97 lakh diesel tubewells and pump sets were proposed to be established by March 2007 under this programme. Against this target, 3.72 lakh private shallow tubewells have been sunk over the last five years. Farmers are given a grant of 30 per cent and a loan of 50 per cent of total costs under this Programme and no discrimination is made between different categories of beneficiary farmers. This is for a standard depth of 4.5 meters. Planning Commission has been approached by the Government to allow additional depth wherever required and enhance the subsidy quantum. Against the distribution target of 5.37 lakh pump sets by March 2006, 3.80 lakh pump sets were distributed by November, 2005. Over the years 2005-06 and 2006-07, there has been a huge increase of 1131 State Tubewell Schemes, 970 of which were completed in 2005-06. (Table 4.7).

Table 4.7: Physical Target and Achievement of Shallow Tubewells

Sl. No.	Year	Physical Target (hect)	Achievement (hect.)	% Achievement
1.	2001-02	33798	4269	12.63
2.	2002-03	23313	43445	190.21
3.	2003-04	160000	107358	67.1
4.	2004-05	160000	172619	107.89
5.	2005-06	160000	109119	68.19
6.	2006-07*	160000	30212	18.75
	Total	697111	467022	67.0

* Achievement during 2006-07 till 31 October, 2006.
Source: NABARD Report, 2007.

On-farm water management programme is being run in all the 38 districts of the State under which shallow pump sets, normal pump sets, and other micro irrigation sources are being arranged. Watershed development programme is being run under which water generating ponds, and other devices of groundwater regeneration are being constructed to develop and improve water utilization capacity in the State. Sprinkler sets and HDPIP pipes are being distributed in the State under the centrally sponsored ISOPOM plan for production of pulses, oilseeds and maize. By March 2006, additional irrigation capacity of 18000 hectares of land was created under this programme. Against the target to create additional irrigation capacity of 2.39 lakh hectares in 2006-07, an additional irrigation capacity of 1.76 lakh hectares was created by September 2007.[6] (Economic Survey 2007)

IRRIGATION POTENTIAL

Of the ultimate potential of 10,888 thousand hectares (major 5220 and minor-5664 thousand), the state had created 7396 thousand hectare of irrigation potential by 2002 which amounted to almost 60 per cent of the ultimate potential. Utilization according to Planning Commission was 5474 thousand hectares or 85 per cent of the potential created (Table 4.8). Actual irrigated area according to Land Use Statistics, however, was only 4539 thousand hect. In 2005-06.

Table 4.8: Ultimate Irrigation Potential and Potential Created upto 2002

(*'000 ha.*)

State	Ultimate Potential			Potential Created			Potential Utilised			Gross Irrigated Area 2001-02
	Major and Medium	Minor	Total	Major and Medium	Minor	Total	Major and Medium	Minor	Total	
Bihar	5224	5664	10888	2680	4716	7396	1715	3759	5474	4539

Reasons for the non-utilization of irrigation potential would vary for major/medium and minor projects. Major/medium projects consist of canals, branch canals, distributionaries, minors, sub-minors, and finally field channels. Sometimes, there is slow progress of sub-minors/ field channels construction where there is a divided responsibility between farmers and government. Even after completion of the project, the problems could be of leveling (a design default), breakages/leakages (normal or farmer created), besides monitoring and supervision problems. This involves canal security, gate operation, canal cleaning and minor repairing, etc. There is the need to take immediate steps to examine this issue command wise and take remedial steps to bring the total potential created under full use.

The Government is now alive to the problem of diesel tubewells and other issues which have led to stagnation in irrigated area. They have recently decided to supply diesel to agriculture at Rs. 10 per litre.* This alone will not be sufficient. Majority of the existing Government tubewells are not working for one or the other reason. A careful study of these non-functional tubewells should be undertaken and remedial measures taken. With the programmes already in hand and further improvement in the working of WUAs, MST programme and appropriate water conservation measures including rain water harvesting—some of the constraints discussed above can be overcome.

The XI Plan has a target of 1.4 million hect. additional potential (Table 4.9). Irrigation potential utilized as in 2005-06 is 3.2 million hect. (Table 4.10). Operational and financial goals for irrigation for

Table 4.9: Irrigation: Additional Potential During the Eleventh Plan

(lakh ha.)

Schemes	*Creation of Additional Irrigation Potential*	*Restoration of Lost irrigation Potential*
On-going schemes	3.19	—
ERM schemes	—	1.87
New schemes	8.80	—
Total	11.99	1.87
Grand Total	13.86	

Source: Bihar Approach to Eleventh Five Year Plan, Vision for Accelerated Inclusive Growth 2006.

* One litre of diesel is equal to around 3.5 units of electricity.

XI Plan are as (Table 4.11). Government may also look for new sources of energy like wind power and bio-diesel from jathropa to overcome this major problem.[7] Having done all this, it can be assumed that out of 2.86 million hect. of unutilized irrigation potential, the state should be able to irrigate around 1.5 million hect. additional area. With rainwater harvesting projects and this additional irrigated area, gross cropped area can be estimated at around 9.0 million hect.

Table 4.10: Minor Irrigation Potential Created Utilised

Minor Irrigation Potential in Bihar	*Area (lakh hectares)*
Surface schemes including traditional ahar and pynes	15.44
Groundwater utilization schemes of open wells, shallow tube wells, deep tube wells	48.57
Total	*64.01*
Minor Irrigation Potential Created	
Groundwater schemes	28.99
Surface water schemes	7.27
Total	*36.26*
Irrigation Potential Utilised in 2005-06	
Surface water including lift and barge schemes	0.49
Surface water ahar, pynes and traditional systems	3.32
State Tubewells	0.25
Private tubewells and irrigation wells	27.64
Total	*31.70*

Source: Bihar Approach to Eleventh Five Year Plan, Vision for Accelerated Inclusive Growth 2006.

Table 4.11: Operational and Financial Goals for Irrigation

Scheme	*No.*	*Operational (lakh ha.)*	*Financial (Rs. Crore)*
Execution of new surface and lift minor irrigation schemes under NABAD RIDF	2250	1.650	310.00
Renovation and restoration of ahar and pyne	8000	4.000	1200
Execution of state electric based tubewells	6000	2.4	390
Restoration of old state tubewells and lift irrigation schemes	4500	2.1	270
Execution of private shallow tubewells including replacement of old tubewells	600000	12.0	1500
Plan establishment			
Total		*22.15*	*3670*

Source: Bihar Approach to Eleventh Five Year Plan, Vision for Accelerated Inclusive Growth 2006.

without any increase in the net sown area by the end of XI Plan. By the year 2020-21, Bihar should be able to make full use of its irrigation resources to achieve a target of 11 million hect. as gross cropped area. (Table 4.12) Immediate steps should be taken by mounting specific studies firstly to locate the total 2.86 million hect. unutilised irrigated area. This should be followed by area specific plans to reclaim this area and also take up additional new projects.

Table 4.12: Projected Irrigated Area 2020-21

('000 hect.)

	2002-03	*2010-11*	*2020-21*
Net irrigated area	3462	4500	6000
Gross Irrigated	4572	6500	8000
Double Cropping	2232	3288	5288
Net sown area	5725	5712	5712
Gross sown area	7957	9000	11000

Larger Outlay for Water is Alright, but Project Efficiency is Key

As in the country as a whole, the present State of irrigation in Bihar, is the result of enormous backlog of deferred maintenance of water, machinery, and large proportion of recurrent budgets in the state being really spent on staff and personnel. Irrigation revenue collections are hardly 6-7 per cent of the dues and cost of collection is double the amount collected. (Table 4.13)

Table 4.13: Irrigation Revenue in Bihar

(In Rs. Crore)

	Demand	*Collection*	*Cost of Collection*
1974-75	11.00	3.69	1.16
1980-81	15.99	4.74	4.06
1985-86	44.23	6.99	7.24
1990-91	55.84	4.94	12.14
1997-98	75.03	9.18	22.21
1998-99	82.21	9.05	23.00
1999-00	90.76	11.16	28.00
2000-01	96.76	10.35	26.50
2001-02	106.35	10.76	19.85
2002-03	121.94	10.51	N.A
2003-04	140.66	10.77	N.A

* Data till 1990-91 refer to undivided Bihar.

Source: Director (Revenue), Department of Water Resources, Government of Bihar.

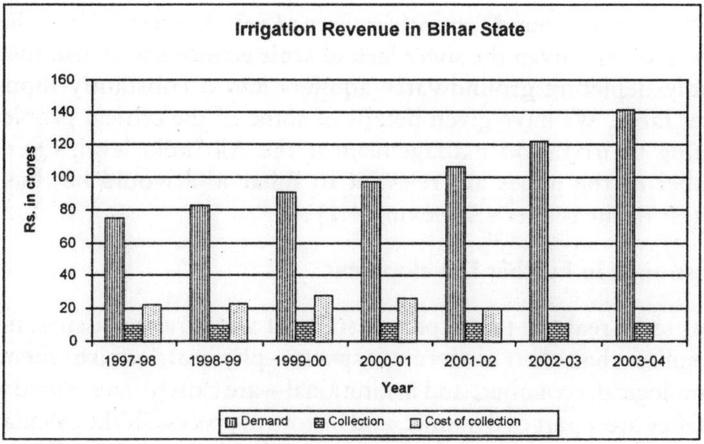

The case of one of WUA, Paliganj was examined by the World Bank. Besides lack of trained staff in irrigation management, the WUA invariably had inadequate resources for properly maintaining the distributary. This is because of a sharp decline in water fee collection as a percentage of the demand.

The ground reality is that there is "endemic corruption" in the field. The fact remains that with insufficient funds for routine operations and maintenance, the water wastages can be relatively high. Generally speaking, there are a surfeit of essentially avoidable distortions in the water economy.

Graph 4.3: Water Fee Demand and Collection—Paliganj WUA

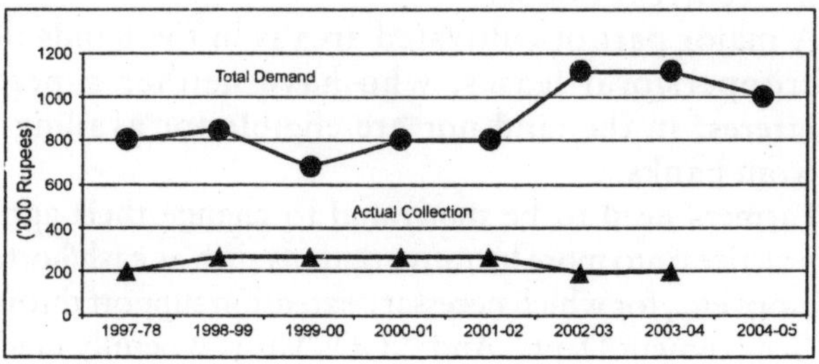

Source: Bihar Agriculture, Building on Emerging Models of Success World Bank.

With over 80 per cent of domestic water supply now coming from ground water, it has led to elaborate and informal water markets.

But "working around", poorly governed water systems imply huge societal costs. Given the sheer lack of scale economies, it also means rapidly depleting groundwater aquifers and a constantly dipping water table. We have given details of some of the critical problems relating to irrigation management at the All India level. Quite a number of the issues are relevant to Bihar and would be thus of interest to the reader. (Annexure 4.1)

Constraints in Further Development

It has to be realized that a proper study of water-related issues, must recognize that their different aspects—physical, environmental, technological, economic, and institutional—are closely interrelated and that they are a part of dynamic, and historical process. If the calculated irrigation potential is to be achieved, besides various issues discussed below, easy and reasonably priced power must get a high priority:

- Although a majority of the inhabited villages in Bihar are electrified, the supply of power is erratic.
- Although a number of petrol/diesel outlets exist in the state, the availability of fuel in rural areas-particularly during cropping season—is inadequate.
- Most of the private drilling rigs available in the State are not registered with SGWIO thereby hampering the utility of their services, particularly in cases where bank finance is involved.
- Major portion of the northern part of the State is affected by recurrent floods.
- A major part of cultivated area is in the hands of share-croppers/oral leases, who have neither a permanent interest in the land nor are eligible for availing finance from banks.
- Farmers need to be motivated to change their agricultural practices into more lucrative options such as cash/horticultural crops etc., for which necessary extension support through State Government Depts. And NGOs/Vas is woefully inadequate.

FLOODS AND DROUGHTS

Climatic conditions in the state are generally considered to be congenial for crop production. Average annual rainfall is 1234 mm.

However, the distribution of rainfall is not uniform. As a result, dry and wet spells are commonly experienced in the state. Every year, Bihar faces vagaries of flood and waterlogging. After bifurcation of the State, Bihar has become the most flood prone area in the country. Total flood prone area of the State is 68.80 lakh hectares which is 73.06 per cent of its total geographical area and 17.2 per cent of the total flood prone area in the country. A substantial proportion of the total cropped area (nearly 41%) gets frequently affected by flood. Flood situation is most severe in the northern plains of Bihar—18 Districts, the heart of Mathila. (Annexures 4.2 and 4.3)

Graph 4.4: Population and Households Affected by Flood in Bihar

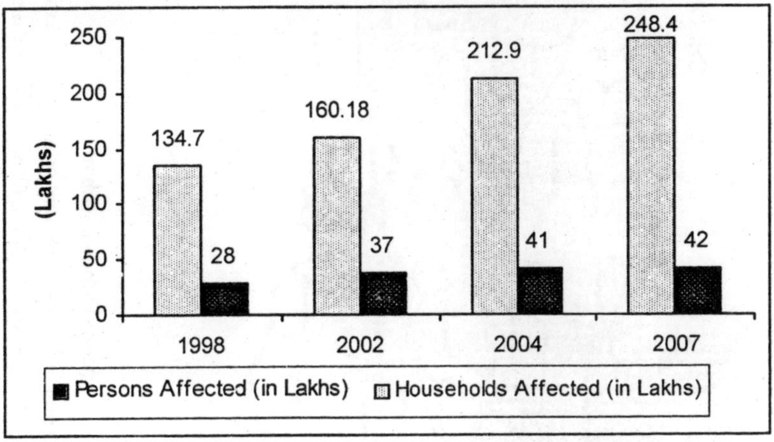

This is because almost all the major rivers in the State enter into Bihar from Nepal in this region. Bed slope of these rivers is very sharp in Nepal and they usually enter the State on plain lands. Because of sudden drop in bed slope, silt brought by the flow of these rivers gets deposited at their base and causes recurring floods. (Table 4.14)

It is a sad story that almost 100 per cent of the gross cropped area in the districts of Araria, Muzaffarpur, Sheohar and 50 to 75 per cent in Begusarai, Katihar, Sheikhpura and Sitamarhi are flood prone. It has been estimated that during the kharif season, about 23 per cent of paddy area remains waterlogged where the adoption of modern technology is either nil or low and consequently the crop yields are low.

Map: Floods in Bihar

Note : Every District in Bihar has flooded to some degree at one time or another. No district can claim to be flood free.

Source: Relief and Rehabilitation Department, Government of Bihar, Central Water Commission, Bihar Media Reports, Daily Reports from the districts. Our source: Economic and Political Weekly.

Table 4.14: Extent of Flood Affected Area in Different Districts

District	*Flood Affected Area*	
	Area ('000 ha)	*Percentage of Cropped Area Affected*
Araria	255	100.00
Muzaffarpur	308	100.00
Sheohar	47	100.00
Katihar	194	70.00
Khagaria	97	74.90
Sitamarhi	123	57.50
Begusarai	99	53.20
Darbhanga	80	40.60
Purnia	119	40.60
Sahrsa	96	44.80
Saran	80	34.00
Sheikhpura	34	56.20
Bhagalpur	51	26.80
E. Champaran	100	28.50
W. Champaran	105	25.10
Gopalganj	51	20.70
Kishanganj	39	22.80
Lakhisarai	33	38.20
Madhepura	24	10.60
Madhubani	32	11.00
Samastipur	43	17.00
Supaul	38	15.30
Vaishali	32	16.40
Munger	9	13.10
Banka	2	1.00
Total	2199	41.40

Source: Collected from Government of Bihar.

Graph 4.5: Human Deaths due to Floods in Bihar

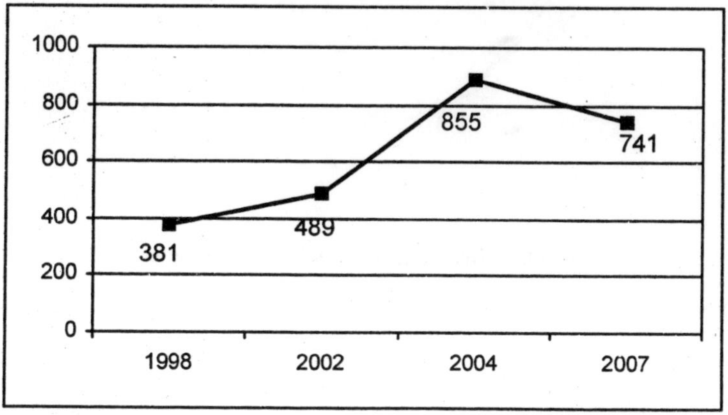

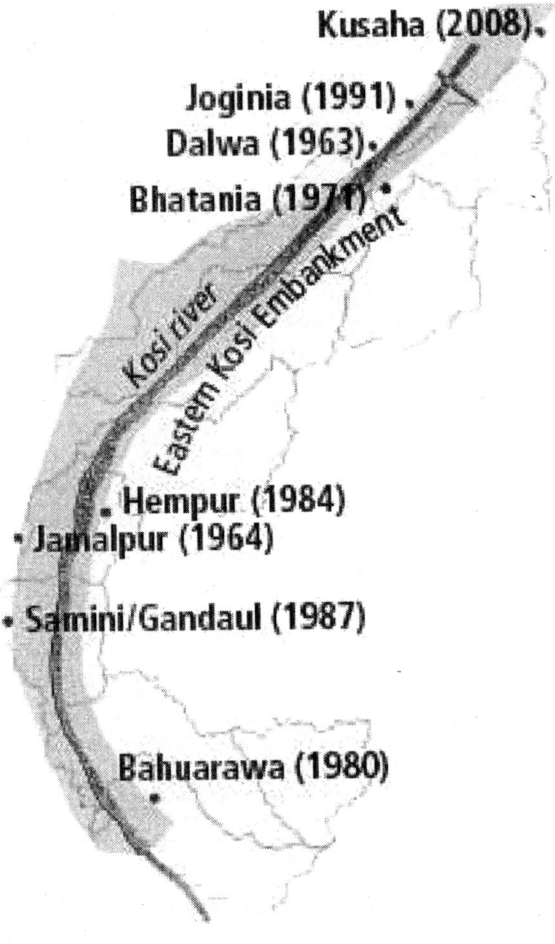

Map: Breaches on Kosi's embankments

Source: Economic and Political Weekly.

Graph 4.6: Crop Area Loss Due to Floods in Bihar

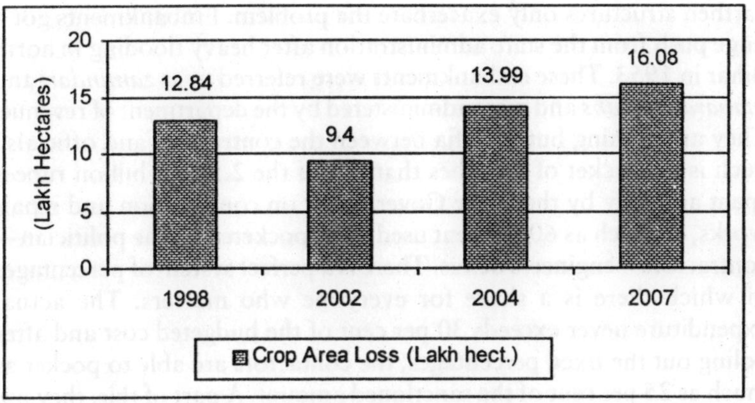

Recurring floods damage agricultural production as much as it plays havoc to various infrastructure. (Graph 4.7) An innovative disaster management system which deals with flood management needs to be put in place. This requires concerted action both at national and international level. At international level, we need to impress upon the Central Government to initiate a dialogue with Nepal for reforestation in the upper catchment area as well as creating irrigation upstream structures both for harnessing benefits of irrigation as well as enormous hydroelectric potential. In addition to this, a technology needs to be evolved to anticipate and contain the enormous damage caused by periodic floods both to crops as well as the vital infrastructure. This requires an integrated plan to be prepared in close partnership with the Government of India.

Graph 4.7: Number of Damaged Houses Due to Floods

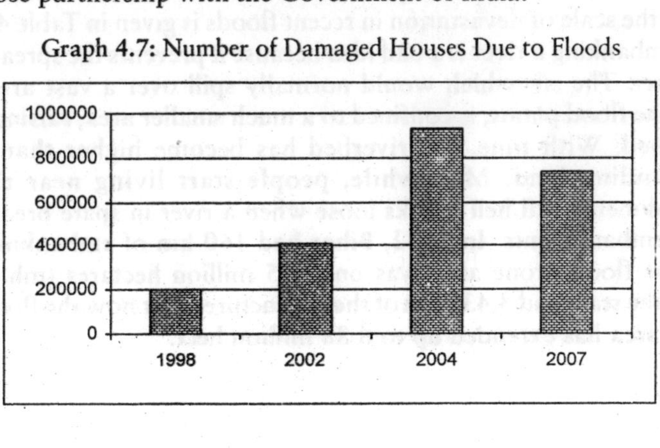

Since independence, successive Bihar Government have sold embankments as an answer to floods, despite warnings that these earthen structures only exacerbate the problem. Embankments got a huge push from the state administration after heavy flooding in north Bihar in 1953. These embankments were referred to as *zamindari* and *maharaji bandhs* and were administered by the department of revenue. They are nothing but a mafia between the contractors and officials.' Such is the racket of breaches that out of the 2.5 to 3 billion rupees spent annually by the Bihar Government on construction and repair works, as much as 60 per cent used to be pocketed by the politician—contractors—engineers nexus. There is a perfect system of percentages in which there is a share for everyone who matters. The actual expenditure never exceeds 30 per cent of the budgeted cost and after doling out the fixed percentages, the contactors are able to pocket as much as 25 per cent of the sanctioned amount. A part of this, they use to finance the political activities of their pet politicians and to get further projects sanctioned. Thus the cycle goes on. [The result is that ...] the contractor's bills are paid without verifying them.[8]

In 1954, two senior officials of the Central Water Power Commission (now Central Water Commission) were sent to China to study embankments along the Hwang Ho. The choice was dim-witted because the embankments on Hwang Ho had been breached 200 times between 1855 and 1954. The British Government had also tried embanking the Damodar in 1854, but by 1869 was forced to shelve the project after it was found infeasible and environmentally dangerous. Dinesh Kumar Mishra has documented the tragedy and extent of the floods in Bihar in a series of articles, books, and papers. Table 4.15 summarises some of his findings for the floods of 1954, 1974, 1987, 2004, and for part of the 2007 flood season. An idea about the scale of devastation in recent floods is given in Table 4.16.

Embanking a river is a bad idea because it prevents the spreading of water. The silt which would normally spill over a vast area to form the flood plains, is confined to a much smaller area, raising the river bed. With time, the riverbed has become higher than the surrounding land. Meanwhile, people start living near these embankments. All hell breaks loose when a river in spate breaches these embankments. In 1952, Bihar had 160 km of embankments and the flood prone area was only 2.5 million hectares (mh.) By 2004, the state had 3,430 km of these structures. But now the flood—prone area has extended up to 6.88 million hect.

Map:

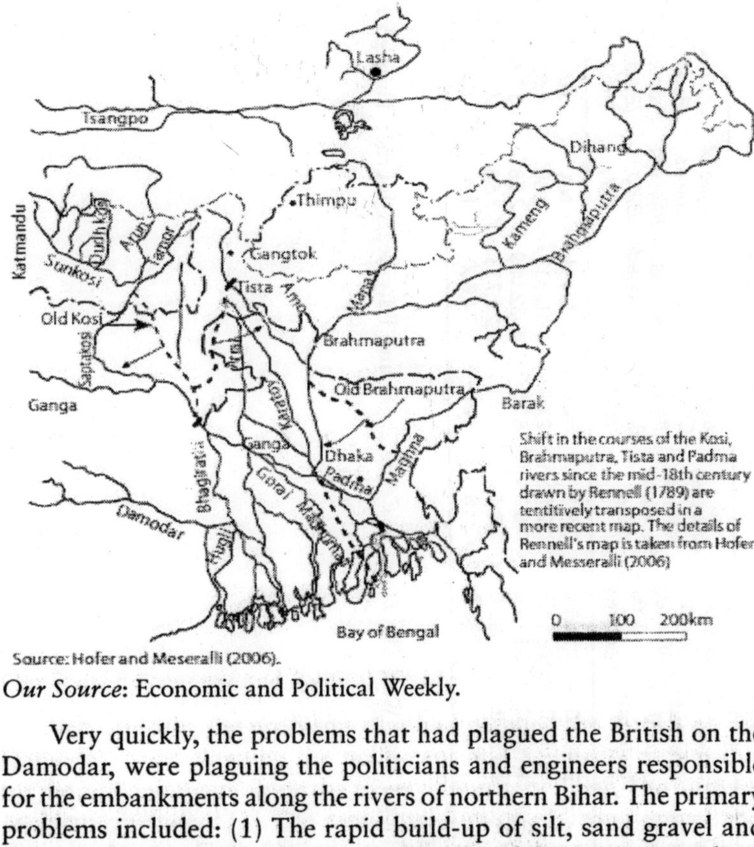

Source: Hofer and Meserallì (2006).

Our Source: Economic and Political Weekly.

Very quickly, the problems that had plagued the British on the Damodar, were plaguing the politicians and engineers responsible for the embankments along the rivers of northern Bihar. The primary problems included: (1) The rapid build-up of silt, sand gravel and stone in the riverbeds, (2) increased height of the mean water level of the rivers, (3) increased waterlogging, and (4) scouring of the earthen embankments and catastrophic breaches.

As long as population pressure on the land remained in equilibrium with the land's capacity to produce, annual floods in north Bihar were a welcome event, even celebrated. It is erroneous to assume that north Bihar is geographically positioned to remain flooded. Conversely, it's the state's arrogance and misplaced faith in engineering that has stopped these rivers from performing their natural task of land building. Without the nurturing role of these rivers, Bihar would never have become the apex centre of knowledge. The collapse of this knowledge culture within the state is an outcome of embankment.[9]

Table 4.15: Flood Losses for Worst floods, Bihar, 1954-2004 and 2007

Selected Year	No. of Districts Affected	No. of Villages Affected	Total Population Affected Millions	Area Affected (1,000 Hectares)	Cropped Area Affected (1,000s Hectares)	Cropped Area as % of total Area*	Value of Crops Affected (Billions Rs.)	No. of Homes Damaged	Value of homes Affected (Billions Rs.)	Loss of Public Property (Billion Rs.)	No. of Human Lives Lost
1954	NA	8,119	7.61	2,460.0	1,596.0	29.7	NA	1,79,451	NA	NA	63
1974	18	NA	16.4	3,182.0	1,751.0	32.5	2.69	5,16,353	0.50	0.80	80
1987	30	24,518	28.7	4,668.0	2,510.0	46.7	6.78	16,82,059	2.58	3.73	1,399
2004	20	9,346	21.3	2,772.0	1,399.0	26.0	5.22	9,29,773	7.58	10.30	885
2007	19	7,565	15.1	NA	1,230.0	22.9	NA	1,74,600	NA	NA	260

* Total Area Estimated at 5,380,000ha.

Source: Adapted from Mishra. The Flood That Was, The Flood to Come, 2005 and Fifty Years Journey Through Foods, 2005.

Prior to the construction of the *bandhs*, when the rivers were in spate, they would gently sweep across the landscape, depositing their burden of silt. As the rivers were channelised, their velocity increased,thus allowing the waters to carry, in addition to silt, sand and gravel and rocks. Initially, this burden was deposited in the river channels, as defined by the levees. This had two effects, First, the fertility refreshing benefits of the silt were lost, and agricultural production fell. Second, the mean level of the rivers rose, as the original rivers bed was filled with debris. Soon, the mean levels of the rivers were higher than the original floodplains. In some areas, the river levels increased to heights of three to four metres above their floodplains, requiring that the levees be further strengthened and elevated.

Table 416: Scale of Devastation Caused by Floods in Bihar

Year	*Number of*			*Human Deaths*	*Crop Area Loss Lakh Hectares*
	Person Affected (in lakhs)	*Households Affected (in lakhs)*	*Number of Damaged Houses*		
1998	134.7	26	199611	381	12.84
2002	160.18	31	419014	489	9.4
2004	212.99	41	929773	885	13.99
2007	248.13	47	736857	741	16.63

Source: Department of Disaster Management, Government of Bihar.

Table 4.17: Physical Progress of Flood Protection Works in Bihar

Upto March	*Length of Embankment (Kilometer)*	*Area Protected (in Lakh) hectares*
1964	1108	12.40
1969	1772	12.40
1974	2192	15.00
1980	3059	26.05
1990	3420	28.68
2004	3430	29.16

Source: Department of Water Resources, Bihar, Patna.

The breaches, themselves, would not be catastrophic, but the increased height of the mean river levels causes them to become catastrophic. River bed of Kosi which has been in the news recently, has risen 12-15 feet or account of silt deposition. Instead of simply spreading out over the floodplain, the rivers now cascade/crash down onto the floodplains below. These cascades carry with them sand, gravel and stone. These are then deposited on the surrounding landscape. In many areas, the accumulation of sand and gravel has made cultivation impossible. These infertile sand and gravel soils are called '*patpar*' by local cultivators. Between waterlogging and lands lost to sand and gravel deposits, cultivators have lost well over 8,00,000 hectares of previously cultivable land. The Government has been taking various steps to control floods. Region wise status of flood protection works is an in Table 4.18

Table 4.18: Region-wise Status of Flood Protection Works in Bihar

(*As on end-2003-04*)
(*Area lakh hect.*)

Items	*North*	*South*	*State*
1. Total Geographical Area	53.80	40.35	94.15
2. Flood Prone Area	44.46	24.34	68.80
3. % of Flood Prone to Geographical Area	82.64	60.32	73.07
4. Length of Flood Protection Embark-ments (in km)	2952	478	3430
5. Protected Area	27.16	2.00	29.16
6. Protected Area as % to Flood Prone Area	61.09	8.21	42.38

Source: Water Resources Department, Government of Bihar.

The focus so far has largely been on short term measures, such as creating embankments and shoring up wells, when long term measures that can go a long way in preventing siltation and heavy erosion, e.g. afforestation and preventing deforestation are ignored. Different Ministries such as water resources and those of environments and Forests rarely work in coordination. The result is that every year the state suffers heavy losses.[10] According to latest National Disaster Management Agency figures, at least 24.5 million were affected by floods during 2007 in 22 districts. Damage to crop, homes and public property is estimated at Rs. 4,000 crore. Another estimate of damage in Bihar presents the picture as in Table 4.19.

Looking at the history of floods and sufferings of the people in North Bihar, the Government since independence has found embankments as the only solution. Rather than serving as a solution, this as discussed here has only aggravated the position. Since Independence, India has sought to engage the Nepalese government in a joint venture, where the rivers that pour into Bihar would be dammed and flooding could be controlled, or at least managed, downstream. After 60 years of meetings with the Nepalese Government, following a multitude of proposals for studies and investigations, nothing concrete has been negotiated between the two Governments.

Table 4.19: Flood Damage in Bihar, 2007

Population affected (lakh)		203.30	
Human lives lost		495	
Affected districts		20	
Villages		9,884	
Lost cattle/livestock		555	
Affected cropped area (ha.)		1,588,000	
Houses damaged		401,906	
Value of damage (Rs. In Lakh)			
	Crop	Houses	Public property
	70,183.64	43,152.92	63,969.44
Total damage		Rs. 177,306 lakh	
Evacuated people		998,479	
Relief Camps		1,288	
Med teams		547	
People in Camps		867,871	
Cattle in cattle camps		349	

Source: Down to Earth, September 15, 2007.

It must be appreciated that, the Nepalese Government has nothing to gain from the construction of dams or barrages for the purpose of flood abatement in India. The only tangible benefit accruing to Nepal might be the electrical power, the proposed dam(s) would generate. They would only inherit a massive maintenance obligation. Every river leaving Nepal carries with it huge burden in the form of silt, sand gravel and stones. If a dam were constructed, the burden of the river would collect on the upstream side of the dam, requiring a massive maintenance effort. If this accumulation of waterborne sediment is not routinely disposed of, the functional life span of the dam is reduced markedly. Even a dam upstream would only be a temporary way out. When a dam-often cited as the best solution to Kosi's floods by some—was first proposed in 1937, engineers hastened to specify that it would silt up in just 37 years given the large sediment deposit Kosi brings along with it as it flows downhill. Moreover, the dam would have to be built in a zone prone to earthquakes and would involve, like any other dam, immense environmental and social costs.

This type of maintenance would be massive, never ending and expensive for either or both governments* . If the proposal is to succeed, the implementation agencies will have to take into consideration all these costs and a proper cost benefit analysis study will have to be undertaken in the first instance. If so, the only solution would be to revert to olden days. First step in this case would be to start a desilting operations. This may be a long and costly exercise, but would appear to be the only permanent solution.

Treatment of catchment should be a pre-condition to building any big dam. Kosi floods and what followed also show complete lack of disaster management in the country. Another fact that environmentalists stress upon was that floods were not a natural calamity, but natural phenomenon. Devastation is what happens when you do not respect nature and take your rivers for granted. During monsoon there will be floods that occur in cycles—a decadal cycle, 25 year cycle or 100 year cycle. For example, last time when Yamuna saw floods was in 1998. Since it runs in decadal cycle, 2008 floods were bound to happen. Nature gives sufficient signals, it is we who ignore them. Floods have to be studied and rivers understood as ecosystems. The country needs a river policy in which floodplains are considered as an integral part of river ecosystem. We also need to a flood-zoning system. For a 10 year cycle, no structure should be allowed in flood plains. For a 50 year cycle, temporary structures can be permitted and for 100 year cycle there can be permanent structures but at one's own risk.[11]

The way out would be to rely on a more diverse catchment—based approach that depends on an extensive network of water-retaining bodies. "That would mean wetlands, much of which is being presently destroyed, lakes, ponds and groundwater recharge to help reduce the water inflow into the Kosi. We would also have to give up the idea that a flood necessarily implies a disaster. It's only human mismanagement that makes it one", (Annexure-4.2)

Drinking Water and Water Quality

In Bihar, a sizable section of people living in flood prone areas have a peculiar problem-excess water during floods and yet not a drop to drink. Not only are people rendered homeless every monsoon, they

---

* *Ibid*, J. Albert Rorabacher.

also face acute drinking water shortage. It was to address this problem that the Megh Pyne Abhiyan (cloud water campaign) was started during 2006. The rainwater falling on the sheets are channeled and stored in a container. In its first phase, the Megh Pyne Abhiyan is targeting about 46,000 people. The objective was rooftop rainwater harvesting for people seeking shelter on higher ground during floods. At such times, inaccessibility to safe drinking water and sanitation are two key problem areas. Launched in north Bihar, the campaign was an attempt to convince people to practice rainwater harvesting. A group initiative launched by four NGOs-Samta, Kosi Seva Sadan, Gramyasheel and the Gogardiha Prakhand Swarajya Vikas Sangh—it was brought together by Eklavya Prasad, who believes in natural resource management.

People living along Manusmara river in Bihar's Sitamarhi district dread *kala pani*—the river water turned dark red or black with effluents. This lifeline now kills their cattle and fish, causes infection and disease and ruins their farms. "People have stopped using river water. It is neither fit for drinking nor for irrigation. The source of pollution is a distillery unit at Pachhaur village in Riga block, not far from the Nepal border. Though a "zero-discharge mill" on paper, Riga Sugar Mill and Distillery located just 10 metres from the river, discharges wastewater into it at night twice or thrice a week. The impact is visible till Runnisaidpur, 40 km down stream, as the river changes it colour.

The river, also known as the old Bagmati, flows in a narrow channel up to Belsand, beyond which it spreads and loses its course, engulfing large chunks of fertile land in Runnisaidpur. The current is negligible here, and discharged effluents tend to accumulate "In the *kala pani*-impacted area, the yield of sesame, mango and jackfruit is badly affected. Polluted water has affected 485 hectares and 100 villages.[12]

Inland Navigation

The World Water Forum (WWF) in its Kyoto (Japan) convention, held in March 2003, recommended that all the future water resources development projects in the world should include the component of inland water navigation. It was suggested that flat bottom sea going vessels up to 3,000 tonnes capacity should be able to ply in major rivers such as the Godavari up to Pochampad (SRSP) dam and the Krishna up to Nagarjunasagar dam, throughout the year. This would be possible when a minimum water depth (draught) of 1.5 metres is maintained all along the river.

Cheap transportation of bulk and finished goods would lead to industrial growth, all along the banks of rivers. industries will come up quickly. Also adequate water can be made available to them. Inland water navigation coupled with industries will create new livelihoods for the landless rural poor throughout the year. This is in tune with the then President A.P.J. Abdul Kalam's Republic Day-message on the need to create additional employment. Inland navigation in the Mississippi was developed several decades ago. Ten developed countries in Europe completed in 1992 a new navigation route connecting the Black Sea and the North Sea through the Rhine-Danube river linkage. India's neighbour, China, has already launched a major navigation programme in the Yangtze river.

It is disappointing to note that practically no effort is being made in India in general and Bihar in particular to develop inland water navigation in any of the water resources projects under construction. On the other hand there is a wrong understanding in several quarters that navigation, as a means of transport, is outdated and not relevant to the modern Jet Age. Interestingly, there are recorded versions (*Encyclopedia Britannica*) that development had resulted in the U.S., Europe and Russia after establishing navigation in the rivers. The modern inland irrigation envisages sea going vessels plying in rivers and is different from the earlier canal navigation. Since over development through inland navigation was already experienced in the developed countries, the Government of India should make this component mandatory for all the future water resources projects in the country. This would also amount to honouring the Kyoto Convention 2003. This is all the more important for Bihar which has a network of 14 rivers and the present road network is rather weak. Due to continuous flow of silt, the river bed has risen, leading to more floods. Desilting of rivers will not only help reducing the flood intensity, but also provide facilities for inland navigation.

Though still in its nascent stages, the Jharkhand Government has initiated an exercise to optimally use its water bodies for commercial gains. Jhakhand Tourism Development Corporation managing director says, "While we have introduced boating facilities in Jamshedpur, we will eventually offer river cruises". Meanwhile, the presence of the Ganges is prompting Bihar to join the bandwagon. We are due to begin the Ganga river cruise by December 2007. We have got a 100—

seater cruise liner to ferry tourists between Gandhi Ghat in Patna and Gandhi Setu. While for tourists we have designed a 30 minute package, for corporates, we will work out tailor—made packages.[13]

INTERLINKING OF RIVERS

An allied solution is the scheme of linking of rivers which has been identified as an important initiative for ameliorating the endemic problems of floods in North Bihar and recurrent droughts in South Bihar. An Expert Committee formed by the State Government has recommended that rather than inter-state linking of rivers, it is important for the State to formulate and implement schemes of Intra-state linking of rivers. The Committee has held that the objective of interlinking of rivers should not be limited to transferring water from surplus to water scarce basins alone. It should also address the entire gamut of issues related to water management which would generate additional irrigation capacity, alleviate problems of floods and drought and also facilitate simultaneous development of indigenous irrigation system. The scheme of interlinking of rivers, while addressing the problems of floods and droughts in different basins, should also provide for augmenting the ever depleting lean season flow of the river Ganges, especially in the part flowing through Bihar. A general consensus has emerged among expert groups in the State on the requirement of intra—state linking of rivers. In the light of recommendations of the Expert Committee, 4 schemes have already been identified in the flood-prone sector and 9 schemes in the irrigation sector. We give below examples of Madhya Pradesh and Tamil Nadu—two states which have benefited from such a programme.

Schemes for interlinking of State rivers have already been completed by Madhya Pradesh (Sardar Sarover Project). A Rs. 4,000 crore plan had been prepared to strengthen the canal system in Punjab. There was a proposal to link various rivers for the optimum utilization of their water in the state. A committee under the chief secretary (Punjab) has been set up in this regard. There was also a programme to provide canals in areas where these did not exist at present.

Link Rivers in Tamil Nadu

The Tamil Nadu Government is planning to link rivers within the

State to divert surplus water to drought-prone regions. A macro analysis has been done by the Public Works Department for taking up proposals to harvest flood water to dry basins. The analysis gave priority to linking the Cauvery with the Vaigai not only to mitigate flooding in the Cauvery basins but also to provide water to dry districts such as Pudukottai, Sivaganga, Ramanathapuram and Virudhunagar. According to a proposal, the existing Kattalai bed regulator across the Cauvery near Karur will be converted into a barrage for flood water storage, regulation and better water management.

For linking the rivers, a canal is to be excavated from the Kattalai barrage to Gundar, connecting Agniar, South Vellar, Pambar, Kottakariyar Vaigai River with Gundar in phases. The total length of the canal will be 255 km. A study of the surplus details of the Mettur reservoir from 1975 to the present found that the reservoir surplussed in 15 years for five to 81 days. Though the dependability is low at 45 per cent, diversion can be attempted even where there is sudden rainfall in the delta region and when there are pre-releases from the reservoir in the river. The pre-released flows in the river up to the regulator can be stored in the Kattalai baarage and can be diverted to the new canal.

The link will not only divert Cauvery flood water but also the waters of Ariyar and Koraiyar during the northeast monsoon to protect the Tiruchi and Srirangam towns. Karur, Tiruchi, Pudukottai Sivaganga, Ramanathapuram and Virudhunagar districts will be benefited by the link, as there is a proposal to fill all tanks now fed through the existing anicut system with the diverted river water. The investigation work would have two phases. The first phase was for studying issues relating to construction of the barrage and canal excavation. The investigation, which is in progress, would be completed by March 2008. The project cost of the first phase is likely to be Rs. 1,000 crore. In the second phase, study on the extension of canal would be taken up. The scheme required 3,684 hectares of land. Of this 2,228 hectares was patta land, 1416 hectares was poromboke and the balance was forest land.

Many other States are initiating these programmes and such projects can be completed without any political complications and inter State issues. This can also slove to a large extent the power problem by setting up small hydro power plants of Bihar which the State needs so badly.

Fig. 4.5: Westward Movement of Kosi

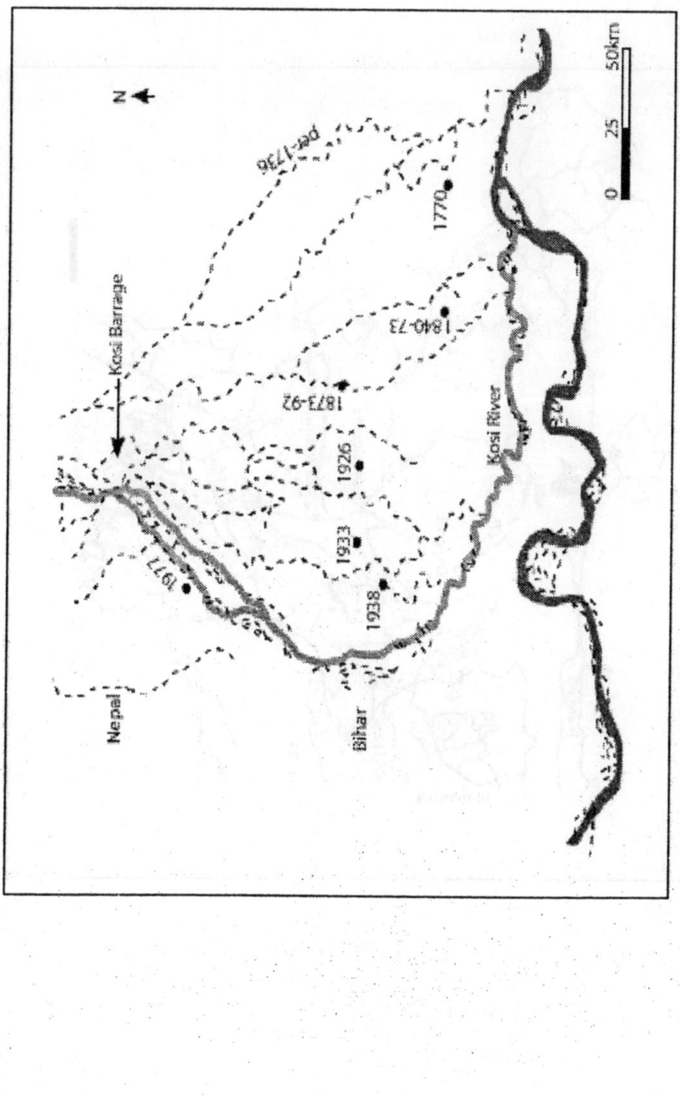

Source: Economic and Political Weekly.

Fig. 4.6 Dynamic River System of Eastern Ganga Plain

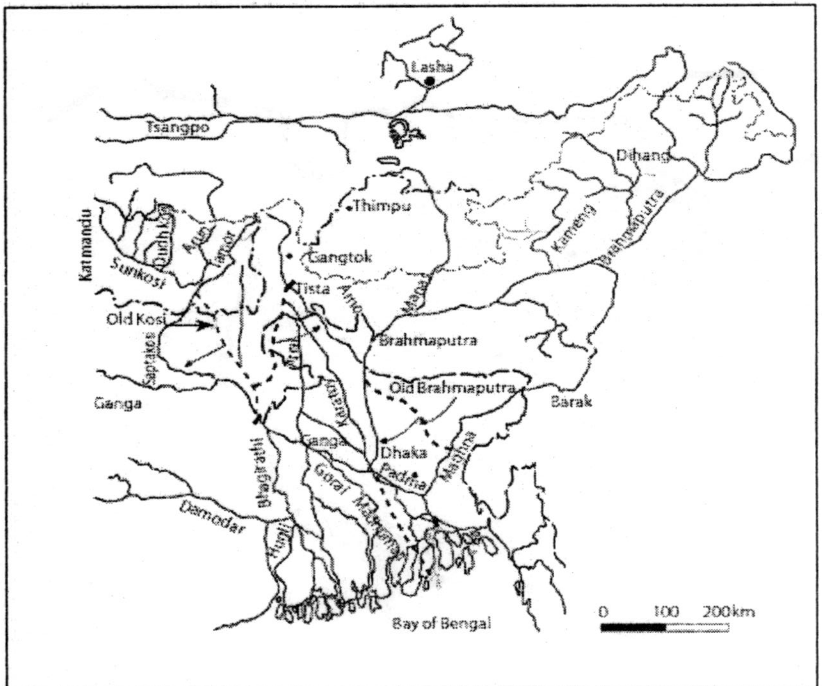

Source: Economic and Political Weekly.

DROUGHTS

Just like floods in North Bihar, there are said to be frequent occurrences of drought in the districts of Munger, Nawada, Rohtas, Bhojpur, Aurangabad and Gaya. A look at irrigation facilities in these Districts would show that but for district Munger where irrigated area is 26.4 per cent, all the other districts have high percentage of irrigated area. Zone-III where these Districts are located has an average annual rainfall of 1104 mm. With such rainfall in the region and most of the districts having irrigated area higher than the State average (Table 4.20), it is a pity that they are still being identified as drought prone. The Planning Commission has already sanctioned watershed development projects for 8 Districts of Southern Bihar—Banka, Jamui, Gaya, Munger, Aurangabad, Rohtas, Bhabua and Nawada. In District Gaya (Barachatt Block) the catchment area was earlier a dense forest, but has been de-forested or degraded in the last 10 years. Due to this reason, volume of runoff is not being checked and water table has gone down, leading slowly the entire area into a degraded unproductive land. Ten years back when these areas were covered under dense forest, water availability was abundant and production was very good. There is the need to take up Watershed Development Projects in all such areas. In the Southern Zone, where groundwater depletion is a cause of concern, projects of ahars, pynes and tanks should be revived and promoted for induced ground water recharge. Other solutions could be, projects with inter-catchment water transfer and injected low-recharge. (Annexure 4.3)

Table 4.20: Extent of Drought Prone Area

District	*No. of Taluks*	*Area ('000 ha)*	*Irrigated area (per cent)*
Munger	4	788	26.4
Rohtas	2	720	84.6
Gaya	2	651	47.6
Aurangabad	1	331	67.9
Bhojpur	2	397	79.4
Nawada	1	249	50.8

Source: Central Water Commission, Govt. of India and Government of Bihar.

It is true that rainfall may be erratic, but Bihar can certainly go for rainwater harvesting programmes in a big way and follow the example of Rajasthan where with less than 400 mm rainfall, the State is in a position to have horticulture by rainwater harvesting like Kund Horticulture Project. We have provided details of two projects—Orissa Watershed Development Mission and Kund Irrigation in Rajasthan. The state can certainly learn from these experiences so as to rid South Bihar from the so called drought problems.

Orissa Watershed Development Mission (OWDM)

The natural resources base of the region which includes Bihar is being increasingly threatened by climatic changes—marked by variable weather changes in rainfall regime, temperature, and increased frequency and severity of natural disasters. Therefore, its protection and regeneration must be our top priority, if we are serious about poverty reduction. The 2007-08 UN Human Development Report focuses on the link that exists between climate change and its impact on the poor. In India, especially, the poorest, are very vulnerable to the impacts of climate change, mainly because the nation's economy is closely tied to natural resources. It adds: "Assisting the poorest communities to increase their income options will be critical."

In Orissa (a state that is the ground zero of many a battle between communities and companies over the control of climate—sensitive resources), a fight has been launched for the regeneration of land and water and improving the livelihood opportunities of the poor. An exhaustive watershed development plan is at the centre of this approach. The Government of India and Britain's Department of International Development (BDID) have sponsored this plan and it is being implemented by the Orissa Watershed Development Mission (OWDM). The Mission's projects target women, tribal and caste groups and its livelihood programmes assess what is needed to strengthen the response of the communities to the mounting impacts of climate change.

Technically speaking, a watershed is an area from which rainwater run-off flows into a single stream. It is around this area that soil and water conservation structures (recharge pits, wells, ponds and bunds) are built and water is harvested where it falls. This harvested water helps in recharging acquifers, leading to better green cover and productivity of farmlands. And, as water becomes the driver for change, other things follow—income generation improves, distress migration drops and education and health become the new talking points. In other words, it is an integrated programme that not only looks at ecological treatment of a particular area but also at improving the human development indices of the villages that fall in that zone.

"Ecological treatment of parched landscapes can play a major role in building a community's capacity to cope with disasters and climate change. Despite a good history of rainfall, this region is still drought-prone. Watershed management can revive old water-harvesting traditions and help communities fight drought and climate change.

For starters, there has been a drop in people migrating out of Nauapada and Bargarh areas. An earlier research conducted by BDID showed that this was a high migration area with mostly Scheduled Tribes and Scheduled Castes moving out in search of jobs. Life of the farmers in the area was divided into two parts: growing paddy in their small patch of land and then moving to Andhra Pradesh for work in the brick kilns, or even Surat to work as labourers. They would get hardly Rs. 5,000—Rs. 8,000 for six months. But there was no option because there was no work available for them here for about six months after the cropping season.

Moving back and forth meant that the children had no access to school and health facilities and, worse, they were getting sucked into an inter-generational poverty cycle. The wheel, started turning when help came from the OWDM's programme. Farmers were taught the technique of cultivating cash crops like sunflower and vegetables. A new onion storage plant helped them hold on to their crop and sell it as and when the prices improved. Today they have stopped the seasonal flight. They are building a new house and their children are in school. The other success story has been the involvement of women. "Who form the backbone of this movement and at many places they are heading the watershed committees. Many have started or joined self-help groups, moving towards economic self-sufficiency. There has been a 50-60 per cent drop in migration thanks to improved farming and other allied income-generation activities. People have found a new reason to stay back. This type of a replicable model in Bihar will provide an immediate help to the poverty driven migrants in the state."

The future looks bright for these areas. But there are challenges: integrating the livelihood schemes with other Government-run schemes, integrating with the market (the OWDM is in talks with private companies for establishing a cold chain for agricultural products) and, most important, keeping the tempo going. The best practices of this movement need to be replicated in other parts of the country—particularly in a State like Bihar—which is most suited for this type of a Mission. Otherwise, these areas will just be small islands of prosperity, a kind of rural special economic zones.

Kund Horticulture Project (Rajasthan)

The Rajasthan Government has envisaged a Rs. 22.43 crore Kund Horticulture Project for 358 Common Interest Groups (CIG) across the state. The project, under the District Poverty Initiative Programme (DPIP), would benefit 3,120 below poverty line farmer families of Churu, Raigarh, Ratangarh, Sardarshahar, Sujangarh, Tara Nagar and Dungarpur districts. DPIP would bear 90 per cent of the cost while the rest would be shared by the GIGS consisting of 10 farmer families each. The project is targeted at poor farmers who have land but lack irrigation facilities. In Churu district itself, they plan to plant 2.5 lakh fruit bearing crops like amla, ber and karana interspersed with plants needing less irrigation like chana, aradu, moth and gwar in the gaps.

According to estimates, a farmer would be able to earn Rs. 3,000–5,000 per month after one year of implementation of the project and Rs. 15,000–20,000 per month after three years of implementation. Under the programme, each CIG member would get a 30,000 litre water kund constructed on an acre of land. A plastic tank of 2,000 litre would be mounted 5-6 feet above the ground. The plastic tank would be filled with water through a mechanical hand pump, which would draw water from the kund. As many as 80 fruit-bearing plants and 30 shady plants like neem, ber, sheesham, etc., would be planted at a gap of 6 x 6 metres. These plants would be irrigated through a pipeline connected to the water tank due to gravity. A plant can get 10 litres of water in an hour through this scientific technique, which keeps a check on wastage of water and ensures equal distribution among plants. The construction of a tank is expected to cost Rs. 42,000 while the total cost including training and barb wiring of the area works out to be Rs. 70,122 per member. The project has shown good results in Churu district. The payback period is of 25 years and it is expected that this technique will not only reduce wastage of water, but increase the yield by optimizing irrigation facilities.

This so-called drought prone region with rain water structure in a districts like Munger where irrigated area today is low, is the potential source of leading the State in medicinal plants, floriculture, seed production of vegetables and dairy development. Marketing facilities are, however, required to be improved for an easy disposal of all types of perishables like horticulture, floriculture and dairy products. It is important to examine the possibility of setting up processing industries for these enterprises and also encourage apiculture to provide some additional income and employment to the small and marginal land holders.

POLLUTED GANGA

Along with the issue of floods and droughts, it is important to note that polluted Ganga is turning into the Death Valley. The Cancer Valley of India lies between Delhi and Patna along the Ganga. A joint study—by Indian and Japanese doctors for over two years, has found that the region has the second highest rate of gall bladder cancer occurrence in the world. The rate is three to four times higher than in any other part of India. The study, conducted by the International Hepato-Pancreato Biliary Association (HPBA) re-affirms an earlier Indian Council of Medical Research (ICMR) finding that gall bladder cancer among women in North India, including Delhi, was the highest in the world—10.6 out of every lakh women. The study has concluded that the abnormally high rate is primarily due to pollution of the Ganga waters and the soil around with heavy metals like lead, cadmium and chromium.

The project, which screened 22,000 people across 60 villages in the Indo Gangetic belt with the help of Mumbai-based International Institute of Population Sciences (IIPS), found the incidence of cancer in the region was between 12 and 20 for every 100,000 people. The corresponding figure for southern India is 0.8 while for Mumbai is 1.2 (in Chile, it is 24 people per lakh.) Three sites were selected in rural Patna, Vaishali and Varanasi districts. The soil and water samples were sent to Industrial Toxicology Research Centre in Lucknow for analysis, while the tissue and hair sample went to the Aichi Cancer Centre in Nagaoya, Japan. High level of cadmium, lead and chromium were found in the sample of soil and water.[14]

Notes

1. Tushaar Shah, 1993.
2. State as Paper on Participatory Irrigation Management Bihar Model, 2005.
3. Total canal irrigated area today is around 1.37 million hect.
4. Bihar Agriculture Building on Emerging Models of "Success".
5. According to Bihar XI Plan, the figure is 32 distributaries aggregating to a total command of 110,549 hect.
6. Economic Survey 2007, Government of Bihar.
7. Some of these issues have been discussed in Chapter on Infrastructure.
8. Water in Nepal/Rivers Technology and Society, Bharti, *The Tribune*, 8/9/08.
9. Sudhirender Sharma, Missing the River for the Dam, *The Economic Times*, 3/9/08.
10. Economic Weekly 1-7 September 2007.
11. Vibha Sharma, Disrespect for Nature led to Kosi Disaster, *The Tribune*, September 1, 2008.
12. *Down to Earth*, February 15, 2008.
13. *The Economic Times*, October 5, 2007.
14. *The Hindustan Times*, February 28, 2008.

ANNEXURE 4.1

Water Problems in India

The urban middle class has come up with "coping strategies given the lack of proactive policy on the water front. These include household storage and pumping system, purchase of bottled water and the like. But the costs are high: several times higher than the average payment to the water utility for example in Delhi. For low-income households, the situation can be far worse. As for industry, there are added costs to bear, as when aquifers are either not available or are exhausted. The way ahead is to opt for high-cost "captive" alternatives, such as reverse osmosis of wastewater or desalination. Given the overall costs of negligence public policy, we do need to comprehensively modernize the systems for allocating funds and monitoring surface and groundwater resources.

As India picks up economic speed, the demand for water would invariably rise. Given the investment backlog in water supply, it suggests all the more need for heightened policy attention. In tandem, what's essential to improve the deteriorating quality of the water-related environment. Fortunately, international experience shows that as incomes rise, environmental quality tends to improve hand in hand. But it would be presumptuous not to have corrective policy in place. For the foreseeable future, there will need to be increased budget support, read money of taxpayers, for irrigation. The idea of reasonable charges—against the backdrop of generally poor services provided by corrupt and inefficient agencies—would naturally be resisted. Which is why any plan for tariffs to better reflect costs, even proper maintenance, must really be the third leg of a triangle in which the first two legs ought to be to improve services in an efficient and accountable manner.

Project management will be the key. The Central budgetary allocation under the irrigation head has quite rightly been handsomely hiked to Rs. 20,000 crore. In parallel, what's required are standardized norms for project management and effective policy implementation. Project delays and cost overrun are endemic to irrigation projects. Besides, Inter-disciplinary team work, so essential in the water sector, seems to be absent pan-India. The ground reality is that the links between academic institutes and water sector practioners are weak or simply non-existent. The result is that academics don't get to look at important practical issues and problems on an ongoing basis. Back-to-back, water managers are more likely not to be kept abreast of the latest technologies. It all needs to change. Fast!.*

According to the Ministry of Water Resources, there are 388 incomplete irrigation projects spread all over the country. Of these, 340 were started before 1992, and 40 before 1974. Till 2003, Rs. 78,449,63 crore had already been spent on them, and an additional Rs. 89,872,72 crore was needed to complete them. Had these projects been completed on time, over 20 Mha of land would have come under irrigation. However, only about 7 Mha worth of irrigation facility has actually been created. Twenty four State Governments were supposed to implement these projects. Since they claimed that they were cash strapped, the Centre announced the Accelerated Irrigation Benefits Programme (AIBP) in 1996 for completing 172 of the more wide-impact projects in two years. Half the money was to be given by the Centre as loans while the respective State Governments were to bear the remaining expenses.

Money Down the Drain	
Incomplete Irrigation Projects	388
Expenditure till 2003	Rs. 78,450 cr.
Additional Funds Needed	Rs. 89,873 cr.
Irrigation Potential Target	20 Mha.
Actual	7 Mha.
Accelerated Irrigation Benefit Programme 1997-2003	
Total Fund Available	Rs. 18,906, cr.
Projects taken up	172
Actually completed	23
Irrigation Potential Target	10 Mha
Irrigation Potential Created	2.8 Mha
Bharat Nirman 2005-2009	
Irrigation Potential Target	10 Mha
Irrigation Potential Created (till September 2006)	2.2 Mha

By 2003, about Rs. 19,000 crore had been released for the programme of which around Rs. 13,000 crore had been spent. Only 23 projects had been completed in seven years. Of the targeted 10 Mha land to be irrigated, work for covering only about 2.8 Mha was completed. Even out of this, only 0.3 Mha was actually being irrigated. The Comptroller and Auditor General (CAG) pointed out that in many

* *The Economic Times*, New Delhi, March 28, 2008.

cases, projects selected for inclusion in the programme were either not taken up, abandoned midway or declared complete when they were not. CAG carried out actual test audit of about Rs. 8,146 crore of the money spent and came up with shocking results. Nearly Rs. 3,000 crore had been misutilised, diverted to other uses, reported as inflated expenditure or parked in bank accounts. It also found that, in order to accommodate requests from states, several ineligible projects were included in the programme.

The Central Government indeed came up with a major new programme—Bharat Nirman—under which irrigation was a key sector. The goal of creating 10 Mha of fresh irrigation potential in four years (2005-06 to 2008-09) was setup. However, results of this special thrust to irrigation have also been disappointing. In about two years of implementation, only about 2.2 Mha worth of irrigation potential has been created, that is 22 per cent of the target. The rest of 78 per cent remains to be created in the next two years.* The picture presented here is for India, but position in Bihar may be worse. The State has to look deep into the issue of project efficiency in all its irrigation projects.

ANNEXURE 4.2

Kosi River

Kosi originates as a small stream in the Eastern Himalayas in Nepal. The river is also called the Sapta (seven) Kosi in Nepal, because of its seven tributaries (the Indravati, Sun Kosi or Bhot Kosi, Tamba Kosi, Likshu Kosi, Doodh Kosi, Arun Kosi, and Tamar Kosi). Characteristics of These rivers are as in Table below.

Table : Characteristics of the Tributary Rivers

River	*Catchment Area km^2*	*Mean Flow M^3/s*	*Specific Flow $M^3/S/km^3$*	*Annual Sediment Load $m^3/sX\ 10^6$*	*Specific Sediment Load m3/s/Km^2*	*Average Gradient%*
Sunkosi	18,800	471	0.025	54.2	2,818	0,0062
Arun	34,300	451	0.01	34.6	9,47	0,0134
Tamor	5,800	347	0.06	29.6	5,016	0,0276
	58,900			118.4	1,920	

The first five tributaries join to form the Sun Kosi, Arun Kosi and Tamar Kosi join it at Triveni in Dhankutta district of Nepal and assume the names Sapta Kosi, Maha Kosi or the Kosi. Trivent is located in the hills, about 10 km. north of Chatra where the river onto the plains. After flowing for nearly 160 miles, it enters the plains of Bihar and joins the Ganga in eastern Bihar. In ancient times, Kosi was known as Kausiki. As per mythology, the river represents Satyavati, the mother of Sage Jamadagni and the elder sister of Sage Visvamitra. After Satyavati passed away, Visvamitra decided to settle on the banks of the Kosi. The place where he stayed is called Kausikasrama and is considered sacred. Many other ancient Hindu sages like Rishyasringa also resided on the banks of Kosi. Figure below gives Kosi River catchment area and characteristics of the Tributers rivers in Table above.

This river, which gathers water from some of the highest mountains peaks in the world, including Everest, enters India in north Bihar. It has changed its course and shifted over 120 km eastwards, going back to a course, it had abandoned more than 300 years ago.

---

* *The Economic Times*, New Delhi, June 7, 2007.

Graph : The Kosi River Catchment Area

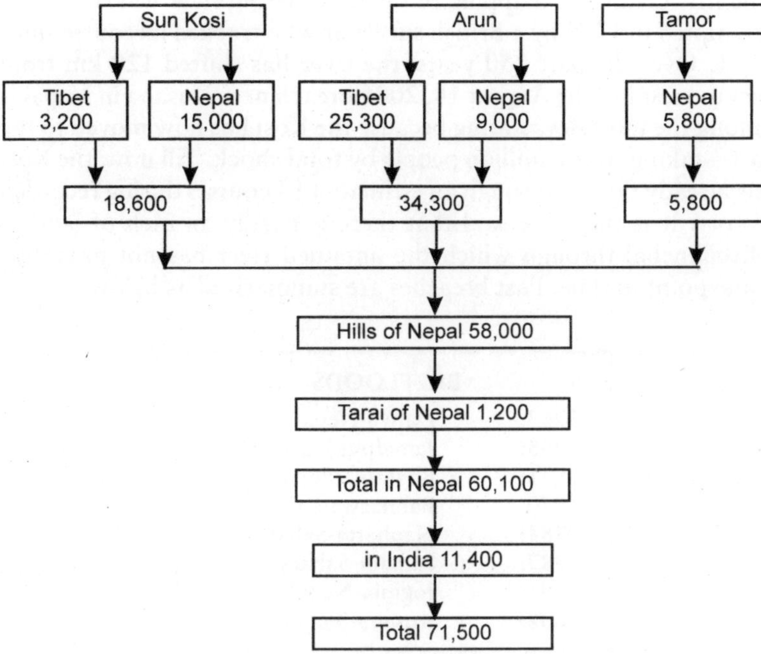

Source: Economic and Political Weekly.

The first creditable mapping of the meandering river was, however, attempted in 1779 by a British surveyor who tracked its course since 1731. Over the past 250 years, the river has shifted 120 km from east to west but the August 18, 2008 breach near Kusaha in Nepal is among the most devastating because the Kosi has now moved to the east—taking over a million people by total shock. Till now, the Kosi has already flowed through an estimated 15 courses during recorded history. It is therefore said that there is hardly an inch of land in Mithilanchal through which the untamed river has not passed at some point in time. Past breaches are summarized as below:-

BIG FLOODS	
1963:	Dalwa-Nepal
1968:	Jamalpur-Darbhanga
1971:	Bhatania-Supaul
1980:	Bahuarwa-Saharsa
1984:	Nauhatta-Saharsa
1987:	Samani-Saharsa
1991:	Joginia-Nepal
2008:	Kusaha-Nepal

When a river changes course, Geographers offer different reasons for this amusing phenomenon. While swiftly flowing down the mountain, the river carries with it mud, sand and boulders. When the river enters the plains, it does not flow or move so fast and is not able to carry the heavy load of sand and boulders. Hence, the sand and boulders get deposited in the bed of the river. In course of time, these deposits lead to the silting of the river Kosi. In other words, the deposits become so big and hard that they prevent the flow of the river. carries 81 million tonnes of silt every year. The river under such conditions is compelled to change its route or course.

The Kosi has a history of westward shift of its stream from Purnia to Supaul over the last 300 years. For a change, the breach in the eastern affiux bund has caused eastward movement of the Kosi water flow, inundating the areas hitherto considered flood safe. Bihar Government sources said the river, which has a heavy load of silt, has been maintaining pressures on the spurs near 12.8 km point of the eastern afflux bund since 1985. This has been forcing the engineers to undertake bund strengthening work every year.

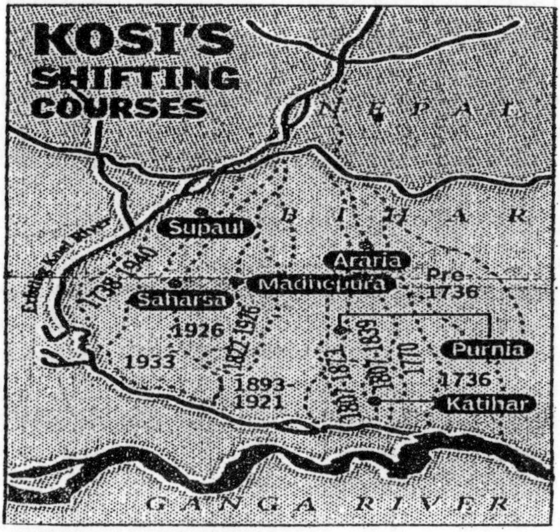

Source: Economic and Political Weekly.

The problem, in its latest form began after the Kosi was embanked in 1950. "The jacketing of the river proved to be disastrous. The river's flow was controlled but there was no way to check the resulting siltation. This has complicated matters". Seen from a satellite, the area looks like a conical fan. Created by hundreds of years of shifting, it is the largest such cone in the world, covering an area of over 15,000 square km. The tip is near Chatra on the Nepal border. The cone is made up of various courses of Kosi and the land in between, which gets submerged during floods.

Experts argue that building embankments allowed too much silt deposition in a shorter stretch, leading to devastation. In other words young Kosi is not being allowed to grow up and settle down. Meanwhile, people have to adjust to a new reality, the Kosi flowing in its new course, redefining fields, roads and boundaries.

There are historical records documenting the floods cause by the Kosi. In 1875, an entire township was washed away by the river. In the 1940s, the river caused extensive floods by changing its course, it is said that before any major flood, the water of the river turns turbid and reddish. The local people also believe that there is a strong wind for three days before the onset of floods.

The first attempt to tame the Kosi began in 1956, after the devastating floods of 1954. In the techno economic survey prepared at that time, clearly the optimum solution was the construction of a 239 metre dam at Barakshetra, about 50 km within Nepal to be backed by a barrage down stream. For several reasons including cost, a deflection of focus due to the Bhakra Nangal Project, the complexities in the construction of multiple structures resulted in shelving this optimum option. It was considered prudent to settle during the interim with merely a barrage at Hanuman Nagar (Birpur). Thus the first credible attempt to tame the river began in 1956 with eastern and western enhancement of 105 and 106 km. respectively, of which about 32 km of the eastern embankment is located in Nepal. The embankments were completed in 1959. The barrage itself at Birpur to regulate water flow was completed in 1964. The indo-Nepalese agreement signed between the two countries which facilitated this project brought benefits to both India and Nepal. The Nepalese side, however continued to question the adequacy of benefits received from the project. Nepal had all along wanted a barrage system up stream of the present which in their perception would have yielded more optimum returns.

There are several questions to be addressed with regard to the future. First, the failure of our bilateral diplomacy with Nepal. We are a lower riparian state. There are well-accepted international obligations on the rights and obligations of upper and lower riparian states. It cannot be anyone's case that the lower riparian state must suffer serious ecological disaster to force the pace of change. Repeated attempts by India to settle this issue during the multiple Indo-Nepalese Commissions and other fora did not receive adequate response.

Second, attempt to engage multilateral institutions like the World Bank and, the Asian Development Bank did not cut much ice with the Ministry of External Affairs, which felt that this would unduly international an issue which lay in the bilateral domain of India and Nepal. This was strange. Thus, while bilateral diplomacy remained stalled in the penumbra of suspicions, multilateral institutions which were keen to help on this project were kept out.

Third, the initiative of the Indo-Nepalese Joint Agreement on Project offices to carry survey and investigation for the preparation of a project report for the Sapta-Kosi and Sun-Kosi high dam is an inadequate response to years of diplomacy given the urgency of the problems. The snail pace with which these talks have proceeded and tardy implementation of some limited agreements or the Maha Kali Treaty, and Agreements on the Bagmati, Kamla and Lal-Bakaya rivers of North Bihar, only heighten our anxiety.

Fourth, the simple fact that apart from the multiplier benefits to agriculture, flood protection and livelihood patterns to both people of

India and Nepal, a substantial part of the hydro-electricity of 3000 MW could be sold by Nepal to India to bring about huge improvement in real income of Nepalese and meet the energy deficiencies of the eastern region of India failed to carry conviction. Will we now at least wake up and impart new meaning and content to the stalled efforts of five decades? There is a new government in Nepal. This may alter the accepted paradigms of diplomacy which we have pursued with Nepal. Nonetheless, the human dimensions of the problems need to transcend the compulsions of contemporary politics. This needs meaningful, credible and perhaps aggressive engagement. We cannot afford to ignore the misery and suffering of millions living in north Bihar and what is perhaps one of the most demographic dense regions in the world.

Some Facts

- The Kosi River, is Kosi in Nepal. It is a trans-boundary river between Nepal and India, and one of the largest tributaries of the Ganga.
- The river, along with its tributaries, drains a total area of 69,3000 square km up to its confluence with the Ganga in India.
- The river basin is surrounded by the ridges separating it from the Brahmaputra in the north, the Gandaki in the west, the Mahananda in the east, and by the Ganga in the south.
- The river is joined by major tributaries, approximately 48 km north of the Indo-Nepal border, breaking into more than 12 district channels with shifting courses due to flooding.
- Over the last 250 years, the Kosi River has shifted its course over 120 kilometers and the unstable nature of the river is attributed to the heavy silt which it carries during the monsoon season.
- The Kosi (The Sorrow of Bihar) is one of the two major tributaries of Ganga in Bihar, the other river being Gandak, draining the plains of north Bihar, the most flood-prone area of India.
- On the positive side, thanks to the silt brought by the river the banks of the river are very fertile for agriculture. The quantity of silt deposited by this river is 37 million tonnes per annum.

ANNEXURE 4.3

Floods in India

Despite massive amounts of money spent on dams, embankments and canals, the floods in India just keep on increasing in frequency and destructive capacity. India is the second-most flood-prone country in the world, after Bangladesh. Nearly 40 million hectares of its land is flood-prone, of which about eight million hectares are flooded annually. The average annual total damage to crops, houses and public utilities during 1953-1995 was about Rs. 9,720 million. The price of human death is incalculable.

Hundred years have passed since 'Article I was published in the December 21, 1907 issue of the *India Mirror*. Still, Article 2' from one of today's leading newspapers *Business Standard* (published on September 5, 2008) expresses the same helplessness. And the same concerns. The only thing that has changed is the frequency and magnitude of floods in India. Obviously, we have not learnt a lesson from the mistakes in the past. What went wrong? To answers this, let us go back to Article-1. India was visited by floods every year. Still, the people who lived here used ingenious, home-spun techniques to manage the floodwaters. Like in Bengal.

Bengal's Genius

Bengal once had an extraordinary system of inundation canals. Floodwater entered the fields through these canals, and irrigated the fields. The water carried rich silt, which fertilized the fields, and fish, which swam through these canals into the lakes and tanks to feed on the larva of mosquitoes. This helped to check malaria in this region. But, unlike Bengla, there are some regions that do not have a tradition to battle floods. Like North Bihar—the local cultivators welcomed the floods.

Bihar's Tale

Floods in the North Bihar plains date back to antiquity. However, they were never considered as a bad phenomenon, rather they were deemed as nature's bounty. Farmers extensively used floodwaters for irrigation. The waters refilled the tanks and wells, and the silt deposits of inundating rivers replenished the natural fertility of the soil and ensured a good winter harvest. Eklavya Prasad of Megh Pyne Abhiyan explains, "The floodwater used to spread across a large area. Thus, its height was low. It would recede quickly, as there were no obstructions. And there was no need for flood control because flood were as and devastating as today".

Flood Affected Area and Flood Damages in India (1953 to 2004)

Item	Unit	Average flood damage	Maximum damage (Year)	Damage during 2004
Area Affected	Million Hect.	7.63	17.50 (1978)	8.03
Population Affected	Million	32.92	70.45 (1978)	34.22
Human Lives Lost	Nos.	1,590	11,316 (1977)	1275
Cattle Lost	Nos.	94,485	618248	63,869
Value of Damage to Crops	Rs. Crores	705.87	4,246.62 (2000)	615.07
Houses Damaged	Million	1.23	3.51 (1978)	1.49
Value of total Damage to Houses, Crops and Public Utilities	Rs. Crore	1,805.18	8,864.54 (2000)	3,336.59

Source: Central Water Commission (FMP Directorate).

Note: Figures from 2003-04 are tentative.

However, with growing population the British rulers began to pressure the locals to move into the fertile areas around the rivers—the floodplains—to occupy and cultivate these rich tracts. As they settled closer and closer to the rivers, the floodwaters that could earlier spread across a larger area, now exploded on human habitat, causing havoc. This changed the entire perception about floods. Floods now became a destructive force for the people living in the floodplains. Post-independence, the government decided to control this menace by constructing dams and embankments.

History: Prophet of the Future

During the British Raj, the colonial rulers did not pay enough attention to flood control measures besides constructing embankments. When they left, there were some 5,280 km of embankments along different rivers, of which 3,500 km were in the Sundarbans in West Bengal and 1,209 km along the Mahanadi in Orissa. In independent India, the First Five Year Plan, which began in 1951, decided to shift from embankments and focus on building dams. Dams were deemed as the 'most effective way of preventing flood damage' (the Plan document). Because embankments contain or restrict floodwater within rivers, but dams hold floodwater in the rivers' upstream and later release it slowly to prevent any flooding downstream.

But before the country could begin any of these plans, there came a spate of severe floods in 1954. All northern rivers flooded simultaneously and caused enormous devastation across Uttar Pradesh, Bihar, West Bengal and Assam. The floods attracted public attention to the inadequacy of control measures. To counter this, the government immediately claimed that it was possible to get rid of this menace. It started a massive upsurge to construct structural measures like embankments and dams.

Debunking Embankments

- *Build-up of floodwaters*: Embankments act like walls and limit the river's natural flood plains, the lands adjacent to rivers. So, a dangerous amount of water builds-up within embanked area.

> *"The Idea that Nature can be Conquered Carries within it the Seed of Human Destruction"*

- *Drainage Congestion*: They cut off the natural drainage of water from flood protected areas into the rivers.
- *Waterlogging*: If the bed of a river rises rapidly after embanking, say due to excessive silt deposition, the water level may reach a point where it would always remain higher than the surrounding area. Thus, the flood-protected area will get waterlogged.
- *Sediment Load*: The sediments, which would have earlier deposited over a wider area of the flood plains, get deposited within the embanked area. This raises the floodwater levels. To stop this, the height of embankments is constantly increased.
- *Reduced Natural Fertility*: Embankments also deprive the floodplains of the rich silt deposits left behind receding floodwaters.
- *River Attacks*: Shifting rivers can erode embankments.
- *Breach Blow*: The main problem is that embankments encourage human occupation of flood plains by installing a false sense of security. So when they break, the floodwaters

literally cascade upon the surrounding area. What makes these "Protectors" tumble down?

According to Professor Vishwanath of National Institute of Disaster Management, "Too much water often forces rivers to change their course, as in the case of River Kosi. The water hits the embankments directly. If there is even a slight fault in their engineering, they break. This is not all. Embankments often dig up their town graves. The silt that gets deposited in the embanked area puts pressure on the structure and erodes it.

The Rashtriya Barh Ayog. Ministry of Water of Water Resource, 1980 report states, "embankments are not a feasible measure of flood protection in cases where the country run off draining into the river is so large as to inundate appreciatively the area protected by the embankments from river spills ... The Report of the Government of India's National Commission on Floods (NCF) mentions that construction of embankments in certain areas c.an lead to increase in flood levels upstream and downstream.

Dam-ed Reality

Even dams, which are expected to provide flood moderation, are often the cause of massive floods. Before monsoon, huge quantities of water are stored in them. When they are nearly full, a lot of water is suddenly released into the river to lower the water levels. If the water is more than the river can take, it overflows in the downstream.

"Dams are actually the cause of many of the floods that visited India in July-August 2006, particularly those in Mahi, Sabarmati, Narmada, Tapi, Chambai, Krishna and Godavari basins", writes Himanshu Thakkar of South Asian Network of dams, Rivers and People (Sandrop) in A Dam-made Disaster. Many of these disasters could have been avoided if the dams had been operated properly. For instance, the 2006 Gujarat floods.

The catastrophe, which killed more than 120 people and rendered 25,000 people homeless, occurred because of the mismanagement of the Ukai Dam on Tapi River. First, the "dam managers" allowed the reservoir to fill past the allotted 'flood storage' point, and then they delayed the release of water. Not only was Tapi River's carrying capacity not taken into account, it's reduced storage capacity due to siltation was also overlooked. All these factors led to the destructive floods in the cities located downstream. This was not a unique case. Dams, including the Bhakhra, Hirakud, Tawa-Bargi, and Damodar, have led to major flood disasters.

Other Factors

Apart from embankments and dams, snowmelt, tropical storms, hurricanes, tsunamis, and undersea earthquakes are the other common causes of floods in India. But the main culprit is, of course, the monsoons. India gets an annual rainfall of 400 million hectare meters. Seventy-five per cent of this is received during the four monsoon months of June to September. As a result almost all the rivers swell up with water and flood due to,

- Inadequate capacity within the banks of the rivers to contain high flows.
- River bank erosion and silting of riverbeds.
- Landslides leading to obstruction of flow and change in the river course.
- Simultaneous flooding in all tributary rivers.
- Poor natural drainage.
- Cyclone and heavy rainfall.

Floodplains are an integral part of the flood regime and natural drainage. They give the floodwaters the room to spread across a large area, which diminishes their damage. But, people are encroaching these lands and building upon them. Constructing houses, roads, railways, canals, urban and industrial centres on these lands block the natural drainage system.

No wonder, therefore, the force of floods has not diminished. It has only increased. Add to this the dangerous mix of climate change, which experts say, is likely to cause heavier monsoons and more devastating storms. So, floodwaters rampage these lands and the lives and livelihoods of the people living on them.

By-Gone Managers

The crisis has deepened also because the natural flood managers, forests and wetlands, are being destroyed systematically. Massive deforestation has aggravated the flashiness of the rivers and their silt load due to higher erosion.

Destruction of wetlands, which act as water sponges and are natural flood moderator, is another spin-off of over exploitation of land. For example, more than a third of the area comprising of lakes, ponds, marshes and watercourses between the Ghaghra and Ganga, which form the wetlands of eastern U.P., have already been lost. The story is the same in the wetlands of Bihar, Bengal and Assam. Degradation of catchment areas and loss of flood plains to urban development and agriculture have accentuated the intensity of floods' states India's Planning Commission's Eleventh Five Year Plan document.

Cursed Kosi

Kosi River is an epitome of how mismanagement causes devastation. The river is one of the largest tributaries of Ganga. It has a steep gradient in its upper Himalayan reaches, but after Chatra, it faces abrupt flattening in the plains. This makes the river shift its course quite regularly. Another key characteristic is its high sediment load. The silt yield of the river is one of the highest in the world.

Excessive soil erosion in its upper region caused by both natural and human factors has increased the silt load of the river. It floods in the monsoon due to extreme rainfall, deposits coarse silt on agricultural lands, and shifts its course causing more damage. The high density of population on the river's floodplains makes each of its move even more deadly. Human interventions have only worsened the situation for the people of North Bihar. The measures to control the inundating river have, instead, spelt doom for them. The recent floods are a stark example of this.

Finding Answers

Kosi's curse finds expression in all the rivers of India, and North Bihar's saga is true of every flood-prone region in the country—from Assam, Bengal, Orissa to U.P. and Rajasthan.

At present, there are 4525 large and small dams in the country, and more than 15,675 km of embankments. As per the Eleventh Five Year Plan, the total area reasonably protected against flood by the end of the Tenth Plan is about 18.22 million hectares. And we are spending stupendous amounts of money on constructing these structures. During the beginning of the Tenth Plan, the Planning Commission outlaid Rs. 4619 crores for State sector to protect 1.93 MH of land against flood. The flooding rivers still claim lakhs of human lives, damage crores of property and infrastructure, destroy agricultural lands and kill livestock. This will continue. Because people will continue to live on the "disaster-prone" floodplains. Because there is no space left in the hinterlands to allow them to move back.

Forecasting Disaster

One solution is, of course, alerting and warning people before the catastrophe actually hit them. This means, developing efficient and accurate flood forecasting and early warning systems in the country. Many people consider these services as one of the most cost-effective measures available today. They help alert people in the flood risk area, give time to the disaster management agencies to move in and prepare for the impending disaster and thus, reduce the damages. In India, Central Water Commission of Government of India is responsible for issuing flood forecasts at 172 stations—145 for water level forecast and 27 for inflow forecast. These stations are located in 14 flood prone states.

The Process

1. The Indian Meteorological Department monitors the catchment area of various rivers, and under flood-like circumstances, informs the flood-forecasting centres.
2. The centres send the 'daily water level and flood forecast bulletins to concerned civil and engineering authorities'.
3. The warnings are then sent to the central control room at state and district headquarters of the "in danger" State Government.
4. From there, the warning is sent to the affected areas. Rescue and relief operations are also organized.
5. National broadcasting agencies like All India Radio, and Doordarshan are informed for wider reach.
6. Revised forecasts are sent as new information comes in.

The effectiveness of forecasts depends on two aspects-accuracy and timeliness. These make sure that there is enough time to disseminate the forecast, and warn people. But in practice, the time lag between flood warning and actual steps taken is huge. In some cases, the warnings are not even considered seriously. Take the case of the recent Bihar flood disaster.

According to Satyanarayan, the Chief state engineer stationed at Birpur near the Nepalese border where the Kosi River breached its embankments, flood management officials had sent four desperate warnings between August 9 and 16. But, the messages went unanswerable because the relevant officer was on leave and there was no one else to substitute for him. Eventually on August 16, warning telegrams were sent to 11 senior officials associated with flood water management in Patna, Bihar. But these, too, were ignored. Two days later the inevitable disaster occurred in Bihar. So, what is the future?

Flood forecasting methods are being improved in the country said, "Professor Vishwanath of NIDM. The government is working with bodies like United Nations Development Programme (UNDP) and United States Agency for International Development (USAID) to make flood forecast more reliable and timely". This includes installing automatic data collection system by sensors, transmission of data by latest techniques of communication and formulation of forecasts using computer-based comprehensive models. Flood preparedness and awareness among the people have a key role to play.

The Real Solution

All these developments will take time, and will still depend on how smoothly the system functions. As for the structural measures, the government still plans to increase the embankments and the number of dams in the country. It plans to effectively monitor the functioning of these structures, and carry out pre-and post-monsoon checks and special repairs. But, there is just one real solution to the problem of floods—people must re-learn to live with floods. We had this wisdom, and we lost it somewhere along the way. But now, we must revive out past lessons.

Real action needs to be taken in the floodplains. Managing these fragile lands adeptly and cleverly is the key to salvation. Disciplined land-use and settlement policies can emerge only if determined government can undertake land reforms and involve the poor who live in these areas in the management of their natural resources. The most crucial step would be to 'flood zone' the various states of the country. It includes demarcation of flood-prone areas and prescribing strict guidelines and land use policies.

However, this is not going to be an easy task. The plains are densely populated and there is extreme shortage of land. So each step will be fiercely opposed. Meanwhile, we can only pray for a calmer Kosi in the coming years. "The vast plains can still have the capacity to support millions of people and there are few signs of irreparable damage but only if they (people) are disciplined, only if they learn to live in harmony with the natural surroundings they have inherited and only if their management systems are equitable and sustainable. We know very little about how to do this but a beginning has to be made.*

---

* *The Times of India*, New Delhi, June 7, 2007.

5

AGRICULTURAL SCENARIO

Though agriculture is the mainstay of the people of the state, in terms of overall development, Bihar lies at the bottom of the ladder. Foodgrains—rice, wheat, jowar, bajra, maize, ragi and barley among the cereal crops and gram, arhar(tur), moong and urad among pulses—occupy nearly 84 per cent of the gross cropped area. In case of pulses, the share of the state in area and production are only 3 and 5 per cent respectively. No doubt the state has only 7 per cent of the country's total area under foodgrains contributing 11 per cent of the total production. It is, however, a matter concern that the production of foodgrain shows a declining trend over the years. (Table 5.1) (Graph 5.1)

Table 5.1: Production of Various Crops in Bihar

(Million Tonnes)

Year	*Rice*	*Wheat*	*Maize*	*Cereals*	*Pulses*	*Oilseeds*
2000-01	5.4	4.4	1.5	11.4	0.6	0.13
2001-02	5.2	4.4	1.5	11.1	0.55	0.12
2002-03	5.1	4.0	1.4	10.5	0.56	0.10
2003-04	5.4	3.7	1.5	10.7	0.56	0.12
2004-05	2.6	3.3	1.5	7.4	0.47	0.14
2005-06	3.5	2.8	1.4	7.7	0.45	0.14
2006-07	5.0	4.3	2.0	11.4	0.45	0.15
2007-08*	3.8	3.5	1.1	8.4	0.37	0.12

Source: Department of Agriculture, Government of Bihar.

* Second Advance Estimate.

Oilseeds cover less than 2 per cent of the gross cropped area of the state and position in the country in area and production is insignificant—less than 1 per cent each. Rapeseed and mustard is the major oilseeds crop here followed by linseed. Although area under oilseeds, particularly rapeseed and mustard and sunflower are small, their productivity levels are the highest in the country. Despite this, area under oilseeds has steadily come down from 153.7 thousand hect. In 2000-01 to 129.5 in 2007-08 (Second Advance Estimate). Changes in Area and production of different crops is as in Table 5.1.

Graph 5.1: Production Trends of Foodgrains and Oilseeds in Bihar

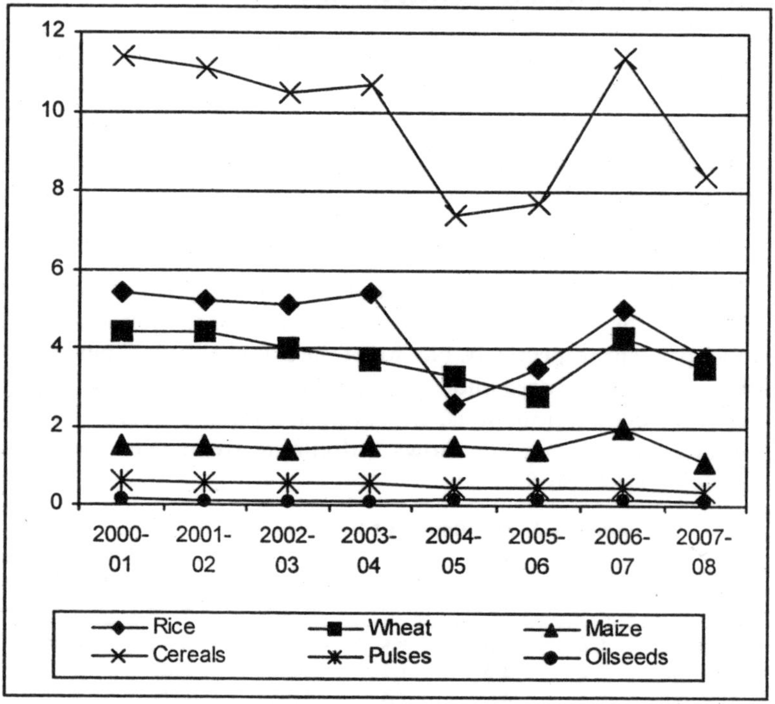

Jute and mesta are the only fibre crop grown in the state and the state's position in area and production in the country is a distant second only after West Bengal. Among other commercial crops, sugarcane is important in the state. It shares slightly less than 2 per cent of the gross cropped area occupying only about 0.1 million hectares constituting less than 2.5 per cent of the country's total. The state's share in the country's production is 4 to 4.5 per cent and ranks 10th among the sugarcane producing states. The place of the state in the agriculture economy of India is shown in Table 5.2

Table 5.2: Production and Productivity of Various Crops in India and Share of Bihar State—2005-06

Crop	All-India Production (In mt.)	Share of Contribution of Bihar (%)	All India Productivity (kg/hae)	State Productivity (kg/hae)
Rice	91.79	3.18	2102	1075
Wheat	69.35	4.67	2619	1617
Maize	14.71	9.25 3rd Position	1938	2098
Pulses	13.38	3.36	598	749
Foodgrains	208.60	4.12	1715	1311
Oilseeds	27.98	0.50	1004	982
Sugarcane	281.17	1.54	66928	42822
Potato	23.9	5.16 3rd Position	17058	8663
Jute	10.84*	12.82 2nd Position	2173	1692

Source: Agriculture Statistics at a Glance 2006.

* Production in Million Bales of 180 kg each.

Poor Yield Rate

One of the principal reasons for dismal performance in agriculture in the state is the poor yield rates for most of the major crops. Average rice yield in Bihar during 2005-06 was only 1.07 tonnes per hectare as against the national average of 2.10 tonnes per hectare. This is the lowest among all the states and UTs in the country, though slightly more than 50 per cent of the area under rice is irrigated. Rice yield rate in Bihar is pitiably low when compared with Punjab and Haryana where it is as high as 3.86 and 3.05 tonnes per hectare respectively. It appears that there has not been much technological upgradation in rice cultivation in Bihar. It may be stated that rice is the staple food of the state and to meet the growing deficit in foodgrains, improvement in the yield rate of the crop is imperative.

Wheat is the second most important cereal crop in the state, which is grown mostly under irrigated condition. According to the latest information, nearly 90 per cent of the area under the crop is irrigated. Yet, the yield rate is very low being only 1.6 tonnes per hectare which is not only below the national average yield of 2.6 tonnes per hectare, but less than half that of Punjab (4.2 tonnes per hectare) and Haryana (3.84 tonnes per hectare). Here also, there is an urgent need for adoption of location specific technology in order to attain a breakthrough in productivity. However, it is interesting to note that performance in respect of maize in the state where it is grown both in kharif and rabi is much better. The state shares 8 to 9 per cent of the country's area under the crop and 10 to 12 per cent of the production. Average yield of this crop is higher at 2.1 tonnes per hectare compared with the national average of 1.9 tonnes per hectare. In case of pulses (total) also, productivity in the state is significantly higher (7.5 quintals per hectare) as compared with the national average of 6 quintals per hectare.

As for oilseeds, the average yield rate in 2005-06 was only marginally lower at 982 kgs. per hectare than that of the national average of 1000 kgs. per hectare. Rapeseed and mustard is the main edible oilseed grown in the state. Groundnut crop is cultivated in a very limited area and its yield rate is almost half that of the country's average. A comparison of yield rates of crops for Bihar compared with India are as in Graph 5.2.

Graph 5.2: Yield of Major Crops in Bihar and India during 2006-07

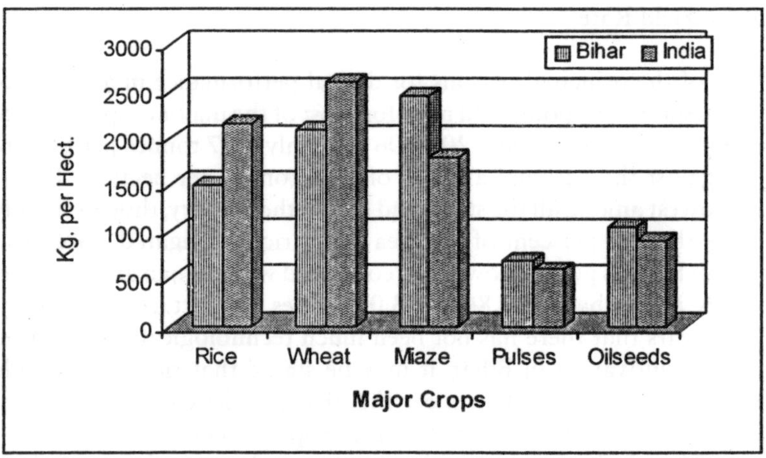

Rice and wheat occupying nearly 70 per cent of the gross cropped area in the state contribute in the range of 5.0 million tonnes (mt) of rice and 4.0 mt of wheat annually. As discussed above, the productivity level remains low despite well-endowed soil, water, climate and human resources. The reasons identified for low productivity of crops in the State are low Seed Replacement Ratio (SRR), low and regionally unbalanced fertilizer use, lack of access to credit (only 20 lakh Kisan Credit Cards have been issued covering only 1.49 per cent of farmers), inadequate mechanization (17 tractors per thousand hectare), small and fragmented holdings, high cost of energisation mainly through diesel for irrigation, weak agricultural extension system, chronic floods (north Bihar) and drought risk (South Bihar), inadequate storage and processing capacity (against estimated 25 lakh MT storage capacity, only 12 lakh tonnes). The agro-climatic condition is congenial for sugarcane. However, any expansion in sugarcane area should be taken up without compromising food security i.e. without diversifying potential wheat growing areas. The issue of promoting maize processing plants needs urgent attention, as the productivity is considerably high even when compared to national average.

Growth Trend

Growth trend in area, production and yield of major crops in the state during the last five years (2000-01 to 2005-06) (Table 5.3). shows that both production and yield of most of the crops in the state has recorded a negative trend which is a matter of great concern. Only in the case of maize growth rate in production has been positive at 0.99 per cent per annum, despite negative area growth rate of (-) 0.59 per cent per annum. This has been possible due to higher annual growth rate of yield by 1.60 per cent.

Table 5.3: Area Production and Yield in Bihar—Major Crops

Area: Thousand Hect.
Production: Thousand Tonnes
Yield: kg./hect/

Year	*Rice*			*Wheat*			*Maize*		
	Area	*Produ.*	*Yield*	*Area*	*Produ.*	*Yield*	*Area*	*Produ.*	*Yield*
2000-01	3656.3	5442.6	1489	2067.8	4438.0	2146	620.5	1497.2	2413
2001-02	3552.3	5202.9	1465	2126.3	4391.1	2065	594.3	1488.3	2504
2002-03	3584.7	5085.5	1419	2130.9	4040.6	1896	603.6	1349.8	2236
2003-04	3578.0	5447.8	1523	2076.8	3688.9	1776	616.4	1473.5	2390
2004-05	3122.6	2472.2	792	2027.6	3263.4	1609	614.4	1465.7	2386
2005-06	3252.4	3495.5	1075	2003.7	3239.0	1617	648.8	1361.1	2098
	Sugarcane			*Potato*					
2000-01	93.5	3987.6	42648	144.6	1400.4	9685			
2001-02	113.4	5211.1	45953	140.2	1412.2	10073			
2002-03	107.3	4520.5	42130	148.2	1476.6	9964			
2003-04	103.6	4285.9	41370	158.7	1538.9	9697			
2004-05	104.2	4111.7	39460	138.7	1062.8	7663			
2005-06	101.3	4337.9	42822	142.3	1232.7	8663			

Source: Directorate of Economics and Statistics, Ministry of Agriculture.

In case of oilseeds, despite a positive growth of 2.21 per cent per annum in the yield rate as against only 0.96 per cent of the national average, production has exhibited a negative growth by 0.59 per cent due to sharp decline in area growth by 2.86 per cent per annum. The decline in area for most of the major crops has caused the decline in both gross and net cropped area. From the above trend analysis it is clear that agriculture in the state is suffering from certain serious disabilities requiring concerted efforts to achieve a turn around. Bihar has a very high concentration of poor population with a relatively higher population growth rate; hence faster growth in agriculture is a prerequisite for overall economic prosperity and reduction of poverty in the state.

Some Issues

State has 38 soil testing labs. However, their current capacity utilization and delivery was very low (about 10%) and therefore urgent steps are required to upgrade it. In order to promote balanced and integrated fertilizer use, soil testing service needs to be strengthened by establishing new labs under KVK/Agri-Clinics, revamping the existing labs and exploring the possibilities of using soil test kits.

Lack of efficient on—farm water management is considered as one of the major constraints for low productivity and poor economic status of the farmers of Bihar. Rainwater harvesting should be a major strategy to meet irrigation needs of the crops. Since, the State has abundant groundwater, Planning Commission, under Rashtriya Sam Vikas Yojana (RSVY) has sanctioned the installation of one million shallow tube wells. Unfortunately, the programme has remained deadlocked with little progress made so far. A call for a revisit of the procedure is required immediately, if the potential is to be quickly channelised to increase production in the State. Bulk finance by NABARD to the State Government under RIDF could be considered. Watershed activities taken by NABARD in 8 districts, though planned systematically are, however, moving very slow, as only Rs. 1.88 crore out of Rs. 10 crore released by the Government of India had been sanctioned so far, against the total project cost of Rs. 60 crore.

The public extension delivery is in real bad shape, against 3000 sanctioned position of the field staff, only about 1500 are in position.

The Department has been promoting para-tech employees as Krishak Salahakar on contractual basis. As the KVKs are established with a major mandate of technology transfer and dissemination, these need to be strengthened with adequate and trained manpower. The State has also taken steps to establish Agriculture Technology Management Agencies (ATMAs) in all the 38 districts.

With the increase in road connectivity, contributed by the schemes taken up under Bharat Nirman and other programmes, the problem of agricultural marketing is expected to ease up in the coming years. However, the State has to improve the infrastructure in the form of godowns and cold storages, etc., in order to provide processing facilities and ensure profitable marketing of their produce by the farmers. The machinery for Minimum Support Price (MSP) has also to be strengthened, especially in respect of those crops which have high potentiality of growth. In order to provide a level playing field for private participation in marketing, the APMC Act which has since been repealed should be substituted by an appropriate one in order to meet the aspirations of the farmers.

The availability of institutional credit is also a problem, as this accounts for only 29 per cent of the total agricultural credit. This calls for thorough restructuring of co-operative structure as recommended by Vaidyanathan Committee Report.

A large tract of the State is waterlogged. There is ample scope to tap the potential of promoting fisheries. A detailed study shall have to be undertaken to examine the technical feasibility and socio-economic viability of both large and small fisheries. The State could also encourage private entrepreneurs to produce fish seeds.

As per NDC Resolution, a National Food Security Mission (NFSM) covering wheat, rice and pulses has been launched in India including Bihar as a centrally sponsored scheme. The scheme will be totally funded by the centre. The machinery for implementing the scheme has been undertaken by a project management team which will supplement the efforts of ATMAs. The scheme was launched w.e.f. 2007-08 with the objective of (i) increasing production of rice (in 22 selected districts), wheat (in 25 selected districts) and pulses (in 11 selected districts) of Bihar through area expansion and productivity enhancement in a sustainable manner; (ii) restoring soil fertility, (iii) provide good returns by improving farm level economy (i.e. farm profits) to restore confidence of farmers of targeted districts. The ultimate objective of the NFSM is to achieve self-sufficiency in foodgrains. The scheme will provide 40.2 per cent of allocation for rice, 34.4 per cent for wheat and 25.4 per cent for pulses for implementation during the Eleventh Plan. It is expected that being one of the major rice and wheat cultivating state, Bihar should get more benefits under this scheme.

Over the Eleventh Plan period, all the SAUs that have yet not received the special grant, would be covered for revamping. The special funds to the selected university will be released in installments and the Oversight Committee will monitor the progress. The High Power Committee at the end of the Plan will study the impact of the special grant.

Agriculture is the core competence of the State. However, spread of the agricultural education has been limited. This requires strengthening of agricultural research, education and extension. In Eleventh Plan, an outlay of Rs. 179 crore has been proposed for Agriculture Research as against the Tenth Plan anticipated expenditure of Rs. 65.99 crore. For the year 2007-08, an outlay of Rs. 31.5 crore is provided. For Eleventh Plan, 2.8 per cent outlay has been allocated for Agriculture and Allied Sectors out of which 10.9 per cent is proposed for Agricultural Research and Education in Bihar.

Key areas requiring focused attention.

(1) There is urgent need to revamp seed production system for appropriate varieties including strengthening of seed certification.
(2) Proper coordination at the district level and below, for all the agriculture-related sectors including research and extension are urgently needed.
(3) With 50 per cent of the cultivable land as flood prone and 50 per cent as drought prone of which only 25 per cent is under irrigation, there is a need to adopt alternative and innovative strategies for agriculture development—with a clear cut distinction between Flood Prone Area (FPA) and Drought Prone Area (DPA).
(4) Bihar also faces shortage of storage facilities, both for fertilisers and foodgrains. There is, therefore, an urgent need for consideration by the Central Warehousing Corporation to provide additional storage facilities. When Bihar's level of marketable surplus increases in the coming years, storage would be a key constraint for enhanced production.
(5) Tremendous opportunities for crop diversification, and

augmenting farm productivity and profits prevail in Bihar due to excellent quality of soil and water. What is important is to infuse efficiency and accountability at all the levels.

(6) The State could avail the programme on rural electrification under Bharat Nirman to make power available to the rural areas.

We have tried in the subsequent pages to analyse crop-wise data at the District level—top and bottom Districts—according to the area under the specific crop. The primary object of this District level study is to identify potential areas for future action. However, we found some instances where inconstancy in the data at the District level could not be reconciled. The readers may take this fact into account while interpreting the data. For example, there appears to be a sudden jump of more than 88 per cent in Rabi maize yield in Patna during 2004-05 and a sudden drop of 50 per cent in Madhepura during 2005-06.

RICE

Bihar with around 3.2 million hect. under rice is grown in all the three seasons viz., autumn, winter and summer and accounts for over 8 per cent of the area in India. While winter rice is grown in all the 38 districts of the state, autumn and summer rice are cultivated mainly in those districts where irrigation facilities are available or where it does not complete with other crops like Jute rapeseed and mustard and other rabi crops. Although it is a major crop in Bihar, where it occupies around 48 per cent of the gross cropped area, it is a sad story that with over 55 per cent of the area having irrigation facilities, productivity levels at around 1.5 tonnes/hec. are much lower than the all India or even West Bengal and Assam levels. This is not all, productivity levels vary widely from year to year. During 2004-05, for example, lowest ever rate at 792 kg./hect. was witnessed. Such variations are much more sharper at the district level as between seasons and time. This is quite evident from the District level study.

Among the cereal crops, rice is the most important crop in a majority of the districts having more than 50 per cent of the cropped area. In Aurangabad, Gaya, Madhubani, Nawada and Sitamarhi, districts, average share of rice in the total cropped area is above 60 per cent, with highest concentration of nearly 80 per cent in Madhubani district. In districts like Katihar, Munger, Saharsa and Vaishali, the average share of area under the crop ranges between 40.3 to 47.4 per cent. In Gopalganj, Samastipur,

Saran-Chhapra and Siwan, the share is between 33.2 and 38.3 per cent. Begusarai among all the district has the least share (9.3 per cent.) Table 5.4. Bihar has all the potentialities to increase productivity levels, if the State takes advantage of modern research as discussed else where in this study. Tamil Nadu, for example, has adopted the System of Rice Intensification (SRI) and has succeeded in achieving higher yields. In addition to achieving higher yields, there is the need for special attention to encourage the production of special rice varieties like 'Organic Basmati' scented rice and varieties such as Katarni of Bhagalpur region and Sugandha in Patna region which have high export demand.

Table 5.4: Average Percentage Share of Area Under Rice in the Total Cropped Area during the Last 4 Decades Ending on 2000-01 in the Districts of Bihar State

Percentage Share	*Name of the Districts with Percentage Share*
Above 60	Aurangabad (62.6), Gaya (64.2), Madhubani (73.9), Nawada (60.7) and Sitamarhi (66.3)
50 to 60	Bhagalpur (52.6), Bhojpur (52.6), E. Champaran (54.5), W. Champaran (53.5), Darbhanga (56.4), Nalanda (57.8), Patna (52.4), Purnea (58.2), Rohtas (53.1) and Muzaffarpur (50.4)
40 to 50	Katihar (47.4), Munger (42.8), Saharsa (47.0) and Vaishali (40.3)
30 to 40	Gopalganj (35.6), Samastipur (35.1), Saran (33.2) and Siwan (38.3)
>30	Begusarai (9.3)

Note: Figure in brackets is the percentage share in the cropped area of the district.

Winter Rice

Winter rice is cultivated in all the districts-important being Rohtas, Madhubani, E. Champaran, Aurangabad, Bhabhua and Bhojpur. These six districts occupied 25 to 30 per cent of the total area of the state. The under winter rice and contributed 33 to 44 per cent in production during the years 2003-04 through 2005-06 (Table 5.5). Further, nearly 70 to 85 per cent of the area was under high yielding varieties (HYVs) in these districts as against 56 to 58 per cent in the state as a whole. These districts have better irrigation facilities 90-95 per cent under HYV and 40-50 per cent under traditional varieties. As against this, only 70 and 16 per cent areas under HYV and traditional varieties respectively, were irrigated in the state as a whole. Naturally, these districts produced more than one-third of the total production of winter rice during these three years.

Table 5.5: Bihar Winter Rice

Sl. No.	Districts	Area (hectares)			Production (tonnes)			Yield/hectares,/tonnes		
		2003-04	2004-05	2005-06	2003-04	2004-05	2005-06	2003-04	2004-05	2005-06
					Top Six Districts					
1.	Rohtas	195387	178135	195296	516983	471832	413327	2.65	2.65	2.02
2.	Madhubani	135156	131724	133256	136088	36193	58711	1.01	0.27	0.44
3.	E. Champaran	107849	124724	132496	150395	38595	115644	1.39	0.31	0.87
4.	Aurangabad	173050	105259	118145	312702	136815	115894	1.81	1.30	0.98
5.	Bhabhau	109797	95185	108086	341697	139200	220023	3.11	1.46	2.04
6.	Bhojpur	105425	95267	102788	223506	38700	182467	2.12	0.41	1.77
					Bottom Six Districts					
1.	Khagaria	15032	11059	11491	14635	545	10814	0.97	0.05	0.94
2.	Sheohar	19587	13729	17950	7795	3968	11396	0.40	0.29	0.63
3.	Begusarai	21832	19600	18763	29364	8553	4535	1.34	0.44	0.24
4.	Sheikhpura	29386	8511	22472	38431	2974	19733	1.31	0.35	0.88
5.	Munger	25974	24617	24716	38980	38810	39679	1.50	1.58	1.60
6.	Arwal	31836	24626	26814	54058	24322	38648	1.70	0.99	1.44
	State Total	2907199	2519734	2607565	4589705	1957683	2841519	1.58	0.78	1.09
	Share of the top six districts (%)	25.1%	29.0%	30.3%	36.6%	44%	32.5%	—		
	Share of the bottom six districts (%)	3.5%	4.0%	4.7%	4.0%	4.0%	4.4%	—	—	1.2%

On the other hand, districts like Khagaria, Sheohar, Begusarai, Sheikhpura, Munger and Arwal had only 3.5, 4.0 and 4.7 per cent of the state's area under the crop in 2003-04, 2004-05 and 2005-06 respectively. These six bottom level districts produced only about 4 per cent of the crop in these three years. Further, unlike the top six districts, total HYV area in these districts was only 45 and 52 per cent, of which only 45 and 52 per cent was irrigated during 2004-05 and 2005-06 respectively. However, quite a significant area under traditional varieties was also irrigated 49 per cent in 2004-05 and 70 per cent in 2005-06. No wonder there the contribution of winter rice production in these six districts was not significant.

Looking at the yield rates in the three years under reference, it is observed that, in 2004-05, there was a significant decline both in the top (except Rohtas) and bottom (except Munger) districts as compared to those of the previous year, presumably due to unfavourable weather conditions. The yield rates were very high at 3.11 tonnes per hectares in 2003-04 in Bhabhua district followed by Rohtas (2.65 tonnes) and Bhojpur (2.12 tonnes). However, in these districts too (except in Rohtas), yield rates sharply declined in 2004-05. The situation was also the same in all the bottom districts in 2004-05, except in Bhabhua where the yield rate was marginally higher than that of the previous year and this higher trend was maintained in the next year too. In 2005-06, both in the top and bottom districts, yield rates improved though these did not reach the 2003-04 level. In fact there was a marginal decline in Rohtas district and further sharp decline in Bagusarai district. The yield rates in the bottom two districts, namely Munger and Arwal, were consistently higher than that of the state's average, hence all efforts should be made to increase the area under this crop in these districts.

Autumn Rice

Autumn rice occupies only about 16-20 per cent of the total area in the state. But the top six districts, sharing 51-54 per cent of area contributed 52-59 per cent production in autumn rice. There are also districts like Buxar, Jahanabad, Nawada, Aurangabad, Sheikhpura, Lakhisarai, Jamui and Banka where autumn rice is not cultivated at all.

On the other hand, the six bottom districts, viz. Rohtas, Bhojpur, Gaya, Patna, Sheohar and Kishanganj, had a nominal share (2 to 2.4) per cent of the state's total area under autumn rice during the years referred to above. The share of irrigated area which was only 5 per cent in 2004-05, increased to 44 per cent in the following year due to sizable extension of irrigated area in Sheohar and Patna districts. In fact, in 2003-04 and 2004-05 autumn rice was not grown in Patna district. But in 2005-06, the crop was cultivated in an area of 1682 hectares, of which 1566 hectares were under irrigation. Even in Gaya, where area under the crop was only 13 hectares in 2004-05, jumped to 1253 hectares in 2005-06. Of this, nearly 57 per cent was irrigated. Contribution of autumn rice of these bottom districts varied between 1.8 to 2.5 per cent during the 3 years. (Table 5.6)

It may be observed, the yield rates of the crop in 2003-04, were above the state average of 1.24 tonnes per hectare in East Champaran, West Champaran and Gopalganj districts and almost at par in Siwan, Rohtas, Bhojpur and Gaya districts. But in 2004-05, due to reported unfavourable weather and rainfall conditions, the yield rates declined in all the districts, except Kishanganj where a marginal increase was recorded.

In 2005-06 also there was no significant improvement in the yield rate in the districts under study. In fact, in West Champaran, Gopalganj, Siwan and Sheohar districts, there was further decline in yield rate, despite increase in irrigated area in some of these districts. In Vaishali, the productivity of autumn rice was highest (1.6 tonne per hectare) though the coverage of area was only 2 per cent and that also unirrigated. The productivity was also higher than that of the state average in Madhepura (1.4 tonnes/hect), Supaul and West Champaran (1.31 tonnes each), East Champaran, Sitamarhi and Saharsa (1.1 tonnes each), Patna, Bhabhua and Purnea (1 tonnes each) and in the rest of the districts, it was at par with the that of the state average or lower than that. The lowest productivity was reported from Samastipur, where it was only 0.3 tonne per hectare. The productivity was also unsatisfactory in Madhubani, Muzaffarpur, Saran and Katihar where it ranged between 0.6 and 0.7 tonne per hectare. It appears that the cultivation of this crop is quite uneconomic in such Districts.

Irrigated area under the crop in the state as a whole was only about 27 per cent. No information was available in respect of coverage of area under high yielding varieties or application of other inputs which help in raising the productivity. Further, in most of the districts, yield rate in irrigated areas was not significantly higher than that of the unirrigated areas. In fact, in a few districts like Saran,

Table 5.6: Bihar Autumn Rice

Sl. No.	Districts	Area (hectares)			Production (tonnes)			Yield/hectares,/tonnes		
		2003-04	2004-05	2005-06	2003-04	2004-05	2005-06	2003-04	2004-05	2005-06
	Top Six Districts									
1.	W. Champaran	60482	60838			85278	78674	1.67	1.40	1.28
2.	E. Champaran	61815	61131	61213	101979	55096	69256	1.46	0.90	1.14
3.	Gopalganj	61274	61612	60672	91437	81605	46121	1.50	1.32	0.79
4.	Siwan	42996	46737	58317	91873	46137	34487	1.22	0.99	0.79
5.	Madhubanii	37220	36084	43765	52557	31336	20423	0.92	0.59	0.61
6.	Samastipur	31888	31545	33617	34390	6660	11012	1.08	0.21	0.35
	Bottom Six Districts									
1.	Rohtas	323	145	224	404	133	213	1.25	0.91	0.96
2.	Bhojpur	1001	1453	453	1255	1326	441	1.25	0.91	0.97
3.	Gaya	1720	13	1253	2124	12	1225	1.23	0.91	0.98
4.	Patna	—	—	1682	—	—	1721	—	—	1.0
5.	Sheohar	5670	6288	3880	4218	6150	3407	0.74	0.98	0.88
6.	Kishanganj	5554	5045	3915	4586	5053	4192	0.83	1.00	1.07
	State Total	583904	554214	533776	688489	505708	502203	1.24	0.91	0.94
	Share of the top six districts (%)	50.8%	53.8%	54.1%	59.0%	58.5%	51.8%			
	Share of the bottom six districts (%)	2.4%	2.3	2.1	1.8	2.5%	2.2			

Siwan, Gopalganj, Muzaffarpur and Purnea, it was lower than that of the unirrigated areas. Needless to say, reasons for such low productivity are required to be investigated.

It may be mentioned that the crop is to face competition from jute in the allocation of area particularly in the jute growing districts of Kishanganj, Saharsa, Purnea, Katihar and to some extent in Supaul, Madhepura and Saran. Area under jute fluctuates from year to year, depending upon price expectations. Variations under autumn rice are thus directly related to these changes in area under jute. The policy makers have to keep in mind this particular phenomenon while planning improvement in the production of autumn rice.

Summer Rice

Summer rice, with nearly 70 per cent of the area is under high yielding varieties, is mostly irrigated. The crop is cultivated mainly in Katihar, Araria, Purnea, Kishanganj and West Champaran districts. Nearly 80 per cent of the crop area is in these 5 districts. Sitamarhi, Supaul, Madhepura, Darbhanga, Madhubani, Samastipur and Saharsa Districts, account for another 18 per cent of the area. The rest of 2 to 3 per cent area is in Khagaria, Bhagalpur, Begusarai, Sitamarhi, East Champaran and Saran. In point of fact, in 2003-04 the crop was not cultivated at all in 10 districts viz. Nalanda, Buxar, Aurangabad, Bhabhua, Jahanabad, Siwan, Sheikhpura, Lakhisarai, Jamui and Banka. In 2004-05, besides these districts, there was no cultivation of summer rice crop in Arwal, Nawada, Muzaffarpur, Sheohar and Munger districts as well.

Six major districts where the crop is cultivated, are Purnea, Katihar, Araria, Kishanganj, West Champaran and Saharsa. The coverage of area in these districts was to the extent of 81 per cent in 2003-04 to 84 per cent in 2005-06 (Table 5.7). Commensurating with the size of area, nearly 82 to 83 per cent state's production of summer rice was also from these six districts.

On the other hand, the six bottom districts viz., Munger, Siwan, East Champaran, Khagaria, Rohtas and Saran, together had a coverage of only 3.1 per cent of the state's total area in 2003-04, which declined to 2.7 per cent in 2004-05 and further to only 1 per cent in 2005-06. In fact, in the latest year area under the crop sharply declined from 1925 hectares in 2004-05 to 294 hectares in Khagaria district presumably due to non-availability of irrigation water. According to the coverage of area, production of

Table 5.7: Bihar Summer Rice

Sl. No.	Districts	*Area (hectares)*			*Production (tonnes)*			*Yield/hectares,/tonnes*		
		2003-04	*2004-05*	*2005-06*	*2003-04*	*2004-05*	*2005-06*	*2003-04*	*2004-05*	*2005-06*
				Top Six Districts						
1.	Purnea	20481	19635	18303	28481	28561	26960	1.37	1.45	1.47
2.	Katihar	39255	36277	35041	65936	53477	46691	1.68	1.47	1.33
3.	Araria	13862	12132	14021	21226	17745	12941	1.53	1.46	0.92
4.	Kishanganj	11110	10919	11507	13473	19290	18578	1.21	1.77	1.61
5.	West Champaran	7660	10496	9461	7992	8631	13027	1.04	0.82	1.38
6.	Saharsa	2615	4298	4457	3809	6537	6506	1.46	1.52	1.46
				Bottom Six Districts						
1.	Munger	328	322	103	436	399	141	1.33	1.24	1.37
2.	Siwan	—	133	133	—	185	181	—	1.39	1.36
3.	East Champaran	903	194	115	914	165	159	1.01	0.89	1.38
4.	Khagaria	1987	1925	294	2664	2346	401	1.34	1.22	1.36
5.	Rohtas	16	174	172	24	222	234	1.50	1.28	1.36
6.	Saran	360	306	273	539	466	377	1.50	1.52	1.38
	State Total	116885	114298	11016	169902	161738	152410	1.45	1.41	1.37
	Share of the top six districts (%)	81.3	82.0	83.6	82.9	83.0	81.8			
	Share of the bottom six districts (%)	3.1	2.7	2.7	2.7	2.3	1.0			

summer rice also declined and was only 1.0 per cent of the state's total in 2005-06 as against 2.3 per cent in 2004-05 and 2.7 per cent in 2003-04, in this District.

Average yield in the top six districts was more or less at par with the state's average at 1.41 tonne and 1.37 tonne per hectare in 2004-05 and 2005-06 rabi seasons. But in 2003-04, it was only 1.38 tonne per hectare as against the state average of 1.45 tonne hect. On the other hand, average yield in the bottom six districts, was significantly higher than that of the state average, –1.49 tonne per hect in 2004-05 and at par in 2005-06. But in 2003-04, it was significantly lower, being 1.36 tonne per hect. only.

Conclusion

It may be stated that among the three seasons, summer rice has the highest yield as the crop is cultivated mainly under irrigated conditions, when the chance of extensive damage of the crop by any natural calamity is much less. If irrigation facility in the state could be increased, the crop can be cultivated in many other areas and thereby there is a possibility of increasing the total production of rice in the state. Presently this crop hardly contributes 4 per cent of total rice production in the state.

The farmers lose their rice crop when fields are submerged by annual floods triggered by rivers, fed by heavy monsoon rains and melting Himalayan glaciers burst their banks. The rice type called Swarna Submergence I, developed by the Manila-based International Rice Research Institute (IRRI), proved to be flood resistant in trials in northern Bangladesh. Normal rice varieties cannot survive being submerged by flood water for more than three days, resulting in huge losses to the farmers. During (2007), when the flood water receded from two farms in which Swarna Sub-I was planted, paddy stood up again, 10 days after it was completely submerged by water. In short, additional irrigation for summer rice and introduction of flood resistant varieties, will go a long way to increase the production of rice in the state.

WHEAT

Wheat which is cultivated in all the districts, is the 2nd most important cereal crop in the state. The state ranks sixth among all the wheat producing states both—in respect of area (8%) and production (5%). The share of the total cropped area in the districts varies from 72 per cent in the erstwhile Madhubani district to 42 per cent in Saran. Of the total area of nearly 2 million hect. under wheat, nearly 75 per cent is under high yielding varieties. But unfortunately seed replacement ratio is only 10 per cent against the required 35 per cent or so. Of this, nearly 93 per cent is irrigated. In the case of area under traditional varieties too, which shares about 25 per cent of the total area, nearly 84 per cent is irrigated.

The six major wheat producing districts in the State are Rohtas, Saran, Siwan, Muzaffarpur, Champaran East and Madhubani. These six districts covered 27.5 per cent of the total area under the crop in 2003-04 and 29 per cent in 2005-06. Average percentage share in area of each district in the state under the crop was 4.23, 4.42, 4.04, 5.61 and 3.39 respectively. Among the districts, Rohtas has the largest area under the crop, being about 11.6 lakh hectares constituting about 6 per cent of the state's total, followed by East Champaran (5%), Siwan (4.8%), Saran-Chhapra (4.7%), Madhubani (4.6%), Muzaffarpur (4.3%), Gapalganj (4.2%) and Nalanda (4.1%). Jamui has the least area of 0.5 per cent of the state's total.

The share of these six top districts in total production was 28.5 per cent in 2003-04 and 30.2 per cent in 2005-06. An analysis of district wise production, indicates that Rohtas district constituting 8.8 per cent of the state's total of 2.8 million tonnes in 2005-06, with a yield rate of 2.1 tonnes per hectare as against the state average of 1.4 tonne per hectare, was the top producer. (Table 5.8) Siwan was the second highest producer with 5.6 per cent of state's total, though in terms of coverage of area its position was third. The yield rate of the crop in this district was 1.7 tonne per hectare—higher than that of the state average. The other large producing districts were Saran-Chhapra (5.6%), Bhabhua Kaisur—(4.8%), Gopalganj and Nalanda (4.7%) each, Buxar (4.6%), Patna and Bhojpur—Arrah (4.5% each) and Muzaffarpur and East-Champaran (4% each). Thus, the share of these 11 districts was 55.7 per cent of the state's wheat crop in 2005-06 and the rest 27 districts produced only 44.3 per cent. Average yield per hectare was also higher than that of state average in Vaishali, Samastipur, Lakhisarai, Sheikhpura, Begusarai and Khagaria districts producing together 13 per cent of the crop from 11 per cent of the total area.

One of the major reason for low wheat yields is its late sowing. In South Bihar, for example, the entire area of late paddy cultivation and transplanting is mostly done after 15th July and carried upto August and harvesting of these crops is in the last week of November or early December. The Wheat/Rabi crops sowing starts after 15th December and completes by middle of January. These practices are being adopted in Rohtas, Gaya, Aurangabad, Nawada, districts. Due to late sowing and harvesting of paddy, the farmers get very short period to grow wheat and other crops. Therefore, the productivity of rabi crops is low. It is, therefore, suggested that concentrated effort needs to be undertaken among farmers for early sowing/transplanting of paddy, at least 20-25 days so that sufficient time would be available for Rabi crops like wheat, mustard, lintel, linseed, pulses to harvest maximum yield.

On the other hand, among the bottom six districts viz., Munger, Lakhisarai, Sheikhpura, Jamui, Banka and Kishanganj, the coverage of area was 6.8 per cent in 2003-04 and 5.6 per cent in 2005-06. and the share in production only 5.4 per cent in 2003-04 and 5.2 per cent in 2005-06.

Table 5.8: Top and Bottom Wheat Producing Districts in Bihar

Sl. No.	*Districts*	*Area (hectares)*		*Production (tonnes)*		*Yield/hectares,/ Kg.*	
		2003-04	*2005-06*	*2003-04*	*2005-06*	*2003-04*	*2005-06*
		Top Six Districts					
1.	Rohtas	128102	115857	322248	243741	2516	2104
2.	Saran	94593	93422	208739	156035	2207	1670
3.	Siwan	84459	95397	181132	160203	2145	1679
4.	Muzaffarpur	90699	85854	129362	111618	1426	1300
5.	East Champaran	91390	98368	143031	105735	1565	1075
6.	Madhubani	82822	92058	65802	62979	794	684
		Bottom Six Districts					
1.	Munger	22780	19181	41423	26752	1818	1395
2.	Lakhisarai	23938	21364	51182	34960	2138	1636
3.	Sheikhpura	22287	15814	39392	28124	1767	1778
4.	Jamui	16972	9604	22120	9437	1303	983
5.	Banka	27157	23311	42013	27807	1547	1193
6.	Kishanganj	27750	22607	22669	17554	817	776
	State Total	2076839	2002221	3688941	2777978	1776	1387
	Share of the top six districts (%)	27.5	29.0	28.5	30.5		
	Share of the bottom six districts (%)	6.8	5.6	5.4	5.2		

Looking at the district-wise yield rate of the crop during these two years, it is observed that both in top and bottom six districts, the yield rates were significantly higher in all the districts in 2003-04 when compared with those of 2005-06 except in Sheikhpura (one of the bottom wheat producing districts) where the yield was marginally lower during 2003-04. Even in the state as a whole in 2003-04, the yield rate was markedly higher (1776 kgs. per hectare) as against only 1387 kgs. in 2005-06. Taking into consideration the agro-climatic conditions of the state, yield rate in Rohtas, Saran, Siwan and Lakhisarai was above 2 tonnes per hectare in 2003-04 which was considered to be quite satisfactory. Even in 2005-06, when there was general decline in the yield rate of the crop in all the districts, Rohtas produced above 2 tonnes per hectare.

In this context, it may be stated that in 2005-06 among all the districts in the state, the highest yield rate of 2.6 tonnes per hectare was achieved in Bhojpur which contributed 4.5 per cent of production with only 2.4 per cent area. The yield rates were also high, being slightly more than 2 tonnes per hectare in Patna, and Bhabhua districts. As the productivity of the crop is high in these districts, an all out effort should be made to bring maximum areas during the rabi season under this crop. On the other hand, the yield rates were dismally poor in Madhubani and Kishanganj districts—below 1 tonne per hectare. In 2005-06, average yield per hectare was also higher than that of state average in Vaishali, Samastipur, Lakhisarai, Sheikhpura, Begusarai and Khagaria districts producing together 13 per cent of the crop from 11 per cent of the total area.

In 2005-06 the yield was lowest in Sheohar, being only 384 kgs. In other districts where the yield rate was very low, are Katihar (458 kgs.), Araria (602 kgs.), Saharsa (689 kgs.), Supaul (803 kgs.), Purnea (847 kgs.), and Madhepura (995 kgs.). These 9 districts together had 15 per cent of the state's total area under the crop producing only 7.5 per cent of the crop. Possibility of improving the productivity of the crop in these districts need to be explored urgently. Or else, efforts should be made to replace this crop with some other more profitable rabi crop like oilseeds and pulses. The yield rate of the crop need to be improved in all the districts in order to achieve a quantum jump in the state's production of wheat. Incidentally, it may be stated that even in the neighbouring West Bengal, a traditionally non-wheat producing state, present yield rate is above 2 tonnes per hectare.

MAIZE

While there is a decline in the area under rice and wheat in recent years, area under maize in Bihar has shown a slight improvement. With more or less stagnating yield levels, the state has produced a little over 1.4 million tonnes of maize. Still Bihar occupies fourth position with 9 per cent of the area and 12 per cent production at the all India level. The crop is cultivated in all the three seasons, viz., autumn, winter and summer, in all the Districts. Maize is an important cereal crop in Bihar and it constitutes nearly 12 to 13 per cent of the total foodgrains production in the state. Bihar is a leading State in maize production. It has multiple uses. The baby corn, sweet corn, pop corn and green cobs are remunerative enterprises for increased income and profit.

Autumn maize has the maximum share of area under the crop—42 to 43 per cent of the total followed by rabi maize (31 to 32 per cent and summer maize (25 to 26%). Though autumn maize has the largest share in area, it contributes only 30 to 31 per cent of the total production, marginally higher than that of summer maize which account for 29 to 30 per cent. Rabi maize has better yield rate than that of autumn maize and thus produces nearly 40 to 41 per cent of the crop in the state.

Both traditional and hybrid maize are grown in all the three seasons. Share of area under hybrid maize in the total maize crop of autumn and summer seasons are 40 to 43 per cent and 70 to 73 per cent respectively and the production of hybrid maize in these two seasons constitute nearly 43 to 45 per cent and 71 to 72 per cent respectively. It appears that Bihar has made use of the potential of hybrid cultivars more than any other state. The crop is now replacing upland rice in Kharif season and wheat in rabi season. Although maize is produced practically in all the Districts, major maize producing Districts are Begusarai (10.5%), Samsatipur (7.6%), Bhagalpur (7.6%), Khagaria (7.4%) and Purnea (6.8%). With rich water resources and available irrigation in the winter and summer seasons, irrigated area under maize increased and so did the yields. Today maize is grown throughout the year in Bihar, and increasingly in rotation with wheat and potato. Winter maize dominated total maize production in the Sate with 42 per cent share. Kharif maize accounts for 32 per cent, and summer maize the remaining 26.5 per cent.

Autumn Maize

During the period from 2003-04 to 2005-06, average share of area under autumn maize in the six large producing district, viz. Saran, Vaishali, Begusarai, Bhagalpur, Purnea and Katihar in the state was between 45 to 49 per cent which contributed nearly 49 to 50 per cent towards the state's production (Table 5.9). On the other hand, the bottom six districts, namely, Bhabhua, Jahanabad, Arwal, Sitamarhi, Supaul and Kishanganj, covered just 1 to 1.5 per cent area. Percentage share of production of maize in these six districts ranged between 0.6 and 1.6.

Table 5.9: Autumn Maize

Sl. No.	Districts	Area (hectares)			Production (tonnes)			Yield/hectares,/(Kgs.)		
		2003-04	2004-05	2005-06	2003-04	2004-05	2005-06	2003-04	2004-05	2005-06
	Top Six Districts									
1.	Saran	17975	16583	18368	22654	31709	22329	1260	1912	1216
2.	Vaishali	17352	17206	15569	20587	31763	25977	1186	1846	1669
3.	Begusarai	40362	40164	40910	76611	60514	58380	1898	1507	1427
4.	Bhagalpur	29353	25109	26200	51402	27755	34894	1751	1504	1332
5.	Purnea	11319	10013	15884	24395	20488	34108	2155	2046	2189
6.	Katihar	9700	13194	18839	23350	27227	31586	2407	2064	1677
	Bottom Six Districts									
1.	Bhabhua	332	338	498	408	436	786	1229	1290	1578
2.	Jahanabad	354	353	308	605	470	126	1709	1341	409
3.	Arwal	250	632	397	428	840	344	1712	1329	866
4.	Sitamarhi	484	210	211	813	374	353	1680	1781	1675
5.	Supaul	1216	1549	259	3564	4711	760	2931	3041	2935
6.	Kishanganj	173	161	151	434	296	313	2509	1838	2070
	State Total	259030	270973	276747	438012	443819	420033	1691	1638	1510
	Share of the top six districts (%)	48.7	45.1	49.1	50.0	49.3	49.3			
	Share of the bottom six districts (%)	1.5	1.2	0.7	1.4	1.6	0.6			

The state average yield rate per hectare varied between 1.5 and 1.7 tonnes. However, in Purnea, Katihar, Supaul and Kishanganj, the average yield rates were significantly higher than those of the state average, being 2.1, 2.0, 3.0 and 2.1 tonnes per hectare respectively. There are a few districts (East Champaran, Saharsa, Madhepura, and Araria) which do not fall either in the top or bottom six districts, but have sizable area under the crop. Since yield rates were quite high there, ranging between 2 and 3.5 tonnes per hectare, extension of autumn maize cultivation may be considered there. However, proper arrangement for marketing and better utilization of the produce are considered a pre-requisites for such planning.

Rabi Maize

As stated earlier, though rabi maize stands second in the coverage of area in the State, its share in production is the highest. No doubt, the crop is cultivated in all the 38 districts of the state, a sizable area is found only in six districts—Muzaffarpur, Samastipur, Begusarai, Khagaria, Madhepura and Katihar. These six districts with nearly 54 per cent of the total area under the crop, contributed 55 to 57 per cent of the total production. On the other hand, the bottom six districts, namely, Patna, Gaya, Jehanabad, Nawada, Jamui and Kishanganj, covered hardly 1 per cent of the total area and contributed barely 1 per cent towards total production. There are a few districts like Bhojpur, Rohtas, Buxar, Aurangabad and Sheikhpura, where area under the crop was negligible. On the other hand, cultivation of the crop was quite considerable in Saran, Siwan, Gopalganj, East and West Champaran, Vaishali, Darbhanga, Bhagalpur and Purnea districts. These nine districts will around 32 per cent of the total area under the crop, had a share of nearly 35 per cent in total production of maize in Bihar. (Table 5.10)

Among the three seasons yield rate of rabi maize is the highest (3 tonnes) on an average. The yield rates as high as 4 and 5 tonnes per hect. were also achieved in Khagaria and Muzaffarpur districts respectively in 2003-04. Even where the coverage of area was low, the yield rates were generally 3 tonnes or so during 2003-04 and 2004-05. In Patna district in 2004-05, the productivity rose almost to 5 tonnes per hectare. However, in 2005-06, the yield rates declined in most of the districts to about 2.6 tonnes per hectare, reportedly due to unfavourable weather condition.

Table 5.10: Rabi Maize

Sl. No.	Districts	*Area (hectares)*			*Production (tonnes)*			*Yield/hectares,/(Kgs.)*		
		2003-04	*2004-05*	*2005-06*	*2003-04*	*2004-05*	*2005-06*	*2003-04*	*2004-05*	*2005-06*
	Top Six Districts									
1.	Muzaffarpur	11749	12218	10146	22687	32353	29941	1931	2648	2951
2.	Samastipur	23836	23964	24938	61521	62019	27232	2581	2588	1092
3.	Begusarai	16840	16849	16733	37384	38702	54750	2214	2297	3272
4.	Khagaria	24745	25792	29868	101380	97881	126521	4097	3795	4236
5.	Madhepura	14487	13536	14976	73391	67274	36991	5066	4970	2470
6.	Katihar	13331	13496	15753	31341	38680	41352	2351	2866	2586
	Bottom Six Districts									
1.	Patna	546	657	678	1421	3220	1599	2603	4901	2641
2.	Gaya	238	151	98	721	458	259	3029	3035	2641
3.	Jehanabad	139	168	134	421	510	354	3029	3035	2641
4.	Nawada	96	136	112	291	413	296	3031	3035	2641
5.	Jamui	226	216	56	742	678	216	3283	3140	3862
6.	Kishanganj	272	271	922	441	681	2384	1621	2513	2586
	State Total	196338	195111	208924	594736	592072	551768	3028	3055	2641
	Share of the top six districts (%)	53.5	54.2	53.8	55.1	56.9	57.4			
	Share of the bottom six districts (%)	0.8	0.8	1.0	0.7	1.0	0.9			

Steps are being taken to increase the area under Rabi Maize in the potential maize growing districts like Begusarai, Khagaria, Patna, East Champaran, Bhagalpur, Munger, Vaishali, under centrally sponsored scheme of ISOPOM. It may then be concluded that this crop has good potential in the state. However, the market-efficiency in the state is too poor for which the producers do not get adequate incentive to bring larger area under the crop.

Summer Maize

Among the 3 seasonal maize crops in the state, summer maize covers only one-forth of the total area. Both hybrid and traditional varieties are grown during the season and hybrid varieties occupy nearly 70 to 72 per cent of the area. Production share of summer maize, however, is only marginally lower than that of autumn maize—nearly 29 to 30 per cent of the total. Again, in the total production of summer maize, the share of hybrid maize is about 71 to 72 per cent. The six major summer maize growing districts—Muzaffarpur, Vaishali, Saharsa, Madhepura, Purnea and Katihar—cover nearly 50 per cent of the total area under this crop, producing nearly 48 to 56 per of the total (Table 5.11).

There are a few districts like Rohtas, Bhabhua, Jehanabad, Arwal, Aurangabad, Sheikhpura and Lakhisarai in the State where summer maize is cultivated for name sake only. However, the bottom level districts, Buxar, Jehanabad, Arwal, Sheohar, Jamui and Kishanganj, do have some area under summer maize cultivation. The six districts together have a share of 0.7 to 0.9 per cent of the toal area and almost the same percentage share in production.

Average yield of summer maize is also very good and significantly higher than that of autumn and only marginally lower than rabi maize, ranging between 2.4 and 2.8 tonnes per hectare. Madhepura had the highest average yield during the years 2003-04 to 2005-06—being 3.9 tonnes per hectare. The yield rates were also very satisfactory and generally above the state average in a few other districts viz. Supaul, Vaishali, Jehanabad and Saharsa.

It may be reiterated that productivity of rabi and summer maize is very good and much above the country's average. Bihar has also the superiority with nearly 60 per cent as against the country's average irrigated maize area of 20 per cent or so, over all the maize growing states in the country. Efforts should be made to increase their coverage by further extending irrigation facility. The need for providing adequate

Table 5.11: Summer Maize

Sl. No.	Districts	Area (hectares)			Production (tonnes)			Yield/hectares,/(Kgs.)		
		2003-04	2004-05	2005-06	2003-04	2004-05	2005-06	2003-04	2004-05	2005-06
	Top Six Districts									
1.	Muzaffarpur	11633	11081	13130	25960	30191	16448	2232	2725	1252
2.	Vaishali	9394	9679	9762	22231	29249	27224	2367	3022	2789
3.	Saharsa	13419	13572	13401	39285	39857	48442	2928	2937	3615
4.	Madhepura	17366	18225	18530	60828	89774	59692	3503	4926	3221
5.	Purnea	17616	16242	14697	40942	40034	26781	2324	2465	1822
6.	Katihar	10402	11917	9930	24990	25771	19239	2402	2163	1937
	Bottom Six Districts									
1.	Buxuar	142	116	158	413	301	382	2908	2599	2419
2.	Jehanabad	185	158	148	516	442	354	2789	2797	2392
3.	Arwal	112	116	113	322	310	268	2875	2672	2374
4.	Sheohar	494	370	279	1115	962	624	2257	2600	2237
5.	Jamui	323	349	164	1077	863	249	3334	2473	1518
6.	Kishanganj	209	192	213	465	421	386	2225	2193	1812
	State Total	161062	160769	163151	440821	454492	389309	2737	2827	2386
	Share of the top six districts (%)	49.6	50.2	48.7	48.6	56.1	50.8			
	Share of the bottom six districts (%)	0.9	0.8	0.7	0.9	0.7	0.6			

price incentive to the farmers and efficient and improved marketing facilities can, however hardly be overlooked.

Bihar with its maize potentialities should also try to take advantage of new researches. US scientist have, for example, developed a way to breed corn that can boost the vitamin A. Vitamin A deficiency is an important cause of eye disease and other health problems in developing countries. Scientist are trying to come up with ways to increase—or "bio-fortify"—levels of specific nutrients in a crop like corn which has precursors to Vitamin A compounds called "provitamins" including beta-carotene—which the body uses to make Vitamin A. There are thousands of different corn varieties, and they differ greatly in provitamin A levels. White corn does not have provitamin A, but yellow varieties have it in varying levels. Last but not the least, maize processing industry may play a bigger role in extending maize cultivation in the state.

Demand Pattern

The demand for maize in Bihar is mainly divided between human consumption and animal feed. Other minor uses, such as starch production, tend to take place outside the state. Nearly 50 per cent of total demand is for home consumption. Maize is used to make gruels and in breads. Maize still constitutes as a major cereal in the consumption basket of most Bihari families, particularly the poorer section of the population. Some of the common corn products are:

- Corn on the cob or bhuna bhutta.
- Makai ki roti in north India, tortillas in Mexico.
- Polenta (corn porridge).
- Corn flakes, a popular breakfast cereal.
- Bakery goods like corn muffins and corn bread.
- Corn oil used in cooking and in manufacture of margarines and salad dressings.
- Corn starch or corn flour used as a thickening agent in cooking.
- Popcorn.
- Corn syrup—used in candies and confectionery.
- As an ingredient in some beer and whisky.

With varied uses of corn (Table 5.12), Bihar can capitalize on its advantage in maize production.

Table 5.12: Current Trends of Maize Utilization in Bihar

Item	Percentage
Food	28
Feed: Poultry	40
Animal	12
Industrial utilization	16
Spoilage in Storage and Transportation	3
Seed	1

Source: Rajendra Agriculture University (2005).

- Demand from poultry expected at 6MT during 2008 against 5.5 MT in 2006.
- 48 per cent of total production is used as poultry feed.
- Farmers increasing area under maize cultivation on good price realization.
- Maize MSP raised from Rs. 540 to Rs. 620.

Demand for maize from the poultry sector is expected to rise by about 10 per cent per annum. Indications are that farmers, encouraged by a good price realization, may increase the area under cultivation. The demand from the poultry business could be approximately 6 million tonnes (2008) against 5.5 MT in 2006. According to Godrej Agrovet vice-president (Integrated Poultry Business. Maize production is less expensive and requires less water which makes it a better choice for farmers. Processing industries for corn oil and starch etc., should also be encouraged.

PULSES

There are two seasonal pulse crops grown in the state—Kharif and Rabi. Bihar is the 8th largest pulse producing state in the country. Pulses in Bihar are grown mainly in unirrigated areas and are very much influenced by the prevailing weather and rainfall conditions.

Kharif Pulses

Kharif pulses (other than tur) shared an area of only around 7 per cent, producing just about 1 per cent of the total crop. Major kharif pulses growing districts are: Khagaria, Bhagalpur, Supaul, Kishanganj,

Araria and Katihar. These six districts had a share of 55 to 63 per cent of the State's total area, under the crop with a production share of 55 to 59 per cent. (Table 5.13)

Table 5.13: Total Kharif Pulses (Other than Tur) 2003-04 to 2005-06

Sl. No.	Districts	Area (hectares)			Production (tonnes)			Yield/hectares,/kg.		
		2003-04	2004-05	2005-06	2003-04	2004-05	2005-06	2003-04	2004-05	2005-06
	Top Six Districts									
1.	Khagaria	3835	3668	3266	2831	2769	2418	738	755	740
2.	Bhagalpur	9620	8883	8533	6458	6806	7228	671	766	847
3.	Supaul	2958	2791	2923	2111	2228	2300	714	798	787
4.	Kishanganj	1982	1426	1439	1405	1179	1265	739	827	879
5.	Araria	4541	8184	5019	3125	3211	3691	688	392	735
6.	Khatihar	4044	4450	5168	2893	3305	3658	715	743	708
	Bottom Six Districts									
1.	Nalanda	115	116	121	69	87	66	600	750	545
2.	Bhojpur	111	90	209	67	76	142	604	844	679
3.	Jehanabad	49	100	60	33	45	43	673	450	717
4.	Arwal	175	144	224	104	85	154	594	590	688
5.	Gopalganj	485	906	268	335	532	181	691	587	675
6.	Lakhisarai	232	215	116	160	112	100	724	521	858
	State Total	49153	46726	46176	34268	34891	34971	697	747	757
	Share of the top six districts (%)	54.9	63.9	57.1	54.9	55.9	58.8			
	Share of the bottom six districts (%)	2.4	3.4	2.2	2.2	2.7	2.0			

The average yield rate which was only 697 kgs. per hectare in 2003-04, increased to 747 kgs. per hectare in 2004-05 and further to 757 kgs./hect. in 2005-06. It is observed that generally in the top six districts, the yield rates were higher than those of the state average. Even in the unimportant districts like Bhojpur and Lakhisarai, yield rates were very good. There are a few districts where kharif pulses are not grown at all or the area is negligible.

Rabi Pulses

In the total pulses (including gram and tur) in Bihar state, rabi pulses occupy nearly 70 per cent of the area and contribute 60 to 65 per cent of the total production. The main rabi pulses grown in the state are summer moong, lentil, khesari, peas, etc., Major rabi pulses growing districts are—Patna, Muzaffarpur, Supaul, Saharsa, Madhepura and West Champaran. These six districts shared an area of 32 to 36 per cent of the state's total area during the triennium ending 2005-06, contributing nearly 36 to 38 per cent (Table 5.14) of the total production. However, wide fluctuation in the productivity of the crop were observed in the different districts mainly due to prevailing weather conditions.

Table 5.14: Total Rabi Pulses (Other than Grams)

Sl. No.	Districts	*Area (hectares)*			*Production (tonnes)*			*Yield/hectares,/Kg.*		
		2003-04	2004-05	2005-06	2003-04	2004-05	2005-06	2003-04	2004-05	2005-06
	Top Six Districts									
1.	Patna	49788	47002	41494	61501	52743	44098	1235	1102	1063
2.	Muzaffarpur	32930	20912	31213	22048	16597	23133	669	574	741
3.	Supaul	30597	26155	28198	14780	13828	15827	483	529	561
4.	Saharsa	22233	23149	24770	8334	12667	9119	375	547	368
5.	Madhepura	22165	21230	20343	9806	12179	12243	442	574	602
6.	W. Champara	14391	18800	18791	9166	15345	5126	637	816	273
	Bottom Six Districts									
1.	Munger	1072	1472	791	726	1155	720	677	785	910
2.	Jamui	2882	2405	1107	2623	1411	1045	910	507	944
3.	Siwan	2984	2907	2060	2239	1789	1655	750	599	579
4.	Sheohar	1563	1943	2116	1176	1305	1044	752	672	493
5.	Begusarai	3009	2508	2413	2442	1798	2034	812	719	843
6.	Gopalganj	3125	2672	2477	1814	1904	1704	580	713	688
	State Total	512572	493944	455272	34856	326362	307395	772	661	675
	Share of the top six districts (%)	33.6	31.8	36.2	36.3	37.8	35.6			
	Share of the bottom six districts (%)	2.9	2.8	2.45	3.2	2.9	2.7			

Though rabi pulses are gown in all the districts in the state, there are the districts like Munger, Jamui, Siwan, Sheohar, Begusarai and Gopalganj, where total area under rabi pulses is quite small. These six districts shared an area of 2.4 to 3.0 per cent during 2003-04 to 2005-06 and contributed around 3 per cent towards state's total production.

The average yield rate of the crop was highest in 2003-04 (772 kgs. per hectare), but declined to 661 kgs per hectare in 2004-05. In

2005-06, there was a marginal gain in the yield rate with 675 kgs. per hectare. It may be highlighted that though area under the crop in the bottom six districts was only 2 to 3 per cent of the total, the productivity in most of the districts was generally higher than that of the state average in most of the years.

Tur

Tur or Arhar, with nearly 14 per cent of cropped area, is an important kharif pulse crop. West Champaran district has the largest area, occupying above 7,000 hectares which is slightly more than one-fifth of the total state area. The other major tur producing districts are Gaya, Samastipur, Siwan, Gopalganj and East Champaran. These five districts together had a share of 28 per cent in the state's total area of 33311 hectares in 2005-06. While West Champaran district had the highest production of above 11 thousand tonnes (25 per cent of the state's total of 44 thousand tonnes in 2005-06), share in the production of the above 5 districts was about 23 per cent. The coverage of area was also quite satisfactory in Bhojpur, Rohtas, Nawada, Vaishali and Begusarai, having more than 1000 hectares under the crop. In other districts, area under the crop was rather small and in many of them it was negligible.

Yield rates of the crop were generally quite good in the state—more than 1 tonne/hect in most of the districts. In fact in 2003-04, Buxar, Gopalganj and Muzaffarpur districts produced around 2.5 tonnes per hectare as against the state average yield of only 1.2 tonnes per hectare. There is, of course, wide variation in the yield rate in the districts from one year to another. Since the average yield of the crop in the state is the highest among the major producing states, an all out effort should be made to bring lager area under tur to boost production and partially mitigate the chronic deficiency in this respect.

GRAM

Gram, grown during rabi is not an important crop in the State. Average area under the crop is around 72 thousand hectares and production 65 thousand tonnes; only about 1 and 1.5 per cent respectively of the country's average area and production. The crop is grown mainly in the dry land areas. Only about 4 to 5 per cent of the area under the crop is irrigated. As such, there is sharp fluctuation in area and productivity from year to year. For instance, in 2003-04, area under the crop was as high as 80.3 thousand hectares, but gradually declined to 62.3 thousand hectares by 2005-06. Similarly production also markedly declined from 78.6 thousand tonnes in 2003-04 to 56.1 thousand tonnes in 2005-06. Along with sharp decline in area under the crop, the productivity also recorded a sharp fall from 978 kgs. in 2003-04 to 832 kgs. in 2004-05. However, in 2005-06 productivity partially improved to 899 kgs. per hectare. Use of bio-fertilisers, say Rhizobium will help increase yield of gram.

Patna, Aurangabad, Rohtas, Bhabhua, Nalanda and Gaya are the major gram producing districts in the state. These six districts share more than 50 per cent of the total area under the crop in the state and 53 to 57 of per cent of the total production (Table 5.15). Of these, Patna is the leading district and the productivity of the crop is also very high. The year 2003-04 was the best for gram crop, when the productivity of these six districts ranged between 918 kgs. per hectare in Rohtas to 1246 kgs. per hectare in Bhabhua district. In Nalanda district, however, the productivity was only 732 kgs. per hectare, which declined to 454 kgs. per hectare in the following year. However, in 2005-06, the productivity was restored almost to the 2003-04, level. Similarly in Gaya district, productivity which was as high as 1210 kgs. in 2003-04 declined by 50 per cent in the following two years. The same was the position with area and production of the crop in this district during the years 2003-04 to 2005-06. Gram which is one of the promising crop of Bihar, needs encouragement particularly in Zone-I after harvest of medium and late rice.

There are a few districts in the state where the crop is grown for namesake only. For example, the bottom six districts—East Champaran, Gopalganj, Supaul, Muzaffarpur, Begusarai and West Champaran, occupy less than one per cent of the area under the crop and also produce insignificant quantity. But in all the district, average yield was almost at par with the state average. As gram has a very good demand in the state as well as in the country and since the productivity of the crop in most of the district is generally higher than that of the country's average, extension of cultivation of this crop should be encouraged. Since the country is chronically deficit in pulses, attempt should be made to increase its production wherever there is any scope of extending cultivation of any kind of pulses crop and Bihar is an ideal state in that way for the expansion of area under gram crop.

Table 5.15: Total Gram 2003-04 to 2005-06

Sl. No.	Districts	Area (hectares)			Production (tonnes)			Yield/hectares,/kg.		
		2003-04	2004-05	2005-06	2003-04	2004-05	2005-06	2003-04	2004-05	2005-06
	Top Six Districts									
1.	Patna	9204	8467	11171	10667	9331	12668	1159	1102	1134
2.	Aurangabad	6689	5455	5843	6348	3851	3950	949	706	676
3.	Rohtas	5188	8641	4862	4763	9116	5129	918	1055	1055
4.	Bhabhua	6990	6653	4587	8710	5196	4839	1246	781	1055
5.	Nalanda	5197	4751	4019	3804	2157	2817	732	454	701
6.	Gaya	8598	4091	3755	10404	2524	2294	1210	617	611
	Bottom Six Districts									
1.	E. Champaran	193	—	96	189	—	86	979	—	899
2.	Gopalganj	355	74	33	347	62	30	977	832	899
3.	Supaul	25	22	4	24	18	4	960	832	899
4.	Muzaffarpur	90	42	78	88	35	70	978	832	899
5.	Begusarai	70	67	56	67	44	48	957	658	856
6.	W. Champaran	41	—	96	40	—	96	976	—	86
	State Total	80322	73278	62363	78585	60960	56061	978	832	899
	Share of the top six districts (%)	52.1	51.9	54.9	56.9	52.8	56.5			
	Share of the bottom six districts (%)	1.0	0.3	0.6	1.0	0.3	0.6			

Chickpea Replaces Other Cash Crops

According to official sources, during 1964-65, chickpea area in northern and north-eastern states was about 51.47 lakh hectares (Punjab-7.44, Haryana-13.10, Uttar Pradesh-26.27 and Bihar-4.66). This has now been reduced to 9.56 lakh hectares (Punjab-0.04, Haryana-1.46, Uttar Pradesh-7.40 and Bihar-0.66) during 2005-06. Fortunately, this reduction in area has been partly compensated by increase in central and southern India.

As chickpea is a cool season crop, the general perception is that it requires cooler and longer winter season and more suited to northern India. However, a major shift in chickpea area from northern India, suggests that if can be successfully grown through out India, provided suitable varieties and production technologies are available.

Andhra Pradesh, for example, was earlier not considered suitable for chickpea cultivation because of mild and short winter season.

However, a chickpea revolution has taken place there Andhra Pradesh during the past 10 years because of the availability of short-duration varieties and production technologies suitable for warmer shot-season environments. The chickpea area in Andhra Pradesh has increased from 1.06 lakh hectares in 1996-97 to 3.94 lakh hectares in 2005-06. The most important development is the enhancement in yield, which has increased from 853 to 1,596 kg. hect. during this period.

Andhra Pradesh now has the highest chickpea yield in the country and it is almost double than the national average. Increase in area and yields have led to seven-fold increase in chickpea production in Andhra Pradesh during the past 10 years. The cultivation of kabuli chickpea, once confined to cooler long season environments, has now extended up to southern India. It started with the development of extra-short duration kabuli variety ICCV 2. Later, other large-seeded short-duration kabuli varieties such as (KAK-2, Vihar and LbeG 7) were developed. Andhra Pradesh has about 20 per cent of India's chickpea area under kabuli type now.

Other Varieties

The farmers have also started growing extra-large seeded (seed size more than 50 g per 100-seed) kabuli chickpea and these are unknown cultivars, which originated from other counties, as no extra large seeded kabuli variety has been released in India. The farmers call these extra-large kabuli varieties by various names such as Dollar and double Dollar. Experiments conducted at the Chickpea Breeding centre at ICRISAT suggest that kabuli type is more tolerant to high temperature stress and perform better than desi-type under late sown conditions, where the crop is exposed to high temperature at reproductive stage. This probably explains why kabuli chickpea cultivation is expanding rapidly in Andhra Pradesh and other semi-arid tropical regions of the world. Bihar which is already among the leading states in gram production, should allocate more area under the crop in its future Plans.

SUNFLOWER

At the all India level sunflower, a major source of edible oil, is grown both in kharif and rabi seasons, and its cultivation has increased significantly over the years. Though Bihar is not a very important sunflower producing state, its acreage recorded a significant increase from mid-nineties. However, of late, it appears to have attained a plateau. In fact, the crop is still either not at all cultivated in many districts or the area under the crop is negligible in some other districts. Despite this, in terms of size of area and production, the state ranks fifth in the country. It is mainly a rabi crop in the state and the area under kharif is quite small being only 13 to 16 per cent of the total area (kharif and rabi together) producing only 12 to 14 per cent of the total crop.

The major producing districts in the state are Purnea, Supaul, Madhepura, Araria, Katihar and Bhagalpur. These six districts together occupied 88 and 93 per cent of the total area under the crop during 2004-05 and 2005-06 respectively and contributed nearly 86 and 88 per cent towards the state's total production. On the other hand, the six bottom districts viz. Samastipur, Gaya, East Champaran, Muzaffarpur, Nawada and Jehanabad, had only 1 per cent of the area and produced approximately 1 per cent of the crop (Table 5.16)

Table 5.16: Total Sunflower 2004-05 and 2005-06

Sl. No.	Districts	Area (hectares)		Production (tonnes)		Yield/hectares/kg	
		2004-05	2005-06	2004-05	2005-06	2004-05	2005-06
	Top Six Districts						
1.	Purnea	4086	5796	5879	7937	1439	1369
2.	Supaul	4454	5676	5787	7070	1299	1246
3.	Madhepura	3506	3561	4459	4582	1269	1246
4.	Araria	2763	2330	4024	3383	1456	1452
5.	Katihar	2916	1630	4186	2289	1408	1404
6.	Bhagalpur	527	1988	741	1568	1406	1441
	Bottom Six Districts						
1.	Samastipur	34	—	45	—	1336	—
2.	Gaya	36	27	48	36	1333	1474
3.	E. Champaran	40	29	54	39	1350	1345
4.	Muzaffarpur	18	31	24	45	1333	1452
5.	Nawada	84	44	113	57	1476	1295
6.	Jehanabad	23	55	32	77	1391	1400
	State Total	21183	22631	29122	30429	1375	1345
+	Share of the top six districts (%)	86.2	92.7	86.1	88.2		
	Share of the bottom six districts (%)	1.1	0.8	1.1	0.8		

It may be observed that though the area under the crop in the state is not very large, productivity of the crop is very high in all the districts. Average productivity of the crop in the State during the last five years was 1396 kgs. per hectare as against the country's average of only 544 kgs per hectare. It is pertinent to suggest that as the agro-climatic conditions of the state are quite favourable for growing the crop, endeavour should be made to extend its cultivation in those districts where the area under the crop is still very small or introduce the crop where it is not cultivated at all. This is quite important in order to bridge the burgeoning gap between the supply and demand of edible oils in the country.

Sunflower in Rice Fallows[1]

Large tracts of shallow lowlands are left fallow after rice harvest in eastern states of Orissa, West Bengal and Bihar. Sunflower (*Helianthus annuus*) during rabi/summer season after harvest of kharif rice in the alluvial tract of Bihar, with limited irrigation is an ideal crop sequence-rice (shallow rooted) and sunflower (deep rooted) as it ensures a rational uptake of nutrients from different layers of the soil. Sunflower forms a competitive profitable crop under rice fallows and will improve the oilseed production of the country and reduce the import bill. Keeping this in view, exploratory trails (both on-station and on-farm) were conducted during 1996-99 at the Central Rice Research Institute, Cuttack to identify suitable varieties and to generate appropriate production technology for growing sunflower under such situations.

Varietal Suitability

Selection of appropriate variety is one of the most important criteria for realizing the desired yield of sunflower. Several sunflower cultivars (varieties and hybrids) were evaluated both under on-station and on-farm conditions. It has been found that sunflower cultivars like KBSH1, BSH 1, PAC 36, PAC 308, ICI 308, TNAU SUF 7, etc., performed very well under such situation with an average seed yield of 12.8-27 q/ha. The yield increase of these cultivars, over standard check Morden, ranged from 50.0 to 71.5 per cent. All these cultivars matured in 86 (morden) to 102 days (Tables 5.17 and 5.18).

Table 5.17: Mean Seed Yield and Maturity Duration of Different Promising Sunflower Cultivars Grown in Rice Fallows of Shallow Lowland (CRRI, Cuttack 1996 and 1997)

Cultivar	*Maturity duration (days)*	*Mean seed yield (q/he.)*	*Per cent increase over check*
ICI 308	102	21.5	74.8
PSCL 5015	99	20.1	63.4
PAC 36	97	19.5	58.5
TNAU SUF 7	102	19.4	57.7
KBSH 1	98	19.1	55.3
BSH 1	94	18.9	53.7
Morden (check)	86	12.3	

Table 5.18: Yield Performance of Sunflower Cultivars in Different Locations of Jagatsinghpur District, On-Farm Trials, 1997

Location	*Seed yield (q/he.)*		
	ICI 308	*KBSH 1*	*Morden*
Puranodapada	26.3 (54.7)	24.8 (45.9)	17.0
Puranodapada	27.1 (62.3)	25.1 (50.3)	16.7
Puranodapada	26.0 (62.5)	24.0 (50.0)	16.0
Joirikoti	25.5 (70.0)	25.0 (66.7)	15.0
Joirikoti	25.0 (95.3)	24.5 (91.4)	12.8
Joirikoti	25.0 (85.2)	24.0 (77.8)	13.5
Mean	25.8	24.6	15.2

Appropriate time of sowing is the most important yield determining factor in this region because of short winter period and sudden rise in air temperature during April onwards. It was observed that the second fortnight of January is the optimum time of sowing for shallow lowland and any delay beyond January end, may cause reduction in seed yield. This is mainly due to increase in atmospheric temperature during the time of seed setting from April onwards. High temperature with dry winds result in poor setting of seeds and yield reduction (Table 5.19).

Table 5.19: Performance of Sunflower Hybrid (KBSH 1) under Different Dates and Fertilizer Levels (CRRI, Cuttack, 1997)

kg./he.

Dates/ Fertiliser level	*Control (no fertilizer)*	*40:18.3:16.6 N.P.K*	*60:13.1:24.9 N.P.K*	*80:17.5:33.2 N.P.K*	*Mean*
January 06	9.5	13.2	17.7	20.4	15.2
January 25	11.4	16.6	19.0	23.7	17.7
February 15	9.5	12.4	16.1	18.9	14.2

RAPESEED AND MUSTARD

Bihar shares nearly 2 per cent of the area and 1.5 per cent of the production of rapeseed and mustard in the country. The crop is grown in all the 38 districts in the state, and the produce is consumed mainly within the state itself as the consumers of the state prefer mustard oil to other edible oils. Major producing districts are West Champaran, Samastipur, Begusarai, Muzaffarpur, Khagaria and Rohtas. These six districts together shared an area of 45 and 46 per cent during the years 2004-05 and 2005-06 respectively. On the other hand, the bottom six districts, viz. Jehanabad, Munger, Jamui, Sheohar, Arwal and Nawada occupied only 2 per cent of the state's area during the two years. (Table 5.20)

Table 5.20: Rapeseed and Mustard 2004-05 and 2005-06

Sl. No.	*Districts*	*Area (hectares)*		*Production (tonnes)*		*Yield/hectares/kgs.*	
		2004-05	*2005-06*	*2004-05*	*2005-06*	*2004-05*	*2005-06*
		Top Six Districts					
1.	W. Champaran	16142	16780	2437	10068	151	600
2.	Samastipur	6145	6284	3638	5121	592	815
3.	Begusarai	4894	4840	6264	9375	1280	1937
4.	Muzaffarpur	3519	3748	4867	4415	1383	1178
5.	Khagaria	2886	3111	2759	6303	956	2026
6.	Rohtas	3811	3001	2771	2128	727	709
		Bottom Six Districts					
1.	Jehanabad	170	207	159	259	934	1251
2.	Munger	345	220	287	148	831	671
3.	Jamui	350	243	84	471	240	1937
4.	Sheohar	198	256	87	73	439	287
5.	Arwal	271	315	155	394	573	1251
6.	Nawada	334	400	124	181	372	453
	State Total	83339	82057	59802	77022	718	939
	Share of the top six districts (%)	44.9	46.0	38.0	48.0		
	Share of the bottom six districts (%)	2.0	2.0	1.5	2.0		

During 2004-05 rabi season, the weather and rainfall conditions were not favourable for the crop and its average productivity declined sharply from the normal of 800 to 900 kgs. per hectare to only 718 kgs.

per hectare. Irrigated area under the crop is insignificant. Hence the crop often suffers from moisture stress. Owing to these factors, production of the crop in 2004-05 in the six top districts was only 38 per cent of the state's total of 59.8 thousand tonnes. On the other hand, 2005-06 rabi season was very favourable for the crop and the total production jumped to 77 thousand tonnes and the share of the six top districts was 48 per cent. The bottom six districts produced only 1.5 and 2 per cent of the total production respectively in those two years.

Looking at the district-wise data, it is observed that the yield rates are very high in Begusarai, Muzaffarpur and Khagaria districts in the top six. In fact, in 2005-06 season, yield rates touched 2 tonnes per hectare in Begusarai and Khagaria districts. Among the bottom six districts, the yield rate in Jamui in 2005-06 recorded an unprecedented level of nearly 2 tonnes per hectare as against only 0.2 tonnes per hectare in 2004-05. The yield rates in Jehanabad and Arwal districts were also very high—more than 1-tonne per hectare in 2005-06. There were a few other middle level rapeseed and mustard crop producing districts like Sheikhpura and Lakhisarai where the yield rates were almost 2 tonnes per hectare during 2005-06 season.

Since there is a serious mismatch between the demand of edible oils and their supply, it will be proper to extend the cultivation of the crop in those districts where the yield rates are consistently high. To achieve this objective, the farmers should be provided with timely credit, quality seeds and other inputs besides providing a guaranteed price for the produce.

LINSEED

Bihar is one of the important producing states of linseed in the country. It is grown mainly for extracting oil and the oil content of the seed varies from 33 to 47 per cent. Linseed oil is an excellent drying oil used in manufacturing paints and varnishs, oil cloth, waterproof fabrics and linoleum. In some parts of the country, it is also used as an edible oil. Lineseed cake is a very good manure and animal feed. Linseed straw produces fibre of good quality and is often used in making paper and plastics.

In this state, area under the crop varies between 28 and 30 thousand hectares. The crop is grown in all the districts of the state. However,

Kishanganj, Araria, Bhabhua, Rohtas and Supaul are the six major producing districts. With almost half of the total area under the crop in the state, they share nearly 48 to 52 per cent of the production. On the other hand, the bottom six districts Begusarai, Lakhisarai, Vaishali, Saran, Khagaria and Nalanda-share only 1.2 to 1.6 per cent of the total production from almost the same share of the area.

Table 5.21: Total Linseed 2004-05 and 2005-06

Sl. No.	Districts	Area (hectares)			Production (tonnes)			Yield/hectares,/Kg.		
		2003-04	2004-05	2005-06	2003-04	2004-05	2005-06	2003-04	2004-05	2005-06
	Top Six Districts									
1.	Aurangabad		5440	4319		4695	3866		863	895
2.	Kishanganj		1458	2399		1076	2147		738	895
3.	Araria		1781	2429		1314	2174		738	895
4.	Bhabhua		1249	2271		1142	2076		914	914
5.	Rohtas		1919	1962		1754	1793		914	914
6.	Supaul		1649	1804		1423	1615		863	895
	Bottom Six Districts									
1.	Begusarai		9	14		8	11		863	815
2.	Lakhisarai		43	37		37	30		863	815
3.	Vaishali		99	50		98	45		988	895
4.	Saran		27	78		23	70		863	895
5.	Khagaria		58	86		50	70		863	815
6.	Nalanda		204	94		186	86		914	914
	State Total		27768	29530		23958	26433		863	895
+	Share of the top six districts (%)		48.6	51.4		47.6	51.7		845	900
	Share of the bottom six districts (%)		1.6	1.2		1.7	1.2		914	870

The yield rates of the crop both among the top and the bottom districts did not vary very much and remained more or less the same at the State's average level of around 880 kilograms per hectare. Since the crop has importance in commercial use and has good demand, and does well on the alluvium loamy soil of the gangetic plain, north-eastern zone of the state is best suited for this. Hence, the extension of cultivation of the crop may be attempted in districts like Begusarai, Lakhisarai, Saran, and Khagaria where area under the crop is still very negligible.

SUGARCANE AND SUGAR INDUSTRY

India is known as the original home of sugar and sugarcane. Sugar has an age old association with this country which is believed to have pioneered production of sugar from sugarcane around the 4th century BC. In fact, the English word 'sugar' is believed to have originated from the Sanskrit word 'sharkara'. Alexander the Great's soldiers, after their visit to India, are reported to have marvelled at the production of honey without the intervention of bees'. It has been clearly established that sugarcane is the cheapest sources of sugar, the cost of sugar production through the cane route can be as low as 40 per cent of that of sugar from the sugar beet route. One of the major producers of sugar through the beet route, the EU, has already begun cutting down on production.

Indians knew the art of making sugar since at least the fourth century. The advent of the modern sugar industry in India dates back only to mid 1930s when a few vaccum pan units were established in the sub-tropical states of Uttar Pradesh and Bihar. Until the mid-1950s, the sugar industry was almost wholly confined to these states. It was only after late 1950s/early 60s that the industry dispersed into southern, western and other parts of northern India. India is the largest consumer and the second largest producer of sugar in the world. Sugar production in India is concentrated in six states namely Maharashra, Uttar Pradesh, Gujarat, Tamil Nadu, Karnataka and Andhra Pradesh which together account for 85-90 per cent of sugar production in the country. The Indian sugar industry is highly fragmented with over 450 mills and no single player having a market share of over 5 per cent. The number of operating sugar mills in the country has increased from 29 in 1930-31 to 412 by 1996-97 and a little above 475 in 2007-08. Some 95 factories are non-operational. Around 60 per cent of the mills are in the cooperative sector, 35 per cent in the private and the rest are in public sector. The sugar industry can be broadly classified into two sub-sectors, the organized sector i.e. sugar factories, and the unorganized into two-sub-sectors, the organised sector i.e. sugar factories, and the unorganized sector i.e. manufactures of traditional sweetener like gur and khandsari.

Traditionally Bihar has been one of the major sugarcane growing States in India. There was a time when Bihar produced 25 per cent of India's sugarcane and was the first state with a prosperous sugar industry. Now with fertile land, enough water for irrigation, the share of Bihar in India, has been reduced to an abysmal 1.7 per cent with a productivity level of 42 tonnes/hect. as against the all India level of above 64 tonnes/hect. Sugar is still the largest agro-based industry. An estimated 5 lakh farmers and their dependants are engaged in the cultivation of sugarcane and approximately another half a lakh unskilled people, including highly qualified and trained technologists work in the sugar industry in the state. It generates considerable employment in the farm sector, directly as well as through ancillary industries and related activities indirectly. There has been a small increase in area and production of sugarcane in 2001-02, but thereafter, there is a stagnancy. The data given in the Bihar Government Eleventh Plan, however, presents a very bright picture for 2005-06 where sugarcane, production is shown as 12995 thousand tonnes from an area of 230 thousand hectares.

Table 5.22: Area, Production and Yield of Sugarcane in Bihar and India

Area ('000 hectares)
Production ('000 tonnes)
Yield (Kg./he.)

Year	*Bihar*			*All India*		
	Area	*Production*	*Yield*	*Area*	*Production*	*Yield*
2000-01	93.5	3987.6	42648	4315.7	295956.2	68577
2001-02	113.4	5211.1	45953	4411.6	297207.8	68577
2002-03	107.3	4520.5	42130	4520.3	287383.0	67370
2003-04	103.6	4285.9	41370	3938.4	233861.0	59380
2005-06	102.0	4346.0	42608	4245.0	278387.0	65580
2006-07	103.0	4249.0	41252	4386.0	283404.0	64615

Source: Directorate of Economics and Statistics.

Till recently, the situation was dismal with just about 9.45 per cent of the cane being utilized for sugar production. Most of the sugar mills had closed down and there was a sense of despondency among those industrialists who were holding out against all odds in an apathetic system. Capitalising on the natural affinity that the state has for sugarcane production, the Bihar Government, after due consultation with experts, recognized that sugar and ethanol production were the need of the hour. The new incentives for the sugar sector allowed for a manifold increase in sugarcane cultivation area. Bihar Government has taken appropriate measures to promote new high-yielding varieties of the crop. Area under sugarcane is to receive coverage under the tube-well irrigation programme.

The industry also needs rejuvenation and new investment if it is to play a dynamic role in the future. Several sugar mills under the state-run Bihar Sugar Corporation had to be closed, as they had not modernized their plant and equipment or retrained their personnel. Those sugar mills that continue to function have a capacity that is far lower than the national average. Bihar has the lowest sugar recovery rate in the country at 9 per cent against the national average 10.36 per cent. Present status of Sugar Industry in Bihar is as in Table 5.23.

Table 5.23: Status of Sugar Industry in Bihar

Sl. No.	*Particulars*	*/ Unit*	*2004-05*	*2005-06*	*2006-07*
1.	Total Area under Sugarcane	Lakh Hect.	2.27	2.30	2.52
2.	Total Production of Sugarcane	Lakh MT	125.82	129.95	143.64
3.	Productivity	Mt/hect.	55.43	56.50	57.00
4.	Cane Crushed	Lakh MT	28.48	44.52	52.04
5.	Sugar Produced	Lakh MT	2.54	4.22	4.52
6.	Recovery Rate	%	9.50	9.49	8.69
7.	Crushing Duration	Days	85	126	150

The revitalization of the closed sugar mills are under way, and expansion proposals for existing Private Sector Sugar Mills have been received. The pro-active governance, fertile land, abundant water resources as well as skilled and unskilled work-force, served to make Bihar the investment destination for sugar industry. The new incentive policies for the sugar sector resulted in 13 green field proposals from sugar majors.

A series of sugar companies had come calling, rekindling hope in a state that had not received any private investment in the past two decades. Coimbatore based Rajshree Sugar proposed to setup three green field sugar mills through an investment of over Rs. 700 crore, one in Mirzapur and two integrated sugar Mills in East Chaparan and Madhubani). Similarly, India Glycols had proposed to set up a 10,000 TCD sugar complex in Saran with an investment of over Rs. 550 crore. Amidst all this, film-maker Prakash Jha's P and M Infrastructure is an exception. It has arranged land on its own for its new sugar mill in Gurawlia of West Champaran district.

Sugarcane production and sugar industry hold great potential in Bihar. Out of 27 Mills established in the State, only 9 are in operation. Application for liquidation of all the 15 Mills of Bihar State Sugar Corporation has been filed before Patna High Court. The State is seeking large scale private investments for new sugar mills in 14 identified viable districts. The State Government is seeking assistance in revival of cane/sugar production in the State.

Sugar Sector Achievements

- Thirteen green field Sugar complexes with an investment of Rs. 3629.64 crore approved.
- Capacity expansion of 8 Sugar Mills with distillery and power co-generation and an investment of Rs. 762.72 crore.
- Payment of cane price arrears of Rs. 8.84 crore to the farmers.
- Sugarcane crushing increased from 2.6 in 2004-05 to 4.5 million tonnes 2005-06.
- Production of sugar increased from 2.5 lakh tonnes in 2004-05 to 4.20 lakh tonnes in 2005-06.
- Power purchase agreement approved by BSEB for purchase of Co-generated power.
- Infrastructure Development Authority to facilitate acquisition of land for industrial purpose.
- More than Rs. 2000 crores for repair of roads in 2005-06.
- Additional Rs. 100 crores allotted for repair and construction of roads in sugarcane growing areas.

Land acquisition is coming in the way of Bihar's economic revival plan, particularly in the rural sector. A number of projects cleared by the State Cabinet are yet to get moving because there is no land available. The worst hit is the sugar industry and it is particularly unfortunate for the potential to turnaround its rural economy. So far, the Government has not been able to arrange land for any of the proposed 14 green field sugar projects and 19 ethanol generation units cleared by the Cabinet.

Farmers are coming up with stiff opposition wherever the Government is proceeding to acquire land and agencies concerned are quietly stepping back as it has been made clear that there will be no forcible acquisition. The government is cautious after the experience of Singur and Nandigram in neighbouring West Bengal. After the initial failure, the Government framed an attractive land acquisition policy in February (2001), offering 50 per cent over and above Government rates plus an additional compensation between 30 to 60 per cent. But despite that the farmers are not willing to give up land. The catch is that the market price of even agricultural land in many areas is much above the Government rates.

New Proposals

There is now a proposal by the Central Government to promote land-share companies. According to this concept, farmers can lease their land to a company, say an agro-processing unit, and become shareholders in proportion to the size of their holdings. The land will be considered as the farmer's equity in the company and he will receive a share of the company's profit. The farmer can also take land on lease, including his own land, from the company and cultivate for a fixed rent. The farmer thus stands to gain from both farming as well as his share in the agro-processing unit. The idea of land-share companies is part of a new policy note prepared by the Working Group 'Land Policy and Globalization, for the Eleventh Five Year Plan. At present, some land-share companies exist, though there is no legal framework. The ITC group grows tobacco in Andhra Pradesh and Pepsico recently started growing potatoes, tomatoes, chillies and rice in Punjab, Maharashtra, Karnataka and West Bengal. The Mittals, in collaboration with UK Bank, Rothschild, have started cultivating several crops in a big way. Narendra Murukumbs of Shree Renuka Sugar Mills, Karnataka is also running on the basis of a shareholding company.

In 2002, the Tamil Nadu Government leased out wasteland to corporates for 30 years to grow crops like cotton, flowers, fruits, vegetables and spices. The Gujarat Government has also leased out over 2,000 acres of wasteland. The Government hopes that the concept of land-share companies—midway between contract farming and direct corporate farming—can ensure faster growth in the farm sector and increase farmers' income. But the fear is that the farmer could be alienated from his land if the firm heads for liquidation. To ensure it doesn't happen, the Government plans to put in some safeguards. One of the several proposals is that access to shares in the land share company should be restricted to farmers and up to 25 per cent of the company's paid up capital can be subscribed to in cash by others, including an agro-processing unit or a trading company.

Another proposal is that the land share company can operate on a commercial basis and the management should be vested in an elected board of directors. Voting rights could be on the basis of cooperative principles—one member, one vote—irrespective of the value of share of each member. Descendants of farmers can inherit shares in the company. A farmer can sell his share to other farmers, but the shares cannot be traded through public issue as there may be a risk of takeover by corporate houses or other entities. In case of liquidation, the shareholders would get back their land rights. Since these variants are being tried in different states, Bihar should have no problem to go ahead with the sugar development programmes.

New Research

Navsri Agricultural University (NAU) in Gujarat has developed and released two high yielding varieties of sugarcane that in test conditions produced up to 150 tonnes of the crops per hectare. Put that against the national average yield of 60-70 tonnes a hectare it makes NAU's Gujarat SC-5 and SC-6 cane variants the best yielding in the country. The canes are also richer in sugar content by 20 per cent, with reported 18 tonnes of sugar per hectare. While SC-5 is an early variety, meaning the crop matures in 10 to 12 months with a yield of 142 tonnes a hectare, the SC-6 is mid-late variant, takes 12-14 months to grow and yields 150 tonnes. With these and other research results, Bihar should be able to obtain much higher yields.

The History of Sugar

Over 2000 years ago in the Pacific Islands, the stalks of a plant were found to contain a sweet liquid. The plant is now known as sugarcane. In the 17th century, another plant was found to contain sugar. It is called sugar beet. It is believed that sugarcane culture began in New Guinea and then gradually spread throughout the South Pacific, Southeast Asia, and India. Thereafter it spread to China and to the ancient Arab world, but sugar remained a scarce luxury in Europe. In the 15th and 16th centuries, however, European explorers and colonizes of the Caribbean and South American regions, brought sugarcane cuttings with them, and once planted, the cuttings thrived in the warm, climate and productive soil. By the year 1600, sugar production in the subtropical and tropical Americas had become the world's largest and most lucrative industry.

Sugarcane culture and sugar production began in the United States with the planting of cuttings in New Orbeans during the 1600s, and by 1689 the first sugar refinery was in operation in New York City. It was not until the 1830s that sugar production became a major

U.S. industry. Sugarcane remained the major source of sugar until the 18th century, when a German Chemist Andreas Margarat, discovered that the sucrose in beets had the same chemical composition as that in sugarcane. When one of his students, Franz Kart Archard, developed a practical method for extracting sugar from beets in 1787, sugar-beet culture spread rapidly for France, Austria-Hungry, and Russia. This expansion led to the construction of the world's first beet-sugar factory in Germany in 1803 and to the further spreading of sugar beet cultivation throughout Europe and North America. Beet sugar now accounts for more than 40 per cent of word's total sugar production.

Sugarcane is cultivated in tropical/semi-tropical climates in countries like Brazil, India, China, Thailand, the Caribbean islands, etc., Sugar beet, on the contrary, is grown mainly in the temperate regions like Europe, Japan, etc. It is also produced in syrup from sources like corn (High Fructose Corn Syrup or HFCS) in countries like the U.S.A.

World's Largest Sugar Producing Countries 2006-07

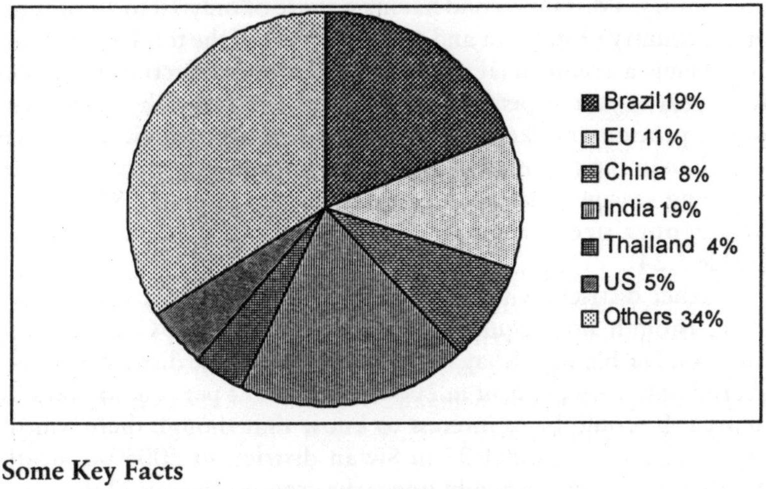

Some Key Facts

- Sugar is produced in more than 100 countries around the world.
- Over 25 nations produce sugar from sugar beets, around 80 from sugarcane, and another eight from both sugarcane and sugar beet.
- World sugar production for 2006-07 was 164 million tonnes;

- As many as 150 countries are engaged in sugar trade mostly as importers;
- World sugar trade accounts for 45-50 million tonnes;
- The closing stock of 41.5 million tonnes sugar for 2006-07 has been the highest in the recorded sugar history;
- Sugar is used as the primary feedstock in as much as 60 per cent of world ethanol production;
- In Brazil, half the total sugarcane output is diverted to ethanol production;
- India's per capita sugar consumption for 2005 stood at 19.6 kg.
- Swaziland's per capita sugar consumption of 97.4 kg. in 2005 is the world's highest, closely followed by Switzerland and Israel at 70.3 kg and 65.7 kg. respectively; and
- Sugar consumption in Rwanda and Congon in 2005, at 1.5 kg and 1.6 kg respectively, is the lowest in the world.

JUTE

After West Bengal, Bihar is the largest raw jute producing state in the country, a distant second having a share of only 16 to 17 per cent of the country's total area and 10-12 per cent of the total production. Jute being a commercial crop, its area and production fluctuates from year to year, depending upon the prices received by the farmers in the previous marketing season. Further, autumn rice competes with jute in the coverage of area as both these crops have the same cropping season. Kishanganj, Araria, Purnea, Katihar, Supaul and Madhepura are the major jute producing districts in Bihar (Table 5.24).

Other districts where the crop is cultivated, in a very limited area, though not regularly, are Begusarai, East Chamaparan, Vaishali, Darbhanga, Khagaria and Saharsa. These districts together shared only 0.06 per cent in 2004-05 and 0.12 per cent in 2005-06 in area. It would be of interest to know that though there was no area under jute in 2004-05 in Siwan district, in 2005-06, nearly 4,000 hectares were brought under the crop producing 37 thousand bales of raw jute—nearly 3 per cent of the state's total production. The six major jute producing districts produced nearly 96 and 97 per cent of the total crop of the state during 2004-05 and 2005-06 respectively.

Table 5.24: Jute 2004-05 and 2005-06

Sl. No.	Districts	Area (hectares)		Production (bales)*		Yield/hectare/ Kgs.	
		2004-05	2005-06	2004-05	2005-06	2004-05	2005-06
		Top Six Districts					
1.	Kishanganj	32983	35698	254519	244333	1389	1232
2.	Araria	34887	31127	308169	289135	1590	1672
3.	Purnea	30490	24202	246291	189179	1454	1467
4.	Katihar	26256	23490	193419	340866	1326	2612
5.	Supaul	11039	9960	143507	34611	2340	1407
6.	Saharsa	5132	4547	76752	61435	2692	1232
		Bottom Six Districts					
1.	Begusarai	3	3	26	29	1564	1758
2.	East Champaran	17	17	148	166	1564	1758
3.	Vaishali	3	70	26	604	1564	1758
4.	Darbhanga	5	—	43	—	1564	—
5.	Khagaria	6	—	52	—	1564	—
6.	Saharsa	53	73	722	986	1564	—
	State Total	140876	133001	1223691	1298565	1564	1758
+	Share of the top six districts (%)	99.9	97.0	95.6	97.0		
	Share of the bottom six districts (%)	0.06	0.12	0.08	0.14		

Note: * = Production in bales of 180 kgs. each.

Source: Government of Bihar.

The yield rate of the crop depends to a large extent on the two crucial operations viz. weeding and retting, besides the quality of seeds and application of other inputs like fertilizers and, pesticides, etc. These two operations again depend on weather conditions at the time of weeding operation and availability of adequate water for retting. Because of all these unpredictable factors, sharp variations in the inter-distinct yield rate are observed. This is amply depicted in the above table.

Note

1. Sanjoy Saha, B.T.S. Moorthy and A. Poonam—Central Rice Research Institute, Cuttack, Orissa-753 006. *Indian Farming*, July 2006.

6

HORTICULTURE

Horticulture and Plantations (H & P) is widely heralded as sunrise sector that provides the dynamic tool for improving economic conditions of the farmers and entrepreneurs, creating diversification opportunities with high value crops, increasing the productivity of land, providing nutritional security, generating employment, ensuring ecological sustainability and enhancing the export earnings. Infrastructure infusion coupled with technological advances and financial interventions have created an enabling environment for attracting private sector investments in this sector. Analysts are of the view that the emergence of Agri-Business ventures in India, is directly correlated to the progress in the plantation and horticulture sector.

India is the second largest producer of fruits and vegetables after China and accounts for close to 10 per cent of world fruit and vegetable production. Horticulture crops are estimated to contribute about 25 per cent to the value of agricultural exports from India. These crops are grown over 11.7 million hectares in India, of which the fruits occupy 43 per cent. (Table 6.1)

The State of Bihar is fifth in fruit production and third in vegetable production and a leading producer of spices like turmeric, ginger, garlic, chilly coriander, etc.,[1] Bihar State is one of the major producers of vegetables and fruits in India, with 12-14 and 6-7 per cent respectively of national production. Area under vegetables in 2004-05 reached 816.6 thousands ha, or 12.9 per cent of the aggregate vegetable area in the country. Similarly, Bihar has 6 per cent of the total fruit area in the country with 276,900 ha. (Table 6.1A). Although fruits and vegetables occupy just 14 per cent of the state's gross cropped area, they hold an important position in the agricultural economy of Bihar. Together, they contribute more than 50 per cent of gross value of agricultural produce.

Table 6.1: Area, Production and Yield of Major Fruit in India—2005-06

Crops	*Area (000 He)*	*% of Total Fruit Area*	*Production (000) Tonnes*	*% of Total Fruit Production*	*Yield Kg/he*
Apple	230.70	18.6	1,739.00	3.50	7,538.00
Banana	529.70	10.7	16,225.00	32.90	30,631.00
Citrus	712.40	14.4	5,996.90	12.20	8,418.00
Grapes	60.20	1.2	1,546.00	3.10	25,686.00
Guava	162.00	3.3	1,685.60	3.40	10,405.00
Litchi	60.00	1.2	368.60	0.70	6,143.00
Mango	1,961.90	39.5	11,160.20	22.60	5,688.00
Papaya	73.00	1.5	2,568.20	5.20	35,181.00
Pineapple	81.20	1.6	1,229.40	2.50	15,140.00
Others	1,092.70	8.0	6,775.30	13.90	

Table 6.1A: Area and Production of Fruits and Vegetables in India

Area—Thousand Hectares
Production—Thousand Tonnes

State	*Fruits*				*Vegetables*			
	Area		*Production*		*Area*		*Production*	
	2003-04	*2004-05*	*2003-04*	*2004-05*	*2003-04*	*2004-05*	*2003-04*	*2004-05*
Andhra Pradesh	656.0	652.0	7580.1	7735.0	248.0	258.4	2882.3	3861.9
Bihar	307.2 (6.6)	276.1 (5.6)	3454.8 (7.6)	2769.5 (7.8)	815.0 (12.9)	816.6 (12.1)	13296.9 (14.3)	13349.1 (13.2)
Karnataka	242.5	255.1	3833.4	4046.9	363.3	367.2	4176.9	4382.9
Maharashtra	1265.0	1346.5	9769.7	10586.1	370.0	372.2	4132.1	4044.4
Punjab	43.6	47.1	628.2	679.5	153.0	158.6	2588.1	2677.4
Tamil Nadu	221.0	224.6	3609.4	4467.6	187.1	215.3	4672.7	6218.3
U.P.	269.4	274.9	2320.4	2912.8	819.5	840.9	14862.0	15792.8
West Bengal	172.9	166.3	2111.6	2128.3	1165.9	1189.0	18510.6	18103.2
All-India	4675.4	4963.8	45644.6	49294.8	6308.9	6755.6	93165.0	101433.5

Note: Figures in bracket for Bihar represent per cent share of all India.
Source: Indian Horticulture Database, 2005.

In the first few years following the bifurcation of the state, area under fruits and vegetables shrank, on an average, by 3 per cent per year. This was largely driven by a decline in area under vegetables by an average of 4.6 per cent per year, which was partially offset by a small average increase of 1.6 per cent in fruits area. Major vegetables include potatoes, brinjal, cauliflower, okra, and tomatoes, while mango, litchi and banana constitute the state's major fruit crops. Major fruit growing districts include Muzaffarpur, Vaishali, East Champaran, West Champaran, Dharbhanga and Samastipur. In the case of vegetables, major districts include Patna, Nalanda, Muzaffarpur, Vaishali and Katihar details about major fruits and vegetable growing Districts as in Tables 6.2 and 6.3.

Table 6.2: Major Fruits and Vegetables Growing Districts in Bihar

District	*Area under fruits per cent*	*Area under vegetables per cent*
Patna	2.67	4.54
Nalanda	2.11	5.82
Rohtas	3.5	1.85
Saran	3.2	2.86
East Champaran	5.71	3.74
West Champaran	4.75	4.08
Muzaffarpur	7.91	5.7
Sitamarhi	3.34	2.38
Vaishali	6.04	6.0
Bhagalpur	4.0	2.97
Darbhanga	6.41	4.35
Madhubani	3.61	3.49
Samastipur	5.53	3.51
Begusarai	2.57	3.66
Purnia	3.2	3.24
Katihar	2.51	5.19
Madhepura	2.35	3.32

Source: Based on data from Government of Bihar.

Bihar's fruits and vegetables (mangoes and tomatoes in particular) yields compare poorly with those of other Indian states, although litchis and okra are exceptions. Potato is another promising vegetable. Since it occupies second place in area, even with low productivity as compared with many States, it had the third position in production in the country. Area and production of horticulture crops in Bihar is as in (Table 6.4).

Growth pattern differs from crop to crop. Potato and brinjal are most important in Zone-II, tomato and cauliflower in Zone-I, Onion and mango in Zone-III and litchi and banana are, however, the fruits of Zones-I and II, Zone-III being more or less deprived of these fruits (Table 6.4). All those residents of Zone-III have to depend for their supplies from North Bihar across the Ganges, involving marketing problems Inland water navigation recommended elsewhere in this study can go a long way to take care of all such and many other problems. Across agro-eco regions, since Zone-II performed better in most of the commodities, it will call for a greater focus.

Table 6.3: Area and Production of Selected Fruits and Vegetables in Bihar

Area ('000 hectares)
Production ('000 tonnes)

	Area ('000 hectares)					Production ('000 tonnes)				
	2000-01	2001-02	2003-04	2004-05	2005-06	2000-01	2001-02	2003-04	2004-05	2005-06
					Fruits					
Banana	27.0	27.2	27.7	28.0	28.0	540.2	544.9	443.5	920.0	920.0
Litchi	25.3	27.7	28.3	28.4	28.4	303.3	221.7	339.0	204.9	204.9
Mango	139.1	139.3	140.0	140.1	140.1	1112.5	1253.5	1540.1	865.5	865.6
Total	268.4	272.3	307.2	276.1	291.12	3237.5	2877.0	3454.8	2769.5	2920.3
					Vegetables					
Potato	148.1	140.8	318.1	318.1	—	1375.9	1432.3	5656.7	5656.7	—
Brinjal	35.5	55.6	54.1	53.7	53.6	110.2	1111.7	1081.9	1073.0	1073.0
Cauliflower	58.9	59.2	59.7	36.5	59.7	942.9	947.3	955.6	598.8	955.0
Okra	57.2	57.4	58.0	56.2	56.2	801.2	860.5	928.7	730.2	674.1
Tomato	46.9	47.0	45.3	46.0	43.0	938.7	751.9	588.4	735.8	601.9
Cabbage	34.9	35.1	36.5	59.7	36.5	557.8	561.4	584.1	955.2	599.0
Total	707.8	578.9	815.0	816.6	487.0	10219.7	8022.9	13296.9	13349.1	7257.8

Note: There is a wide difference in area and production of vegetables particularly during 2005-06 as compared with previous years. This is because State sources data for vegetables differ widely from National Horticulture Board.

Source: Data for 2005-06 from Economic Survey, Government of Bihar and earlier one from National Horticulture Board.

Table 6.4: Growth Trends in Major Horticultural Crops in Bihar (1990-91 to 1999-2000)

(Per cent Change)

Crops	Zone-I		Zone-II		Zone-III		Bihar	
	Area	Production	Area	Production	Area	Production	Area	Production
Potato	-.337	-0.43	3.40*	3.26*	1.03	-0.70	1.10*	0.49
Brinjal	2.41*	5.70*	9.14*	13.78*	-2.25*	-2.98*	4.06*	7.16*
Tomato	7.79*	10.65*	3.90*	5.05*	1.85*	0.31	5.28*	6.19*
Chillies	-1.62*	1.71*	-1.22	-0.32	-1.78	2.78	-1.46	-0.57
Cauliflower	7.42*	5.80*	5.57*	3.23*	5.14*	2.56*	6.93*	4.68*
Onion	-1.71*	-0.22	3.51*	6.14*	7.40*	8.14*	2.31*	4.59*
Mango	2.50*	-3.89*	-0.79	-2.39*	10.75*	7.37*	3.72*	-1.61*
Guava	-1.02*	-6.32*	2.74*	-1.27*	-3.51*	-14.51*	-0.83	-6.91*
Litchi	4.66*	8.72*	9.70*	9.99*	—	—	6.28*	9.83*
Banana	-1.20	12.28*	11.66*	27.87*	—	—	5.54*	23.04*
Jackfruit	5.59*	3.76*	4.45*	2.47*	5.03*	-8.95*	5.12*	2.47

Note: * Indicates statistical significance at 5 per cent level.

Source: Base data are from Directorate of Statistics and Evaluation, Government of Bihar.

Latest available data for 2005-06 regarding Horticulture shows (Graph 6.1 and 6.2) that productivity of major fruits and vegetables is actually higher than the all India levels. Among the fruits, productivity levels of pineapple are as high as 62 per cent. Similarly among vegetables, brinjal, cabbage and onion yields are much higher in Bihar.

TRENDS AND PATTERNS OF AGRICULTURAL DIVERSIFICATION

Bihar has a diversified agriculture. Fruits and vegetables constitute nearly 50 per cent of the gross value of agricultural produce (Table-3.16). The state produces about 2.8 million tonnes of fruits and 13.3 million tonnes of vegetables annually (as of 2004-05). Among fruit crops, mango occupies about 48.4 per cent of total area, followed by litchi (9.7%), banana (9.5%), guava (9.5%), pineapple (1.4%) and others remaining area. Potato occupies nearly 18 per cent of the total area under horticulture crops in the state. Among other important vegetables, cauliflower (9.8%), brinjal (9.2%), tomato (7.8%), cabbage (5.8%) and onion (2.3%) are important. During the past two decades, share of area under fruits and vegetables in the state increased from 4.3 in early 1980's to 9.2 per cent in early 2000.

Graph 6.1: Yield of Major Fruits in Bihar and India, 2005-06

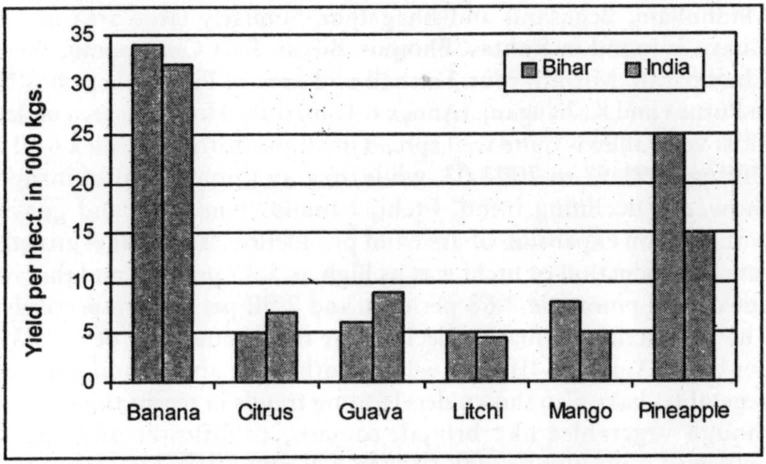

Graph 6.2: Yield of Major Vegetables in Bihar and India, 2005-06

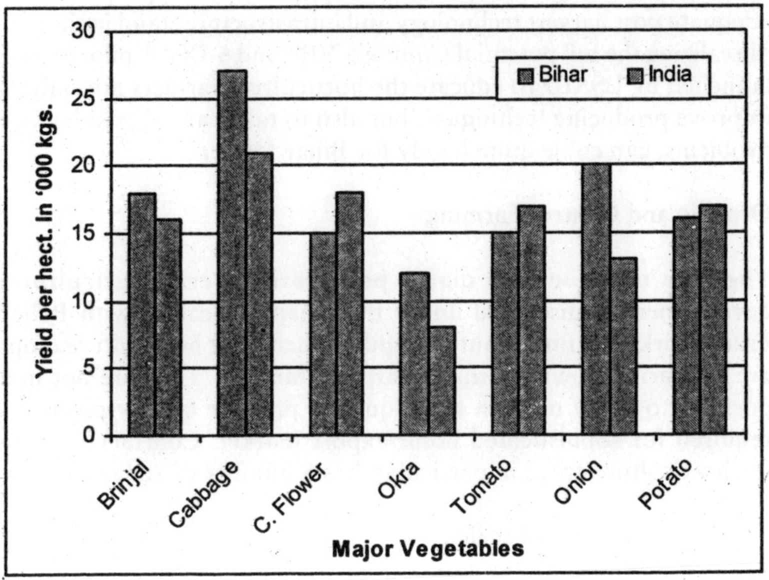

Litchi is concentrated mainly in the districts of Muzaffarpur, Vaishali, East Champaran, West Champaran, Katihar, Sitamarhi, Purnea and Samastipur, while area under Banana is concentrated in Muzaffarpur, Vaishali, Samastipur, Darbhanga, Madhepura, Madhubani, Begusarai and Bhagalpur. Similarly large area under Guava is found in Rohtas, Bhojpur, Buxar, East Champaran, West Champaran, Muzaffarpur, Vaishali and area of Pineapple is mostly in Purnea and Kishanganj. (Annex 6-I and 6-II). However, area under most vegetables is quite well spread in all the districts (Annex 6-III). During 1991-92 to 2002-03, while area and production of mango showed a declining trend, litchi, banana, pineapple and guava witnessed an expansion of area and production. The annual growth rate of production of litchi was as high as 5.35 per cent and that of guava and pineapple 2.68 per cent and 2.90 per cent respectively. The production of mango declined at the annual rate of (-) 3.35 per cent (Annex 6-III and 6-IV). Both area and production of vegetables have also shown decelerating trends in recent times, even though vegetables like brinjal, tomato, cauliflower and onion witnessed a positive growth (Annex 6-V and 6-VI). On the whole, the state has a tremendous potential for productivity growth of fruit and vegetable crops. But unfortunately absence of appropriate and adequate post harvest technology and infrastructure stand in the way of realizing the full potential (Annex 6-VIIb and 6-IX). A pilot project launched by USAID to educate the horticulture farmers not only to improve producing techniques, but also to take care of post harvest problems, can come quite handy for Bihar farmers.*

Organic and Control Farming

The role of these two major policies in overall agricultural development is discussed under the Chapter dealing with Policy Framework. It is important to highlight here that horticulture crops are primarily grown by small/marginal farmers. They are not in a position to apply modern techniques to produce quality products-required for sophisticated home/export market. Contract farming for horticulture crops in particular, has a number of advantages as:

- Economics of scale.

---

* For details refer to Chapter 12 on Replicable Models.

- ➢ Quality products fit for a sophisticated home/export market.
- ➢ Reduction in wastages.
- ➢ Production of quality products for the processing industry.
- ➢ Marketing margins are reduced in favour of both the producer and the consumer.
- ➢ Application of organic farming, which helps to produce high value and quality produce, becomes easy.
- ➢ A win situation both for the farmer and the consumer.

Marketing: A production pattern with relatively more perishable fruits and vegetables, will increase the need for effective marketing linkages supported by modern marketing practices, including introduction of grading, post-harvest management, cold chains, etc. Further, fruits and vegetables grown in Bihar are mostly of table rather than processing variety. There is, therefore, need to upgrade the production technology and adoption of modern post-harvest technology on a large scale for giving dynamism to this sector. Marketing Channels as between vegetables do not show any uniform pattern, while major part of tomatoes and cauliflower are disposed of at the village level, most of potato and cabbages are marketed at the primary market. (Table. 6.5)

Table 6.5: Spatial Allocation and Marketing Channels of Selected Vegetables in Bihar

(*Percentage*)

	Spatial Allocation					*Marketing Agencies*				
Vegetables	*Village door level*	*Primary market/ rural halt*	*Main market halt*	*Outside market yard*	*Total*	*Village merchant/ itinerant trader*	*Katcha arathiya/ agent of the wholesaler/ commission agent*	*Wholesaler/ Commission agent*	*Cooperative Institutions*	*Total*
Potato	16.91	46.73	19.16	17.20	100	25.83	41.05	33.12	—	100
Tomato	44.45	38.0	8.38	9.17	100	33.39	49.43	15.95	1.23	100
Brinjal	36.34	46.70	6.89	10.07	100	40.44	43.59	14.97	1.00	100
Cabbage	39.25	49.04	10.08	1.63	100	27.05	59.55	12.07	1.33	100
Cauliflower	42.21	45.96	8.88	2.95	100	44.63	43.17	11.58	0.621	100
Ladyfinger	29.04	67.07	0.77	3.12	100	33.43	62.27	3.89	0.41	100

Source: Prasad, J. (1999).

POST HARVEST TECHNOLOGY

Wastage of fruits and vegetables persist due to lack of post harvest facilitates. It is estimated that this leads to a colossal loss of more than Rs. 50,000 crore per annum—nearly 30 per cent of the country's production. Since the 100 per cent tax deduction provided for cold chain infrastructure has not worked, Ficci has on behalf of the industry—now suggested that a 10 year tax holiday should be offered to encourage investments in this crucial link. The incentives would be justified since it would also improve quality and boost exports, apart from checking losses. The apex chamber has argued that boost to food processing would result in higher incremental economic activity, as compared to other segments and create significant employment.

Farmers would get better prices and the government could reduce food as well as fertilizer subsidies. Pre-cooling facilities, cold storages, refrigerated trucks and handling facilities should get the benefit so that farm produce reaches processing units and retail chains without waste. The investment needed in cold storages alone is estimated to be Rs. 15,700 crore. The food processing industry is pushing hard for sops, but so far has not met with success. While policy-makers are aware of the situation, bureaucrats are keeping their fingers crossed to see if the Government goes for an integrated approach rather than providing piecemeal incentives to some sections.

PROCESSING INDUSTRY

According to Annual Survey of Industries, 2002-03, the total value of agro-based industries in Bihar (excluding textiles) calculated to Rs. 2005 crore. Bihar shared only 0.6 per cent of the total value of output of agro-based industries in the country. Based on a rough estimate, Bihar can produce about 5 to 6 per cent of the total agro-based industrial products in India which will enable the state to become one and a half times of its present size.

If value addition to food is an indicator of the economic development of a country, India is a step away. A KPMG-Ficci report on the state of the food processing (FP) and agri-business sector has projected an estimated 9-12 per cent growth for this segment in the near future. Of this, fruit and vegetable processing, currently only around a meagre 2 per cent, is expected to shoot up to 10 per cent by

2010 and 25 per cent by 2025. Overall value addition in food products is projected to grow from the current 8 per cent to phenomenal 35 per cent by end 2025. The FP sector, the report maintains, has the potential to attract 2.7 bn euro in investments by 2015. Higher sales growth, increased earnings of companies, rising exports of agri-processed foods and government policy initiative, according to the marketing consultants, have already set the stage for this buoyant performance in the sector.

Table 6.6: Size of Selected Agro-based Industries in Bihar (ASI 2002-03)

NIC Code 1998	*Industry Group*	*Output value (Rs. Crore)*		*Share of Bihar (Percentage)*
		India	*Bihar*	
15	Food/Beverages	168565	1295	0.8
16	Tobacco Products	15649	418	2.7
17-18	Textiles/Textile Products	103450	4	0.0
19	Leather/Leather Products	10309	77	0.7
20	Wood/Wood Products	2522	22	0.9
21-22	Paper/Printing/Publishing	30228	188	0.6
	Total	330723	2005	0.6

Around 30 per cent of the total produce valued at Rs. 30,000 crore, is wasted due to poor post-harvest and storage facilities. This is equivalent of fruit consumption in U.K. Hardly 2 per cent of fruits and vegetables are processed and preserved in the country as against around 70 per cent in Brazil, USA and Philippines, 80 per cent in South Africa, 85 per cent in Malaysia and 30 per cent in Thailand. This is because a large chunk of the fruits and vegetables grown in the country are consumed at home. The Union Agriculture Ministry is keen that at least 30 million tonnes of fruits and vegetables, constituting nearly 25 per cent of the productions, should be processed by 2025 to avoid spoilage losses and provide higher returns to growers. Horticulturists in India feel that changing food habits, a cash rich and upwardly mobile urban middle class and growing health consciousness could give a boost to the popularization of processed fruit and vegetable products.

Horticulture including vegetables cover around 14-15 per cent of the gross cropped area of Bihar. However, the spatial distribution of Fruit and Vegetable Processing Industry (FVPI) is skewed and does not bear relationship with the availability of raw material. Bihar accounts for a very low share of the processing units. FVPI in the state is very underdeveloped and faces many constraints. Post-harvest technology, particularly for fruits and vegetables is still primitive. The huge loss amounting to 25-30 per cent of the total fruits and vegetables produced in the State occur annually on account of poor methods of harvesting and transport. A range of other factors impede the development of food processing product varieties unsuitable for processing, and the slow development of vertical linkages in the farm and food industries. Throughout India, taxes and duties ranging 21-23 per cent are levied on various branded food items. In Bihar changes are underway that would both reduce tax rates on branded food products and shift their burden by being changed into VAT. A recent study shows that the quality and quantity of marketing outlets in Bihar is among the worst in the country. Furthermore, a World Bank analysis of some commodity supply chains has shown that up to 40 per cent of the margins along the value chains are lost due to wastage and transportation. (Table 6.7)

Table 6.7: Post Harvest Losses in Fruits and Vegetables in Bihar

Fruits	*% loss*	*Vegetables*	*% loss*
Mango	25-30	Cabbage	17-27
Banana	15-25	Cauliflower	41-47
Papaya	30-50	Potato	15-29
Litchi	20-36	Tomato	19-31
		Onion	19-31

Source: Verma, V.S., 2002.

Although most fruits and vegetables are sold fresh in Bihar, there is some processing activity taking place which is at a low scale. A number of food processing plants (63) are registered under the Food processing order. About 40 per cent of them are not in working condition. Five of them have a processing capacity of more than 250 MT. Current processing activities are known to include: canning of pineapple, jam and pickles from mango, tomato sauce, litchi juice and pulp, and lime for squash and pickles. Increased processing will not only improve market access to producers, but would also help utilize the low grade fruits and vegetables products, thus reducing losses. Important fruits in the state are mango and lichi and potato as the main vegetable. This accounts for 39 per cent of the area and over 42 per cent of horticultural production in the state. Production of mushroom among vegetables and floriculture is also coming up in the State. There are scattered examples of entrepreneurs having achieved success and others following them in the area. Potential agricultural processing commodities in Bihar are as in Table 6.8.

Table 6.8: Horticulture Products having Potential for Processing in Bihar

S. No.	*Cereals*	*Fruits*	*Vegetables*	*Spices*	*Others*
1.	Paddy	Litchi	Cauliflower	Chilli	Milk
2.	Maize	Mango	Tomato	—	Fisheries
3.	Ragi	Fruits like Guava, water-	Potato	—	Makhana
4.	—	melon and jamun	Cabbage	—	—

- *Litchi*—As much as 70 per cent of litchi produced in India comes from Bihar. Litchi is mostly grown in Muzaffarpur, Vaishali, Bhagalpur and Champaran districts. Pre-cooling and packing industries and juice and pulp industries have good scope in these districts to support the crop related activity.
- *Mango*—It is extensively grown in Rohtas, Saran, Sitamarhi, Dharbhanga, Samastipur, Madhubani and Champaran districts. Ample scope exists for pulp and juice industries to be set up under the sector as a supportive activity.
- *Banana*—Widely grown in the State with Vaishali, Bhagalpur, Katihar, Samastipur, Muzaffarpur and Madhepura being the important districts. Industries which indulge in production of quality banana chips, banana sauce and baby milk food based on banana have excellent opportunities in these area.
- *Guava*—Grown in Rohtas, Bhagalpur, Nalanda, Buxar and Kaimur. Industries for jelly making, juice extraction or pulp making can come up in these districts.
- *Makhana*—Dharbhanga and Madhubani districts offer good scope for processing of this unique crop produce of Bihar.

Agri-Export Production Zones (AEPZ)

Export of horticultural produce has been adopted as a priority, but has been slow to develop. Establishment of Agri-Export Promotion Zone (AEPZ) for litchi: To give a boost to agricultural exports from the country, GoI has adopted the concept of Export Oriented Agriculture Units. In AEPZs, there is a focused effort for promotion of agri-Exports by pooling the resources available under various

Ministries of GoI and State Governments and promoting a particular crop in a zone/area in which that crop is well adopted. The concept of "Contract Farming" has also been introduced in these areas. An Agri-Export Zone was created in 2002 by the GoB under the APEDA scheme. The scheme covers litchi and other fruits, vegetables and honey in the districts of Muzaffarpur, Samastipur, Hajipur, Vaishali, East and West Champaran, Bhagalpur, Begusarai, Khagaria, Sitamarhi, Saran and Gopalganj.

Its role was to promote the export of fruits, vegetables, and other agricultural products grown in the state. To date just 7 functional plants have been registered under the scheme. The Department of Food Processing Industries, Government of India, estimates that it will invest or facilitate investment of Rs. 69.7 billion in the processing industry in Bihar—4.65 per cent of the total investment in the sector in the country over the next ten years. Bihar State Export Corporation Ltd has been identified as the Nodal Agency for implementation of the scheme. Four export oriented units for processing of litchi have been supported under the programme. In 2002-03, 293 tonnes of litchi was exported from Bihar, of which 252 tonnes was in processed form while 41 tonnes was exported as fresh. However the export potential is yet to be fully tapped and the progress is very slow in this direction. A potential of Rs. 214.81 crore has been estimated for this sector for the year 2007-08.

There is also the potential of processing banana as fibre, which has a lot of potential applications in industry. Banana plant, by virtue of its gigantic nature, produces a lot of biomass. This is re-circulated into the soil for enrichment and in most cases, it goes as a waste. With versatile utilities of banana fibre, the waste can be effectively put to use and it can be bring about a lot of change to the banana farming community.

A Farmer in Vaishali[2]

A progressive small scale farmer of Vaishali district, Bihar, cultivates elephant foot yam (commonly called as Ool in North India) in his three-acre field. Without stopping with just growing the tuber crop, he has also taken up value added products by creating 33 different varieties of recipes from it such as sweets, pickle mouth-freshner, and brewing powders similar to the tea. Elephant foot yam is a tuber crop commonly used as a vegetable and for preparation of pickles and sauce. The tuber is commonly called as Jimikand or Ool in North India, Sooran in Gujarat and Maharashtra, Kand in Andhra Pradesh, Karnal Kilangu in Tamil Nadu and Suwarnagatti in Karnataka, and is cultivated all over the country.

This value addition has increased the marketability and has demonstrated immense product potentials of this tuber. He normally harvest 50-60 tonnes from a hectare of land and earns a gross income of Rs. 2 lakhs in about 10 months. A number of farmers in the state have adopted Ool farming as a means to supplement their income. This farmer provides marketing support to the other yam farmers at a margin of 5-10 per cent through contract farming. Besides the contract agreements, there are individual farmers who have taken up Ool cultivation and sell the products to local traders.

Punjab Model

A fruit and vegetable processing plant at Hoshiarpur built at a cost of Rs. 38 crore, is the only plant in the world, which can process all types of fruits and vegetables produced in India. The plant has the capability to switch from processing of one fruit to another within four hours. Meanwhile, the government's move to usher in contract farming in the state got a boost with 5,500 farmers opting for the scheme.

The Government approved a blue print for setting up a network of agri-processing infrastructure on a scale capable of eventually turning the state into the biggest producer and exporter of processed fruit and vegetable products in the world. Under the scheme, farmers who lease out their land for setting up fruit orchids, would get a lease amount of Rs. 8,000-12000 per acre with a 2 per cent hike every year. The owners of the land and the horticulture development council the organizers of the project—will have a 50-50 profits and risks sharing arrangement after five years. The land, with an established orchard, will be handed over to the farmer after 12 years. The orchid will have a life span of 30 years.

STORAGE

Capital Investment Subsidy Scheme (CISS) for Construction/ Expansion/Modernisation of Cold Storages and Storages for Horticultural Produce was introduced by Government of India, to promote modern cold storage so as to avoid spoilage of perishable horticulture produce and other items like dairy, meat, fish, chicken etc.; provide remunerative price to the farmers and providing business opportunities to cold storage owners through out the year by establishment of multipurpose cold storages. Nearly one third of our horticulture produce, especially fruits and vegetables are wasted, mainly on account of poor cold storage and other storage facilities the county also experiences fluctuation in prices of horticulture produce, particularly potatoes and onions. The loss of fruits and vegetables and spices due to inadequate post harvest handling, transportation and storage ranged between 25 and 40 per cent of the total production (NAMBARD).

Considering the annual production of perishable horticulture produce at 152.50 lakh MT in the State during 2002-03, the existing post harvest handling and cold storage capacity is inadequate. Assuming, 30 per cent of the marketable surplus produce is to be stored, an additional cold storage capacity for approximately 14 lakh MT is required which translates into 281 additional cold storage units of 5000 MT capacity each, with a capital investment of Rs. 562 crore, is required. At present, 27 districts out of 38 are having cold storage facility, which can take care of only about 5 per cent of the total production of fruits and vegetables. Post-harvest handling facilities and cold storage units are required to be established in the remaining 10 districts. Additional facilities may be created in those districts where the number of cold storages and their capacity are inadequate. Rural godowns have also vast potential in the State in view of the fact that majority of the farmers are small and marginal who command little holding capacity.

Despite the vast potential available under the sector in Bihar, full exploitation of the available potential is hampered due to the following factors:

- Lack of awareness among prospective entrepreneurs about CISS.
- Most of the produce stored in cold storages in Bihar is potato; GoI's objective of increasing storage capacity through multi-produce cold storages is not achieved.
- Irregular power supply necessitates usage of generators, which increases the operational costs substantially.
- Poor road network, particularly in the rural sector hampers

movement of farm produce.

- Delay in obtaining clearance from various Government agencies.

Bihar has only 5 per cent of the total storage capacity in India. Of the 238 cold storage units available, 187 are used for potatoes. Thus 77 per cent of the capacity of these cold storage units are used to store potatoes alone, leaving only 23 per cent for other purposes. The majority (92%) of the State's cold storage facility is owned by the private sector and the rest by the cooperatives (8%). State has been adding to its storage and preservation capacities over last few years. Under the Plan of cold storage construction, capacity expansion and modernization in the State (with financial assistance from Central Government), permission has been given for construction of 42 new cold storages since 2000, out of which 22 cold storages have started functioning and the remaining ones are under construction. (Table 6.9). The State Government may identify potential areas and draw up a plan for construction of specialized storage facility for fruits like litchi, mango banana, etc., and their processed products. This is possible only if an uninterrupted power supply to the Cold Storage unit is ensured by the Bihar State Electricity Board.

Table 6.9: Overall Scenario of Cold Storages in Bihar

Sl. No.	*Description*	*Nos.*	*Capacity*	
			Cubic Meter	*Metric Ton*
1.	Total Cold Storages	215	2359060	778550
2.	Total in Private Sector	203	2182461	721850
3.	Total in Co-operative Sector	12	176599	56700
4.	Total in Public Sector	Nil	Nil	Nil
5.	Total for Potato	165	1781469	586912
6.	Total for Multipurpose	50	577591	191638
7.	Total for Fish and Marine Products	Nil	Nil	Nil
8.	Total for Milk Products and Others	Nil	Nil	Nil
9.	% share of Pvt. Sector on total	94.40	—	92.70
10.	% share of Co-op Sector on total	5.60	—	7.30
11.	% share of Public Sector on total	0	—	0
12.	% share of Potato on total	76.75	—	75.40
13.	% Occupation for Multi purpose	23.25	—	24.60
14.	% share of fish, Marine and Others	0	—	0

Source: Government of Bihar.

The importance of the cold chain as a part of the infrastructural set up in Bihar is yet to be highlighted as it plays a vital role in the agricultural economy of perishable commodities and an aid to orderly marketing. Cold chain sector will be one of the major agro-based industries in the state classified under 'Priority Sector'. With the increase in quantum of vegetable and fruits consumption among the common people, due to its availability throughout the year, it has become necessary to store more vegetables to meet this continuing demand. This process will usher in a new era of increased storage space for these commodities in years to come and in this context cold storage units will come in handy for support of this situation. In order to encourage marketing the above crops, the Government has formulated a scheme for construction of cold storages in the surplus growing areas of the State. The scheme has a provision for grading and packing of seasonal fruits and vegetables. The banks are required to finance the scheme so as to help the farmers to export the horticultural products.

Constraints: Farmers' Level

Lack of Awareness Among the Farmers

The farmers are not adequately guided on the socio-economic importance of various horticulture crops/activities, system of cultivation, linkages and facilities available from various agencies and marketing etc. Poor extension service is perceived to be one of the reasons for this.

Poor Supply of Quality Planting Materials

Quality planting materials of selected crops and improved crop varieties are reported to be not readily available at a competitive price in many districts. Improper functioning due to lack of infrastructure like water and skilled man power in the Government nurseries are the reasons quoted for poor production and supply.

Poor Market Infrastructure and Market Intelligence

There are no organised market centers in the State, to facilitate convenient marketing of perishables by the farmers at a competitive price. There is no nodal agency to provide market intelligence/ information to guide the farmers in organising their production and deciding a suitable market channel.

Inadequate Extension Support

Due to delay in filling up of various posts in the Department of Horticulture, the required extension support and technical guidance are not made available to the farmers. It is reported that many potential districts are not provided with District Horticulture Office. Even in those districts where the office is opened the services of regular DHOs and other field and extension staff have not been provided.

Lack of Coordination

Line departments, banks and beneficiaries are found to be working in isolation without necessary coordinated effort to harness the potentials available in the sector.

Poor Road Communication Facility

The communication network/surface road facility is poorly maintained, especially during and after the rainy season. This comes in the way of speedy transport of perishables like banana, litchi, tomato, etc, resulting in huge avoidable loss (quantity as well as quality) to the farmers.

Banking Sector

Slackness in Extending Investment Credit

Most of the financing banks are not evincing interest in extending credit support to this sector. Long gestation, lack of manpower at the branch level, documentation/security problem, poor/no eligibility, etc. are the reasons attributed for poor lending to this sector.

Lack of Adequate Technical Staff at Bank Branches

Commercial banks and Regional Rural Banks are having good network of branches. However, they do not have adequate trained technical staff to identify the potential activity and guide the farmers to take up crop development with credit support.

Non-availability of Original Registered Land Document

Due to the inordinate delay in issuing the original land documents (in those cases where there is transfer of property) the beneficiaries are unable to mortgage the land to banks as security, at the time of availing term loan. This is coming in the way of flow of credit, despite good demand in crops like tea.

NATIONAL HORTICULTURE MISSION (NHM)

Government of India launched the National Horticulture Mission in April 2005 for a period of seven years (2 years of Tenth plan and 5 years of Eleventh plan) to provide a comprehensive and holistic growth of the sector in the country, aiming to double horticulture production by 2011-12. The objective of the scheme is to improve the production and productivity of horticulture crops by harnessing the potential of the region. Special emphasis is to be given to "Low Volume, High Value, and Less Perishable Horticulture Crops". Through this Mission, a horticulture based farming system is to be developed, thereby providing viable and ample opportunities for employment, especially for women, besides improving the productivity of land. The programme in Bihar is being implemented by State Horticulture Department. Initially 16 districts were identified. (Table 6.10).

Table 6.10: Districts Identified for Horticulture Development

Sl. No.	*District*	*Mango*	*Litchi*	*Banana*	*Guava*	*Pine-apple*	*Veg. Seed Prod.*
			CLUSTER I				
1.	W. Champaran	***	***	***	***		
2.	E. Champaran	***	***		***		
3.	Muzaffarpur	***	***				
4.	Samastipur	***					
5.	Vaishali	***	***		***		***
6.	Darbhanga	***	***		***		
			CLUSTER II				
7.	Araria					***	
8.	Kishanganj					***	
9.	Purnea		***	***		****	
10.	Katihar			***		***	
			CLUSTER III				
11.	Bhagalpur	***	***		***	***	
12.	Banka	***					
13.	Munger	***					
			CLUSTER IV				
14.	Patna						***
15.	Gaya						***
16.	Nalanda						***

Source: Planning Commission, Government of India.

Progress Under NHM

- ➢ 10 progeny and 70 block nurseries identified for strengthening during 2005-06.
- ➢ Rs. 3.5 crore released to District Mission Committees as the first installment.
- ➢ Planting material of Mango graft (35,000), Guava layers (5,000), Litchi graft (50,000), and Citrus graft (25,000) to be available for plantation in the rainy season.
- ➢ Rest of 22 districts may either be included under NHM or old schemes under Macro Management.

A number of constraints have been observed in the way of organized development of Horticulture Sector in the State. Poor extension service is perceived to be one of the main reasons. Farmers are not adequately guided on the socio-economic importance of various horticulture crops/activities, system of cultivation, linkages and facilities available from various agencies and marketing etc. Quality planting materials of selected crops and improved crop varieties are not readily available at a competitive price in many districts. Improper functioning due to lack of infrastructure like water and skilled man power in the government nurseries are the reasons quoted for poor production and supply. There are no organized market centers in the State, to facilitate convenient marketing of perishables by the farmers at a competitive price. Wastage of fruit and vegetables perists due to lack of post-harvest facilities. It is estimated that this leads to a colossal loss of more than Rs. 50,000 crore per annum—nearly 30 per cent of the country's production. Since the 100 per cent tax deduction provided for cold chain infrastructure has not worked, FICCI has suggested on behalf of the industry—that a 10 year tax holiday should be offered to encourage investments in this crucial link.

SPECIFIC FRUITS AND VEGETABLES

Banana

Banana is one of the major fruit crops in Bihar. While Zone-III hardly produces any banana, it is primarily concentrated in Zone-II. Growth pattern of banana in the State is as in (Table 6.11). Production of banana there has the highest growth among all the fruits, 23.04 per cent during 1990-91 to 1999-2000 and 76 per cent during 2000-01 to 2005-06. New varieties of banana—Tissue Culture (TC)—are now available. Farmers of Theni, Chinnamanur, and Uthamapalayam regions of Tamil Nadu grow the crop in a minimum of 2-3 acres and are fully satisfied with the returns. These farmers mainly grow Tissue Cultured (TC) Grand Naine banana variety.

Table 6.11: Area, Production and Productivity of Banana in Bihar

Year	*Area ('000he.)*	*Production ('000 mt)*	*Productivity (mt./he.)*
1991-92	23.4	350.5	15.0
2000-01	27.0	540.2	20.0
2001-02	27.2	544.9	20.0
2003-04	27.7	443.5	16.0
2004-05	28.0	920.0	32.9

Grand Naine is popular because it has a relatively good shelf life, has attractive golden yellow colour at maturity and is internationally acceptable—both as a fresh fruit and in processed form. With an initial investment of Rs. 80,000–85,000 per acre these farmers get a yield of 38-40 tonnes per acre which amounts to Rs. 1,50,000–1,60,000 per acre as gross income and a net income of Rs. 70,000–75,000 per acre annually. Farmers are advised to go in for two rotations, which give them a net income of about Rs. 1 lakh each per acre. Thus in about 27-30 months, with intensive cultivation, farmers can get a net income of Rs. 2.25 to 2.50 lakhs per acre. The farmers have also standardized the harvesting and packing methods locally. This value addition has ready a market in metros like Mumbai and Kolkata where consignments have gone from Tamil Nadu. Surely Bihar has a better geographical advantage.

Mango

As the dominant fruit in Bihar, mango occupies 7.1 per cent of the total fruits' area and accounts for 7.5 per cent of national production. Area production and yield of mangoes is given in Table 6.12. Bihar is the fourth largest producer of mango in India after Andhra Pradesh, Uttar Pradesh and Karnataka. It is grown in all districts with Darbhanga, Samastipur and Muzaffarpur leading others. During 1991-92 to 2004-05, production of mango has declined from 1462.3

to 865.6 thousand tonnes driven by a decline in both yield and area. Bihar's yield of mango during 1991-92, at 10 mt/hect. has declined by 61 per cent to just 6.2 mt/hect. by 2004-05. Mango production is constrained by a number of problems. Orchards are neglected and are in need of rejuvenation, cultivation technology is old and outdated. There is a need to improve orchards establishment and management following modern techniques. For this, a strong extension system is needed to transfer modern technology to farmers and establish demonstration plots. Planting material is inadequate in terms of quality which is further affecting quality and yields. Moreover, the varieties grown are often not preferred on the export markets.

Table 6.12: Growth Pattern of Mango in Bihar

Year	*Area (he)*	*Production (mt.)*	*Productivity (t/he.)*
1991-92	146232	1462320	10.00
1992-93	147059	1764708	12.00
1993-94	150527	1806324	12.00
1994-95	149490	1793152	12.00
1995-96	150512	1806144	12.00
1996-97	151755	910410	6.00
1997-98	148595	1783140	12.00
1998-99	143686	1772232	12.33
1999-2000	138500	1742601	12.58
2000-01	139068	1112544	8.00
2001-02	139282	1253547	9.00
2002-03	139510	1255590	9.00
2003-04	140009	1540099	11.00
2004-05	140100	865600	6.20

Source: National Horticulture Board, Government of India.

In Bihar, high density orchards can be taken up in long gestation mango varieties like Langra (Malda), Chausa with normal spacing (80-100 trees/ha) but high density orcharding is achieved by palnting medium duration crops like guava, papaya and anola in between mango, thereby increasing the plant population by 4 times. These crops are removed at the age of 15 years, by which time the mango comes to commercial bearing. Mango crop, especially of a variety like Amrapali has since been established under high density plantation as the stature of the plant is dwarf and canopy is relatively small. More and more farmers are now coming forward to adopt this technology to step up production resulting in better returns at the farm gate level.

Most of Bihar's mango is consumed within the state, with some sent to neighbouring states of West Bengal, U.P. and Delhi. Production seasonality could give Bihar an advantage to send the Mango to states like Madhya Pradesh and Maharashtra, where production peaks in the months of April-May compared to June-July in Bihar. Price comparison with two adjacent markets, Jharkhand and Uttar Pradesh reveals that Bihar prices are competitive with those of Jharkhand, but not Uttar Pradesh.

India exports both fresh and processed mangoes. Fresh mango is a significant export product for India. It has now been accepted in USA and Japan after a gap of many years. Preliminary analysis shows that, despite the significant infrastructural difficulties in Bihar, its mango can compete with African (Senegal) and South American (Mexico, Brazil) product. A more in depth analysis that accounts for variety and quality difference and seasonal advantage through early and late varieties is needed to complement this finding. Processing mango is apparently a well-established and successful Indian export, and one that is diversified over several lucrative markets. There is a potential for counter-seasonal exports to fill gaps left by (southern hemisphere) African and South American producers.

Litchi

India accounting for about 20 per cent of global production is one of the largest producers of litchi in the world, with 205 thousand tonnes produced over an area of 28 thousands ha. (Table 6.13). It leads all states with 56 per cent of total national production and 47 per cent of the national area. West Bengal ranks a distant second. Litchi is exceptionally well suited to Bihar physical environment in terms of temperature, altitude, soil conditions and water availability. Major litchi growing districts which are primarily located in the Northwestern region of Bihar include Muzaffarpur with 17.8 per cent of total state production followed by Vaishali (13.2%), Sitamarhi (7.95), East Champaran (6.9%) and West Champaran (5.5) per cent.

Litchi fruit is a native of southern China. The deliciously flavoured, sweet, juicy and pearly white flesh of the fruit is a rich source of vitamin C. Its growing areas have well-drained, sand loam or clayey loamy soil of high fertility suitable for its cultivation. The frost free climate and comparatively low temperature during winters are favourable for flowering and bud growth. The crop is propagated commercially by air layering and also by grafting. The success rate with air layering is normally more than 90 per cent while grafting technique is yet to be standardized. Good preparation of the planning site, 1-2 months before planting helps in better establishment of the saplings and ensures good growth during the early phase.

About 100 plants are required per hectare to be planted at a spacing of 9×9 mts. During the first two years after planting, intercrops such as vegetables, cereals, fruit crops such as papaya and gooseberry can be grown. The crop comes to flowering from March to mid-April. The duration of the flowering period varies from 27-38 days according to the variety grown. The fruit has short shelf-life harvesting at right maturity, packaging, temperature maintenance during storage, and transport are critical factors. If properly managed this can prolong the storage and shelf-life of the fruits to as long as six weeks. It is a normal practice with litchi growers to generally lease out their standing crop to contractors, usually for one year just before the flowering season. The contractors negotiate individually with growers regarding price of the orchard, considering age of trees, maintenance of orchard proportion of fruit bearing trees and earlier record of fruit bearing, Generally, 25 per cent of the total price is paid to farmers at the time of contract and the remaining 75 per cent is paid during the time of harvest.

Table 6.13: Growth Pattern of Litchi in Bihar

Year	*Area (he)*	*Production (mt.)*	*Productivity (t/he)*
1991-92	19115	191150	10
1992-93	19769	200000	10.12
1993-94	20753	249036	12
1994-95	21680	266700	12.3
1995-96	22508	277000	12.31
1996-97	23537	282444	12
1997-98	23441	201282	8.5
1998-99	24421	293052	12
1999-2000	24905	298896	12
2000-01	25272	303264	12
2001-02	27708	221664	8
2002-03	28071	336852	12
2003-04	28254	339048	12
2004-05	28400	204900	7.2

Source: National Horticulture Board, Government of India.

Litchi deteriorates very rapidly after harvest, and the onset of a characteristic brown colour renders the product unmarketable. Fumigation is typically used as a post harvest treatment to reduce browning. Fruit cracking after harvest is also a problem. Losses along the supply chain are quite significant at 22 per cent of production. The Agricultural and Processed Food Products Export Development Authority (APEDA) has drawn up a series of steps to enhance litchi quality, detailing procedures for production, inspection, harvest, peeling, de-stoning, sorting and grading, receipt at pack house, acceptance, and transport. Pre-cooling and transportation in refrigerated vehicle is essential to improve the post harvest life and quality of the fruit.[3]

Several factors have limited the expansion of litchi production in Bihar. These include a narrow cultivar base, poor arboriculture, high incidence of insect pests, poor post-harvest management, and shortage of planting material. At present the litchi economy in the State is plagued by a number of problems such as lack of good planting material, processing and marketing facilities. Due to perishable nature of the fruit, unorganized marketing chain and failure to meet international export standards. Just because of these problems export potential of this fruit has not crossed even one per cent levels of the State's production. The immediate need of the hour, is to provide improved planting material, training in scientific, cultural practices and harvesting methods including prevention of post harvest losses, grading and improved transportation, pre-cooling and processing facilities for the litchi growers.

Priorities for new varieties include a longer shelf life and smaller seeds. Main litchis varieties grown in Bihar are Shahi and China. The very short shelf life combined with a very compact ripening and harvesting season are major hurdles to effective marketing which result in leaving a very short processing season. Fruit cracking is also a major limiting problem for the cultivation of litchi. Furthermore, litchi is highly susceptible to a variety of pests and diseases. In the absence of proper extension service, and the lack of proper orchard management by the brokers, no testing is done regarding the safety (pesticide residues) and/or quality of the fruit.

In recent years, there have been attempts to export litchis from India. Though some export has taken place, packaging and lack of quality standards remain impediments to increased exports. Export of litchi during 2003-04 was 393 MT representing 82 per cent of the total exports of the country. Litchi and makhana grown in Bihar have good export potential. Present level of export of these crops and value realization are found to be far from satisfactory. To address these issues and promote litchi export, APEDA has setup an Agriculture

Export Zone (AEZ) in various districts, such as Muzaffarpur, Samstipur, Hajipur, Vaishali, East and West Champaran. Price spread in the different sales mode in litchi marketing under both the Channels shows wide differences (Table 6.14).

Table 6.14: Price Spread in Different Modes of Sale in Litchi Marketing

Sl. No.	*Particulars*	*Channel-I (Pre-harvest Contract)*		*Channel-II (Self-marketing)*	
		Price (Rs/kg)	*% to Total*	*Price (Rs/kg)*	*% to total*
1.	Price realised by the producer	10.50	17.50	20.00	33.33
2.	Cost for PHC/grower	2.50	4.17	8.00	13.33
3.	Pre Harvest Contractor (PHC)	15.00	25.00	—	—
4.	Commission agents price	28.00	46.67	28.00	46.67
5.	Commission agent's margin	6.00	10.00	6.00	10.00
6.	Stockist price	34.00	56.67	34.00	56.67
7.	Cost and margin of stockist	7.00	11.67	7.00	11.67
8.	Retailers price	41.00	68.33	41.00	68.33
9.	Retailers cost	7.00	11.67	7.00	11.67
10.	Retailers margin	12.00	20.00	12.00	20.00
11.	Consumers Price	60.00	100.00	60.00	100.00
12.	Price spread	49.50	82.50	40.00	66.66

Source: Mysore, Sudha (Undated).

It was sometime in mid nineties that the sulphering process was used for the first time to export litchis to the European market. There are two main varieties of litchis—China and Shahi—which are grown in Muzaffarnagar District. China being a profuse bearer and having a lesser percentage of rejection, is preferred for export although its seeds are bigger. Three Companies Godrej Agrovet Ltd., R.K. Kedia and Sharven Enterprizes Pvt. Ltd.,—are engaged in exports.

Importers have been satisfied by the quality of Indian products. But the European market is quite difficult at the time when temperate fruits and Thai litchis are there in the market. Not only are the exports from Thailand done on a much larger scale (shipping by sea allows greater quantities at a lower cost) but the quality is of a higher level. As the litchis from both countries have been sold on very restricted market prices were high and homogeneous. An idea about cost of exports of litchis from India to Europe is as in (Table 6.15) These results cannot be used as a reference for larger scale exports because on the European market, litchi is a Christmas fruit and consumers may prefer temperate fruits in summer time. The potential market may prove limited under such conditions. India's litchis obtained a good price thanks to the small quantities sold on the market. On the other hand, exporters will have to go by sea to reduce their costs and increase their quantities. An equilibrium has to be found between demand and supply.

Table 6.15: Cost Structure of Litchi Exports from India to Europe

Sl. No.	*Particulars*	*Total Cost (Rs/kg)*	
		Sea	*Air*
1.	Cost up to harvest	1.06	1.06
2.	Export packing (labour)	3.00	3.00
3.	Export packing boxes	3.00	3.00
4.	Transport upto Mumbai	15.00	15.00
5.	Freight Charges	12.75	98.00
6.	Commission	10.00	10.00
	Sub total	44.81	130.06
7.	Overheads	12.00	12.00
8.	Total cost	56.81	142.06

Source: Mysore, Sudha (Undated).

Establishment of cold chain is an essential step in the development of most of Bihar's food industry, but for litchi it is a pre-requisite for access to markets. Existing stakeholders in the litchi chain have not, to date, invested in this cold chain and have reported that they have no plans to do so. While the lack of investment in new trees puts a time limit on Bihar's dominance of the domestic litchi market, lack of cold chain investment denies large-scale access to international markets.

Several factors have limited the expansion of litchi production in Bihar. These include a narrow cultivar base, poor arboriculture, high incidence of insect pests, poor post-harvest management, a shortage of planting material. Priorities for new varieties include a longer shelf life and smaller seeds. The main litchis varieties grown in Bihar are Shahi and China. The very short shelf life combined with a very compact ripening and harvesting season are major hurdles to effective marketing and provide for very short processing season. Fruit cracking is also a major limiting problem for the cultivation of litchi. Further more, litchi is highly susceptible to a variety of pests and diseases. In the absence of proper extension service, and the lack of proper orchard management from brokers, no testing is done on the safety (pesticide residues) and/or quality of the fruit.

Litchi is exceptionally well-suited to Bihar's physical environment. Its wholesale price in Bihar is far below that of potential foreign competition as well as that of competing provinces. A form of vertically-coordinated system has evolved in Bihar which primarily servers the purpose of security of supply and credit provision throughout the chain. The system could be developed further to include information exchange and quality control issues. Litchi is widely acknowledged as an excellent prospect for Bihar's entry to international export markets and much current investment is targeted at promoting and developing that outcome.

Makhana

Makhana is another important horticulture crop of Bihar. Gorgon nut or makhana is an aquatic weed which grows in shallow water bodies of north Bihar. The seeds (which are small in size and white in colour) are used for making different food items. North-eastern Bihar accounts for nearly 90 per cent of the global makhana production. This is cultivated in about 5000 hectares in the entire country, 90 per cent of which falls in Bihar. Contribution of different districts in the production of Makhana are as in Table 6.16. Its production is the specialty of North Bihar Districts of Madubani, Darbhanga, Saharsa, and Motihari of Zone-I and II which have monopoly on makhana (water fruit) production and processing. A rough estimate indicates that the produce worth Rs. 20 crores transported across the State only by rail from Darbhanga, Madhubani, Samastipur and Katihar Railway Stations. Hitherto neglected, this enterprise requires due planning in terms of production, processing and marketing facilities. Makhana, obviously a potential source of strong economy for trading farmers, may be exported in sizeable quantities.

Projects for production of makhana, litchi and mangoes are currently being promoted by NABARD under RSVY in the districts of Muzaffarpur, Darbhanga, Samastipur and Madhubani. About 30 per cent of the total area in the State is planned to be diverted from other crops towards horticulture (Economic Survey). During last two years, 2005-06 and 2006-07, plans for horticulture development are being run in the State with 100 per cent grant from the Central

Government under National Horticulture Mission (NHM). The districts left out by the NHM are being covered under the Chief Minister's Horticulture Mission. This plan consists of expansion of area under fruit orchards, commercial flower cultivation, cultivation of medicinal plants, bee-keeping, integrated pest management, training of cultivators and officers, post-harvest management, etc., This plan is to be implemented in the public as well as private areas.

Table 6.16: Contribution of Different Districts in the Total Production of Makhana in Bihar State

Sl. No.	*District*	*% Share*
1.	Madhubani	40
2.	Darbhanga	25
3.	Saharsa	20
4.	Katihar	7
5.	Purnea	5
6.	Champaran	3
	Total	100

Source: ATMA (Undated), Makhana—Mithilanchal ki ek pahchan.

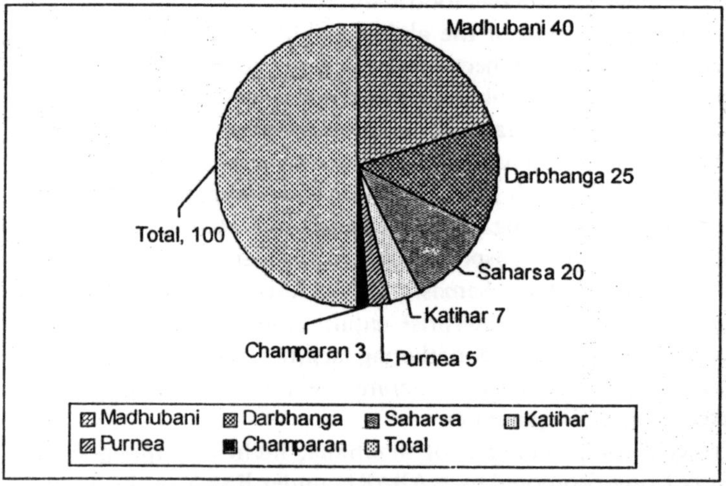

Despite the significant economic importance of makhana production, farmers are still cultivating the crop with traditional seeds and methods. Rajendra Agricultural University, Pusa, Bihar, which is located in a makhana growing area, is yet to come out with any

improved package of practices or a new variety for increasing production in the State. Though the Indian Council of Agricultural Research had established the National Research Centre for Makhana in Darbhanga in 2002, it is still in its infancy and yet to reach the farming community with any new technology. If farmers in Bihar properly exploit this crop, they will have the potential to produce makhana worth more than Rs. 400-500 crores.

Potato

Potato is the single largest vegetable crop grown in Bihar. During 2004-05, the state produced 6 per cent of national potato crop, compared to 44 per cent in Uttar Pradesh and 30 per cent in West Bengal. Total cropped area under potato of these three states represents 71 per cent of the total area sown to potato in the country. Bihar shares 11 per cent of the national area. Its potato production has been stable during the last decade growing at a modest rate of 1.7 per cent per annum, driven entirely by a similar growth in area. The crop is grown in every district of Bihar. It is estimated that nearly one quarter of the area under vegetable production in the state is allocated to potatoes alone. The most commonly grown potato variety in Bihar is the Kufri Sandhuri and Kufri Lalima. The crop is usually grown in the winter season and adapts well to the state soil and physical climatic conditions. Although potato yield during the last 2-3 years is stagnant at 17.8 mt/hect., it has increased significantly from 10.7 mt/hect. in 1991-92. (Table 6.17).

The data for the past five decades (1949-50 to 2002-03) in the Nalanda and Muzaffarpur districts of Bihar has shown a positive growth rate in terms of area (2.77%), production (3.73 per cent) and yield (0.97 per cent) of potato production in the state, but it is far below the national values. Low yield of potato in the state has been attributed to poor adoption (69 per cent) of improved agronomic practices by farmers. The major constraints in adoption of improved technologies have been identified as non-availability of good quality of improved seed varieties at affordable price, inadequate cold storage space, and lack of knowledge about the improved package of practices, particularly about plant protection. Besides, farmers have also reported a number of infrastructural and institutional bottlenecks like, poor road network, erratic power supply for irrigation, poor density of regulated market, fragmented landholdings and poor availability of desired quality of fertilizers on time. The study has suggested that

Government should address these problems of farmers for the development of potato production in the state.[4]

Table 6.17: All-India Area, Production and Yield of Potato 1970-71—2005

Area—Million hect. Production Million tonnes yield kg./hect.

Year	Area	Production	Yield
1970-71	0.48	4.81	9976
1980-81	0.73	9.67	13258
1990-91	0.94	15.21	16254
1991-92	1.03	16.39	15904
1992-93	1.05	15.23	14458
1993-94	1.05	17.39	16610
1994-95	1.07	17.40	16262
1995-96	1.11	18.84	16973
1996-97	1.25	24.22	19376
1998-99	1.21	17.65	14587
1999-2000	1.32	23.61	17886
2000-01	1.34	24.71	18440
2001-02	1.22	22.49	18434
2002-03	1.21	23.92	19806
2003-04	1.34	23.27	17887
2004-05	1.32	23.63	17923
2005-06	1.40	23.91	17058

Source: Agricultural Statistics at a Glance, Directorate of Economics and Statistics, Ministry of Agriculture, Government of India, New Delhi.

Potato is one of the few vegetable crops for which APMC market arrivals constitute a large share of total sales in Bihar, though at 23 per cent it is still small by the standard of other states. Moreover, this exceeds combined percentage arrival of all other vegetables crops, estimated at 6 per cent. One of the reasons why potato arrivals are higher than that of other vegetables is the continuous demand for potato by cold storage owners. However, cold storage space being limited, farmers sell their products to brokers, near or at the mandis.

Most of the potato produced in Bihar is used for table purposes. A negligible quantity is used for processing. Recently two varieties—Chip Sona-1 and Chip Sona-2—have been introduced in areas around Patna. They have a high dry matter, low sugar, less processing loss with high recovery ratio. These varieties are thus most suitable for processing. Every effort has to be made to increase their production. Two supply chains for potatoes are represented below, reflecting the seasonality of potato production. Channel I reflects the peak producing season and Channel-II is the off season. Market participants are the same under the two channels with the exception of the cold storage activity for the off season supply chain.[5]

- *Channel I*. Peak season (Feb.-March); Farmer wholesaler-retailer-consumer.
- *Channel II*: Off season (July-August); Farmer—cold storage—wholesaler-retailer-consumer.

SEED SPICES

Indian spices are as old as human civilization and the Vedas, ancient literature and travelogues are replete with innumerable references to them. Spices trade in India started with the Arabians, who enjoyed the monopoly till the 15th century. From the beginning of the 16th century, the Dutch began to control the Indian spices trade, till the British emerged as a strong rival to them. Even though spices exports are low in volume, thy are high value crops. That is, 94 per cent of the total production of spices in India is domestically consumed and the remaining six per cent is exported. But share in the world trade of spices is 47 per cent in volume and 25 per cent in value terms. India exported 37.37 million tonnes of spices valued at Rs. 3,675.75 crore during the year 2006-07 which went to all.

The state of Bihar is naturally endowed with vast fertile plain lands and congenial sub-tropical climate for raising seed-spices. A number of seed-spices like coriander (*Dhania*), fenugreek, (*Methi*) fennel (*saunf*), omum (*Ajwain*), nigella (*Magrail*), dill, turmeric and chillies are grown in Bihar. Champaran (West and East), Muzaffarpur, Samastipur, Bagusarai are leading coriander growing districts. But maximum acreage under fennel and omum is in Begusarai and Bhagalpur districts. Fennel, coriander and omum are commercially cultivated in Bhagalpur, Munger, Begusarai, Samastipur and Muzaffarpur districts. However, cultivation of nigellas is confined to the districts of Rohtas. East Champaran and West Champaran and Bhojpur. Major problems in seed-spices cultivation have been identified in Bihar which hampered the interest of growers and reduced the productivity and quality of the produce. They are:[6]

1. Lack of high yielding varieties.
2. Standardised agro-technology.
3. Plant protection measures.

4. Improved knowledge of processing and grading.
5. Organised marketing system.
6. Organised spices extension programme to disseminate recent technology to the farmers.

Table 6.18: Area and Production of Spices in Bihar, 2004

Spice crops	*Area (he.)*	*Area (% to total spices)*	*Production (t)*	*Production (% to total spices production)*
Turmeric	3,968	26.31	7,326	36.35
Ginger	942	6.25	1,327	6.58
Garlic	2,972	19.71	3,533	17.53
Chilli	7,181	47.62	7,967	39.53
Total spices	15,081	100.00	20,153	100.00

Source: Kumar, A. (2005).

Future Prospects

There is tremendous scope for increasing seed-spices productivity in the State. But efforts in right direction should be made. Considering the need of spices research in the State to develop its cultivation, Rajendra Agricultural University started a research project for the improvement of spices with the limited financial assistance of the Government of Bihar. But due to lack of fund no adequately qualified staff was appointed in this project and this adversely affected the smooth functioning of the scheme. As such, inspite of some very landable achievements, the output and the quality of the work has not matched fully the requirement of the spices growers of Bihar.

Spices improvements project has developed coriander variety R&D. 44, fenugreek variety R.M.-16 and fennel variety R.F.-17 and has standardized date of sowing seed-rate, spacing N.P.K. and plant protection measures. Results of on-farm trials conducted in the farmers fields are testimony for the brighter future of seed-spices industry in Bihar. Both for coriander and fenugreek yield results were 30-35 per cent higher than the local varieties.

To increase seed-spices productivity further, varieties should be bred for different agro-situations such as for Tal and Diara area for rainfed conditions and for other normal conditions. It is also essential to breed varieties for various socio-economic groups of the farmers. For meeting the requirement of seeds of high yielding varieties, ICAR and Spices Board, Cochin should provide separate fund for seed production to the Agri. Universities. There is an urgent need to replace present Spices marketing system which is very defective. In this connection, Spices Board, Cochin is the only body which can start spices purchasing centers in the spices growing districts of Bihar. This is to be done in collaboration with the Bihar State Agri. Produce marketing and development Board, Patna. This will facilitate growers to get good price for their produce. State Government of Bihar should take necessary steps to implement it immediately. Steps should also be taken to export spices in semi processed and packaged form in the beginning and processing the same later.

For better spices extension programme, there is a great need to organize an extension set up in the State. There is no separate Directorate of Horticulture, but it is looked after by the Deptartment of Agriculture. There is a need for a separate Deptartment of Horticulture with its offices at District and Block levels. These officers will augment the horticulture development programme in the State by working as liaison officers for research organization and growers of various horticulture crops including spices.

MEDICINAL PLANTS

According to the botanists, the dividing line between the medicinal and aromatic plants is rather thin, while there may be some which can be distinctly identified as being of pure medicinal value, there are a large number of species which have both the characteristics. The National Commission on Agriculture (1976), in their report mentioned that there were 1300 species of essential oil bearing plants, but the commonly known ones were sandal, lemon and ginger-grass palmarosa, vertiver (khus), geraniun, rose, celery, clove, jasmine, cinnamon, caraway, eucalyptus, lavender, ocimum (tulsi), anisc, artemisia, etc. Sandal is, however, the main contributing plant, lemon-grass being the next in importance. Others are palmarosa, encalytus and keora. Other sources mention that nearly 8000-odd species of native medicinal plants are known to the Indian people and these are referred to by more than 2,00,000 local names, in different parts of the country. In this list we may further add Neem (*Azadirachta indica*), Bilva (*Aegle marelos*), the Durva Grass (*Cynodon dactylon*), Brahmic (*Bacopa monnieri*), Haldi (*curcuma longa*), Jeera (*Cuminum* cyminum), Dhania (coriandrum sativum), Amla (*Plyllanthus emblica*) and Katha (*Acacia cateclur*).

India has the oldest, richest and most diverse cultural traditions in the use of medicinal plants. The Indian system of medicine has reportedly identified 1500 medicinal plants of which 500 are mainly used in the preparation of medicines.

- About 90 per cent of India's traditional herbs are being traded. India's share in the global complementary medicines market valued at US $ 62 billion is only 0.3 per cent out of which 70 per cent comes from the export of raw materials.

The use of herbal medicines (*kastha aushadhis*) in multifarious ways is quite considerable in India. Even in modern medicines, many of the plant products are finding increasing use. In fact, all over the world there is now increasing evidence of the importance in use of herbal medicines. The global market for traditional remedies has shot up to $ 70 bn. today from less than $ 20 bn just five years ago. But, at present, India has a rather small turnover of around Rs. 5,000 crore in medicinal and aromatic plants and its share in the global market is only around 1 per cent. However, a resurgence in the study and usage of medicinal plants has been observed in the recent past. Production, consumption and international trade in medicinal plants and phyto-medicines are growing. Needless to say, India has a good opportunity in expanding trade of these products. The scenario can be changed by putting to use the country's vast knowledge base and infrastructure in medicinal plants with supportive policy formulation and proper linkages, the sector also offers tremendous scope for crop diversification, waste land development and promotion of agro-based industries.

Taking into account huge demand for herbal drugs and green medicines in the herbal drug industry, the production of medicinal plants has become indispensable. It is also the only source of health care for more than 80 per cent of rural population where modern drugs are not available due to high prices.

Medicinal and Aromatic Plants (MAP) in Bihar

The varied agro-climatic conditions in the state offer scope for commercial cultivation of a variety of medicinal plants that are in demand, including crops like neem and pyrethrum which are sources of bio-pesticides. Of late, growing of medicinal plants is gaining momentum and there is an urgent need to standardize the package of practices for commercially important medicinal plants suitable for cultivation in the state. Commercial cultivation of medicinal and aromatic plants is already in vogue in the state of Bihar. The important species include Periwinkle (Catharanthus), Gloriosa, Coleus, Phyllanthus, Aswagandha, Annota, Citronella, Palmarosa and Geranium. A majority of these crops are being exported in bulk form. During Tenth plan, Medicinal and Aromatic plants have come to focus with setting up of National Medicinal Plant Board and constituent State Medicinal Plant Boards. Majority of the medicinal herbs and aromatic crops are annuals or biannuals. However, under contract farming system, banks are funding working capital and term loans with a duration of 3-5 years. Most of Indian exports are in raw farm. China is one country which imports from India unprocessed MAP and exporting to other countries by adding value. Bihar has the potential to produce many of these plants and start processing industry for MAP.

Under Agri-Export Zones and contract farming, this new innovation of funding as term loan for activities hither to eligible for only short term funding has helped in capital formation in MAP sector. This has also helped to bring into focus funding of specialised horticulture, which could not have received the attention as crop loan. In Bihar, an innovative model is adopted for financing herbal plants like stevia, where the initial cost of planting material is high (Rs. 0.60 lakh/acre) and the same can be multiplied by farmers in their own farms. A model has been recently introduced for financing only the targeted area under an area development plan with some more crops under the category in addition to Stevia. This model not only gives a chance to farmers to assess the production quality and marketing performance of the crop but also rationalised the risk of bankers and investors. This model also insures against the ever falling prices of planting material with expansion of production base. This model can be adopted for financing term loans to other herbal plants like mint, lemon grass, etc.

MUSHROOM

Mushroom, a fungus fruit body, is considered a delicious food all over the world because of its taste, aroma and structure. It is full of nutrients, low in calories, high in vitamins, and minerals. It is mainly used for preparing soups. Unlike other crops, cultivation of mushrooms needs no land and these can be grown in a small 5 x 5 feet room without windows. The fungus grown in India broadly falls into two categories: guchhis and white button mushrooms. Guchhis, also known as truffles and morels, are rated as the best quality and high priced in the world. These are found in the hilly regions of Kashmir Valley, Himachal Pradesh, Sikkim, Darjeeling, West Bengal,

Arunachal Pradesh, Pohri Garwal and Kuman. On the other hand, the white button mushroom is a cultivated variety and its commercial production in India commenced in 1961 when Indian Council of Agricultural Research (ICAR) in collaboration with the Department of Horticulture, Himachal Pradesh, took up its cultivation in the State. Later on, its cultivation spread to other states like Haryana.

A Profitable Proposition

Mushroom cultivation is capable of recycling agricultural wastes and generating employment and income in rural areas without competing for land resources. It has high nutritional value, contains more protein than vegetables, and mineral content is reported to be more than that of meat and fish. Since most of the Indian population is vegetarian, mushroom can provide the required protein. Mushroom is considered to be a delicacy with no party complete without a mushroom dish, which is eaten by one and all. Besides, one can also have mushroom soup, mushroom pickles, etc. Its cultivation offers following advantages:

(a) Farmers can obtain high returns of Rs. 6.20 lakh in 5 months from a small plot of 1,000 sq.yds. by investing say, Rs. 4 lakh.
(b) No arable land is needed, as it is grown inside rooms, on various agricultural wastes like paddy/wheat straw, dry leaves, sugarcane baggasse, corn cobs, and animal/poultry bird droppings (excreta).
(c) The produce (mushroom) has good export market and unlike floriculture, it is low volume high value commodity and does not require to be marketed fresh as international trade is in processed form, packed in cans.

Mushroom production is labour and management-intensive. Specialty mushrooms are not a "get rich quick" enterprise. On the contrary, it takes a considerable amount of knowledge, research, planning, and capital investment to set up a production system. One must also be prepared to face sporadic fruiting, invasions of "weed" fungi, insect pests, and unreliable market prices.

Mushroom production is completely different from growing green plants. They do not contain chlorophyll and therefore depend on other plant material (the "substrate") for their food. The part of the organism that we see and call a mushroom is really just the fruiting body. Unseen is the mycelium—tiny threads that grow throughout the substrate and collect nutrients by breaking down the organic material. This is the main body of the mushroom. Generally, each mushroom species prefers a particular growing medium, although some species can grow on a wide range of materials.

These are the steps in mushroom production—a cycle that takes about 15 weeks (time varies by species) from start to finish.

- ➢ Choosing a growing medium,
- ➢ Pasteurizing or sterilizing the medium,
- ➢ Seeding the beds with spawn (material from mature mushrooms grown on sterile media),
- ➢ Maintaining optimal temperature, moisture, and other conditions for mycelium growth and the conditions that favour fruiting (This is the most challenging step.)
- ➢ Harvesting, packaging, and selling the mushrooms,
- ➢ Cleaning the facility and beginning again.

The substrate on which mushrooms fruit, must be sterilized or pasteurized in order to destroy any fungal and/or bacterial competitors. To produce spawn, the grower should inoculate a pasteurized medium, usually grain, with the sterile culture of a particular mushroom species. After the culture has grown, it is called spawn. Producing spawn requires exacting laboratory procedures.

> Many mushroom suppliers sell several kinds of spawn. Mushroom farmer at the starting time should take advantage of this selection in early trials to determine which species grow best on available materials. Eventually, learning to produce spawn would reduce the cost of production. While the mycelium is growing—and until it fully occupies the substrate—the mushroom farmer typically manipulates the growing environment to favour mycelial growth. The atmospheric conditions are then changed to initiate "pinheads," and then to complete fruiting. For example, in oyster mushroom production under closely controlled conditions, the grower lowers the temperature to enable (CO_2) to grow room to initiate fruiting. Each species has specific requirements for its stages of development.

Prospects in Bihar

Mushroom cultivation in Bihar is not very popular as yet. There are, however, scattered examples of individual entrepreneurs having achieved success and others are following them in and around Patna. Because of its profitability; particularly for landless and small/ marginal households, there is an urgent need to encourage its cultivation and marketing on commercial lines.

FLORICULTURE

It is well said that "apart from the satisfaction of being surrounded by colours all the time, growing flowers commercially can bring in rich monetary rewards as well." A flower is one of the best gifts one can give to anyone anytime. Saying it with flowers will never go wrong. Indeed, it never goes out of fashion. Floriculture comprises cultivation and trade of flowers (fresh and dried), pot/nursery plants, seed and bulb production foliage and green live plants and their economic products (micro propagation and extraction of essential oils) like scent, oils as well as medicines. Flowers like rose, carnation, chrysanthemum, tulip and lily dominate the market for cut flowers. Dracaena of the green plants, begnia and kalanchoc of flowering plants also occupy important position in the world trade. Floriculture today is a recognized scientific discipline and a growing industry.

Floriculture comprises cultivation and trade of flowers (fresh and dried), pot/nursery plants, extensive research and creation of healthier/stronger seed varieties, and bulb, production of foliage and green live plants and their economic products (micro propagation and extraction of essential oils) like scent, oils as well as medicines. Flowers like rose, carnation, chrysanthemum, tulip and lily dominate the market for cut flowers. Dracaena of the green plants, begnia and kalanchoc of flowering plants also occupy important positions in the world trade. Floriculture today is a recognized scientific discipline and a growing industry. World over, more than 140 countries are engaged in the cultivation and trade of flowers.

Thanks to the diversity of climatic and soil conditions, the flora of India in general and Bihar in particular, is extremely rich, ranging from temperate to tropical flowering plants. In good olden days floriculture in the country was more of an art than a science; however, with the increasing application of scientific techniques in the field of plant improvement, floriculture developed into a full-fledged scientific discipline. Floriculture is now included in the curriculum of many institutions of higher learning in the country and a large number of private firms are engaged in floriculture research and seed production. In fact, there exists immense potential in India to export orchids, roses, gladioli and other distinctively Indian flowers. With its varied climate, Bihar can take advantage of this opportunity.

With the declaration of floriculture as an extreme focus area and identified as a sunrise industry with 100 per cent export oriented status by the Ministry of Commerce, Government of India, this industry has acquired a special position in the basket of India's export commodities. Commercial floriculture industry in India began only in the late 1980s. With the economic liberalization, the Government identified floriculture as one of the major areas for export. Several entrepreneurs set up export-oriented units, mainly around Bangalore, Pune, Nashik and Delhi.

Current Trends

India's floriculture business is worth Rs. 9 billion. With rich and varied climate conditions and abundance of cheap labour, it is most suitable for floriculture. Organised floriculture in India is only 15 years old. Of the 110,000 hectares under floriculture in the country, only 500 hectares is under organized floriculture. A large number of projects came up in the 1990s. But many of the older ones have not been able to survive. Karnataka, Tamil Nadu, Andhra Pradesh, West Bengal, Maharashtra, Rajasthan, Uttar Pradesh, Delhi and Haryana are showing interest in floriculture.

Floriculture has not yet been taken up seriously in Bihar. There are, however, stray examples of some enterprising farmers in Dist. Bodh Gaya. One farmer has invested Rs. 1.6 lakh per acre for production of Safed Musli, cut flowers, rose plantation, gladiolus and lemon grass. He is supposed to save Rs. 400 per kg. from Safed Musli alone. Government is now planning to give a subsidy of 30 per cent through National Medicinal and Aromatic Development Board. This will go a long way to encourage the spread of such projects in Bihar.

Floriculture is a viable profitable alternative for the new generation of farmers. For instance, the Punjabi farmers have taken up floriculture seriously. A progressive farmer in Punjab, had gone to Holland to make himself familiar with the technical know-how of the business. He is now able to export Asiatic and Oriental Liliums to South-east countries. VSL group, which has interest in the mining sector has developed a 25 acre floriculture facility in Hampi, which had only historical important once. The VSL project is unique as it seeks to promote hydroponic form of agriculture. Hydroponic, or soil less agriculture, is popular in the West as conventional agriculture has resulted in soil contamination through rampant use of pesticides and chemicals. It is estimated that the hydroponic technique of flower production can yield about 20 per cent more output than conventional floriculture.

History of Flowers

The history of flowers is older than of humans. Flowers are ubiquitous. Even the cotton clothes we wear come from flowering plants.

Flowers appeared on earth about 130 million years ago, during the Crustacean period. Humans are believed to have existed for a mere two hundred thousand years. Once flowers took firm root about 100 million years ago, they quickly diversified into some 250,000 species.

Flowering plants have the botanical name of Angiosperm. Angiosperms enclose their seed in fruit, and each fruit contains one or more carpels. Carpels are hollow chambers that protect and nourish the seeds.

The oldest fossil of an angiosperm was discovered in China in sediments that date back to 130 million years. It was so primitive that it had not developed the lovely flower that is typical of a flowering plant. It is still considered a flowering plant since it had carpels enclosing seeds that grew into fruits

The first angiosperms are believed to have evolved in areas where there were ecological disturbances like floodplains and volcanics. They were slow growing and had short life cycles. Thus, they matured and reproduced and multiplied much faster than trees. As they seeded quickly, they also evolved faster than other plants that were their competitors.

Scientists believe that petals evolved 30 to 40 million years after the first angiosperms evolved. Once this happened, the angiosperms had the distinct advantage over all the other plants, as they attracted the birds and the bees, which, in turn, pollinated them.

The world must have started bustling with activity with butterflies and bees and insects swarming everywhere. When this happened some 70 to 100 million years ago, the number of flowering species on Earth exploded.

> By the time the first flowering plant appeared, plant-eating dinosaurs had been around for a 100 million years. Dinosaurs were eating these and when the dinosaurs became extinct, another group of animals took their place—the mammals, which dispersed the angiosperm fruits, nuts and many vegetables. Now, the flower kingdom and the human race depend on each other.

Some Problems

Though the quality of Indian cut flowers has been gaining popularity globally, domestic flower growers may not be able to meet the growing domestic demand owing to lack of re-plantation activities. Re-plantation of cut flower plants is not taking place in India as importing the planting materials is too expensive. High freight rates add to the crisis. Freight rate in India is 50 to 70 per cent higher than of Kenya—India's closest competitor in cut-flower market. There is htefly menace in flower crops. Spiralling whitefly causes severe damage to all types of field crops including flowers. The insect is always found on the lower surface of leaves. Young and adult insects suck the plant sap, which leads to premature falling and yellowing of leaves. Quality of flowers gets reduced. In young plants the infestation leads to retarded growth and poor flowering.

India faces a major challenge in terms of infrastructure and awareness among small producers. Producers have small and fragmented holdings. Global rose output, for example, rose by over 8 per cent in 2006 due to increased acreage in Ethiopia, Ecuador and Kenya. Indian Exporters are losing even traditional markets in Asia-Pacific like Singapore. Research studies have been carried out by the Tata Economic Consultancy Services and A.F. Ferguson & Co. They have identified several constraints facing floriculture.

Potentialities in Bihar

Floriculture in Bihar has yet to make a start. The State has the necessary agro-climatic conditions available and the National Horticulture Mission is mandated to support this important activity. Since Bihar will be starting more or less from scratch, it has the advantage of learning from the experience of others and take care of the problems mentioned above. Whether it is contract farming or any other production system, Bihar can launch floriculture in the state on scientific lines and create necessary desired infrastructure to enter the export market from day one. Floriculture can also be taken up by some enterprizing farmers even on small farms. We quote here the example of a landless farmer in village Malti (Ranchi).

The landless son of a labourer with nothing but a matriculation certificate degree and a ready supply of mettle, has found his fortune in flowers. In fact, so impressive has his turnaround been that many farmers in the small villages of Malti in Ranchi have taken to floriculture. The magic transpired in 2000, when the staff of the Wold Bank—sponsored nursery and the Indian Council of Agriculture Research's subsidiary, HAP, helped the young man cultivate flowers like rose, jasmine, gladioli, marigold and gerbera. He leased land, dug up a well, bought a diesel pump to irrigate it and planted 200 marigold saplings.

Today, the scent of flowers fills his life and the flow of money is more than visible. He zips around on a Hero Honda bike, sports a watch, watches Doordarshan on his own TV, and of course, nets profit in his newfound venture. Last year (2007), for instance, over 5,000 plants of roses, marigolds, gladioli us and gerbera fetched him a net profit of Rs. 40,000. This year his target is higher. Be hopes to save Rs. 60,000 by the end of this year. Little wonder that he is a source of inspiration for a growing number of farmers at Malti who have abandoned their meagre vegetable farms for flowers. One reason for this change is higher and quicker profits. A vegetable usually takes three months to grow, but marigold starts flowering within a month.

SERICULTURE

Sericulture is the art and technique of silk production which dates back to the second century BC. It fits very well in India's rural structure. Most of the sericulture activities do not involve hard labour. It provides attractive returns with minimal investment. The cocoon crops have short gestation period. Farmers can have five to six crops per annum. Bivoltine silk refers to silk reeled from silk-worm which lays eggs twice a years. This type of silk is qualitatively superior to multivoltine variety. It also commands a good premium in the international market.

India produces a variety of silks called mulberry, 'tassar', 'muga' and 'eri' based on the feeding habit of the cocoons. The sericulture industry employs over 7 lakh farm families. The world's production of silk stands at 95,858 tonnes of which China produces 68,000 tonnes. India is a leading producer and consumer of silk in the world. Karnataka accounts for nearly half of the silk produced in India. Together with Andhra Pradesh and Tamil Nadu, it accounts for over 80 per cent of the nation's annual silk output which stands at about 15,000 tonnes. Bihar is no where on the silk Map of India, but is known for 'tassar' silk produced in Bhagalpur area. In the year 2000-01, silk production in India was placed at around 15,000 tonnes, while demand stood at 22,000 tonnes, leaving a deficit of 7000 tonnes. At present, the deficit is around 9000 tonnes as demand increased to 25,000 tonnes while production stands at 16,000 tonnes. Bulk of the deficit is being met through imports mostly from China.

Double-hybrid cocoons could be the new segment of focus for the Indian biyoltine sericulture industry. A double-hybrid is a product achieved by multiplying two hybrids and are commonly found in most sericulture practicing nations bivoltine production based on double hybrids is currently about 10 per cent, which can be increased to 50 per cent in the near future. Double-hybrids benefit the seed producers also.

The Indian Government has secured technical assistance from the Japanese International Cooperation Agency (JICA) to promote bivoltine silk production. The first five years (1991-96) were spent in developing the Bivoltine Sericulture Technology Development Project. The second phase of the project (1997 to 2002) is called the Programme for Popularising Bivoltine Sericulture Technology' (PPBST). It focused on issues related to silkworm race maintenance, silkworm seed production, silkworm rearing and training. India has been able to evolve more nutritional and better varieties of mulberry leaves through interaction between sericulturists and JICA. The project has been successful. It achieved some major breakthroughs like rearing silkworms using the 'shoot rearing' technique. This method has helped to reduce the number of feeding from four to two. The worms are fed the entire shoot instead of merely the leaves. Since Bihar is more or less a new entrant, it should adopt the latest technology so as to be the leader in Silk production. The producers will need technological, marketing and credit support. Distt. Bhagalpur offers great potential for silk development. Intervention through DIPP/VSE schemes could provide the needed help.

The Central Silk Board (CSB) has also taken an initiative in developing new equipment to help the spread of bivoltine sericulture.

These include the 'rotary mountages', which helps to increase cocoon production compared to the conventional chandrike mountage. The Kanchi silk is beautiful, sensuous and graceful. But, the units here are facing a crisis situation. There has been a rise in the price of zari used for weaving silk sarees. This pushed up the cost of sarees, and sales declined by 40 per cent in the festival season of 2006. The wavers are demanding a 30 per cent rise in wages. Of course, the price of zari can come down if the state withdraws sales tax.

A project aimed at promoting organic sericulture is slowly gaining ground in key silk producing states of South India. This is another area where Bihar can take initiative. Organic sericulture focuses on organically cultivating mulberry and use of material like sunnhemp, a fibre crop which can be used for paper and pulp production or as green manure. Given the falling soil fertility levels, organic farming can enhance both production and productivity. An acre of land requires about 10 kg. of sunnhemp. One kg. costs about Rs. 12. About 10 kg. can fix nitrogen equivalent to Rs. 1,000 per acre. The green manure helps to reduce consumption of nitrogenous fertilizer up to 25 per cent in the very first year of use. Bangalore-based Central Sericulture Technological Research Institute (CSTRI) has experimented with natural dyes and is thinking of even commercialsing the product. Experts say that the real challenge lies in identifying and developing more natural dyes. Besides lac, natural dyes have been produced using a host of materials like *neem* and even *amla* (Gooseberry).

APICULTURE

Evolution of bees and of plants they pollinate has provided reciprocal influence ever since the origin of angiosperms. The available records suggest that honey bees have functioned as cross-pollinators extensively on geological time-scale much before the primitive man started domestication of wild edible plants or selection of varieties suited to his own choice. Though bees may not play such an important role as in the case of self-pollinated or wind-pollinated crops, but significant increases in the yields can be obtained by harnessing the services of bees insect-pollinated crops. In achieving better performances of some of the pulses, oilseeds, vegetables, fruits and orchard crops, their role may not be trivial. Beekeeping has already developed in many advanced countries as an essential component of mixed farming like dairy, poultry and other occupations.

Honey is a sweet, viscous, golden coloured liquid that is "manufactured" by different species of honey bees, such *as Apis mellifera*, from the nectar of flowers. It is stored in wax combs and used by the bees as a food source to feed both immature larval stages and adult bees. Approximately 120,000 bees are required to produce 1 kg. of honey and a single bee will produce only about a tablespoon of honey during its lifetime.

Honey is also the oldest sweetener known to humans and has been an eagerly sought commodity since prehistoric times. Primitive societies hunted for trees where bees stored honey so that they could obtain a supply of the previous sweetener. Eventually beekeeping or apiculture, the science and art of raising bees in artificial hives, where honey can easily be harvested evolved and is a major agricultural industry today. Modern beekeeping not only provides a supply of a natural sweetener, but more importantly makes large number of bees available when agricultural crops that depend on insect pollination for production of fruits, nuts and seed are in flower. This is an important agricultural industry because about one-third of the human diet is derived from insect pollinated plants, and honeybees are responsible for 80 per cent of the pollination.

Beekeeping (apiary) is an age-old tradition in India as well as in Bihar and its activity is valuable which can provide food, cash income and employment to a subsistence economy. In India beekeeping is practiced in about four lakh villages and provides part time employment to 2.41 lakh persons. With the advancement of technology, emerging trade prospects and wider scope for their use as pollinating agent, honeybees offer scope to adopt them for entrepreneurship. Diversity of bee flora and varied agro-climatic conditions in Bihar offers enormous potentialities for profitable beekeeping. The importance of honey bee is normally associated with the production of honey and wax, but it performs another very major function from the point of view of crop production, i.e. pollination.[7]

Apple orchadists of certain apple growing areas such as Kullu, Manali, and Rohru in Himachal, for example, are now making extensive use of both indigenous and exotic honey bees in pollination. This is because in the event of declining population of cross pollinating agents, apple trees have started showing symptoms of bearing less fruits resulting in low productivity. To arrest this phenomenon, apiculture is becoming popular. As a result of this and many more been development activities, Himachal Pradesh has not only witnessed a phenomenal increase in the number of bee colonies from 580 in 1971 to 56,626 in 2004-05, but the honey production has equally increased from 1.3 to 1,547 tonnes, giving an overall compound growth rate (CGR) of 28.31 per cent per annum over the same period.

The primary object of modern apiculture should be to secure higher crop yields, and honey and wax will automatically come as a by product. Needless to say, honey production generates employment for the rural people, and can be an additional source of income to the farming community. India represents as one of the potential beekeeping areas in the world, having all the different agro climatic zones ranging from tropical, subtropical, intermediate and temperate to arid ones. The flora required for beekeeping is available round the year and provides sufficient raw material in the form of pollen and nectar to honeybees for production of honey and other products of commercial value. All this makes it best fit for both migratory and non-migratory beekeeping. Beekeeping has gained importance as an agricultural activity and has been inextricably linked with rural life. Beekeeping is a form of animal husbandry that deals in scientific method of conservation and rearing of bees for honey, beevenom, beeswax, pollen, royal jelly and pollination of crop plants. Beekeeping had been a traditional household occupation in our society for its natural boons of yielding nutritive honey and other products. Beekeeping shows an increases in a farmer's income by nearly 50 per cent under favourable conditions; it requires little capital outlay, no land, and is not labour intensive. It is particularly suited as an income-generating activity for women, young people, and the landless.

Beekeeping has thus the potential in Bihar to spin wealth through increase in quality and quantity of crop produced by pollination services, hive rentals and honey and other bee products production. Beekeeping should be encouraged since apart from improving economic status of the farmers, it adds to the aesthetics of the environment. We must not allow the vast resources of bee flora to go waste rather it should be used to provide food, nutrition, income and employment to small farmers, the landless, women and children who can easily do beekeeping, and particularly educated unemployed youth who take up beekeeping on a large scale and act as a locus of developing beekeeping in their respective areas.

Apiculture is not new to Bihar. Bihar is one of the leading producer of apiary honey in the country. Due to lack of knowledge on scientific honey extraction, farm level processing, branding and marketing, the farmers are unable to reap the full benefits of apiculture. This is hindering large scale expansion of this potential activity. Districts of Muzaffarpur, Vaishali, Sitamarhi, East and West Champaran, Madhepura, Katihar and Begusarai are the main honey production districts. Approximately 3,900 MT of honey was produced in the state out of total country's production of 8,400 MT in 2002. Bihar is the only state producing litchi honey on commercial scale, and it should rank at par with some of the premium honey in world market such as the Block Forest Honey of Germany and the Spanish Orange Honey in terms of its uniqueness and quality. Honey yield with the Italian honeybee species is the highest in Bihar as compared to other states with a production rate of 40 and 60 kg. honey/hive/year under stationary and migratory bee keeping respectively. The processing industry has not kept pace with the increase in beekeeping in the state with no large processing unit within the state. There are handful of processing units in the organized sector with most of the processing in the unorganized sector—Mirzanagar Gramodyog Samiti being the main one.

Beekeeping has a tremendous potential in Bihar. Taking the acreage of litchi as about 24,000 ha., Bihar can support 2,40,000 bee colonies on Litchi alone. The crops of Mango, Guava, Mustard, and Rapeseed can help the bee colonies develop for the Litchi honey flow. In the world market the demand for honey is around one million tonnes. There is an immense possibility for India to increase its export share from 7,000 tonnes to three lakh tonnes and Bihar can lead the way to capture this market. At a conservative estimate, given the existing cropping pattern and area under different crops suitable for beekeeping, Bihar has the potential to produce about 40,000 MT of honey with an estimated value of Rs. 200 crore annually and creation of employments for more than 6,00,000 persons.

Note

1. The discussion here is based on the data provided by the National Horticulture Board (NHB) in their publication Indian Horticulture Data Base. While estimates for fruit production are very near by NHB and Government of Bihar (GB) there is a wide difference in these two sources—This being 13.35 million by NHB during 2004-05 and 7.3 million by GB for 2005-06. respectively of national production. Area under vegetables in 2004-05 reached.

2. *The Hindu*, March 13, 2008.
3. Bihar Agriculture Building on Emerging Models of "Success", World Bank.
4. N.R. Kumar and N.K. Pandey, "Indian Institute of Vegetable Research", Varanasi, Uttar Pradesh 'Division of Social Science, Central Potato Research Institute, Shimla, Himachal Pradesh, Constraints in Adoption of Improved Potato Production Technologies in Bihar, *Agriculture Economic Research Review*, Volume. 21, 2008.
5. Dr. C.P. Thakur, *Bihar's Development,* by 2015, 1996.
6. For more details refer to Dr. C.P Thakur, *Bihar Development* by 2015.
7. *Indian Farming*, May 2007.

ANNEXURE 6.1

Vegetable Production in Bihar (Area in Hectares, Production in MT, Productivity in MT/Hect.)

Name of vegetables	2003-04			2004-05			2005-06			2006-07
	Area	Production	Productivity	Area	Production	Productivity	Area	Production	Productivity	Estimate Area
(1)	(2)	(3)	(4)	(5)	(6)	(7)	(8)	(9)	(10)	(11)
Cauliflower	59726	955616	16	59701	955216	16	59778	938515	15.7	59867
Cabbage	36509	584144	16	36513	598813	16.4	36602	578312	15.8	36649
Onion	48309	917871	19	48759	975180	20	49837	1011691	20.3	50108
Tomato	45260	588380	13	45987	735792	16	46028	727242	15.8	46171
Chillies	38278	459336	12	38070	456840	12	38134	442354	11.6	38312
Brinjal	54096	1081920	20	29143	582860	20	53697	1030982	19.2	53807
Okra (*Bhindi*)	58045	928720	16	53651	1073020	20	56232	714146	12.7	56295
Bottle-gourd	25205	40328	16	56173	730249	13	29213	566732	19.4	29318
Sponge-gourd	33321	466494	14	33606	470484	14	33684	458102	13.6	33808
Ridge-gourd	7749	46494	6	8078	48468	6	8096	46957	5.8	8114
Bitter-gourd	8484	50904	6	8424	50544	6	8508	49346	5.8	8604
Pomted-gourd	4594	45940	10	4594	45940	10	4642	47813	10.3	4710
Cowpea	11751	70506	6	11582	69492	6	11636	86106	7.4	11723
Others	61049	976784	16	59685	835590	14	59742	955872		59862
Total	492376	7213437		493966	7628488		495829	7654170		497348

Source: Department of Agriculture, Government of Bihar

ANNEXURE 6.2

Production of Major Fruit Crops in Bihar 2005-06

Name of Dostrict	*Share of districts in total Fruit Production (In Per cent)*								
	Mango	*Guava*	*Pine-apple*	*Coconut*	*Banana*	*Lemon*	*Litchi*	*Other Crops*	*Total*
					Zone 1				
Saran	2.95	2.82		0.54	2.40	2.83	3.10	3.82	2.66
Siwan	1.56	2.30		0.64	1.98	2.70	3.26	3.60	1.96
Gopalgamj	1.69	1.96		0.12	1.85	2.19	3.50	2.65	2.82
E. Champaran	5.62	6.40	2.73	1.70	2.72	8.85	6.34	3.30	4.51
W. Champaran	4.76	5.94	3.03	1.30	2.46	9.68	5.38	4.02	4.10
Muzaffarpur	7.27	5.45		3.42	18.11	3.87	26.33	4.21	10.84
Sitamarhi	4.05	2.53		1.09	1.98	1.66	7.03	2.35	3.04
Sheohar	1.84	1.09		0.53	0.85	0.91	3.11	1.48	1.40
Vaishali	6.43	4.74		2.33	10.68	3.45	13.03	4.08	7.34
Darbhanga	9.63	2.23		3.42	6.07	4.89	2.81	4.26	6.49
Madhubani	4.70	1.67		3.11	3.22	4.04	2.44	4.00	3.62
Samastipur	7.72	2.14		2.48	7.34	4.82	3.85	4.11	6.15
Begusarai	2.98	1.83		1.82	2.94	2.50	1.74	3.44	2.69
Total	**61.20**	**41.09**	**5.75**	**22.51**	**62.60**	**52.39**	**81.91**	**45.33**	**45.33**
					Zone 2				
Purnia	2.04	1.33	33.28	15.33	3.02	2.65	4.20	3.31	4.12
Araria	0.37	0.60	4.54	5.72	1.30	1.66	1.17	1.65	1.22
Kishanganj	0.45	0.70	34.56	5.56	2.06	1.57	1.15	2.60	2.56
Katihar	2.37	1.54	4.04	6.81	2.20	1.80	4.86	3.56	2.73
Saharsa	1.95	2.35	9.63	15.28	3.64	3.53	1.40	3.17	3.38
Madhepura	1.51	2.25	8.19	1104	3.80	2.68	0.70	2.72	2.93
Supaul	0.92	1.48		6.64	1.76	1.27	0.35	1.32	1.41
Khagaria	1.15	1.25		1.92	2.29	1.37	0.99	2.11	1.57
Total	**10.76**	**11.51**	**94.25**	**68.30**	**20.07**	**16.53**	**14.82**	**20.44**	**20.44**
					Zone 3A				
Munger	0.72	0.76		0.54	.091	1.20	0.63	1.30	0.81
Shekhpura	0.51	0.40		0.31	0.40	0.44	0.27	0.63	0.44
Lakhisarai	0.32	0.36		0.11	0.35	0.35	0.13	0.43	0.31
Jamui	0.61	0.62		0.36	0.65	0.87	0.49	1.35	0.65
Bhagalpur	5.67	2.47		1.49	3.88	5.70	1.60	3.51	4.14
Banka	3.97	1.01		0.60	2.09	2.57	0.15	1.81	2.48
Total	**11.78**	**5.63**		**3.40**	**8.28**	**11.13**	**3.26**	**9.04**	**9.04**

(Contd.)

(*Contd.*)

Name of Dostrict	*Share of districts in total Fruit Production (In Per cent)*								
	Mango	*Guava*	*Pine-apple*	*Coconut*	*Banana*	*Lemon*	*Litchi*	*Other Crops*	*Total*
				Zone 3B					
Patna	2.36	3.59		1.83	1.74	3.11		4.72	2.226819
Nalanda	1.63	4.85		0.40	1.28	2.02		3.69	1.705075
Bhojpur	3.22	6.40		1.05	0.97	2.17		2.88	2.282361
Buxar	2.20	4.95		0.41	0.61	1.59		1.60	1.541268
Rohtas	3.09	10.38		0.27	0.81	1.81		1.72	2.293889
Kaimur	1.70	4.14		0.12	0.60	1.28		1.26	1.2452
Gaya	0.63	2.12		0.13	0.62	1.89		2.58	0.847979
Jehanabad	0.13	0.77		0.14	0.40	0.87		1.23	0.357154
Arwal	0.12	0.67		0.09	0.29	0.59		0.87	0.271226
Nawada	0.58	1.74		0.04	0.88	2.23		2.40	0.908722
Aurangabad	0.59	2.16		0.50	0.85	2.41		2.24	0.908722
Total	**16.26**	**41.77**		**4.98**	**9.05**	**19.95**		**25.19**	**25.19**

ANNEXURE 6.3

Percentage Share of each district in total area of important fruits in the State (2003-04)

District	Mango	Guava	Litchi	Lemon	Banana	Pin-apple	Coconut	Other Fruit	Total
Patna	2.64	3.68	0.15	3.20	1.92	0.00	2.13	4.76	2.62
Nalanda	1.86	5.16	0.00	2.26	1.42	0.00	0.55	3.92	2.10
Bhojpur	3.19	6.72	0.05	2.28	1.02	0.00	1.43	3.08	2.81
Buxar	2.30	5.28	0.00	1.73	0.66	0.00	0.55	1.66	1.98
Rohtas	3.98	11.59	0.00	1.98	0.96	0.00	0.38	1.99	3.46
Kalmur	2.32	4.83	0.00	1.46	0.71	0.00	0.19	1.50	1.90
Gaya	0.79	2.21	0.00	2.08	0.75	0.00	0.17	3.06	1.11
Jehanabad	0.15	0.77	0.00	0.90	0.46	0.00	0.19	1.29	0.39
Arwal	0.14	0.65	0.00	0.56	0.30	0.00	0.12	0.90	0.29
Nawada	0.73	1.78	0.00	2.52	1.10	0.00	0.06	2.75	1.06
Aurangabad	0.76	2.22	0.00	2.79	1.09	0.00	0.68	2.78	1.17
Saran	3.53	2.82	3.53	3.11	2.21	0.00	0.49	4.09	3.16
Siwan	1.66	2.33	3.72	3.06	2.46	0.00	0.64	4.00	2.25
Gopalganj	2.02	1.96	4.00	2.41	2.36	0.00	1.26	2.83	2.28
E. Champaran	6.46	5.85	6.28	9.45	2.95	2.96	2.14	`3.47	5.63
W. Champarna	4.90	5.56	5.02	9.46	2.49	3.25	1.64	3.77	4.69
Muzaffarpur	6.82	4.91	25.02	3.61	17.37	0.00	3.78	3.94	8.67
Sitamarhi	3.66	2.53	7.22	1.57	2.06	0.00	1.32	2.28	3.31
Sheohar	1.81	1.08	3.33	0.88	0.86	0.00	0.62	1.44	1.62
Vashali	5.76	4.38	12.15	3.19	9.63	0.00	2.63	3.62	6.00
Bhagalpur	5.14	2.29	1.60	5.47	3.68	0.00	1.88	3.28	3.97
Banka	4.39	1.06	0.17	2.76	2.30	0.00	0.82	1.91	2.86
Munger	0.74	0.73	0.69	1.06	0.80	0.00	0.68	1.18	0.79
Sheikhpura	0.53	0.41	0.30	0.46	0.42	0.00	0.42	0.62	0.48
Lakhisarai	0.33	0.36	0.15	0.36	0.36	0.00	0.15	0.43	0.31
Jamui	0.66	0.67	0.56	0.92	0.67	0.00	0.49	1.42	0.73
Khagaria	1.10	1.21	0.94	1.27	2.05	0.00	2.12	2.02	1.33
Darbhanga	9.15	2.09	2.72	4.33	6.12	0.00	3.75	3.97	6.33
Madhubani	4.20	1.61	2.72	3.76	3.31	0.00	3.41	3.94	3.57
Samastipur	7.33	1.96	3.59	4.15	6.59	0.00	2.73	3.66	5.47
Begusarai	2.79	1.65	1.73	2.17	3.02	0.00	2.19	3.45	2.56
Purnia	1.61	1.17	4.11	2.28	2.90	35.01	12.97	3.05	3.20
Kishanganj	0.47	0.68	1.28	1.48	2.09	31.67	6.31	2.61	1.76
Katihar	1.85	1.46	4.74	1.55	1.93	4.25	6.51	3.19	2.50
Saharsa	1.70	2.18	1.52	3.47	3.63	9.75	13.09	2.87	2.85
Madhepura	1.32	2.13	0.83	2.84	4.02	8.09	9.50	2.44	2.33
Supaul	0.81	1.45	0.57	1.59	1.96	0.00	5.27	1.24	1.27

Source: Based on data collected from Government of Bihar, Patna.

ANNEXURE 6.4

Annual Compound Growth Rate of Area under Various Fruits by State (1992-93 over 2003-04)

States/UT	*Apple*	*Banana*	*Citrus*	*Grape*	*Guava*	*Litchi*	*Mango*	*Papaya*	*Pine-apple*	*Sapota*	*Others*	*Total*
Andhra Pradesh	0.00	1.65	20.00	-2.79	4.77	0.00	5.10	39.12	0.00	10.47	-14.84	6.06
Arunachal Pradesh	3.32	9.61	0.00	0.00	0.00	0.00	0.00	0.00	7.64	0.00	-36.63	6.23
Assam	0.00	0.53	8.67	0.00	2.85	0.25	16.55	2.72	0.00	0.00	-22.13	0.68
Bihar	0.00	1.36	-0.42	0.00	0.82	3.94	-0.53	-6.97	4.25	0.00	4.50	0.84
Chhattis-garh	0.00	0.00	0.00	0.00	0.00	0.00	0.00	0.00	0.00	0.00	0.00	0.00
Delhi	0.00	0.00	0.00	0.00	0.00	0.00	0.00	0.00	0.00	0.00	-4.83	-14.90
Goa	0.00	1.78	0.00	0.00	0.00	0.00	1.39	0.00	-4.18	0.00	-4.24	-1.10
Gujarat	0.00	2.31	10.80	0.00	5.65	0.00	7.47	0.00	0.00	12.85	12.20	7.97
Haryana	0.00	0.00	4.86	0.00	11.06	0.00	5.62	0.00	0.00	0.00	9.82	7.28
Himachal Pradesh	2.74	0.00	0.00	0.10	-1.90	8.86	2.66	5.88	0.00	0.00	-1.37	3.14
Jammu & Kashmir	3.04	0.00	2.73	1.64	0.00	0.00	1.79	0.00	0.00	0.00	-22.94	-424
Jharkhand	0.00	0.00	0.00	0.00	0.00	0.00	0.00	0.00	0.00	0.00	0.00	0.00
Karnataka	0.00	2.84	0.00	5.21	-2.69	0.00	3.63	-1.11	-3.76	3.77	-8.98	1.24
Kerala	0.00	8.27	0.00	0.00	0.00	0.00	1.15	2.92	8.40	0.00	0.00	-7.06
Madhya Pradesh	0.00	-2.18	25.51	-18.46	-7.04	0.00	-10.57	-4.98	0.00	0.00	-13.02	-3.56
Maharashtra	0.00	0.95	0.00	6.52	11.78	0.00	12.75	0.00	0.00	18.51	-2.01	7.14
Manipur	0.00	-5.43	31.57	0.00	1.55	0.00	-5.17	31.81	5.09	0.00	-17.80	2.58
Meghalaya	0.00	-0.57	0.00	0.00	0.00	0.00	0.00	2.08	1.07	0.00	0.00	-4.93
Mizoram	0.00	1.90	38.62	0.00	10.83	69.86	3.57	2.19	-0.12	0.00	-12.74	5.41
Nagaland	-4.57	-1.52	8.23	0.00	1.92	8.31	19.62	63.91	7.00	0.00	-3.18	5.13
Orissa	0.00	-4.45	13.29	0.00	8.57	14.25	7.46	0.10	5.95	30.35	1.63	4.61
Punjab	0.00	0.00	-6.76	-5.36	2.49	-3.27	-7.66	0.00	0.00	0.00	-8.04	-6.16
Rajasthan	0.00	0.00	0.00	0.00	0.46	0.00	-1.16	1.76	0.00	0.00	-12.15	0.74
Sikkim	0.00	0.00	0.00	0.00	0.00	0.00	0.00	0.00	0.00	0.00	-22.47	2.42
Tamil Nadu	0.00	2.66	5.06	1.01	7.30	0.00	6.80	0.00	5.51	0.00	1.42	5.02
Tripura	0.00	3.23	-3.35	0.00	-9.83	-18.29	-4.47	0.00	-0.02	0.00	-4.52	-4.95
Uttaranchal	-16.17	0.00	-13.69	0.00	0.00	-10.83	26.33	0.00	0.00	0.00	-14.91	-10.89
Uttar Pradesh	0.00	4.28	-2.40	0.00	-0.87	-11.75	0.38	-2.70	0.00	0.00	-11.88	-0.84
West Bengal	0.00	3.77	0.00	0.00	12.75	6.88	1.87	4.27	2.86	0.00	-9.19	1.30
Andaman Nicobar	0.00	3.29	0.00	0.00	0.00	0.00	8.77	1.17	0.00	0.00	34.49	8.09
Chandigarh	0.00	0.00	0.00	0.00	0.00	0.00	0.00	0.00	0.00	0.00	10.76	-2.21
Dadar & Nagar Haveli	0.00	0.00	0.00	0.00	0.00	0.00	0.00	0.00	0.00	0.00	-0.11	-0.11
Daman & Diu	0.00	0.00	0.00	0.00	0.00	0.00	-1.21	0.00	0.00	0.25	3.93	0.46
Lakshadwep	0.00	-1.56	0.00	0.00	0.00	0.00	0.00	1.52	0.00	0.00	25.28	3.46
Pondicherry	0.00	1.03	0.00	0.00	12.07	0.00	4.94	10.00	0.00	18.71	-8.79	1.20
All India	0.10	1.84	4.29	4.13	3.30	0.10	3.63	3.65	2.99	10.59	-5.48	1.68

ANNEXURE 6.5

Annual Compound Growth Rate of Production of Fruits by State (1992-93 over 2003-04)

States/UT	*Apple*	*Banana*	*Citrus*	*Grape*	*Guava*	*Litchi*	*Mango*	*Papaya*	*Pine-apple*	*Sapota*	*Others*	*Total*
Andhra Pradesh	0.00	1.00	9.91	-5.08	4.82	0.00	0.93	75.18	0.00	10.47	4.39	4.50
Arunachal Pradesh	-0.99	5.59	10.85	0.00	0.00	0.00	0.00	0.00	9.49	0.00	0.00	5.22
Assam	0.00	0.91	3.35	0.00	2.74	4.59	19.97	2.02	1.43	0.00	-26.67	0.21
Bihar	0.00	-0.89	-2.63	0.00	2.68	5.35	-3.35	-10.61	2.90	0.00	2.92	-1.00
Chhattis-garh	0.00	0.00	0.00	0.00	0.00	0.00	0.00	0.00	0.00	0.00	0.00	0.00
Delhi	0.00	0.00	-12.72	-5.17	0.00	0.00	0.00	0.00	0.00	0.00	-8.73	-13.41
Goa	0.00	4.05	0.00	0.00	0.00	0.00	-7.61	0.00	-3.90	0.00	0.77	-1.92
Gujarat	0.00	2.28	5.79	0.00	1.09	0.00	3.83	0.00	0.00	10.49	8.03	4.44
Haryana	0.00	0.00	6.14	-8.16	12.88	0.00	4.40	0.00	0.00	0.00	9.38	6.93
Himachal Pradesh	2.24	0.00	5.56	0.00	6.25	23.14	5.12	-13.89	0.00	0.00	16.32	3.99
Jammu & Kashmir	3.61	0.00	-1.95	-2.79	0.00	0.00	-7.67	0.00	0.00	0.00	-17.03	2.26
Jharkhand	0.00	0.00	0.00	0.00	0.00	0.00	0.00	0.00	0.00	0.00	0.00	0.00
Karnataka	0.00	0.89	0.54	1.00	1.26	0.00	4.63	3.84	-2.16	-1.78	-0.09	1.55
Kerala	0.00	-0.44	0.00	0.00	0.00	0.00	2.19	2.54	5.97	0.00	0.00	-8.02
Madhya Pradesh	0.00	-0.78	5.30	-1.31	-7.13	0.00	-10.57	1.61	0.00	0.00	-5.73	-1.41
Maharashtra	0.00	2.85	14.04	11.38	15.18	0.00	6.52	0.00	0.00	22.96	11.97	6.89
Manipur	0.00	5.22	10.29	0.00	7.65	0.00	7.62	21.15	2.47	0.00	-8.45	3.27
Meghalaya	0.00	-0.49	0.00	0.00	0.00	0.00	0.00	2.08	1.40	0.00	0.00	-4.06
Mizoram	0.00	0.92	2.69	0.00	10.67	79.64	0.61	-4.61	-4.30	0.00	6.20	2.33
Nagaland	-6.02	27.16	14.49	0.00	25.89	5.42	50.17	54.06	26.23	0.00	8.16	21.74
Orissa	0.00	-2.16	8.84	0.00	11.21	10.44	2.99	5.80	1.75	3.87	5.92	3.67
Punjab	0.00	0.00	-1.94	-5.20	8.31	1.63	-2.76	0.00	0.00	0.00	-4.90	-1.80
Rajasthan	0.00	-11.34	6.59	-12.19	11.11	0.00	6.19	9.06	0.00	0.00	7.96	6.92
Sikkim	0.00	0.00	-7.19	0.00	0.00	0.00	0.00	0.00	0.00	0.00	29.57	-8.19
Tamil Nadu	0.00	5.42	-19.21	-2.68	-5.08	0.00	8.02	0.00	2.93	0.00	22.73	5.41
Tripura	0.00	9.27	-6.72	0.00	-5.44	4.58	-18.82	0.00	9.95	0.00	4.22	3.52
Uttaranchal	-15.89	0.00	7.18	0.00	0.00	2.54	6.02	0.00	0.00	0.00	3.84	-0.08
Uttar Pradesh	0.00	9.94	-1.77	0.00	-2.18	-10.13	8.37	3.60	0.00	0.00	-22.85	4.08
West Bengal	0.00	9.66	13.53	0.00	10.57	14.29	3.49	5.59	4.35	0.00	-5.78	4.51
Andaman Nicobar	0.00	-4.21	0.00	0.00	0.00	0.00	9.00	-3.30	0.00	0.00	0.00	0.22
Chandigarh	0.00	0.00	7.73	0.00	12.79	5.24	2.92	0.00	0.00	0.00	-0.58	3.33
Dadar & Nagar Haveli	0.00	0.00	0.00	0.00	0.00	0.00	0.00	0.00	0.00	0.00	1.75	0.00
Daman & Diu	0.00	5.24	0.00	0.00	0.00	0.00	0.00	0.00	0.00	0.00	-12.67	0.73
Lakshadwep	0.00	7.12	0.00	0.00	0.00	0.00	0.00	8.81	0.00	0.00	39.18	11.10
Pondicherry	0.00	4.25	-19.88	0.00	17.35	0.00	5.33	0.00	0.00	13.62	-1.33	3.93
All India	1.44	2.43	6.66	6.69	4.06	6.20	3.28	10.33	3.15	8.00	-1.13	3.21

ANNEXURE 6.6

Annual Compound Growth Rate of Major Vegetables Crops by State (1992-93 over 2003-04)

States/UT	Brinjal	Cabbage	C. Flower	Okra	Peas	Tomato	Onion	Potato	S. Potato	Tapioca	Others	Total
Andhra Pradesh	1.24	1.92	0.00	0.89	-9.88	7.02	3.70	10.26	-3.07	-2.36	3.95	3.69
Arunachal Pradesh	0.00	0.00	0.00	0.00	0.00	0.00	0.00	-2.36	0.00	0.00	10.96	1.82
Assam	-1.80	2.01	0.97	-1.69	-4.10	1.32	2.03	1.93	0.00	0.00	11.09	2.14
Bihar	-1.66	-0.59	-1.54	-2.54	-1.79	-1.57	-12.12	-8.92	0.00	0.00	-0.84	-4.08
Chhattisgarh	0.00	0.00	0.00	0.00	0.00	0.00	0.00	0.00	0.00	0.00	0.00	0.00
Delhi	0.00	0.00	0.00	-5.94	-14.42	-9.56	-24.64	-26.34	0.00	0.00	1.59	-0.83
Goa	0.00	0.00	0.00	31.80	0.00	0.00	0.00	0.00	0.00	0.00	0.00	66.29
Gujarat	9.45	7.33	-15.22	18.34	0.00	13.53	1.96	9.75	0.00	0.00	12.40	6.42
Haryana	9.44	-1.95	0.00	11.01	0.71	11.75	17.12	2.00	0.00	0.00	10.26	9.25
Himachal Pradesh	5.27	4.08	9.83	0.00	0.00	10.69	-2.92	-1.43	0.00	0.00	-6.08	1.54
Jammu & Kashmir	0.00	0.00	0.00	0.00	0.00	0.00	0.00	0.00	0.00	0.00	-31.23	-19.67
Jharkhand	0.00	0.00	0.00	0.00	0.00	0.00	0.00	0.00	0.00	0.00	0.00	0.00
Karnataka	-3.81	0.00	0.00	-3.59	0.88	1.03	6.11	3.95	-11.98	0.00	-6.89	-1.83
Kerala	-22.32	0.00	0.00	-6.73	0.00	0.00	0.00	0.00	0.00	-2.76	0.00	-6.30
Madhya Pradesh	0.00	0.00	0.00	0.00	0.00	0.00	3.06	0.58	-21.85	0.00	1.79	-4.85
Maharashtra	3.84	5.33	7.64	0.00	0.00	3.89	3.76	6.24	9.39	0.00	12.75	8.06
Manipur	14.31	5.11	3.08	0.00	6.17	11.03	0.00	20.44	0.00	0.00	5.90	10.76
Meghalaya	0.00	4.01	4.54	0.00	7.71	17.81	8.91	-0.11	0.00	0.00	1.75	4.64
Mizoram	0.00	38.50	4.38	-1.30	-2.01	12.49	54.45	-1.49	0.00	0.00	-21.37	-11.04
Nagaland	41.49	-0.59	0.00	0.00	38.74	51.65	63.46	-3.67	30.28	0.00	-37.22	-2.10
Orissa	0.00	-2.57	-0.21	-1.11	-3.12	9.69	-23.62	-6.05	0.00	0.00	-10.11	-2.30
Punjab	4.24	0.00	9.25	8.36	0.00	5.01	-11.61	4.09	0.00	0.00	3.58	5.52

(Contd.)

(*Contd.*)

States/UT	Brinjal	Cabbage	C. Flower	Okra	Peas	Tomato	Onion	Potato	S. Potato	Tapioca	Others	Total
Rajasthan	0.99	0.00	0.00	3.72	17.58	1.48	6.36	4.66	-7.56	0.00	0.51	4.74
Sikkim	0.00	0.00	0.00	0.00	0.00	0.00	0.00	2.68	0.00	0.00	25.44	5.94
Tamil Nadu	0.00	2.37	11.55	7.49	77.25	3.21	1.70	-1.74	0.00	0.00	16.72	13.45
Tripura	0.00	0.00	0.00	0.00	0.00	0.00	0.00	6.62	0.00	0.00	-1.67	-0.12
Uttaranchal	-0.48	-4.33	-9.96	0.00	2.08	-4.18	-9.24	-0.13	0.00	0.00	-1.75	-1.32
Uttar Pradesh	0.00	0.00	0.00	0.00	0.00	0.84	-3.87	1.68	-1.92	0.00	2.36	4.12
West Bengal	5.82	8.11	7.28	2.92	23.33	10.61	0.00	4.80	0.00	0.00	11.93	7.89
Andaman Nicobar	0.00	0.00	0.00	0.00	0.00	0.00	0.00	0.00	0.00	0.00	-5.81	0.00
Chandigarh	0.00	0.00	0.00	0.00	0.00	0.00	0.00	0.00	0.00	0.00	10.12	-10.45
Dadar & Nagar Haveli	0.00	0.00	0.00	0.00	0.00	0.00	0.00	0.00	0.00	0.00	29.15	29.15
Daman & Diu	0.00	0.00	0.00	0.00	0.00	0.00	0.00	0.00	0.00	0.00	32.40	14.74
Lakshadwep	0.00	0.00	0.00	0.00	0.00	0.00	0.00	0.00	0.00	0.00	-2.01	-2.01
Pondicherry	4.00	0.00	0.00	1.30	0.00	-10.26	0.00	0.00	0.00	0.00	8.61	6.36
All India	7.30	0.90	1.52	1.75	5.16	5.01	1.22	0.66	6.19	2.76	0.93	1.99

ANNEXURE 6.7

Percentage Share of Various Vegetables in Total area under Vegetables by District (2003-04)

District	Cauli-flower	Cabbage	Onion	Tomato	Chilli	Brinjal	Lady-finger	Pumpkin	Ridge Gourd	Jhingani	Bitter Gourd	Parwal	Bodi	Other
Patna	22.83	13.86	15.41	13.81	17.13	12.08	19.38	3.02	2.21	1.14	1.25	1.31	1.32	14.32
Nalanda	13.88	7.66	15.60	9.40	19.00	40.16	13.65	2.14	1.69	1.00	1.23	0.44	1.73	11.67
Bhojpur	14.99	9.55	17.96	16.76	9.47	16.69	24.41	4.71	2.99	1.88	2.75	1.45	1.91	11.77
Buxar	11.73	10.14	19.27	17.65	9.73	14.93	24.35	6.36	4.07	2.03	2.34	1.03	2.47	10.94
Rohtas	20.65	9.30	20.14	12.70	9.55	16.47	23.56	4.93	3.44	1.58	1.80	0.82	1.48	12.29
Kaimur	17.61	9.80	19.70	15.09	9.39	17.66	21.32	5.60	3.84	1.45	1.54	0.64	1.99	12.51
Gaya	19.07	7.56	14.80	10.53	15.24	17.46	19.74	7.17	5.45	2.73	1.96	0.44	1.56	12.70
Jehanabad	15.63	9.87	15.31	27.60	11.12	20.85	17.77	4.83	3.90	1.43	1.67	0.81	2.10	6.82
Arwal	18.62	10.60	13.47	21.47	13.63	19.07	19.30	5.66	4.17	1.53	2.24	0.88	2.28	6.13
Nawada	18.55	10.15	13.43	10.77	12.24	16.92	31.18	6.27	4.42	1.42	1.00	0.39	1.70	10.99
Aurangabad	16.46	8.53	12.83	31.29	8.36	14.91	23.04	5.31	3.82	0.97	1.01	0.42	1.49	11.42
Saran	20.75	10.30	9.22	15.24	11.80	18.98	22.21	5.67	5.58	1.34	0.93	1.51	1.48	12.80
Siwan	18.75	8.29	9.52	16.89	13.16	18.69	23.70	6.00	5.27	1.08	1.03	0.75	1.60	13.19
Gopalganj	18.52	10.18	7.96	18.74	10.08	13.81	18.73	17.87	3.52	1.17	1.10	0.67	1.47	10.31
E. Champaran	15.12	12.45	18.71	13.89	8.93	13.52	18.07	8.04	7.24	1.06	1.45	0.88	1.78	14.99
W. Champaran	20.32	11.48	15.35	14.35	10.60	14.51	16.09	7.99	6.33	1.07	1.36	1.01	1.82	14.47
Muzaffarpur	19.52	14.57	13.02	18.91	9.62	14.96	14.71	6.91	5.92	1.25	1.25	0.87	1.76	13.98
Sitamarhi	15.86	11.15	15.91	18.93	8.88	15.17	16.53	8.63	7.50	1.35	0.80	0.23	0.96	14.54

(Contd.)

(*Contd.*)

District	Cauli-flower	Cabbage	Onion	Tomato	Chilli	Brinjal	Lady-finger	Pumpkin	Ridge Gourd	Jhingani	Bitter Gourd	Parwal	Bodi	Other
Sheohar	19.11	12.88	15.43	17.07	9.64	15.87	16.02	8.62	6.68	1.48	0.88	0.37	1.39	13.10
Vaishali	24.17	11.72	7.55	18.34	8.37	15.93	18.26	6.14	6.89	1.42	1.43	1.57	2.03	12.67
Bhagalpur	6.06	4.96	6.50	6.51	3.60	6.92	9.23	4.23	7.77	2.17	1.59	0.82	1.62	14.32
Banka	7.13	5.15	6.86	6.10	4.49	7.89	8.88	3.60	5.91	1.54	1.41	0.52	2.55	12.60
Munger	4.37	4.30	8.30	3.89	10.86	6.37	9.55	4.09	6.56	1.74	1.73	2.37	2.16	8.69
Sheikhpura	4.54	3.48	9.68	3.29	15.41	5.95	6.38	4.06	5.80	2.35	2.01	1.55	2.77	7.36
Lakhisarai	5.16	4.45	5.34	4.36	12.80	4.95	7.51	3.47	8.22	2.73	2.71	2.97	3.39	10.00
Jamui	5.90	4.51	7.71	4.49	10.57	6.59	8.77	3.65	7.00	1.38	1.43	1.22	2.40	8.86
Khagaria	5.09	4.23	5.29	7.02	6.00	9.49	11.31	2.41	4.68	0.95	0.87	1.44	1.84	11.44
Darbhanga	12.57	5.06	3.26	4.21	4.81	7.40	6.23	3.51	9.52	1.48	1.52	0.43	4.96	11.44
Madhubani	9.84	5.78	4.03	4.59	7.50	7.73	7.07	2.30	7.49	1.61	1.78	0.53	1.41	14.01
Samastipur	10.64	6.41	4.04	5.35	4.11	7.77	6.82	3.35	6.94	1.14	1.37	0.86	1.54	15.89
Begusarai	6.11	3.61	7.07	7.18	3.90	9.69	9.44	3.53	8.03	0.88	0.87	0.88	1.47	12.19
Purnia	7.71	4.27	6.98	3.58	5.56	4.25	4.81	3.79	9.45	1.79	2.05	1.19	2.54	21.84
Araria	6.66	4.29	12.11	6.14	3.72	5.03	6.37	2.86	7.41	2.30	2.58	0.88	2.16	14.04
Kishanganj	6.93	4.29	11.02	7.39	4.49	4.23	6.33	3.81	7.28	2.28	1.74	0.44	2.08	12.42
Katihar	7.17	4.76	10.68	3.50	4.86	4.55	4.68	4.05	7.61	1.70	2.71	1.15	5.10	15.44
Saharsa	8.24	4.21	2.75	5.45	2.99	6.63	6.20	6.21	11.82	3.15	3.63	1.97	4.94	10.88
Madhepura	6.49	6.47	3.58	5.27	2.76	6.23	4.74	4.79	12.12	2.57	4.21	1.68	5.36	13.74
Supaul	6.49	4.56	3.14	5.92	3.50	6.30	8.82	4.54	11.66	1.94	3.07	1.38	5.64	12.10

ANNEXURE 6.8

Percentage Share of each district in state total area of important vegetables (2003-04)

District	Cauli-flower	Cabbage	Onion	Tomato	Chilli	Brinjal	Lady-finger	Pumpkin	Ridge Gourd	Jhingani	Bitter Gourd	Parwal	Bodi	Other	Total
Patna	8.49	8.55	7.37	6.62	9.95	4.87	7.42	2.86	1.06	2.25	2.29	4.08	1.75	3.19	2.02
Nalanda	6.89	6.31	9.95	6.02	14.74	21.62	6.97	2.70	1.08	2.62	3.00	1.85	3.05	3.46	4.03
Bhojpur	2.28	2.40	3.50	3.28	2.25	2.75	3.81	1.82	0.59	1.51	2.05	1.85	1.03	1.07	1.23
Buxar	3.21	1.73	2.55	2.35	1.57	1.67	2.58	1.67	0.54	1.10	1.19	0.89	0.90	0.67	0.84
Rohtas	3.16	2.36	3.96	2.50	2.28	2.73	3.71	1.92	0.68	1.28	1.36	1.05	0.80	1.13	1.24
Kaimur	1.78	1.64	2.55	1.96	1.48	1.93	2.21	1.44	0.50	0.77	0.76	0.54	0.71	0.76	0.82
Gaya	4.35	2.85	4.33	3.09	5.43	4.31	4.63	4.15	1.60	3.29	2.19	0.84	1.26	1.73	1.85
Jehanabad	1.08	1.13	1.36	2.46	1.20	1.56	1.26	0.85	0.35	0.52	0.57	0.47	0.51	0.28	0.56
Arwal	0.79	0.75	0.74	1.18	0.91	0.88	0.85	0.61	0.23	0.35	0.47	0.32	0.35	0.16	0.35
Nawada	3.39	3.08	3.16	2.54	3.50	3.36	5.87	2.92	1.04	1.37	0.90	0.59	1.10	1.20	1.48
Aurangabad	3.48	2.99	3.48	8.52	2.76	3.41	5.01	2.85	1.04	1.08	1.05	0.75	1.12	1.44	1.71
Saran	4.89	4.02	2.79	4.63	4.34	4.85	5.39	3.39	1.69	1.67	1.07	2.98	1.23	1.80	1.91
Siwan	3.91	2.87	2.55	4.54	4.29	4.22	5.08	3.18	1.42	1.19	1.06	1.31	1.18	1.64	1.69
Gopalganj	4.59	4.18	2.53	5.98	3.90	3.71	4.77	11.25	1.12	1.53	1.34	1.38	1.29	1.53	2.01
E. Champaran	4.61	6.30	7.33	5.46	4.25	4.47	5.67	6.24	2.85	1.70	2.18	2.24	1.92	2.73	2.48
W. Champaran	6.76	6.34	6.57	6.16	5.51	5.24	5.51	6.77	2.72	1.89	2.23	2.81	2.15	2.88	2.70
Muzaffarpur	8.98	11.11	7.70	11.21	6.91	7.46	6.96	8.08	3.51	3.05	2.83	3.36	2.86	3.85	3.73
Sitamarhi	3.11	3.63	4.01	4.79	2.72	3.23	3.34	4.31	1.90	1.40	0.78	0.37	0.67	1.71	1.59
Sheohar	2.02	2.25	2.09	2.32	1.59	1.82	1.74	2.31	0.91	0.82	0.46	0.33	0.52	0.83	0.86

(Contd.)

(*Contd.*)

District	Cauli flower	Cabbage	Onion	Tomato	Chilli	Brinjai	Lady finger	Pumpkin	Ridge Gourd	Jhingani	Bitter Gourd	Parwal	Bodi	Other	Total
Vaishali	11.45	9.20	4.59	11.19	6.19	8.18	8.90	7.39	4.21	3.55	3.32	6.21	3.41	3.59	3.84
Bhagalpur	2.24	3.04	3.08	3.10	2.08	2.77	3.51	3.97	3.70	4.24	2.89	2.54	2.12	3.16	3.00
Banka	1.23	1.47	1.51	1.35	1.21	1.47	1.57	1.57	1.31	1.40	1.19	0.75	1.55	1.30	1.39
Munger	0.70	1.15	1.72	0.81	2.73	1.11	1.58	1.67	1.36	1.48	1.37	3.19	1.23	0.84	1.31
Sheikhpura	0.31	0.39	0.85	0.29	1.64	0.44	0.45	0.70	0.51	0.85	0.67	0.88	0.67	0.30	0.55
Lakhisarai	0.23	0.33	0.31	0.25	0.90	0.24	0.35	0.40	0.48	0.65	0.60	1.12	0.54	0.27	0.37
Jamui	0.78	0.99	1.31	0.77	2.19	0.95	1.20	1.23	1.20	0.96	0.93	1.35	1.12	0.70	1.07
Khagaria	1.34	1.84	1.78	2.37	2.46	2.70	3.05	1.61	1.58	1.32	1.13	3.17	1.71	1.85	2.13
Darbhanga	6.73	4.49	2.24	2.90	4.02	4.29	3.43	4.78	6.57	4.18	4.00	1.92	9.40	3.66	4.34
Madhubani	4.24	4.13	2.23	2.55	5.05	3.61	3.13	2.52	4.16	3.66	3.78	1.90	2.15	3.61	3.50
Samastipur	4.66	4.66	2.27	3.02	2.81	3.69	3.07	3.73	3.92	2.64	2.96	3.17	2.40	4.16	3.55
Begusarai	2.77	2.71	4.12	4.20	2.76	4.76	4.40	4.06	4.69	2.12	1.93	3.36	2.37	3.30	3.68
Purnia	3.10	2.85	3.61	1.86	3.49	1.86	1.99	3.88	4.90	3.80	4.06	4.02	3.62	5.25	3.26
Araria	1.31	1.40	3.06	1.55	1.14	1.07	1.29	1.43	1.88	2.39	2.49	1.45	1.50	1.65	1.59
Kishanganj	1.40	1.43	2.85	1.92	1.41	0.92	1.31	1.95	1.89	2.43	1.72	0.74	1.49	1.62	1.63
Katihar	4.47	4.92	8.54	2.81	4.73	3.07	3.00	6.40	6.11	5.58	8.29	5.98	11.25	5.75	5.05
Saharsa	2.61	2.21	1.12	2.22	1.48	2.27	2.02	5.00	4.82	5.26	5.65	5.21	5.54	2.06	2.57
Madhepura	2.66	4.40	1.89	2.78	1.77	2.77	2.00	4.99	6.40	5.57	8.50	5.77	7.78	3.37	3.32
Supaul	0.98	1.14	0.61	1.15	0.83	1.03	1.37	1.74	2.27	1.55	2.28	1.74	3.02	1.09	1.23

ANNEXURE 6.9

Annual Compound growth rates of Production of Vegetables (2002-03) over 1992-93) by State

States/UT	*Brinjal*	*Cabbage*	*C. Flower*	*Okra*	*Peas*	*Tomato*	*Onion*	*Potato*	*S. Potato*	*Tapioca*	*Others*	*Total*
Andhra Pradesh	8.53	11.90	0.00	-1.34	-10.82	7.02	9.67	8.85	0.89	-4.54	4.51	5.59
Arunachal Pradesh	0.00	0.00	0.00	0.00	0.00	0.00	0.00	236.97	0.00	0.00	-3.29	0.19
Assam	-3.20	2.44	2.52	0.00	-1.09	2.88	3.28	4.29	0.00	0.00	9.96	3.46
Bihar	-4.69	-1.71	-0.75	1.75	-4.36	-0.08	-18.21	-12.44	0.00	0.00	-6.76	-6.48
Chhattisgarh	0.00	0.00	0.00	0.00	0.00	0.00	0.00	0.00	0.00	0.00	0.00	0.00
Delhi	0.00	0.00	0.00	-7.83	-11.63	-8.94	-30.89	-23.58	0.00	0.00	2.82	-1.56
Goa	0.00	0.00	0.00	61.41	0.00	0.00	0.00	0.00	0.00	0.00	0.00	71.05
Gujarat	8.14	5.03	4.97	20.25	0.00	13.57	4.04	7.25	0.00	0.00	19.21	8.50
Haryana	8.39	-3.41	0.00	4.91	-3.35	5.39	12.92	3.11	0.00	0.00	8.86	7.14
Himachal Pradesh	5.40	5.95	12.92	5.46	4.05	15.77	-16.78	2.64	0.00	0.00	1.01	5.00
Jammu & Kashmir	0.00	0.00	0.00	0.00	0.00	0.00	0.00	0.00	0.00	0.00	-19.12	-7.74
Jharkhand	0.00	0.00	0.00	0.00	0.00	0.00	0.00	0.00	0.00	0.00	0.00	0.00
Karnataka	-2.66	0.00	0.00	-1.48	2.20	4.33	-8.53	-13.17	-15.74	-14.32	2.35	-2.26
Kerala	-29.23	0.00	0.00	8.12	0.00	0.00	0.00	0.00	0.00	-0.59	0.00	-1.30
Madhya Pradesh	0.00	0.00	0.00	0.00	0.00	0.00	4.54	1.06	0.53	0.00	10.18	-1.13
Maharashtra	1.17	4.75	9.29	10.19	0.00	4.20	-2.51	-9.09	7.82	0.00	12.88	2.94
Manipur	31.95	4.59	1.92	11.13	7.02	9.60	0.00	10.03	0.00	0.00	8.74	7.04
Meghalaya	0.00	10.59	5.63	0.00	6.68	25.51	7.29	-0.54	0.00	0.00	2.83	3.61
Mizoram	0.00	42.48	-0.70	14.66	0.15	10.51	30.46	-4.14	0.00	0.00	-11.83	-7.60

(*Contd.*)

(*Contd.*)

States/UT	Brinjal	Cabbage	C. Flower	Okra	Peas	Tomato	Onion	Potato	S. Potato	Tapioca	Others	Total
Nagaland	30.90	-5.94	0.00	0.00	55.86	40.91	66.12	-2.30	51.57	0.00	-32.16	1.62
Orissa	0.00	3.79	-7.90	-1.39	-2.21	10.79	-23.60	-5.21	0.00	0.00	-9.45	-0.83
Punjab	3.27	0.00	7.70	7.43	0.00	4.69	-9.39	5.35	0.00	0.00	1.03	4.77
Rajasthan	1.19	0.30	0.00	1.45	4.23	-0.97	2.73	1.06	-13.18	0.00	2.25	1.71
Sikkim	0.00	0.00	0.00	0.00	0.00	0.00	0.00	-2.77	0.00	0.00	17.64	1.33
Tamil Nadu	0.00	7.31	5.52	3.12	80.40	-4.37	1.89	-4.56	0.00	0.00	17.74	15.98
Tripura	0.00	0.00	0.00	0.00	0.00	0.00	0.00	5.94	0.00	0.00	-0.61	1.17
Uttaranchal	13.02	10.41	5.23	8.87	2.69	2.80	-11.09	-9.04	0.00	0.00	-1.69	-2.98
Uttar Pradesh	0.00	-8.36	-21.11	0.00	3.82	21.34	-5.97	5.93	-0.25	0.00	25.06	5.89
West Bengal	3.28	5.83	4.67	4.80	15.67	3.41	0.00	3.40	0.00	0.00	-1.30	2.96
Andaman Nicobar	0.00	0.00	0.00	0.00	0.00	0.00	0.00	0.00	0.00	0.00	-1.54	1.11
Chandigarh	0.00	0.00	0.00	0.00	0.00	0.00	0.00	0.00	0.00	0.00	23.32	-11.09
Dadar & Nagar Haveli	0.00	0.00	0.00	0.00	0.00	0.00	0.00	0.00	0.00	0.00	-0.04	-0.04
Daman & Diu	0.00	0.00	0.00	0.00	0.00	8.93	0.00	0.00	0.00	0.00	-1.81	14.26
Lakshadwep	0.00	0.00	0.00	0.00	0.00	0.00	0.00	0.00	0.00	0.00	4.81	4.81
Pondicherry	10.83	0.00	0.00	6.34	0.00	-5.55	0.00	0.00	0.00	0.00	13.68	10.54
All India	4.43	2.15	0.52	2.70	3.29	4.52	-3.02	1.34	8.82	6.78	0.52	1.79

7

LIVESTOCK

Total population of livestock in Bihar during 2003 was 27.16 million, which included cattle 10.7 million, buffalo 5.1 million and goat 9.5 million (Table 7.1). Further, the total number of poultry was 13.97 million.

Table 7.1: Total Number of Livestock Population in Bihar, 2003

(Thousands)

	Nos.000'	*Per cent*
Cattle	10729	39.5
Buffaloes	5743	21.1
Sheep	382	1.4
Goats	9490	34.9
Pigs	672	2.5
Horse and Ponies	117	0.4
Mules	4	0.0
Donkeys	23	0.1
Total	27162	100

Source: Basic Animal Husbandry Statistics, Ministry of Agriculture.

Bihar's livestock sector is crucial not only in terms of its contribution to rural income, but also for the section of the population to which this income goes. The sector in Bihar accounted for approximately one-quarter of the total value of agricultural output in TE 2002-03. The activity is concentrated among landless households and those with marginal holdings of less than 1 hectare of land. Approximately 60 per cent of rural households in Bihar own cattle, 35 per cent buffalo, and 31 per cent sheep and goats (NSSO 2003). Of all rural households owning cattle and/or buffalo in Bihar, more than three-quarters are either landless or have less than 1 hectare of land. Sheep and goats tend to be even more concentrated among landless and marginal rural households. (Table 7.2).

Table 7.2: Livestock Ownership Based on Size Class of Operational Holding in Bihar (2003).

Size class of operational holding (ha.)	No. of households (000)	No. of Households with			% of Households with			% share of total Households with		
		cattle (000)	Buffaloes (000)	S&G ('000)	cattle	Buffaloes	S&G	cattle	buffaloes	S&G
<=0.04	5082	759	428	747	14.9	8.4	14.7	18.8	17.8	42.7
0.04-1	5238	2371	1410	842	45.3	26.9	16.1	58.7	58.7	48.2
1-2	943	622	410	124	65.9	43.5	13.1	15.4	17.1	7.1
2-4	329	224	115	32	68.0	34.8	9.6	5.5	4.8	1.8
>4	80	66	37	5	82.4	46.9	6.1	1.6	1.6	0.3
All	11673	4042	2400	1749	34.6	20.6	15	100	100	100

S & G—Sheep and Goat

Source: World Bank Report and NSS 59th Round, 2003.

During 1990-91 to 2001, livestock population in the state decreased at the rate of—0.10 per cent. In the 1970's and 80's the annual growth rates of livestock were 1.23 and 2.44 per cent respectively (Appendix Table 7.1). The livestock population per one million population was 740 as against the national average of 617 during 2000-01 Appendix Table 7.2. But the quality and production of livestock in the state is reported to be poor. Total number of crossbred cattle per lakh population in Bihar was only 0.32 as against 2.55 in the country. Punjab, Haryana, Jammu and Kashmir, Karnataka, Himachal Pradesh, Tamil Nadu have comparatively more number of crossbred cattle.

However, per capita goat population in the state was comparatively much higher than the national average. Marginal farmers and landless labourers keep mainly goat for supplementary income. Looking at the economics of goat rearing (Table 7.3), it is advisable that programmes relating to goat development are given high priority in the livestock sector. Appendix Tables 7.3 and 7.4 give district-wise growth rates and per cent shares of livestock in Bihar (Annexure 7.3). There is drastic change in the growth pattern of livestock production in the last 5-7 years. Milk is the most important livestock product in Bihar, accounting for approximately 50 per cent of the total livestock output, followed by meat (24%) and other livestock products. (Table 3.16). During the 1990's milk production in the state had grown at an annual compound rate of 0.29 per cent only, while poultry meat production grew at the rate of 0.24 per cent per annum. The states of Bihar, Madhya Pradesh and West Bengal had the lowest growth Rate of Milk production during this period. (Table 7.4).

Table 7.3: Economics of Different Livestock Enterprises.

Item	*Dairy Cow Unit*	*Dairy Buffalo Unit*	*Goat Unit*	*Sheep Unit*
1. Capital Investment (Rs.)	19,500	24,500	37,000	18,200
2. Fixed Costs (Rs.)	8,125	9,900	12,439	5,855
3. Variable Cost (Rs.)	10,889	11,092	23,532	3,290
4. Total Costs (Rs.)	19,014	20,992	35,971	9,145
5. Gross Income (Rs.)	27,800	32,580	65,000	13,889
6. Net Income (Rs.)	8,786	11,588	29,029	4,744
7. Investment Recovery period (Years)	3	2	2	4
8. Input Output Ratio	1:1.46	1:1.55	1:1.81	1:1.31
9. Fixed Cost Output Ratio	1:3.42	1:3.29	1:5.23	1:2.37
10. Variable Cost Output Ratio	1:2.55	1:2.94	1:2.76	1:4.22

Table 7.4: Annual Compound Growth Rates of Milk and Poultry Meat in Various States-During 1991-2001.

(Percentage)

State	Milk		Poultry	
	Growth	Rank	Growth	Rank
Andhra Pradesh	0.678	4	0.439	7
Assam	0.508	8	0.216	11
Bihar	0.290	16	0.243	10
Gujarat	0.524	6	0.563	5
Haryana	0.400	13	0.160	13
Himachal Pradesh	0.467	11	0.530	6
Jammu and Kashmir	1.138	1	0.430	8
Karnataka	0.947	2	0.572	4
Kerala	0.633	5	-0.352	15
Madhya Pradesh	0.276	17	-1.058	16
Maharashtra	0.497	10	0.182	12
Orissa	0.833	3	0.717	3
Punjab	0.503	9	-1.059	17
Rajasthan	0.405	12	0.898	1
Tamil Nadu	0.389	14	0.773	2
Uttar Pradesh	0.509	7	0.248	9
West Bengal	0.369	15	-0.222	14
All India	0.377			

Institutional Infrastructure

The overall sorry state of affairs of the livestock sector can be explained by the fact that the number of institutions providing extension services to the livestock sector have remained stagnant over the years. (Table 7.5).

Table 7.5: Number of Institutions in Animal Husbandry Department, Bihar.

Sl. No	Item	Based on Livestock Census 1982		Based on Livestock Census 2003			Growth over the period (in %)
		2001-02	2002-03	2003-04	2004-05	2005-06	
1.	No. of Artificial Insemination Institutions	1401	1401	1401	1401	1401	0.00
2.	No. of Veterinary hospitals	39	39	39	39	39	0.00
3.	No. of veterinary dispensaries	814	814	814	814	814	0.00

Source: Animal Husbandry and Fisheries Department GOB.

There are 851 veterinary hospitals in the state, out of which 39 are sub-division level veterinary hospitals, 783 are first class hospitals and another 29 travelling veterinary hospitals. These institutions play an important role in prevention and treatment of animal diseases in the state. There has been an effort to strengthen and modernize these institutions. Of these 851 hospitals, a total of 352 are being run either in rented accommodation or in dilapidated buildings. Further, while there are 38 districts in the state, there are only 29 animal husbandry districts and 9 animal husbandry divisions in the state. An important decision has been taken to create 9 new veterinary districts and one new division to deepen the work of the department. As regards piggery, there is no pig breeding farm or bacon factory in the Government sector. Goat and Piggery activity is financed only under Government sponsored programmes like SGSY. There is only 1 Animal Health and Production Institute and 4 frozen semen Banks in addition to the above.

According to the 2003 livestock census, there were approximately 10 million heads of cattle in Bihar, only 10 per cent of which were cross-bred. It is a pity that there is no cold storage facility for any of the livestock products like fish, meat or milk. Livestock products being perishable commodities, cold storage facilities are extremely necessary. Though Bihar has one of the largest livestock population in the country, there is only one meat producing and processing unit in the state. Very little amount of the meat and poultry products is scientifically produced, processed and packaged. With growing urbanization and quality consciousness, the demand for quality livestock products is developing rapidly in the national and international markets. The state should strive hard to tap these opportunities.

Growth Rates

Despite all this sorry state of affairs, milk production has more than doubled (106%) over the period 2001-2007. With base as 2000-01, when production was only 2489 thousand tonnes (Table 7.6), the growth will calculate to 119 per cent. Meat production grew by only 12 per cent over the same period. (Table 7.6). The only explanation for this happy situation would appear to be the impact of dairy cooperatives which provided the necessary services.

Table 7.6: Production of Livestock Products in Bihar (2001-06)

Sl. No.	Item	Based on		Based on		Based on	
		2001-02	2002-03	2003-04	2004-05	2005-06	2006-07
1.	Total milk (in 000, litres)*	2632	2869	3175	4743	5060	5450
2.	Eggs (in crores)	74	74	78	79	100	94.55
3.	Meat (in 000, tonnes)	156	173	173	176	175	—
4.	Wool (in lakh kg.)	4.24	3.62	3.94	3.78	2.2	2.31

Source: Economic Survey, Government of Bihar, Animal Husbandry Department GOB, 2008.

Note:* This should actually be '000' tonnes.

Organic Production

Small and marginal farms cannot sustain themselves. But introduction of fee-range and organic goods farming are showing that money can be made by growing specialty products that consumers are willing to pay more. French farm in Gurgaon, Roger Langbour, for example, raises thousands of free-range Peking and Muscovy ducks on feed that is free of pesticide and antibiotics. On an early morning visit, white ducks sit placidly on the ground in a large enclosure with wire fencing. Elsewhere turkeys and even a small number of quail and pheasants are pecked at the ground. Langbour who started the three acre farm 14 years ago, has opened the duck market in India. Now he also raises pigs, and sells his products to five-star hotels, expatriates and wealthy Indians, garnering strong weekly sales worth around $ 8500. Another recent start-up, Heritage Health Food, delivers boxed organic food grown on 80 acres of leased land. Livestock sector within agriculture is the one that is going to be the next big thing. It is an un organised sector there's a lot of opportunity.

The two-year-old firm makes 200 to 300 deliveries a week, priced at between Rs. 250 to 500, depending on weight, and the company expects to start turning a profit by this year 2008. Recent studies have shown that if you want to achieve growth, it can be through high-value products and not through cereals alone. They include fruits, vegetables, milk, poultry products and fish. Organic farming as an integrated system of production using local inputs and recycling of bio-mass is ideally suited for livestock products.

CATTLE DEVELOPMENT PROGRAMME IN BIHAR

Ministry of Rural Development, Government of India had sanctioned in February 2004 a "Programme for Generation of Rural Employment and Sustainable Livelihoods through Cattle Development in Bihar". The programme has a project period of 5 years and aims at production of high-producing animals thereby leading to higher milk production which will provide additional income to the poor farmers of the state. The programme is being monitored by NABARD and implemented by BAIF (a Pune based NGO) in 13 districts viz., Bhojpur, Buxar, Patna, Banka, Siwan, Saran, Vaishali, Begusarai, Lakhisarai, Gopalganj, Muzaffarpur, Samastipur and Munger. One hundred Cattle Development Centres (CDC) have been opened in these districts for providing services such as Artificial Insemination (AI), Pregnancy Diagnosis, deworming and supply of mineral supplements. Each CDC covers around 1500-2000 breedable cattle population in order to make them more productive through better breeding and feeding.

Constraints

- Low productivity of existing livestock and large number of unproductive animals which are a drain on the feed and fodder resources.
- Shortage of veterinary health clinics and breeding centres.
- Inadequate extension support to farmers and entrepreneurs.
- Limited availability of good quality animals.
- Problem in procurement of day old chicks within the State due to the absence of adequate number of hatcheries.
- Poor road connectivity as well as erratic power position in the districts pose a hindrance in the establishment of layer farms and hatcheries.
- Absence of modern abattoirs for wholesome meat production and efficient utilization of slaughter house by products.
- Reluctance of bankers to extend necessary credit support to the entrepreneurs.

PROCESSING INDUSTRY

Over and above the local demand, milk is now being exported to the neighbouring States during flush period. Bihar Government Eleventh

Plan proposes to increase capacity of existing plants and setting up new plants at strategic locations. Four new dairy plants with 2 lakh litres capacities are proposed to be set up at Biharsharif, Purnea, Chhapra and Mohania. There is also a proposal to set up a 30 MT per day powder plant and ultra-heat treatment (URT) plant with septic packaging facilities to conserve surplus solids during flush seasons for use when milk supply is low. The aim is to supply packaged milk to milk deficit areas like Madhubani and Darbhanga within the state and the north eastern states.

At the same time, Government of Bihar is planning to produce clean milk with total quality Management—ISO and HACCP certification. To achieve these objectives, the State XIth Plan is starting at the village level to improve the quality of raw materials, apart from processing, packaging and handling. Quality with cost control will be the key challenges for dairy plants in the next plan. Farmers will be encouraged to produce clean milk by maintaining hygiene at their doorsteps. Incentives will be given to them through dairy cooperatives for hygienically better quality milk.

Bihar should also take advantage of the new technologies developed by the NDRI and encourage the private sector to set up such industries at suitable places in Bihar to take advantage of value addition of surplus milk during the flush period and create new employment facilities.

The National Dairy Research Institute (NDRI) is engaged in preparing health promoting dairy products. They are producing 16 dairy items, which take into consideration all health norms. The institute has thus prepared Arjuna Ghee which has functionalities and regulating properties to create resistance against heart diseases and blood pressure. The developed ghee has been found sensorily similar to the market ghee. The Arjuna Ghee has been found to be four times more stable to oxidative deterioration as compared to the control ghee. Thus, Arjuna Ghee could be consumed instead of normal ghee in order to reduce the risk of CVD and to boost up immune system of the body.

Likewise, a process has been developed for preparation of low cholesterol ghee. Through this process, ghee with 80-90 per cent low cholesterol could be prepared. Probiotic dahi has also been manufactured by NDRI. This has multiple health benefits like hypecholestrolemic, hypotensive, immunomodulatory, antidiabetic and antioxidant. NDRI has developed a technology to manufacture dietetic burfi by reducing fat level to about half of the conventional burfi and completely substituting sugar with permitted artificial sweeteners. Shelf life studies have revealed that vaccum package dietetic burfi could be stored without spoilage for 12 days at 30°C and for 40 days at 5°C They have standardized the technology for the manufacture of probiotic Edam Cheese. Animal feeding trials of this cheese has demonstrated intestinal implantation of the bifidobacteria, higher body weight, 65 per cent decrease in faecal coliforms and reduction in activity of Beta-glucuronidase. This cheese is an excellent carrier of beneficial microbial sultures. Ice cream market in India was growing at an impressive rate of 15-20 per cent per annum. Relatively high calorie content of ice cream has often been a cause of avoiding it for many health conscious consumers. As a result of awareness of the harmful effect of excess fat on human health, there has been increasing demand for a low fat and low calorie product.[1]

Another important activity of the NDRI relates to transfer of technology. The institute has transferred technologies relating to tea/coffee whiteners, whey based soups and drinks, mozzarella cheese, processed cheese, cheese spread, paneer etc., directly to the industry and entrepreneurs. NDRI has another feather in its cap when it entered into an agreement with the National Research Development Corporation (NRDC), New Delhi in respect of commercialization of technologies developed by the institute. Several technologies have thus been assigned to NRDC for transfer to the user industry/entrepreneurs. Long life paneer and herbal ghee technologies have already been transferred through NRDC for commercial exploitation. A few other technologies are also being processed along these lines. These include low cholesterol ghee and acidowhey.

Reliance Industries

To facilitate this type of high level processing activity, Bihar can approach a company like Reliance which has already entered this area. Reliance Industries which have been working on plans to enter the country's Rs. 40.000 crore branded dairy sector over the past two years, has kicked off the venture by foraying into the lucrative but low-margin liquid milk category. Reliance Retail has begun testing liquid milk, branded Dairy Pure, in Hyderabad. Unlike other FMCC products, margins in liquid milk are very low—at just 3-5 per cent. The biggest challenge for players operating in this sector is to work out efficiencies in logistics. It has tied up with south-based dairy operator. The liquid milk has been rolled out in four variants at a pricing structure that pits RIL's brand against other competing liquid milk brands in the state such as Mother Dairy. Vijaya and a host of other private players. However, Gujarat Milk Marketing Cooperative Federation (GMMCF)'s Amul, which leads the country's Rs. 15,000 crore branded packaged milk market, does not have a presence in Andhra Pradesh. In fact, Amul liquid milk is not yet present in the southern states.

White Flood

- RIL has signed up with Punjab Government for milk procurement. It plans to buy 7 lakh litres every day from 1.5 lakh farmers.
- The company also plans to support milk cooperatives in U.P. and Bihar, but no agreement with cooperatives has been finalized so far.
- Reliance Retail has planned 5,000 collection centres in U.P. and Bihar. It is working on a private label to make deeper inroads into daily sector.

Future Outlook

Since Reliance Company is already having plans to support Dairy Cooperatives in Bihar, one could expect another Anand and Amul in Bihar. With some improvement in the breeding programme along with feed facilities, there is full justification for milk production to increase by 100-200 per cent during the XI and coming Plans.

DAIRYING

A study of milk production during 1991 to 2001 at the all India level shows that Bihar had the lowest growth rate, being second from bottom with Madhya Pradesh being at the bottom (Table 7.3). One of the reasons for this condition as already discussed could be the very low position of the number of cows per lakh of population 0.32 as against 2.55 at the all India level. Recent developments in dairying, however, show that Bihar is endowed with a large potential of dairy development. This can be gauged from the fact that milk production in the State has more than doubled during the last 5 years. Only around 15 per cent of milk is marketed through the organized sector (cooperatives and private dairy) and during the flush season, the excess milk available in the State is sent to the neighbouring States of West Bengal, Assam and Uttar Pradesh. The total bovine population in the state as per 2003 Census is as in Table 7.7.

Table 7.7: Total Bovine Population in Bihar.

Sl. No.	*Type of Animal*	*Number of Animals ('000)*
	COWS	
1.	HF	169.9
2.	Jersey	586.3
3.	Other CB cows	443.0
4.	Improved Desi cows (Tharparkar, Shahabadi, Bachchaur)	324.6
5.	Indigenous	8946.5
	Total	10470.2
	BUFFALOES	
6.	Murrah	311.7
7.	Indigenous (local)	5432
	Sub-total	5743
	Total Bovine (Cows + Buffaloes)	16235.9

Source: NABARD Report, 2007.

These numbers are much higher than those given by the Animal Husbandry Dept. Milk production in Bihar is going up steadily (Table 7.8). This is primarily due to new policy under which there is a reasonable increase in the number of cross bred cows—over 34 per cent in two years, and productivity increase in yield levels (over 43%) over the same period. Even in the case of non-descript cows, productivity levels increased by over 31 per cent. On the other hand, productivity levels of buffaloes have not shown any improvement (Table 7.8). This has resulted in the growth of per capita availability 80 to 100 grams/day in a period of 3 years—no doubt far below the all India level of around 250 grams a day (Table 7.9). All this growth is demand driven with a very high income elasticity of demand for milk 2.36 for rural and 1.21 for urban areas (Table 7.10).

Despite all this increase in milk production and productivity, present levels are much below than those achieved in Punjab, Gujarat, Kerala and Uttar Pradesh and all India. (Table 7.11). With Economic Survey estimates of 5 million tonnes milk production in 2005-06 and 5.5 million in 2006-07, productivity level of milk in Bihar has gone up appreciably.

Table 7.8: Milk Production in Bihar.

		2000-01	2001-02	2002-03	2003-04
Milk Production ('000 tonnes)		2489	2664	2869	3180
Cows	C.B.	99	96	125	157
	N.D.	651	698	763	854
Total		750	794	888	1011
Buffalo		1460	1565	1649	1824
Goat		279	305	332	345
Total		2489	2664	2869	3180
No. of Animal (in Milk) ('000 Nos.)					
Cows	C.B.	55.2	52.3	61.9	74.2
	N.D.	1093.0	1154.5	1218.3	1288.3
Average yieid Kgs. Per/day					
	C.B.	4.915	5.038	5.527	5.78
	N.D.	1.633	1.657	1.715	1.81
Buffalo (in Milk) ('000 Nos.)		1194.3	1228.6	1298.4	1365.00
Yield kgs/day		3.350	3.489	3.479	3.65
Production ('000 tonnes)		1460	1565	1649	1824.00
Goat ('000 Nos.)		6016.2	6380.0	6587.4	6646.00
Yield kgs/day		0.127	0.131	0.138	0.14
Production ('000 tonnes)		279	305	332	345.00
Per Capita availability gms/day		80	88	92	100.00

C.B. = Cross Breed
N.D. = Non-Descript
Source: Department of Animal Husbandry and Dairying, Ministry of Agriculture.

Table 7.9: Average Milk and Egg Production Per Lakh Population in India.

State	Year	Milk Production ('000t/lakh)	Egg ('000/lakh)
Andhra Pradesh	1991	5.77	97.81
Andhra Pradesh	2001	8.86	109.18
Assam	1991	3.18	81.53
Assam	2001	3.74	76.20
Bihar	1991	4.25	23.35
Bihar	2001	5.61	27.42
Gujarat	1991	12.62	19.88
Gujarat	2001	15.66	20.76
Haryana	1991	27.72	66.69
Haryana	2001	30.21	56.01
Himachal Pradesh	1991	12.5	15.18

(*Contd.*)

Table 7.9: (*Contd.*)

State	*Year*	*Milk Production ('000t/lakh)*	*Egg ('000/lakh)*
Himachal Pradesh	2001	14.43	15.41
Jammu and Kashmir	1991	8.45	75.93
Jammu and Kashmir	2001	13.69	69.90
Karnataka	1991	7.74	50.54
Karnataka	2001	14.29	57.07
Kerala	1991	7.55	92.65
Kerala	2001	11.24	71.11
Madhya Pradesh	1991	9.11	22.37
Madhya Pradesh	2001	12.65	15.09
Maharashtra	1991	7.71	62.76
Maharashtra	2001	9.58	56.30
Orissa	1991	1.81	46.77
Orissa	2001	2.67	56.93
Punjab	1991	36.69	124.97
Punjab	2001	47.68	62.75
Rajasthan	1991	12.88	8.79
Rajasthan	2001	14.2	9.88
Tamil Nadu	1991	8.96	65.56
Tamil Nadu	2001	11.12	87.67
Uttar Pradesh	1991	8.94	9.45
Uttar Pradesh	2001	12.11	8.86
West Bengal	1991	6.03	74.73
West Bengal	2001	6.92	56.46
All India	1991	8.58	47.36
All India	2001	10.75	44.17

Table 7.10: Income Elasticity of Demand for Various Consumption Goods in Bihar.

Commodity	*Rural*	*Urban*	*All*
Rice	0.37	0.11	0.33
Other Cereals	0.50	0.28	0.47
Milk	2.36	1.21	1.85
Animal Production	0.76	0.50	0.69
Vegetables	0.86	0.59	0.76
Fruits	2.06	1.55	2.00
Other Foods	1.18	1.15	1.20
Educ-Healths	1.41	1.62	1.82
Other Non-Foods	1.13	1.32	1.22

Source: Bihar Agriculture Building on Emerging Models of "Success", World Bank.

Table 7.11: Average Animal Productivity in KG/day in Selected Indian States, 2003-04

kg/day

State	Cross Breed Cattle	Indigenous Cattle	Buffaloes
Punjab	8.7	2.8	6.3
Gujarat	8.1	3.1	4.0
Kerala	7.0	2.6	6.2
Uttar Pradesh	6.1	2.4	4.3
Bihar	5.8	1.8	3.7
Orissa	5.5	0.5	2.6
All India	6.5	1.9	4.2

Source: Ministry of Agriculture, Department of Animal Husbandry.

Dairy Development in Bihar

The Dairy development activities in the State are taken care of by the Dairy Development Directorate and the Bihar State Co-operative Milk Producers' Federation (COMFED) in the non-operation flood and operation flood regions, respectively. The Dairy Development Directorate is the implementing agency for the Integrated Dairy Development Programme in the State. Apart from this it is also implementing a Special Component Programme for clean milk production under which training and inputs are provided to the farmers. The COMFED which is operating on the Anand pattern in the State has five affiliated milk producers' co-operative unions, namely Vaishali Patliputra Milk Union, Patna; Barauni Milk Union, Barauni; Tirhut Milk Union, Muzaffarpur, Mithila Milk Union, Samastipur and Shahabad Milk Union, Arrah. These five milk unions cover 6500 villages in 24 districts of the State.

Dairy development in Bihar is in a growth momentum. COMFED which came into existence in 1983, was the implementing agency of Operation Flood in Bihar (following the Anand model) and is the apex organization for dairy co-operative societies and unions. Today COMFED is the largest player in dairy business in the state and has succeeded in developing an integrated supply chain for liquid milk and other dairy products. The achievements of COMFED include outreach, increasing milk production and procurement, product and market leadership, provision of support services, increased income for producers, and social, economic, and political empowerment for its members.

More than 5123 District Cooperative Societies (DCS) have been formed in Bihar with a total membership of 2.54 lakh farmers—

more than any other cooperative in India. District Cooperative Society (DCS) at present covers 6750 villages although the average number of farmers per society is smaller in Bihar than in West Bengal and Orissa. It procures, processes and markets around 500 liters daily. The Social and gender decomposition of the DCS suggests that COMFED has been able to reach the poorest sections of the population. Looking at the success of Dairying Cooperatives, those Districts where such cooperatives are not there, farmers are clamouring for their establishment. Building on the success of the COMFED there should be further efforts to expand the milk catchment areas and also raise the number of supplies within the existing catchments.

Since milk the only commodity, which has to be collected twice a day, every day of the year, cooperatives are the only logical system for the dairy industry. About 85 per cent of the dairy industry in US is run by cooperatives. So is in Denmark and Autralia. No other system will work for milk. It is, therefore important for Bihar to strengthen the boon which is already there and enter the Dairy map of India like Anand. However, much remains to be achieved through improvements in animal productivity, quality of milk produced, and processed, and improvements in COMFED's transport, cooling and processing capacity, besides strengthening the institutions of artificial insemination, veterinary centre etc. Attention will have to be paid to these aspects as well.

Dairying is an important enterprise in Bihar, providing supplementary income, employment and nutrition to thousands of small rural households. The socio-economic impact of COMFED on its member farm households have been positive. Field studies have shown that on an average, cooperative member households have been better off than non-member households. First, it is estimated that there is 7.6 per cent higher number of milch animals with member households than that of non member household. Second, member households market nearly 15 per cent more milk than non member households. Finally, the annual net dairy income per households was comparatively higher than that of non member household, providing member households with additional income to allocate towards other operational and consumption expenses or savings.[2]

According to a plan being discussed by the Central Government after almost 36 years of Operation Flood coming into being, the dairy cooperative sector in the country is all set to take up yet another initiative to compete with private players. The National Dairy Development Board has proposed a Rs. 15,000 crore national dairy plan to strengthen dairy cooperative sector in the country. This plan is being chalked out after studies by the plan panel. To meet the challenges put forth by private players and to help produce 180 million tonne of milk by the year 2021—such a move is needed. In order to fulfil the demand for milk, production would have to be doubled with the help of village level milk unions. For the plan, a consortium has been formed, where National Bank for Agriculture and Rural Development (NABARD) will raise money and NDDB will provide knowledge and technical—based assistance. With Bihar Dairy Co-operatives being already strong, they will be able to enjoy maximum benefit under the programme.

The four basic components of livestock development are breeding, feeding, disease control and management. Although progress in cross breeding of cows is satisfactory, this is not in keeping with the requirements, because so far, milch animals are being procured from surplus, neighbouring states and distributed to the farmers, but little effort has been made to become self-sufficient in good crossbreeds for distribution. The Government should provide inputs and subsidies to farmers to set up breeding units in milk-sheds. Institutional finance will play a crucial role in this. Real issue will, however, be best to enhance the genetic potential of the largely indigenous breeds of cattle in Bihar through appropriate selection programme and/or cross breeding. There should be an institutional arrangement for providing inputs and subsidies to farmers for setting up breeding units in milk sheds. In fact, institutional finance should play a crucial role to alleviate the situation.

It is now appreciated that livestock wealth and its quality are crucially depended on breeding infrastructure available to livestock farmers in the state. And here artificial insemination centres are of key importance as their availability not only facilitates breeding, but also goes a long way in monitoring changes in the quality of livestock. As a step towards strengthening of livestock breeding programme, out of 1401 artificial insemination centres sanctioned in the state, 422 centres have been made functional. Further, these institutions are to be provided with internal technology and structural facilities and are to be strengthened under National Animal and Buffalo Breeding Programme being run with hundred per cent contribution from the central government. Efforts are being made to operationalise the rest of the centres. While 2.38 lakh animals were artificially inseminated in the year 2006-07, 78 thousand animals were inseminated till September 2007-08 (Table 7.12)

Table 7.12: Achievements of Livestock Breeding and Health Programmes

Sl. No.	*Work Type*	*2006-07*	*2007-08*
1.	No. of artificially inseminated animals (in lakh)	2.38	0.78
2.	Animals Treated (in lakh)	24.96	13.69
3.	Sterilisation (in lakh)	2.21	0.92
4.	Immunisation (in lakh)	4.41	29.38
5.	Free distribution of fodder seed (in Qntls)	177.45	146.89

Source: Animal Husbandry Department, Government of Bihar.

As suggested by ICAR, improved germ plasma for frieswal cattle can increase milk yield by 5 per cent. It is worth mentioning here that the state co-operative milk producers federation has made significant improvement in the dairy sector in some areas. Building on the success of COMFED, there should be further efforts to expand the milk catchment area and also raise the number of suppliers within the existing catchments. Besides, much remains to be achieved through improvement in animal productivity, quality of the milk produced and processed and improvement in COMFED's transport, cooling and processing capacity, aside from strengthening the institutions of artificial insemination, veterinary care, etc.

Buffalo milk yield at less than 4 kg per day is much lower than even that of West Bengal. This needs to be looked into. Low yielding animals are a drain on the scarce feed and fodder resources. Improving the quantity and quality of feed and fodder is also essential. This should be coupled with an effective extension system on animal husbandry management and improved access to veterinary services. We may quote here the example of Rajasthan where, with around 320 mm rainfall, small farmers have succeeded to supply enough green fodder for the dairy animals by practicing rain water harvesting. In Bihar where average annual rainfall even in the highly drought prone area is over 700 mm, appropriate rain water harvesting methods should go a long way to put more income in the hands of small/marginal farmers and also add to milk production in Bihar.

The breeding services in Bihar are provided by the State Department, AI centers of Dairy Cooperative Societies, the mobile cluster AI services of COMFED and 100 AI centres of BAIF operating in 13 districts of Bihar. A private enterprise, the Patna Animal Development Pvt. Ltd., a group of private practitioners are also involved in delivering AI services in the State. They provide training on AI and supply frozen semen. There are 1401 AI centres in the State and efforts are now on to make all these AI centres functional. As per the estimate of the State Government 6800 AI centres are required in the State as against which only 1000 centres are fully functional.

Further, in order to improve the quality of livestock it is necessary to have easy access to institutions for providing extension services. But unfortunately their number have remained stagnant at least during the last 5 years. For example, the number of artificial insemination centres, veterinary hospitals and veterinary dispensaries have remained stagnant at 1401,39,814 respectively in the State since 2001-02. Besides, inadequacy of the institutions, the services available from these institutions are grossly inadequate in terms of availability of doctors and other technicians, medicines and quality semen.

For maintenance of the improved herd and for obtaining higher output, they are to be provided nutritious feed. But economic hardship of the livestock owners in general forbids them to provide recommended cattle feed. Non availability of the feed at a reasonable price is also a serious impediment.

Next important issue is to pay due attention to feed requirements and animal health. Around 80 per cent of the cost of milk production depends upon the cost of feeding the animals besides the labour and management costs. The state has three cattle feed plants in the cooperative sector with total installed capacity of 260 MTs per day. The plant at Patna is working at almost full capacity. This would need to be expanded to meet the growing demand of balanced and protein-enriched feed. The capacity of this plant is proposed to be augmented from 100 to 200 MTs per day. Given here are two very interesting successful experiment in Rajasthan.

Another Rajasthan Example[3]

Since availability of sufficient fodder is always a problem, the experience of a village Laporia in Rajasthan is relevant to the Bihar situation. Dairy in Laporiya works even in severe drought because it is connected to the common grazing land and by reclaiming and regenerating these lands. In these villages a fascinating technique has been evolved, called the '*chauka*' system, to trap the little rainfall they get and improve the grasslands. The villagers dig rectangular trenches—less than 1 feet deep—to temporarily hold rainwater before it flows into the next trench and then the next and so into the tank. With this system in place, the village common land has become a grand water collection area. The aim is to make the entire village a rain collection system, to recharge the aquifer and withstand drought. In neighbouring Sihalsagar village, every bit of land has been re-crafted—villagers have dug three big *nadis* (ponds), 25 small ponds and made chauka in their grazing land. They have helped change the life of the poor. Bihar can certainly take advantage and follow such examples. Every field has a bund: every drop of rain is trapped and harvested. As a result, the village has water even when its neighbours do not. Since work began on water conservation, the village has never seen bountiful rain. But it still has some water.

Duni Village Experiment

Women in Duni village of Tonk district in Rajasthan have taken a unique initiative of establishing a bulk milk chilling unit of 5,000—litre capacity to augment their milk supply under the banner of Maitree Mahila Mandal. The group, selling milk in the rural areas, now plans to supply it to Mother Dairy in Delhi. Maitree Mahila Mandal was established as a women's group by the Self-Reliant Initiative for Joint Action (SRIJAN) working in more than 60 villages of Deoli block in Tonk district recently. It has encouraged dairy activities for women and introduced weekly savings and interlending, providing liquid cash to women on a regular basis. Getting advantage of the World Bank-sponsored District Poverty Initiative Programme (DPIP), has been formed today a federation of 165 groups comprising 1980 women spread over 60 villages. Women from Below Poverty Line families, purchasing two buffaloes each on a cost-sharing basis with the State Government paying 80 per cent of the price, have collected more than Rs. 20 lakh in savings.

The Maitree women now have about 2,500 Murrah grade buffaloes for which they have got a subsidy of Rs. 3 crore from DPIP. After forming local clusters of 10 to 15 livelihood groups each and gradually federating themselves into the Mahila Mandal, women have so far sold over Rs. 40 lakh worth of milk. The new bulk milk chilling unit of 5,000—litre capacity, adding to an existing capacity of 2,000 litres, at Duni was commissioned during 2007. Creation of Maitree was an inspiration for all women. The need for the new plant emerged as the Maitree women were facing the constraint of chilling capacity to preserve milk. They could only cool 2,000 litres of milk which was uneconomical to transport and market.

POULTRY

Poultry primarily dealing with egg production has just 1.6 per cent share in the livestock value of the State (Table 3.16). Details about egg production by fowls and ducks as given in Table 7.12, 7.13 show that total egg production has increased from 717.7 million in 2000-01 to 1000 million by 2005-06, an increase of around 40 per cent over a period of 5 years. This is when the average yield of an improved layer is below 190 eggs per annum against a potential of over 350 eggs. Productivity level of Deshi birds is less than 90 per bird per annum.

Table 7.13: Egg Production by Fowls and Ducks in Bihar State.

('000)

		2000-01	*2001-02*	*2002-03*	*2003-04*
Egg Production (Million No.s)		717.7	741.6	737.0	740.2
Fowls	Deshi	3286	3249	3170	3077
	Improved	3455	3716	3794	3912
	Total	6741	6965	6964	6989
Ducks	Deshi	436	451	406	413
	Improved	—	—	—	—
	Total	436	451	406	413
No. of Layes ('000 Nos.)	Deshi	3735.64	3662.00	3643.40	3420.94
	Improved	1855.90	1958.01	2039.6	2129.7
Average yield/Annum (No.)	Deshi	87.97	88.71	87.00	89.94
	Improved	186.15	189.80	186.00	183.71
Production (Lakh Nos.)	Deshi	3286	3249	3170	3077
	Improved	3455	3716	3794	3912
No. of Duck ('000 Nos.)	Deshi	296	306	310.10	304.53
	Improved	—	—	—	—
Average Yield/Annum (No.)	Deshi	147.10	147.21	135.00	135.55
	Improved	—	—	—	—
Production (Lakh Nos.)	Deshi	436	451	406	413
	Improved	—	—	—	—
Per capita availability (Nos./Annum)		8	9	9	9

Poultry Development

With poultry population in Bihar being the lowest among all the States on a per capita basis, development of this sector should have immense potential in the State. However, it continues to be primarily a back yard activity. Broiler farming is quite popular in the State and small broiler farms are concentrated near cities and towns. As regards table eggs, it is estimated that only 10 per cent of the requirement in the State is met by the production within the State. Eggs are mostly obtained from Andhra Pradesh. Although the demand for table eggs in the State is very high, the farmers are not evincing interest in commercial egg production mainly due to the longer gestation period of income generation, high feed cost, high overhead expenses due to smaller size of units, and poor power supply position. Lack of awareness on scientific management of poultry birds as well as perception of high risk in the activity by the bankers is also a constraint in the development of the sector. However, of late, the State Government is focusing on promoting layer farming. As a step in this direction, the Government has accorded agriculture status to poultry farming.

Presently, the Day Old Chicks (DOCs) are mostly being supplied from outside the State and also to a limited extent by some private hatcheries within the State. The State Government has planned to set up modern hatcheries in the districts of Patna, Muzaffarpur, Kishanganj and Purnea on the Pattern of Venkateshwar Hatchery. The modern hatcheries will be set up under the centrally sponsored Low Input Technology Scheme. Once the modern hatcheries are set up, the State Government will train rural farmers in the latest technology of poultry farming. The vaccines for Ranikhet and Fowl Pox are being manufactured at Animal Health and Production Institute, Patna. Poultry feed is available from the food plant at Central Poultry Research Institute, the milk unions of COMFED on no profit basis and also from certain private feed manufactures. There is one feed analysis centre located at Animal Health and Production Institute, Patna. The major ingredients such as maize, etc. required for formulation of poultry feed is available in abundance in the State and hence manufacture of feed by setting up feed plants in the State will considerably reduce the feed cost and improve the profitability of the activity, thereby encouraging the farmers of the State to take up this activity on commercial lines.

Poultry Revolution to Boost Economy of Rural India

Kegg Farms are among the top poultry developers in the country and have already earned the distinction by indigenously breeding and licensing India's first pure line at a time when the entire industry was importing grand-parent stocks from the US and Canada for breeding their lines. Since then there are several big farms producing their own pure line. Then, in 1991, came economic liberalization. The country was swarmed with foreign company representatives looking to come into India and tap the huge poultry potential. They discovered through market research that there was a huge Indian-poultry-specific niche market in the rural areas, where the global interests had no stake.

They decided to be part of the huge rural poultry makeover for more reasons than just survival. It was also a change to transform the rural countryside through poultry development. Industrial poultry production in India is now among the largest in the world and Kegg Farms are the fourth biggest egg producers and the fifth biggest poultry meat producers in the world. But the industrial poultry revolution totally by-passed the rural sector. Now, the Government aims at doubling food production in the country and the poultry sector is one fulcrum of the planned second green revolution.

A way had to be found by which villagers are able to get higher production through existing traditional poultry keeping methods. "For that a new specie of bird is to be introduced that would produce more than that of ordinary birds and rearing would be inexpensive". The white industrial poultry bird was abhorred, it was the brown bird that was considered healthier and hardier. This is where the Kuroiler, a hardy, village specific genetic product branded by Kegg Farms, will work wonders. Egg production from the Kuroiler, compared to the *desi* bird, is closer to 150 as against only 90. The Kuroiler male weighs 3.5. kg. compared to only 1 kg of the *desi* male. The female weighs 2.5 kg. against 0.9 kg of the *desi* female. Kegg Farms currently services over seven lakh village households in 11 states.

For 2007, they had targeted an increase to 10 lakh households. During 2006, they supplied over 7 million Kuroilers and this year, they are targeting 20 million bird supplies. Buoyed by their success, numerous small parties are now engaged in the production and distribution of coloured birds in rural areas. There is the potential to create wealth to the tune of over Rs.7,000 crore in 20 million rural households, not to speak of the nutritional security poultry provides. Growth in rural poultry production will also minimize the diversion of human food (mainly coarse cereals which are consumed in rural areas) to poultry production as chicken feed. The saving has been estimated at seven million tonne a year since the Kuroiler birds are hardy and can survive on household mulch and kitchen waste with only a supplementary feed of broken grain and farm waste.

Poultry Directorate, ICAR have propagated two breeds of poultry-Vanaraja and Gramapriya in the major Indian states including Bihar. Vanaraja is a dual type (meat and egg producer) and Gramapriya is predominantly an egg-laying bird. Both these birds need brooding during the initial 4/6 weeks of age, during which compounded balanced feed, artificial heat and vaccination is given which protect the bird from common diseases. These birds start laying from 5.5/6 months of age after attaining the desired body weight (2.5 to 3 kg.) and continue to lay up to 2.5 to 3 years of age depending on the quantity of natural food base and health management. The more the quantity of natural food base available, the minimum will be the requirement of supplementary feed. Males of these birds of Vanaraja and Gramapriya attain the desired body weight (1.5 kg.) at 8 and 11 weeks of age, respectively. Experience from all over India shows that these birds could withstand harsh environment and utilizing natural food base, scavenging their feed on insects, earthworms and fallen grains etc.—gave 120-180 eggs per year. Male birds are sold at 10-15 weeks giving 1.5 to 2.5 kg. body weight and culled females also give 2.0 to 2.5 kg body weight at the end of their productive period.[4] Along with the improved Kuroiler bird and the new ICAR Vanaraja and Gramapriya one can expect a revolution in the village poultry development, which the Eleventh Plan is encouraging. Since there is tremendous gap in demand -supply of eggs in Bihar some enterprising, farmers/ companies should be encouraged to start commercial farms for egg and broiler production.

The purpose of giving details of poultry development in India is that Bihar—a state where there is a huge gap between supply and demand—has to take advantage of the research already available and adapt the same to meet local conditions. Since Bihar has a very successful dairy cooperatives, normal poultry will get a big fillip, if poultry cooperatives are established. This will go a long way not only to improve rural nutritional levels, but also generate additional income. Free range poultry is most suitable for small/marginal and landless farmers. Poultry development in Bihar has been taken up as a priority area for generating additional employment and strengthening the rural economy which is crucially dependent on livestock income. Three poultry development sectors in Patna,

Bhagalpur and Muzaffarpur are working towards organizing poultry business, and scientific development of poultry farming and providing poultry farmers with high breed chicken. There was a production of 1.07 lakh eggs and 0.08 lakh chicken in these sectors in 2006-07. There is a proposal for each one of these poultry sectors to be further strengthened by a low-input technology sector to be established under a plan, sponsored by the Central Government. Under the Venture Capital Fund Plan, there is a plan to make loans available through banks at low rates (equal to agricultural loans) for establishment of dairy and poultry units. Under this plan, 50 per cent of the loan would be interest free. Since the Government is ceased with the problem, our poultry targets have been fixed much above the normal so as to reach near self-sufficiency.

FISH PRODUCTION

Bihar is blessed with vast and varied fisheries and aquaculture resources. The State occupies third position in inland fish production (2.66 lakh tonnes) after West Bengal (9.88 lakh tonnes) and Andhra Pradesh (6.80 lakh tonnes) in 2003-04. These resources are in the form of rivers, reservoirs, lakes, mauns, chaurs, irrigation canals, ponds and community tanks. However, despite such natural resources and fish as a highly preferred food item, aquaculture and open water fisheries resource remain highly underutilized. During the 1990's, the annual growth rate of fish production worked out to 6.58 per cent per annum. As against this, annual consumption is 4.5 lakh tonnes, nearly half of which is imported Table 7.14. With 60-65 per cent of people consuming fish, that also at a lowest level of per capita consumption (less than 1 kg./year.), the shortfall is met by imports from Andhra Pradesh. Although WHO consumption norm is @11kg./capita/annum, the demand can reach over a million tonnes by 2020 at just 2 kg. per capita per annum.

Out of a total of 408 thousand hect. of water area in the composite state, Bihar is endowed with a major portion—316 thousand hectare. Of this, Zone-I has 37.3 per cent of this area, Zone-III 33.5 per cent and the balance 29.7 per cent is in Zone-II. Table 7.15. In addition, there are about 500 thousand hectares, of paddy/crop area under deep water, which can be exploited for fisheries development. This implies that there is a potential of enhancing fish production in the entire state. Further, water area may increase in future due to increase in irrigation facilities and construction of ponds and reservoirs. A large number of ponds are also expected to be constructed under various government programmes including Fish Farmers Development Agency. Almost all the districts of Bihar except flood prone areas can be considered suitable for fisheries development. However, inspite of immense potential for horizontal and vertical expansion of aquaculture in Bihar, owing to availability of vast and varied fresh water resources mainly tanks and ponds, less than one third of the available water resources have been developed so far. Other aquatic resources (makhana, singhara, water chichorh, etc.) in the state are yet to be exploited scientifically. In proportion to the area, Zone-I is responsible to produce 46 per cent of fish. Next in production is Zone-II (28.4%) followed by Zone-III (25.5%) Table 7.16.

Table 7.14: Fish Production in Bihar.

Year	Fish production from all sources ('000 tonnes)		Fish seed production (Lakhs)	
	Target	Achievements	Target	Achievements
1994-95	250.00	195.37	6000.00	3225.56
1995-96	250.00	239.58	6000.00	3322.26
1996-97	250.00	249.78	6000.00	4419.64
1997-98	225.00		6000.00	4054.85
1998-99	255.00	236.68	6000.00	4513.80
1999-00	280.00	254.74	5540.00	4491.06
2000-01	310.00	222.16	5750.00	3585.70
2001-02	260.00	240.40	3575.00	3299.57
2002-03	275.00	261.00	3940.00	3679.90
2003-04	276.00	266.49	4500.00	3471.14
2004-05	300.00	267.51	4500.00	318.21

Source: Department of Fisheries, Government of Bihar

Table 7.15: Zone-wise Permanent and Seasonal Water Area in Bihar.

(Lakh he.)

	Permanent Area	Seasonal	Total
Zone-I	0.80	0.38	1.18 (37.3)
Zone-II	0.59	0.33	0.92 (29.2)
Zone-III	0.58	0.48	1.06 (33.5)
Bihar	1.97	1.19	3.16 (100)

Source: Singh and Prasad, 2001 AERR.
Note: Figures in parenthsis are percentages.

Table 7.16: Zone-wise Fish Production in Bihar ('000 tonnes).

	T.E. 1993-94	T.E. 2001-02
Zone-I	51.49 (40.12)	92.30 (46.09)
Zone-II	39.50 (30.78)	56.92 (28.42)
Zone-III	37.35 (29.10)	51.04 (25.49)
Bihar	128.33 (100)	200.27 (100)

T. E. = Triennial Ending

Source: Directorate of Statistics and Evaluation, Government of Bihar.

Note: Figures in parenthesis indicate percentage to the total state fish production.

Zone-I is the leader in area, production and growth rate of fish. This is followed by Zones-II and III, growth rate in Zone-III being just the half of Zone-I (Table 7.17).

Table 7.17: Growth Trends in Fish Production in Bihar Across Zones (1991-92 to 2001-02)

Zones	CRG (%)
I	8.10
II	6.25
III	4.41
Bihar	6.58

CRG = Compund Rate of Growth

Source: IHD Report, 2003.

Bihar has bountiful natural resources of water in the form of over 43,000 ponds and tanks, spread throughout the state, majority of them being under the control of the state Fisheries Dept. They constitute 66,336 hectare of ponds and tanks and 9930 hectare of ox-bow-lakes. In addition Bihar has 35,000 hectare of water logged area and many major rivers with total length of 66336 km. There are about 2 lakh hectares of permanent water area and 1.3 lakh hectares of seasonal water area in the state (Table 7.18). However, a significant area under tanks and ponds is in a derelict and semi derelict condition. Of all this, only 27,000 hect. of water area have been developed for fish culture. This calls for appropriate intervention and investment to renovate these water bodies. Besides this, there is about 5 lakh hectares of paddy area under deep water condition, which can be exploited for fisheries. Crops and animal husbandry can easily be integrated with fish culture. Considering the increasing input and labour cost, integration of fish culture with rice culture or several kinds of livestock production (pigs, poultry, ducks and geese, other livestock) is important.

Table 7.18: Water Area as per Satellite Survey Done by CIFRI

(Hectares)

Between 0.5-10 he.	43197.41
Between 10-50 he.	26862.41
Between 50-100 he.	13275.98
Between 100-500 he.	24796.72
Between 5000-1000 he.	11713.16
Above 100 he.	12229.22
Total water area	132074.49

Apart from these, there are numerous ponds and tanks, having water area of less than 0.5 ha., with a total water area of 15,000 ha suitable for fish culture.

Major Rivers

Ganga, Kosi, Sone, Burhi Gandak, Punpun, Kamla Balan, Kareh are the major rivers flowing through the State.

Major Reservoirs of Bihar

Reservoir fisheries potential is relatively poor compared to neighbouring States of U.P. and Jharkhand. The total number of reservoirs in Bihar is 29. All the reservoirs of the State are of either small or medium category. Some important reservoirs are as follows:

Badua (Banka District)	680
Chandan (Banka District)	648
Orhni (Banka District)	576
Phulwar (Patna District)	571
Nagi (Jamui District)	204

All the reservoirs are small or medium in size. But for 3 of them, all the 26 are under the control of Water Resources Department. The contribution from State reservoirs in total fish production is very small, having an average productivity level of only 5 kg. ha. Based on the nutrient status of these reservoirs *vis-à-vis* scientific technologies available in the country, production levels of 30 kg./ha. in large; 50 kg./ha in medium and 200-100 kg. ha. In small reservoirs could easily be achieved by judicious and systematic efforts. There is also a large waterlogged area of 5.6 lakh hect. This should be developed in collaboration with the National Fisheries Development Board. Since, there is a huge potential available in the State to meet the fish requirement, a Mission approach may be adopted. It will go a long way to provide additional employment as well nutrition in the rural areas.

Fish Hatcheries

Availability of quality fish seed is the most critical input for success of fish culture. In order to supply adequate, pure and quality fish seed, establishment of infrastructure i.e. hatchery, nursery and rearing farms are urgently required. Hence the main bottleneck in commercialisation and industrialisation of fish culture is want of adequate quality fish seed. There are no functional hatcheries in the Government Sector and only one of the two hatcheries in the cooperative sector has been functional. It has been estimated that 26 hatcheries exist in the private sector all of which are in the districts of Bettiah, Motihari, Sitamarhi, Darbhanga, Madhubani and Samastipur. However, one hatchery per district is the minimum requirement in the state (Table 7.19).

Table 7.19: Fish Hatchery and Seed Farms in Bihar

Crop Hatchery		
Corporate Sector	2	Raghpur (Sitamarhi) Rampatti (Madhubani)
Govt. Sector	1	Kaimur (Nalanda)
Private Sector	26	Scattered in various Districts
Fish seed farms		
Government Sector	121	Scattered in various Districts
FFDA sponsored	33	In various Districts

Source: Bihar Agriculture Development, Opportunities and Challenges, Planning Commission, 2008.

Of the total 121 fish seed farms in the State, only 10 per cent are in working condition. The total seed production in the State was only about 345 million against a demand of nearly 600 million. Here we quote the example of an enterprising farmer of Daiya Kharwa village in Madhubani district, Bihar, who established his own hatcheries on his own two ponds of five acres. He has improved and expanded the infrastructure on scientific lines and rented 21 ponds in and around his village. At present the total water area of his rented ponds is about 55 acres which is sufficient to multiply the spawn into fry (small fishes) of saleable size. He grows mainly Rahu Catla, common carp, grass carp and mrigal breeds. During the last two months he has sold fish seeds worth about Rs. 30 lakh and earned a net profit of about Rs. 15 lakhs and has repaid all the private loans and cleared the bank dues (within two years).[5] The State Government should identify enterprising farmers, provide them necessary infrastructure facilities and technical know how. This involvement of private farmers will go a long way to meet the fish seed shortage. Basic data on the present status of Bihar fisheries is as in Table 7.20. Shortfall is mainly imported from West Bengal to meet the demand.

Table 7.20: Trends in Fish Production, Income from Jalkars, and Other Initiatives of Fisheries Deptt.

	2001-02	2002-03	2003-04	2004-05	2005-06	2006-07
Fish production from all sources ('000 MT)	240.00	261.00	266.49	267.51	279.00	169.50 (upto Dec. 06)
Seed Production (in lakhs)	3299.57	3679.90	3471.14	3182.11	3449.49	2866.55
Income from Jalkar settlements (in lakh Rs.)	296.39	321.61	371.14	382.95	448.44	309.83 (upto Dec.)
Number of Private fish farmers who have been given training in fisciculture	861	1192	5063	1475	134	1400 (in process)
Loans accepted for maintenance/renovation of privately owned ponds (in Rs. lakhs)	4.00	14.70	35.08	32.84	73.98	74.55
Free housing plan for fishermen (at the rate of 1 per family)	117	56	243	244	205	405

Aquaculture

Bihar is blessed with aquaculture resources in the form of freshwater tanks and ponds which cover water area of over 69,000 hectares. Freshwater aquaculture in the State is being promoted through District level Fish Farmers Development Agencies (FFDAs) which provide technical, financial and extension support to the fish farmers. The FFDAs have so far brought under scientific fish farming over 26,000 ha. in various districts with an average fish productivity of around 2.2 tonnes/ha.

The share of aquaculture to total fish production in Bihar is 40 per cent. Indian major carps and three Chinese carps contribute to almost the entire production from aquaculture. Though no systematic farming of other candidate species is carried out, some of the other species like scampi, act fishes (magur and singhi), different species of murrel, are also being cultured.

Bihar has vast potential for commercial scampi farming. In 2003, MPEDA has shown on experimental basis, the production of scampi to the tune of 1500 kg./he. four months in polyculture with carp. Bihar may be considered as the 'sleeping giant' for fresh water prawn farming. Realising its importance as a candidate species for culture and export, it can be taken up on large scale, for monoculture or polyculture practices. Available water resources are as in Tables 7.21 and 7.22.

Table 7.21: Inland Fishery Resources of Bihar

Resources	*Extent (hect.)*
Ponds and tanks	69,000
Oxbow lakes	9,000
Chaurs (waterlogged areas)	35,000
Rivers and Canals	3,200 kms

Source: Bihar's Agriculture Development, A Report of the Special Task Force on Bihar, Government of India.

Table 7.22: Water Area as Per Satellite Survey Done by CIFRI

Size Range	*Total Water Spread area (hect)*
Between 0.5-10	43197.41
Between 10-50	26862.41
Between 50-100	13275.98
Between 100-500	24796.72
Between 500-1000	11713.16
Above 1000	12229.22
Total water area	132074.49

Source: Bihar's Agriculture Development, A Report of the Special Task Force on Bihar, Government of India.

Socio-economic Aspects

The total fisherman population in Bihar is about 49.59 lakh which is largest among all other states of India. About 50 per cent (23.01 lakh) of the total fisher population is constituted by the children of various age groups. But of this huge population, only 37.079 lakh are active fishermen of which 25,503 lakh are male and 11.576 lakh female. The fishermen population density in the state is 52.66 per square km. The data regarding the literacy of fishermen is not available, but it is assumed that the literacy rate is much below the Bihar average of 47.53 per cent. The total number of fishermen co-operatives in the State was 1125 in 2000 as per the Co-operative Department, Government of Bihar. However, most of the Fishermen Cooperative Societies (FCSs) are not functional. Most of the fishers' villages in the State lack basic amenities like housing communication, drinking water, electricity, health and sanitary facilities (Table 7.21).

Considering that Bihar is plagued with a very low per capita income and protein intake, fishery development in open waters assumes greater significance from social and economic perspectives. Fisheries will also play a pivotal role in increasing fish availability in the State and providing gainful employment to its rural poor. However, the State has a long way to go towards optimum utilization of this important resource.

Policy Issues

Lack of appropriate leasing policy over the years has been one of the major factors of ineffective utilization and development of these water bodies for aquaculture. Though certain positive initiative have been taken by the Government in the form of the Bihar Fish Jalkar Management Act. 2006 for short and long term leasing of these community water bodies, it needs effective implementation at Gram Panchayat level. Community approach for integrated development shall help development of aquaculture in these ponds.

For the development of inland waters, fishing rights must be with the State Fisheries Departments. The State Government has decided to give ponds under Fisheries Department for a lease periods of 5-10 years. This will provide incentive for the lease holder to invest in long term development of the pond and adopt scientific fish farming. The lease amount should be rationalized with lease duration of 12 years or more and lease rent fixed according to productivity and level of multiple uses. Specific policies should be framed to lease out different sized ponds of fishermen co-operatives, group of fisheries or individual fisherman, SHGs, private entrepreneurs in order to have efficient aquaculture management to exploit its full production potential. In additional to this, fisheries graduate can be given preference for taking the pond on lease.

The policy for promotion and development of aquaculture in the State would include (i) tariff realization in order to give parity with agriculture sector, (ii) more vibrant FFDAs, (iii) self-sufficiency in quality and disease free seed and feed and (iv) legal framework. The strategy would include (i) scientific stocking, (ii) horizontal expansion of farming area, (iii) integrated farming, (iv) contract farming, (v) species diversification, (vi) transfer to departmental fish farm to private entrepreneurs, (vii) development of approach road to aquaculture clusters, (viii) empowerment of farmers and fisherwomen and utilization of technical expertise of fisheries graduates, and (ix) increase in productivity.

Capture fisheries which have a good potential, need to be explored. There is vast resource in seven major rivers—Ganga, Kosi, Sone, Bagmati, Gadak and Punpun, but only 27 thousand hect. of water area has been developed.

Since vertical and horizontal expansion of aquaculture is possible, co-operative sector in fish can be strengthened like dairying. This can help to give an annual growth rate of 10-15 per cent so as to achieve self sufficiency.

Constraints

There are a few intricate issues such as, presence of carnivorous fishes, tree stumps, dense macrophytes, poor fishery management practices, unregulated destructive types of fishing, sick Fisheries Cooperative Societies, etc., which should be taken care of for raising production from present abysmally very low level. Some other measures that should be put in place are strict enforcement of management rules, observance of closed season, providing training and fishing tools to fisheries, intensive, extension practices and observing ethics of responsible fisheries.

The current state of disappointing fisheries and aquaculture development in the State can be attributed mainly to the underutilization of aquaculture resources, unscientific management of water bodies and lack of entrepreneurship. The major challenge, therefore, is to develop the aquaculture resources for productive, sustainable fisheries to meet the local demand and enhance the nutritive and livelihood security to farmers as well as consumers. Poor institutional setup, almost non-existing extension services, lack of adequate resources and infrastructure facilities, devoid of conducive policy environment, lack of professionalism among fisheries personnel, fragmented social setup and poverty and illiteracy among the primary producers, are the major issues. The challenge is how to negotiate with these impediments, where many of them do not fall under the purview of the department of fisheries and do not employ a practical and effective management strategy for utilization of available resources. It is necessary to work hard for exploring appropriate approaches and develop field tested models to make this food production sector play a greater role in poverty reduction, food and nutritional security, sustainable rural livelihood development revenue and overall contribution towards betterment of the socio-economic conditions of farmers and fisher farmers.

Major constraints for fisheries development are:

- Fish seed.
- Leasing of Government bodies for aquaculture.
- Weak Fish Cooperatives.
- Lack of Soil and Water testing facilities.
- Lack of special thrust initiative by Banks.

Notes

1. *Dairy Planner*, July 2007.
2. Bihar Agriculture Building on Emerging Models of "Success", World Bank.
3. Sunita Narain, *When Markets, Do Work, Down to Earth*, June 15, 2007.
4. *Indian Farming*, July 2007.
5. *The Hindu*, September 25, 2008.

ANNEXURE 7.1

Table 7.1: Annual Growth Rates of Livestock in Different States

State	*Year*	*Cattle*	*Buffalo*	*Sheep*	*Goat*	*Livestock*
Andhra Pradesh	1971-81	0.43	2.01	-27.07	2.86	-1.37
Andhra Pradesh	1981-91	-0.80	-1.68	34.86	-2.81	0.83
Andhra Pradesh	1991-01	-2.55	3.87	11.71	3.70	4.42
Assam	1971-81	1.99	1.85	-3.55	1.53	1.67
Assam	1981-91	-2.78	-4.00	-6.39	-2.80	-0.14
Assam	1991-01	5.43	5.90	12.19	7.89	3.47
Bihar	1971-81	-0.85	4.46	2.80	9.20	1.23
Bihar	1981-91	3.01	-1.44	2.23	-1.47	2.44
Bihar	1991-01	-1.08	1.95	-3.12	-2.72	-0.97
Gujarat	1971-81	2.00	3.40	7.73	0.49	3.78
Gujarat	1981-91	-0.28	1.72	-1.50	2.63	0.63
Gujarat	1991-01	0.80	2.89	-0.03	0.30	1.30
Haryana	1971-81	-2.75	1.00	0.08	-5.58	0.05
Haryana	1981-91	0.89	-0.64	-0.69	-0.69	0.38
Haryana	1991-01	-4.96	6.68	-1.11	-2.07	0.87
Himachal Pradesh	1971-81	0.42	1.58	0.88	-0.08	0.67
Himachal Pradesh	1981-91	-0.04	1.33	-0.10	0.54	0.46
Himachal Pradesh	1991-01	-1.03	1.87	1.56	0.97	-0.08
Jammu & Kashmir	1971-81	1.18	1.89	6.71	6.11	3.89
Jammu & Kashmir	1981-91	2.42	2.42	8.07	5.99	3.77
Jammu & Kashmir	1991-01	0.19	3.87	-1.01	2.03	1.91
Karnataka	1971-81	1.31	1.09	0.54	-1.31	0.86
Karnataka	1981-91	1.55	1.82	0.96	3.05	3.01
Karnataka	1991-01	-3.18	-0.63	3.25	-1.75	-1.95
Kerala	1971-81	0.93	-2.97	-1.29	2.67	0.82
Kerala	1981-91	0.42	-0.57	45.70	3.52	1.30
Kerala	1991-01	-3.75	-16.85	-33.51	-8.07	-5.55
Madhya Pradesh	1971-81	-2.07	1.40	-2.90	2.16	0.98
Madhya Pradesh	1981-91	0.53	3.94	-0.40	0.99	1.02
Madhya Pradesh	1991-01	-0.34	-0.22	-2.48	2.20	0.35
Maharashtra	1971-81	1.24	2.28	2.27	2.02	1.97
Maharashtra	1981-91	0.48	2.13	-0.11	1.62	1.04
Maharashtra	1991-01	-0.53	1.52	1.72	1.28	0.42
Orissa	1971-81	1.57	-0.44	4.10	5.73	2.43
Orissa	1981-91	-0.69	-1.50	-3.71	-2.49	-2.35
Orissa	1991-01	1.43	2.38	27.36	-8.25	-27.72
Punjab	1971-81	-1.07	2.04	2.42	-2.40	0.89
Punjab	1981-91	-0.91	1.04	-1.04	-2.62	-0.25
Punjab	1991-01	-3.04	1.55	-6.77	-5.35	-0.90
Rajasthan	1971-81	0.72	2.74	4.99	2.62	2.45
Rajasthan	1981-91	-2.13	1.60	-1.05	-0.78	-0.72
Rajasthan	1991-01	-0.90	3.21	-1.96	1.05	0.30

ANNEXURE 7.2

Table 7.2: Number of Livestock per Lakh Population (1991)

State	Crossbreed Cattle	Indigenous Cattle	Cattle	Buffaloes	Yarks	Thuns	Sheep	Goats	Horses	Mules	Donkey	Camels	Pigs	Livestock
Andhra Pradesh	0.95	20.51	21.46	17.94	0.00	0.00	15.27	8.49	0.02	0.00	0.08	0.00	1.27	64.53
Assam	1.62	48.70	50.32	4.77	0.00	0.00	0.74	17.18	0.09	0.00	0.00	0.00	6.78	79.88
Bihar	0.25	29.05	29.30	7.08	0.00	0.00	2.23	23.09	0.15	0.01	0.04	0.00	1.49	63.39
Gujarat	0.82	23.09	23.91	18.51	0.00	0.00	7.13	14.90	0.05	0.00	0.28	0.22	0.36	65.36
Haryana	3.24	13.37	16.61	33.99	0.00	0.00	8.11	6.22	0.39	0.19	0.57	1.00	4.01	71.09
Himachal Pradesh	5.88	39.43	45.31	14.67	0.10	0.00	22.52	23.40	0.27	0.33	0.15	0.00	0.10	106.85
Jammu & Kashmir	13.01	37.12	50.13	11.96	0.54	0.00	48.33	29.00	2.00	0.31	0.31	0.03	0.20	142.81
Karnataka	1.96	39.23	41.19	13.29	0.00	0.00	16.98	19.66	0.04	0.00	0.10	0.00	1.19	92.45
Kerala	7.44	7.46	14.90	1.25	0.00	0.00	0.13	7.82	0.00	0.00	0.00	0.00	0.58	24.68
Madhya Pradesh	0.39	53.99	54.38	15.11	0.00	0.00	1.58	15.87	0.14	0.02	0.11	0.02	1.38	88.61
Maharastra	3.46	30.56	34.02	10.62	0.00	0.00	6.00	19.39	0.08	0.00	0.14	0.00	0.73	70.98
Orissa	2.15	47.41	49.56	5.50	0.00	0.00	6.58	17.70	0.00	0.00	0.00	0.00	2.09	81.43
Punjab	11.10	8.73	19.83	40.96	0.00	0.00	3.59	3.71	0.27	0.11	0.25	0.29	0.69	69.70
Rajasthan	0.35	33.32	33.67	22.19	0.00	0.00	35.97	44.06	0.07	0.01	0.58	2.16	0.73	139.44
Tamil Nadu	4.91	19.86	24.77	7.51	0.00	0.00	15.61	16.94	0.02	0.00	0.11	0.00	1.79	66.75
Uttar Pradesh	2.19	20.27	22.46	17.60	0.00	0.00	2.11	11.49	0.22	0.09	0.24	0.03	2.55	56.79
West Bengal	1.92	32.95	34.87	2.02	0.00	0.00	2.97	28.31	0.02	0.00	0.00	0.00	1.91	70.10
All India	2.35	29.20	31.55	12.99	0..01	0.02	7.83	17.78	0.13	0.03	0.15	0.16	1.97	72.62

(*Contd.*)

Table 7.2: (*Contd.*): Number of Livestock per Lakh Population (2001)

State	Crossbreed Cattle	Indigenous Cattle	Cattle	Buffaloes	Yarks	Thuns	Sheep	Goats	Horses	Mules	Donkey	Camels	Pigs	Livestock
Andhra Pradesh	1.29	16.96	18.26	16.63	0.00	0.00	16.78	8.98	0.01	0.00	0.06	0.00	1.29	62.01
Assam	1.54	32.33	33.87	3.05	0.00	0.00	0.35	11.37	0.05	0.00	0.00	0.00	4.53	53.22
Bihar	0.32	33.59	33.91	8.10	0.00	0.00	2.70	27.89	0.17	0.01	0.04	0.00	1.27	74.09
Gujarat	0.98	18.38	19.36	18.03	0.00	0.00	6.19	12.58	0.04	0.00	0.21	0.19	0.57	57.17
Haryana	5.15	9.42	14.57	29.28	0.00	0.00	7.74	5.88	0.30	0.21	0.38	0.60	4.25	63.21
Himachal Pradesh	6.55	32.15	38.70	13.32	0.12	0.00	19.24	20.80	0.23	0.32	0.14	0.00	0.12	92.99
Jammu & Kashmir	13.62	26.31	39.93	9.90	0.42	0.00	39.87	23.45	1.77	0.26	0.29	0.05	0.15	116.09
Karnataka	3.45	25.44	28.89	11.65	0.00	0.00	21.34	13.00	0.04	0.00	0.07	0.00	1.08	76.07
Kerala	7.56	2.06	9.62	0.43	0.00	0.00	0.01	6.18	0.00	0.00	0.00	0.00	0.34	16.58
Madhya Pradesh	0.37	40.14	40.51	13.81	0.00	0.00	1.36	13.44	0.11	0.01	0.10	0.02	0.78	70.14
Maharastra	3.91	24.84	28.75	9.66	0.00	0.00	5.36	18.19	0.07	0.00	0.11	0.00	0.90	63.04
Orissa	2.82	39.83	42.65	4.29	0.00	0.00	5.45	17.83	0.00	0.00	0.00	0.00	1.86	72.08
Punjab	10.41	4.61	15.02	35.13	0.00	0.00	2.48	2.36	0.19	0.10	0.13	0.17	0.55	56.13
Rajasthan	0.47	26.76	27.23	21.91	0.00	0.00	32.71	38.06	0.05	0.01	0.42	1.50	0.68	122.57
Tamil Nadu	8.42	13.31	21.73	6.58	0.00	0.00	12.63	15.41	0.03	0.00	0.10	0.00	1.46	57.94
Uttar Pradesh	1.54	13.15	14.69	13.94	0.00	0.00	1.40	8.65	0.16	0.06	0.18	0.02	2.30	41.40
West Bengal	1.59	28.64	30.23	2.09	0.00	0.00	2.48	26.53	0.03	0.00	0.00	0.00	1.36	62.72
All India	2.55	22.72	25.27	11.43	0.01	0.02	7.31	15.60	0.10	0.03	0.11	0.12	1.69	61.69

ANNEXURE 7.3

Table 7.3: Annual Compound Growth Rates of Livestock by District

District	Year	Total Cattle	Total Buffaloes	Total Sheep	Total Goat	Total Livestock	District	Year	Cattel	Buffaloes	Sheep	Goat	Livestock
Aurangabad	1971-81	4.48	24.73	1.31	3.72	4.16	Muzaffarpur	1971-81	-9.32	23.46	-5.16	5.95	6.61
Aurangabad	1981-91	-1.79	-17.46	3.73	3.75	-0 47	Muzaffarpur	1981-91	13.16	-16.61	10.80	1.55	11.04
Aurangabad	1991-01	1.88	28.66	-4.26	-2.01	1.67	Muzaffarpur	1991-01	5.42	28.00	8.81	-1.77	-6.06
Begusarai	1971-81	1.32	1.62	2.59	5.93	2 97	Nalanda	1971-81	-1.11	-1.91	1.58	3.03	-1.11
Begusarai	1981-91	1.27	1.31	2.43	1.58	0.70	Nalanda	1981-91	3.76	4.96	3.45	4.44	4.86
Begusarai	1991-01	4.46	7.07	25.36	4.57	5.89	Nalanda	1991-01	3.32	-0.97	1.83	-0.25	-0.62
Bhagalpur	1971-81	1.50	2 87	-3.15	5.45	3.63	Nawada	1971-81	2.58	8.66	3.55	5.33	3.93
Bhagalpur	1981-91	1.09	0.08	8.50	2.04	0.06	Mawada	1981-91	0.03	-5.25	1.48	2.16	-0.23
Bhagalpur	1991-01	-1.87	8.08	-11.61	1.49	4.19	Nawada	1991-01	0.32	7.99	1.50	-3.09	0.23
Bhojpur	1971-81	-4.06	7.69	-3.12	-1.92	-2.66	Patna	1971-81	-4.32	-7.86	9.59	6.82	-3.85
Bhojpur	1981-91	6.95	-4.40	8.47	9.70	6.53	Patna	1981-91	7.25	11.73	-4.11	0.63	7.85
Bhojpur	1991-01	-6.56	11.44	-1.96	-568	-3.53	Patna	1991-01	9.74	-1.88	-2.13	-0.50	2.98
Champaran (East)	1971-81	12.77	0.20	-4.23	-2.59	-0.58	Purnea	1971-81	-1.75	-0.82	-9.83	2.39	2.74
Champaran (East)	1981-91	-9.01	2.75	8.73	10.47	4.30	Purnea	1981-91	4.44	3.80	16.54	5.09	0.93
Champaran (East)	1991-01	15.32	4.49	4.92	-8.25	-0.11	Purnea	1991-01	-3.75	0.29	-3.64	-4.75	-0.22

(Contd.)

(*Contd.*)

District	Year	Total Cattle	Total Buffaloes	Total Sheep	Total Goat	Total Livestock	District	Year	Cattle	Buffaloes	Sheep	Goat	Livestock
Champaran (West)	1971-81	1.10	0.72	-8.31	11.53	18.95	Rohtas	1971-81	2.73	1.65	3.32	4.55	6.24
Champaran (West)	1981-91	1.50	2.21	14.60	-3.52	-12.82	Rohtas	1981-91	-0.12	1.28	1.71	2.92	-2.39
Champaran (West)	1991-01	-0.94	0.68	-2.09	-7.79	9.45	Rohtas	1991-01	-0.18	-2.19	-12.62	3.71	3.27
Darbhanga	1971-81	-13.12	-2.02	-9.94	7.07	2.20	Saharsa	1971-81	-3.26	-1.68	0.51	5.47	-4.45
Darbhanga	1981-91	18.1	5.08	16.68	0.50	1.47	Saharsa	1981-91	6.06	4.71	4.55	2.03	8.53
Darbhanga	1991-01	-16.06	1.87	-16.53	1.13	0.13	Saharsa	1991-01	-6.99	-4.25	11.98	-4.79	-9.28
Gaya	1971-81	-5.49	1.81	1.90	5.01	-2.75	Samastipur	1971-81	6.94	-0.40	9.11	4.19	11.72
Gaya	1981-91	8.57	1.12	3.12	2.47	6.63	Samastipur	1981-91	-4.05	3.37	-3.69	3.27	-7.18
Gaya	1991-01	-5.42	0.50	-5.66	-0.72	-4.29	Samastipur	1991-01	5.34	'3.96	8.92	1.10	11.30
Gopalganj	1971-81	2.67	33.43	2.84	7.50	0.55	Saran	1971-81	0.68	4.18	5.22	-0.34	1.79
Gopalganj	1981-91	-0.06	-22 84	2.18	0.10	3.13	Saran	1981-91	1.92	-1.18	-0.12	7.65	1.88
Gopalganj	1991-01	4.26	37.77	2.55	0.19	-0.07	Saran	1991-01	-4.73	-1.47	4.52	-3.08	-5.52
Katihar	1971-81	1.88	-2.17	2.95	4.97	3.26	Sitamarhi	1971-81	2.48	2.13	-5.04	5.89	3.98
Katihar	1981-91	0.72	5.24	2.08	2.51	0.42	Sitamarhi	1981-91	0.12	0.81	10.66	1.62	-0.27
Katihar	1991-01	1.29	-4.17	-3.76	1.03	1.87	Sitamartii	1991-01	2.86	2.36	-10.30	-3.03	1.97
Madhubani	1971-81	-3.31	2.33	-2.60	4 59	0.16	Siwan	1971-81	-2.31	-2.08	-3.47	-0 03	3.10

ANNEXURE 7.4

Table 7.4: Percentage Share of each District to total Livestock in the State

District	Crossbred Calle	Indi-genous Cattle	Total Cattle	Buffa-loes	Total Sheep	Total Goats	Total Pigs	Total Horse & Ponies	Total Mules	Total Donkeys	Total Camels	Total Livestock	Total Cocks	Total Duck	Total Poultry
Araria	1.53	0.32	0.40	0.74	1.07	0.47	0.29	0.58	1.27	2.51	0.00	0.50	0.34	0.17	0.33
Aurangabad	0.90	3.30	3.14	3.38	11.04	2.70	2.37	0.21	3.02	1.53	0.00	3.13	2.03	2.01	2.03
Banka	1.94	1.94	1.94	1.94	1.94	1.94	1.94	1.94	2.59	1.94	2.04	1.94	1.94	1.94	1.94
Begusarai	2.83	2.83	2.83	2.83	2.83	2.83	2.83	2.83	1.68	2.83	2.92	2.83	2.83	2.83	2.83
Bhagalpur	4.50	2.50	2.64	2.59	0.17	4.14	2.58	3.34	1.94	1.31	0.00	3.04	2.44	1.98	2.40
Bhoipur	4.95	1.17	1.42	4.09	7.59	1.43	1.84	3.50	2.82	9.98	11.08	2.04	1.17	0.92	1.15
Buxar	3.48	0.90	1.07	3.12	4.64	0.84	0.97	2.19	0.38	11.77	5.54	1.45	0.74	0.35	0.70
Darbhanqa	0.47	1.76	1.68	3.39	0.18	2.52	0.79	0.60	25.00	0.13	0.00	2.20	2.43	2.94	2.48
East Champaran	4.75	4.75	4.75	4.75	4.75	4.75	4.75	4.75	1.82	4.74	4.66	4.75	4.75	4.75	4.75
Gaya	1.20	5.23	4.96	5.36	4.17	5.26	13.75	6.50	0.72	3.62	0.00	5.39	5.06	0.93	4.71
Gopalganj	2.59	2.59	2.59	2.59	2.59	2.59	2.59	2.59	1.55	2.59	2.62	2.59	2.59	2.59	2.59
Jarnui	1.69	1.69	1.69	1.69	1.69	1.69	1.69	1.69	1.54	1.69	1.75	1.69	1.69	1.69	1.69
Jehanabad	1.82	1.82	1.82	1.82	1.82	1.82	1.82	1.82	0.00	1.82	1.75	1.82	1.82	1.82	1.82
Kaimur	1.55	1.55	1.55	1.55	1.55	1.55	1.55	1.55	0.00	1.55	1.46	1.55	1.55	1.55	1.55
Katihar	0.40	3.28	3.09	1.37	1.55	5.30	1.76	0.85	0.05	0.05	0.29	3.36	6.81	14.28	7.44
Khagaria	1.54	1.54	1.54	1.54	1.54	1.54	1.54	1.54	1.84	1\54	1.46	1.54	1.54	1.54	1.54

(*Contd.*)

Table 7.4: (*Contd.*)

District	*Crossbreed Cattle*	*Indi-genous Cattle*	*Total Cattle*	*Buffa-loes*	*Total Sheep*	*Total Goats*	*Total Pigs*	*Total Horse & Ponies*	*Total Mules*	*Total Donkeys*	*Total Camels*	*Total Livestock*	*Total Cocks*	*Total Duck*	*Total Poultry*
Kishanqanj	0.01	2.15	2.01	0.87	0.06	3.35	0.49	0.09	0.79	0.00	11.66	2.11	5.94	4.69	5.83
Lakhisarai	2.00	0.81	0.89	1.03	0.02	1.10	0.86	1.84	0.00	0.02	0.00	0.97	0.33	0.23	0.32
Madhepura	1.84	1.84	1.84	1.84	1.84	1.84	1.84	1.84	4.31	1.84	1.75	1.84	1.84	1.84	1.84
Madhubani	4.31	4.31	4.31	4.31	4.31	4.31	4.31	4.31	1.37	4.31	4.37	4.31	4.31	4.31	4.31
Munger	1.37	1.37	1.37	1.37	1.37	1.37	1.37	1.37	4.52	1.37	1.46	1.37	1.37	1.37	1.37
Muzaffarpur	4.52	4.52	4.52	4.52	4.52	4.52	4.52	4.52	2.85	4.51	4.66	4.52	4.52	4.52	4.52
Nalanda	2.86	2.86	2.86	2.86	2.86	2.86	2.86	2.86	2.18	2.86	2.92	2.86	2.86	2.86	2.86
Nawada	2.18	2.18	2.18	2.18	2.18	2.18	2.18	2.18	1.17	2.18	2.33	2.18	2.18	2.18	2.18
Patna	14.45	9.88	10.18	5.13	4.06	2.34	5.84	2.15	4.75	3.50	6.12	6.61	2.24	0.98	2.13
Purnia	0.71	3.31	3.13	3.54	0.06	5.16	3.06	12.76	2.20	0.18	0.00	3.82	5.08	5.15	5.08
Rohtas	2.95	2.95	2.95	2.95	2.95	2.95	2.95	2.95	2.95	2.96	2.92	2.95	2.95	2.95	2.95
Saharsa	1.82	1.82	1.82	1.82	1.82	1.82	1.82	1.82	1.82	1.82	1.75	1.82	1.82	1.82	1.82
Samastipur	4.12	4.12	4.12	4.12	4.12	4.12	4.12	4.12	4.12	4.12	4.08	4.12	4.12	4.12	4.12
Saran	3.92	3.92	3.92	3.92	3.92	3.92	3.92	3.92	3.93	3.92	3.79	3.92	3.92	3.92	3.92
Sheikhpura	0.63	0.63	0.63	0.63	0.63	0.63	0.63	0.63	0.63	0.63	0.58	0.63	0.63	0.63	0.63
Sheohar	0.62	0.62	0.62	0.62	0.62	0.62	0.62	0.62	0.62	0.62	0.58	0.62	0.62	0.62	0.62
Sitamartii	3.22	3.22	3.22	3.22	3.22	3.22	3.22	3.22	3.22	3.22	3.21	3.22	3.22	3.22	3.22
Siwan	3.27	3.27	3.27	3.27	3.27	3.27	3.27	3.27	3.26	3.27	3.21	3.27	3.27	3.27	3.27
Supaul	2.11	2.11	2.11	2.11	2.11	2.11	2.11	2.11	2.10	2.10	2.04	2.11	2.11	2.11	2.11
Vashali	3.27	3.27	3.27	3.27	3.27	3.27	3.27	3.27	3.27	3.27	3.21	3.27	3.27	3.27	3.27
West Champaran	3.67	3.67	3.67	3.67	3.67	3.67	3.67	3.67	3.67	3.67	3.79	3.67	3.67	3.67	3.67
State Total	100	100	100	100	100	100	100	100	100	100	100	100	100	100	100
All India share	3.77	7.32	6.89	5.37	0.70	7.01	6.17	18.75	3.21	2.56	0.05	5.80	23.22	4.67	4.08

ANNEXURE 7.5

Table 7.5: District-wise Livestock Population per Lakh Human Population

District	Crossbreed Cattle	Indigenous Cattle	Total Cattle	Buffaloes	Total Sheep	Total Goats	Total Pigs	Total Horse & Ponies	Total Mules	Total Donkeys	Total Camels
Araria	654.15	1961.33	2615.48	1936.60	231.72	1990.79	126.47	43.28	6.38	21.46	0.00
Aurangabad	416.54	21486.54	21903.08	9463.25	2569.92	12289.22	1115.86	17.02	16.31	14.10	0.00
Banka	1066.11	15071.43	16137.53	6480.59	538.56	10516.39	1091.21	184.58	16.67	21.25	0.45
Begusarai	1140.98	16130.09	17271.07	6935.77	576.40	11255.04	1167.85	197.58	7.95	22.76	0.47
Bhagalpur	1923.29	15138.69	17061.98	6746.07	35.90	17456.59	1130.80	247.10	9.73	11.13	0.00
Bhojpur	2177.29	7262.90	9440.20	10946.32	1686.13	6203.72	827.68	266.25	14.54	87.53	1.96
Buxar	2310.19	8442.05	10752.24	12580.56	1556.01	5500.42	658.02	251.24	2.95	155.62	1.48
Darbhanga	133.55	7086.38	7219.93	5858.10	25.60	7057.15	231.04	29.56	83.22	0.73	0.00
East Champaran	1091.42	15429.55	16520.97	6534.57	551.40	10766.27	1117.14	188.99	4.91	21.75	0.43
Gaya	340.21	20993.14	21333.35	9263.07	598.58	14743.75	3999.47	319.76	2.40	20.52	0.00
Gopalganj	1091.64	15432.39	16524.03	6635.78	551.47	10768.25	1117.35	189.04	7.65	21.76	0.44

8

INPUT MANAGEMENT

Agricultural growth is a function of the level of technology available in the concerned region. This involves seeds, fertilizers and irrigation which are the major inputs in agricultural production. A sustained increase in agricultural production and productivity is dependent on the continuous development of improved varieties of crops and supply of quality seeds to the farmers. Mechanization, agrochemicals and credit facilities are the other major inputs. These were the basic ingredients of the first Green Revolution which triggered the stagnant agriculture. If Bihar has to lead the Second Green Revolution as the cherished goal of the Government, we have to look to these components and critically examine the ground situation in each case.

SEED

One of the major ingredients of the first green revolution in the country was the introduction of new High Yielding Variety (HYV) seeds, particularly of wheat and rice, along with the usage of proper doses of fertilizers and water with an appropriate price policy as a catalytic agent. Waves of this revolution though belatedly, also hit Bihar state which would be evident from the fact that within a period of slightly less than 20 years (TE ending 1982 to TE ending 1999), nearly 68 per cent of the rice, 91 per cent wheat and 75 per cent of maize area were covered by HYV seeds. This further improved to 74 per cent for rice 92 per cent for wheat and 77 per cent for maize in the TE 2005. (Table 8.1).

Zone-III has the highest coverage of area under HYV for rice and maize. Wheat has the highest coverage at over 90 per cent at the State level with Zones-I and III even higher than this. There is all the same a much higher growth in respect of rice in all the zones. (Table 8.2). Joint efforts of ICAR-DOAC in taking the improved technologies on the shelf to the tiller through the compact Frontline Demonstration—a unique experiment launched in 1990-91 played a major role in the expansion of HYV area under rice during 1990s. Since 1990-91, a large number of demonstrations have been conducted every year with the active support and co-operation of Agricultural Universities and Department of Agriculture. The experiment while disseminating knowledge, has facilitated large scale farmer participation and has enabled them to choose varieties/production package suited to their farm situations.

Table 8.1: Trend Share of total Crop-Specific Area under HYV in Bihar (HYV)

Crop	*Year*	*Area (%)*
Rice	TE 1982	24.7
	TE 1990	38.3
	TE 1999	67.9
	TE 2005	73.9
Wheat	TE 1990	76.9
	TE 1999	90.6
	TE 2005	92.0
Maize	TE 1990	62.4
	TE 1999	75.4
	TE 2005	77.1

Source: Bihar Development, October 2006.

Table 8.2: Zone-wise Trend in Share of total Area under HYV in Bihar (HYV Area as % of Crop Area)

Crop		*Zone-I*	*Zone-II*	*Zone-III*	*Bihar*
Rice	TE ending 1982	23.70	30.72	20.55	24.65
	TE ending 1990	24.52	27.11	64.40	38.25
	TE ending 1999	63.31	55.68	77.10	67.93
Wheat	TE ending 1982	77.58	71.93	62.25	70.44
	TE ending 1990	78.99	74.58	75.97	76.95
	TE ending 1999	93.47	85.05	91.68	90.61
Maize	TE ending 1982	56.84	57.09	56.22	56.92
	TE ending 1990	72.77	55.31	60.48	62.38
	TE ending 1999	71.88	76.90	82.38	75.42

Source: Dte. of Statistics and Evaluation, Government of Bihar.

The major problem however, relates to extremely low seed replacement rates. Certainly this leads to losing the potentiality of the seed, hence low productivity levels. Graph 8.1 presents the gap between demand and supply. The HYV coverage alone is not sufficient to help raise productivity, unless the seeds are of the required quality accompanied by proper doses of fertilizers and assured water. Seed treatment and replacement of seeds at regular intervals are sine qua non in the process. Unfortunately, seed replacement rate for most of the crops in the state is one of the lowest in the country. The latest position as available indicates that by 2005-06 kharif season, the replacement rates of paddy, maize, pulses and oilseeds were 12, 50, 8 and 5 per cent respectively. In case of rabi season, the replacement rate of wheat alone was available which was only 11 per cent (Table 8.3). All out efforts have to be made to enhance the replacement rates of the crops, particularly of wheat, rice and maize. In addition to reviving the Seed Corporation, farmers should also be involved to produce certified seeds of recommended varieties.

Graph 8.1: Seed Demand and Supply Scenario in Bihar.

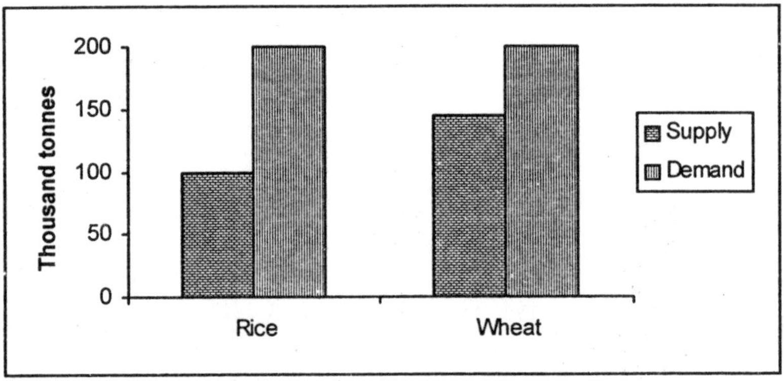

Source: Department of Agriculture, Government of Bihar.

Realising that management of seed supply is an important component of any programme of agricultural development, Bihar State Seed Corporation was revived during 2007 and production of seed is now being undertaken on 45 state managed farms which remained inoperative for some years. More than 1100 agriculture graduates have been trained in the State. A decision has been taken to use the seed produced by agri clinic for public sector programmes. Extension system has also neglected these enterprises so far and there is little awareness among the farmers regarding improved varieties.

Table 8.3: Seed Replacement Rates of Different Crops

Name of Crops	2003-04		2004-05		2005-06		2006-07
	Target	Achievement	Target	Achievement	Target	Achievement	Target
Kharif Crops							
Paddy	8	6.8	10	10	11	12	15
Maize	40	30	50	40	50	50	60
Pulses	10	6	10	7.5	10	8	14
Oilseeds	5	2	5	3	5	5	
Rabi Crops							
Wheat	10	8.1	10	9	15	11	15
Maize	—	—	—	—	50	—	80
Pulses	5	1.2	10	7.5	15	—	10
Oilseeds	25	20	30	25	30	—	30
Vegetables	25	20	60	50	60	—	60

Source: Economic Survey, 2006-07, Government of Bihar.

Unless the seed sector is revamped, technological dynamism cannot come about. To take care of such problems, Kisan Samman Yojana was launched to identify and acknowledge enterprising farmers in the State and using them as extension agents. This plan is unique and innovative in respect of extension reforms programme.[1]

New recommended seed varieties under various conditions in different parts of the State are as follows:

Rice

(i) Rajendra Mahsuri, Rajendra Sweta and Swarna in flood prone and sub-merged areas.
(ii) Short duration and drought tolerant varieties such as Vandana, Tulasi and Rajashree, in water stress areas.
(iii) Hybrids PHB-71, KRH-2, Pusa RH-10, Pant H-1, ARRH-2 and DRRH-2 in normal situation.
(iv) Bacterial blight resistant varieties such as Ajaya, IR-64.
(v) Application of zinc sulphate in zine deficient areas.

Wheat

(i) Varieties like K-0307, HD 2733, HD 2824, HP 1761, PBW 443, HUW 468, K 9107, NW 1012 in irrigated timely sown situation;
(ii) DBW 14, NW 2036 HW, NW 1014, HD 2643, HP 1744 in irrigated late sown situation.
(iii) Salinity/alkalinity tolerant varieties such as KRL 19 and KRL-1.

Maize

(i) QPM hybrids, HQPMI, Shaktiman 1, Shaktiman 2, Shaktiman 3 and 4,
(ii) Winter maize as a better alternative to wheat
(iii) Inter-cropping of maize with potato, pea and other vegetables.

Summarising the position, it can be concluded that to ensure 4.0 per cent growth in agriculture in the country, agriculture in Bihar has to grow at 7 per cent. Due to the financial constrains of the

State, Centre has to share the cost incurred on rationalization of agriculture extension system and make investments in setting up seed banks. The State Government sought to emphasise that agriculture in the State has suffered because of low public sector investment in infrastructure and devastating floods. Technology fatigue has resulted flattening the yield potential over the last five years or so. In many cases the gap between the yield potential of existing varieties and actual realization was 40-50 per cent, while in one case it was 300 per cent. We ought to organize the longer-term restructuring of our research effort to address all issues relating to technology fatigue.

At the same time, private seed growers are to be encouraged and timely payments are made to them by the National Seed Corporation (NSC). This will go a long way in helping this programme as farmers are enthusiastic for it taking up. South Bihar is suitable for seed production of rabi pulses. Seed Processing Plants need to be established under private sector by providing 25 per cent back ended credit link subsidy under Seed Production Programme. There is nothing wrong in the short term to import the required quantities of quality seed to reach the desired levels of seed replacement during the Eleventh Plan itself.

FERTILISERS

Fertiliser consumption in the state has been increasing in the post-bifurcation period. Consumption of nutrient has gone up from 61.2 kg. in 1993-94 to 94 kg. hectare in 2001-02 and 149.17 kg./hectare in 2006-07 (Table 8.4). This trend of increasing use of fertilizers is likely to be maintained in 2007-08 also, as total fertilizer consumption in the kharif-2007 has been 15.11 per cent more than the corresponding figure for kharif-2006. Consumption of fertilizers could have gone up further, but for the supply demand gap. (Table 8.5).

Consumption during the rabi season is generally higher than that of the kharif season which is a normal phenomenon even in other states. (Table 8.6). Total consumption of fertilsers during 2004-05 was 732 thousand tonnes as against 695 thousand tonnes in 2003-04. The consumption ratios of N.P.K. which were 24.3:7: 1 in 2003-04 improved to 6.8: 3: 1 in 2005-06. (Table 8.7). This presents an unbalanced use as against the ratio of 4:2:1.

Table 8.4: Consumption of Nutrient in Bihar

Year	*Nutrient Consumption (kgs. hect.)*
1993-1994	61.20
1994-1995	62.50
1995-1996	65.00
1996-1997	68.00
1997-1998	69.00
1998-1999	72.00
1999-2000	78.50
2000-2001	85.00
2001-2002	94.00
2002-2003	96.00
2003-2004	87.50
2004-2005	99.78
2005-2006	129.32
2006-2007*	149.17

* Estimated

Source: Fertilizer Association of India.

Table 8.5: Demand, Availability and Consumption of Major Fertilizers in 2007-08

(*Lakh MT.*)

Fertiliser	*Kharif 2007*			*Rabi 2007-08*		
	Demand	*Availa-bility*	*Consump-tion*	*Demand*	*Availability (till Jan.'08)*	*Consumption (till Jan.'08)*
Urea	9.00	8.53	7.84	11.0	8.55	7.69
DAP	2.50	1.102	1.04	2.5	2.15	1.86
MOP	1.00	0.302	0.15	1.00	0.69	0.64
NPK	2.00	1.09	0.81	2.00	0.69	0.68

Source: Department of Agriculture, Government of Bihar.

Although fertiliser consumption increased in all the zones during this period. There was wide variation in the level of its use across zones/ districts. It was as high as 104 kg./ha in Zone-III and 69 kg./ha in Zone-I during. (Table 8.7). Across different districts fertilizer consumption was as high as 167 kg./ha in Patna followed by Nalanda (159 kg./ha) and Khagaria (142 kg./ha). Gaya, Nawadah, Begusarai, Saran, Purnea and Rohtas are other districts, which use high levels of fertilizer. Consumption levels of fertilizer in these districts were almost similar to those in Punjab, Haryana and other agriculturally developed states. The lowest fertilizer consumption was observed in Madhubani (18 kg./ha) followed by Lakhisarai (27 kg./ha) and Supaul (31 kg./ha). Other districts, which use very little fertilizer, include Kishanganj, Bhabhua, Saharsa, Gopalganj etc. (Table 8.8)

Table 8.6: Consumption of Fertilisers Season-wise in Bihar During 2003-04 and 2004-05

('000' tonnes)

Nutrient	2003-04			2004-05			2005-06			% increase decrease
N	Kharif	Rabi	Total	Kharif	Rabi	Total	Kharif	Rabi	Total	
N	302.74	320.84	623.58	290.13	328.67	618.80	333.59	475.12	808.71	+29.7
P_2O_5	12.34	33.51	45.85	16.11	54.79	70.90	69.56	108.86	178.72	+289.8
K_2O	3.45	22.22	25.67	3.24	38.74	41.98	26.76	50.61	77.37	+201.4
Total	318.53	376.57	695.10	309.48	422.20	731.68	430.21	634.59	1064.80	+53.18

Source: Fertilizer Statistics of India.

Table 8.7: Zone-wise Fertilizer Consumption in Bihar

(kg. per he)

	Nitrogen			Phosphorous			Potash			Total (N+P+K)		
	1982	1991	1998	1982	1991	1998	1982	1991	1998	1982	1991	1998
Zone-I	17.59	37.76	51.75	4.21	12.10	12.98	2.29	5.19	4.52	24.10	55.04	69.25
Zone-II	11.12	29.53	57.13	2.51	14.13	12.40	1.58	6.70	8.71	15.21	50.37	78.24
Zone-III	18.67	60.32	87.95	2.99	18.41	13.76	1.48	8.63	2.16	23.15	87.36	103.87
Bihar	16.57	41.56	63.99	3.24	14.60	13.01	1.74	6.69	5.23	21.54	62.84	82.23

Source: Fertilizer Statistics, Fertilizer Association of India (Several Volumes) Year refers to Triennium ending.

Table 8.8: District-wise Fertilizer Consumption Per Hectare in Bihar (as of 2004-05)

District	Fertilizer Consumption kg. per he. (NPK)	District	Fertilizer Consumption kg. per he. (NPK)
Patna	214.99	Darbhanga	138.60
Nalanda	143.11	Madhubani	28.04
Bhojpur	174.28	Samastipur	176.34
Buxar	95.78	Begusarai	219.12
Rohtas	106.04	Munger	51.48
Bhabhua/Kaimar	61.87	Lakhisarai	67.66
Gaya	140.06	Sheikhpura	131.07
Jehanabad	154.55	Jamui	164.91
Nawada	82.77	Bhagalpur	211.44
Aurangabad	89.70	Banka	34.24
Saran	91.65	Saharsa	56.74
Siwan	78.42	Supaul	19.18
Gopalganj	37.21	Madhepura	90.42
Muzaffarpur	164.68	Purnea	169.05
East Champaran	127.33	Kishanganj	35.23
West Champaran	169.99	Araria	79.68
Sitamarhi	74.03	Katihar	118.73
Sheohar	11.94	Khagaria	281.84
Vaishali	174.88	State total	116.58

The proportion of farmer households using farm inputs in the winter season exceeds those using them in the kharif season. More farm households use fertilizer, improved seeds, and pesticides in the rabi (winter) season versus the kharif season. This is primarily due to the availability of irrigation in the winter season. For instance, 43 per cent of farmer households use improved seeds in the rabi season compared to an all India level of 34 per cent. In the kharif season the reverse is true: just 34 per cent use improved seeds, compared to the national average of 46 per cent (Table 8.9).

There are complaints that allocation of fertilizers is delayed and that substantial quantities are smuggled to Nepal. Along with NPK, Bihar should also apply micro-nutrients as well as organize fertilizer. We have recommended bio-gas plants for villages needs of power. The slurry coming from these plants will serve as a valuable source of nutrients.

Table 8.9: Fertilizer Consumption and Use of Farm Resource 2003-04.

	Fertilizer Consumption	*Per cent of farmer households using*									
		Fertilizer		*Organic manure*		*Improved Seeds*		*Pesticides*		*Vet. Services*	
	Kg./ha.	*Kharif*	*Rabi*	*Kharif*	*Rabi*	*Kharif*	*Rabi*	*Kharif*	*Rabi*	*Kharif*	*Rabi*
Orissa	41.4	76	15	65	10	19	10	41	10	31	8
Madhya Pradesh	55	59	52	40	39	29	21	28	23	19	14
Bihar	80.5	59	91	32	34	34	43	40	46	19	18
India	89.8	76	54	56	38	46	34	46	31	30	22
West Bengal	122.4	89	72	54	43	63	58	80	65	32	29
Uttar Pradesh	126.7	78	88	48	56	48	53	39	35	26	21
Andhra Pradesh	136.8	81	30	69	28	68	30	71	30	41	30
Punjab	184	54	55	36	35	44	46	52	52	66	71

Credit

In a situation when we are dealing with a poverty ridden peasantry, agriculture has to be made a profitable avocation. This is possible only by raising productivity levels as in other States like Punjab. We have already discussed various steps required to achieve this objective. All this will call for the need to place either sufficient financial resources or physical inputs in the hands of the farmer to enable him to adopt modern techniques of farming. Credit is a critical input for the development of agriculture and allied activities. The key task is to ensure a convergence among credit availability and effective credit delivery system in keeping with adequate credit absorptive capacity of the farmer. Mere availability of credit does not ensure its productive use and increased production/value addition. This is all the more important in the case of small/marginal farmers. An analysis of the credit situation at the all India level presents a paradoxical situation. Even as the credit target of Rs. 225,000 crore for 2007-08 seems to be on track, a significant problem has emerged—the number of farmers being provided credit by the banking sector has actually gone down. An indication that the sharp fall in farmer accounts may be hurting small and marginal farmers is that most of the decline in accounts has been seen in co-operative banks that provide small loans to small borrowers—from 2.25 crore accounts in 2004-05 to 1.89 crore in 2006-07.

Graph 8.2: Credit Flow to Agriculture

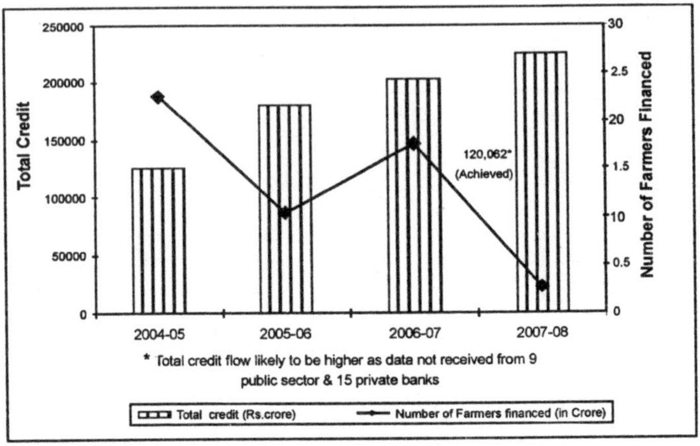

While production for market is not yet the predominant characteristic of agriculture in Bihar, its dependence on the market for input is quite large. This dependence on the market for inputs in an agriculture, where marginal and small holdings together account for 96.50 per cent of total number of holdings and 67.36 per cent total owned area (NSS) report (491), makes agricultural credit a critical input in agricultural production. However, agricultural credit scenario is not very encouraging. (Table 8.10).

Table 8.10: Bank Branches Per Lakh Population by District (as of 2001)

State/District	*Bank branches per lakh population*	*State/District*	*Bank branches per lakh population*
Patna	7.90	West Champaran	5.30
Nalanda	5.40	East Champaran	5.10
Bhojpur	5.80	Vaishali	4.70
Rohtas	5.60	Darbhanga	5.60
Gaya	6.10	Madhubani	5.00
Jehanabad	5.30	Samastipur	5.20
Aurangabad	5.30	Saharsa	4.90
Nawada	5.10	Araria	4.30
Saran	5.20	Purnea	4.50
Siwan	5.40	Madhepura	4.80
Gopalganj	5.00	Kishanganj	4.80
Muzaffarpur	6.10	Katihar	4.90
Sitamarhi	4.90	Bhagalpur	5.70
Sheohar	NA	Munger	5.40
Begusarai	5.60	Khagaria	4.70
Bihar	5.60	India	7.30

While only 1.3 per cent of cultivators in Bihar availed institutional credit during 2002-03, the corresponding number for Punjab and India as a whole was 24.2 per cent and 10.1 per cent respectively. Also, average amount of credit disbursed to cultivators during the year was found to be a meagerly sum of Rs. 322 against 12,118 for Punjab and all India average of Rs. 2647. This will also explain that the number of indebted households in Bihar (33%) was lower than the all India figure of 48 per cent (Tables 8.11 and 8.11A)

Table 8.11: Number of Rural and Farmer Households Indebted in Bihar State

(in '000)

	Bihar	*All India*	*Bihar %*
Number of Rural Households	116853	1478988	7.9
No. of farmer households	70804	893504	7.9
No. of indebted farm hhs	23383	434242	5.4
% of farmer hhs indebted	33.0	48.6	

Source: Agriculture at a Glance 2005.

Graph 8.3: Proportion of Cultivators who availed Institutional Credit During 2002-03

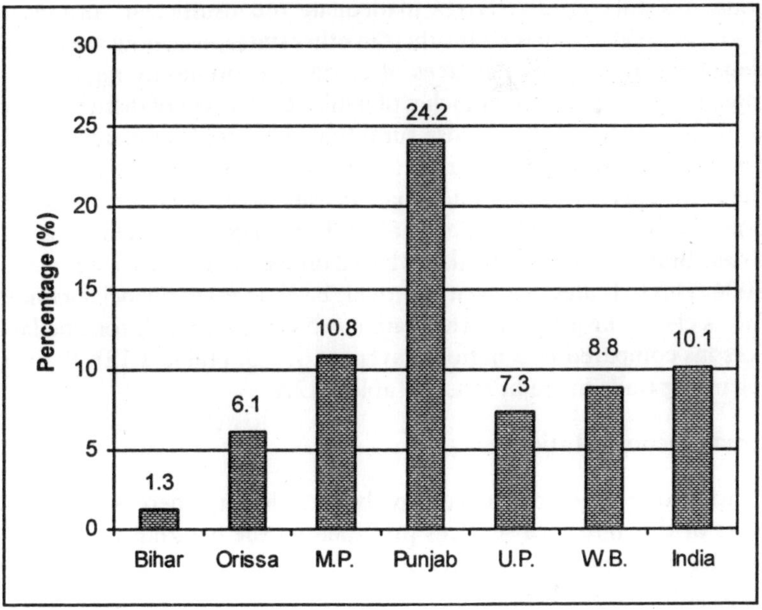

Table 8.11A: Incidence of Indebtedness Based on Size of Land Possessed

State	*% of marginal indebtedness Fhhs (upto 1.0 ha. of land)*	*% of small indebted Fhhs (1.01 to 2.00 ha)*	*% of semi-medium indebted Fhhs (2 01 to 4.00 ha)*	*% of medium indebted Fhhs (4.0 to 10.00 ha)*	*% of large indebted Fhhs (+ 10.00 ha.)*
1. Uttar Pradesh	71.3	17.4	7.8	3.4	0.3
2. Maharashtra	36.0	26.2	23.3	12.2	2.4
3. Madhya Pradesh	33.0	27.1	23.1	13.0	3.9
4. Rajasthan	43.9	19.8	17.8	14.1	4.5
5. Karnataka	50.7	22.8	15.9	9.3	1.2
6. Andhra Pradesh	55.7	21.8	15.1	6.6	0.7
7. Bihar	86.9	9.2	2.8	0.7	0.6
8. West Bengal	88.7	8.5	2.4	0.4	0.0
9. Punjab	53.3	15.8	17.0	11.8	2.2
10. Orissa	70.3	20.6	7.3	1.7	0.0
All India	61.0	18.9	12.5	6.4	1.2

FHHs = Farmers Households

Source: Agricultural Statistical at a Glance 2005.

Though over the years there has been a significant improvement in the availability of institutional credit in the agriculture sector in the state, it is still considered to be inadequate and insufficient. As a result, Farmers in Bihar like their brothers in other states, even now are heavily dependent on informal sources of credit at exorbitantly high rate of interest. Out of the total branches of banks, 65 per cent of them branches were in rural areas. Bank wise rural branches are; Commercial Banks 59.6 per cent, RRBs-86.5 per cent and Co-op Banks—17.66 per cent of the total. There is a gross inadequate density of financial services access points in Bihar which has a total of 45,103 villages. It would translates to one branch in every 13 villages based on the total branch estimates of 2006. This indicates that even to provide basic access to financial services, this is clearly insufficient. The state has 5.6 bank branch for one lakh persons compared to a national average of 7.3. (Table 8.12). Position regarding trade in credit is as in Table 8.12A.

Credit Deposit Ratio

In addition to the presence of low branch density, there is the added issue of the nature of services provided by the banking system as a whole. The extreme poverty level of the state is reflected in the aggregate deposit base.

Table 8.12: Trends in Credit and Deposit

(Rs. crore)

Bank Category	*2003-04*		*2004-05*		*2005-06*	
	Deposits	*Advances*	*Deposits*	*Advances*	*Deposits*	*Advances*
Commercial Banks	3,06,405.0	78,296.3	3,40,562.0	96,657.7	3,91,770.0	1,19,880.0
Regional Rural Banks	44,215.4	12,533.7	54,751.5	17,959.9	61,820.0	22,250.0
Co-operative Banks	7,620.8	5,208.6	7,635.3	5,694.4	7,750.0	5,950.0
Total	3,58,241.0	96,038.6	4,02,949.0	1,20,312.0	4,61,340.0	148,080.0

Table 8.12A: Trends in Agriculture Credit—Bihar (1972 to 2007)

	Dec. 1972	*Dec. 1982*	*Dec. 1992*	*Dec. 2002*	*Dec. 2007*
No. of Accounts ('000)	26	684	2,228	712	1,593
Amount Outstanding, (Rs. Crore)	7	215	987	1,487	6,647
Per cent to State's total Bank Credit	3.7	22.2	25.2	22.7	24.2
Per cent to All-India Agriculture Credit	1.5	3.8	4.9	2.3	2.9

Source: RBI, Banking Statistics, Basic Statistical Returns of Scheduled Commercial Banks in India, in 2007 (Vol. 36) and earlier issues.

The average deposit per individual is only Rs. 55,000 against the national average of Rs. 1,95,000. The situation of advances is even worse for Regional Rural Banks and Co-operative Banks where total amount of advances for the year 2005-06 is only Rs. 22,250 and Rs. 5,950 respectively (Table 8.13). However, this trend is moving in a positive direction over the years. The Credit Deposit Ratio in the state is also very poor as compared to the national average. In case of Commercial Banks and Regional Rural Banks the CD Ratio in Bihar is 31 and 36 respectively. The performance of Primary Agricultural Credit Society (PACS) also reflects a similar state. (Table 8.13). While there are almost 6,000 PACS in Bihar, only 19 per cent of them posted a profit in the year 2004-05, as compared to the national average of 43 per cent. Credit Deposit Ratio in the State in 2005-06 was 30 per cent, up from 21 per cent in 2001-02. Though the Bank lending to agriculture in the country reached the level of

Rs. 952 million in 2004, up from Rs. 249 million 1995, Bihar's share of that amount declined from 5 per cent to 2 per cent.

Table 8.13: Credit Deposit Ratio March 31, 2006

(*Percentage*)

Banks	*Bihar*	*National*
Commercial Banks	31	70
Regional Rural Banks	36	54
Co-operative Banks	77	100
Total	32	73

Kisan Credit Card (KCC) Scheme started in 1999—is one important medium for increasing agricultural credit. At present, there are 1.04 crore landholdings in the State, but till 2005-06, only 14.5 lakh KCCs have been distributed against the target of 30.5 lakhs. To meet the target of crop loans of Rs. 10042 crore in 2006-07, KCCs have to play a very important role. Although total disbursement under KCC has increased almost 2.5 times, from Rs. 342 crore in 2002-03 to Rs. 815 crore in 2005-06. It still channels only a miniscule proportion of the total credit volume for crop loans. The coverage of farmers in the State has not, however, been uniform over the years (Graph).

Graph 8.4: Percentage of Farm Households Covered Under KCCs in Bihar.

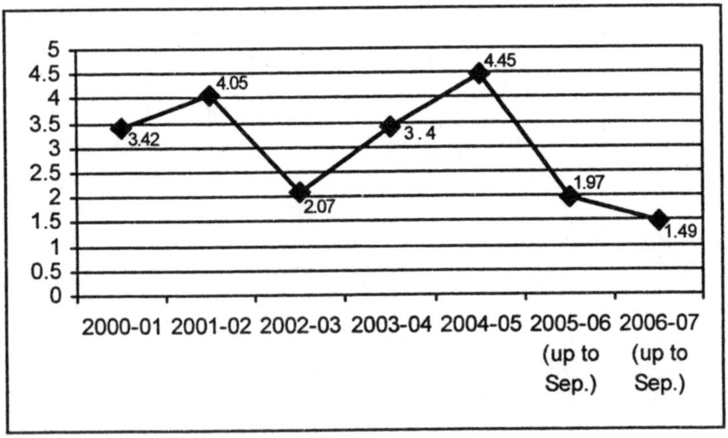

In the area of micro-finance, more than 17,000 Self-Help Group had been financed by banks up to 2005 through their priority lending schemes in 38 districts. The target for 2006-07 is to raise the number to 24,000 SHGs and advance Rs. 48 crore through the scheme. As on March 31, 2007, a cumulative number of 72638 SHGs were credit linked to banks and the total credit flow to these SHGs worked out to Rs. 202.98 crore.

Institutional Credit

The flow of credit for agriculture in the state from the institutional sources from 2003-04 till 2005-06 (Table 8.14). Shows that total institutional credit in the agricultural sector in the state recorded an increase by almost 82 per cent over the period. This may be attributed to the following developments:

(1) *Kisan Credit Card Package*—Effective instrument to the Commercial Banks and the RRBs to enter the crop loans segment.
(2) *Million Shallow Tubewells Programme* (MSTP)—A convenient platform for the CBs and the RRBs for increasing investment credit.
(3) Growing agriculture sector since the late 1990s supported financing for Farm Mechanisation in Bihar.

Table 814: Agency-wise Flow of Credit for Agriculture in Bihar State.

(Rs crores)

Year	*Commercial Banks*	*Co-operative Banks*	*RRBs*	*Total*
2003-04	792.44	199.72	204.87	1197.03
2004-05	1325.06	293.75	431.30	2050.11
2005-06	1489.33	234.61	450.09	2174.03

Source: SLBC, Bihar.

Despite this improvement in absolute term, the availability of credit is far short of requirement. For example, the credit need of the state was estimated at Rs. 11341 crore in 2006-07 (Government of Bihar, Econ. Survey, 2007), Rs. 10042 crore for crop loans, Rs. 78.22 crore for the Macromode (ISOP AM plan), Rs. 60 crore for micro-irrigation, Rs. 18.50 crore for agri-clinics and Rs. 1142 crore for horticulture mission etc. But credit target for agriculture set by bankers was only Rs. 3732 crore, which was only one-third of the assessed requirement.

CBs, RRBs and Co-operative Banks meet on an average 68,21 and 11 per cent of the requirement respectively. Further, credit portfolios are mainly restricted to four activities namely (i) Kisan Credit Cards, (ii) Million Shallow Tubewell Programme (MSTP), (iii) Farm Mechanization and (iv) Sampoorna Grameen Swarozgar Yojana (SGSY).

Per hectare availability of institutional credit was very low by any yardstick in most districts. Average cost of cultivation per hectare in the case of paddy and wheat, comes out to about Rs. 9000 per crop, and judged from this angle, even districts like Patna, Muzaffarpur and Samastipur, which have relatively better availability of credit, do not get their full credit requirement met. The ratio of credit to cost of cultivation works out to less than 5 in most cases. This is particularly relevant because small and marginal farmers in Bihar do not have invisible surplus of their own.

At the top of that institutions concerned with agriculture and rural development have become more or less non-functional. We have no hesitation to say that high level of institutional credit at cheap rates has not led to any increase in agricultural production, but to supply related inflation, if at all. Not only that the distribution of this credit is inequitable; how much of this is used for productive purposes is also a question mark. The experience from Punjab shows that this cheap credit with subsidies led to the easy purchase of tractors, consumption of drugs and liquor and expenditure on ostentatious marriages and such other unproductive purposes. Nearly 44 per cent of farm households meet their credit requirement from money lenders (NSSO 2003). In this respect, Bihar is second only to Tamil Nadu where it is 52 per cent. On the uses of loans, only 33 per cent of farm households contract loans for capital and current agricultural expenses. Typically consumption, marriages and ceremonial expenditures account for the largest share of loan. NSSO data also indicates that, on an average, the amount of outstanding loan per farm household in Bihar is among the lowest in the country. The experience of micro-credit in Bangladesh is an important lesson for others to follow where normal or even a little more than normal interest rates are charged, the borrower has not to give any proof of his credit worthiness and sureties are not needed. Even a beggar can borrow. But the recovery is stated to be nearly 99 per cent without any coersion. Some of the basic problems which inhibit institutional credit flow in agriculture are:

> The problem is not peculiar to Bihar. The position is not very much different at the all India level. While the institutional credit

for agriculture has expanded after bank nationalization, from Rs. 1,865 crore in 1971-72 to Rs. 1,80,486 crore in 2005-06 and to Rs. 2,03,296 crore in 2006-07, the credit needs of agriculture have not been fully met and an overwhelming number of farmer households in India have not been able to borrow from the formal institutional system. According to the National Sample Survey Organization's (NSSO) situation Assessment Survey of Farmers (2003), 48.6 per cent of farmer households were indebted, of which only 57.7 per cent were indebted to institutional agencies. In other words, only 28.04 per cent of farmer households were indebted to institutional agencies. There is huge financial exclusion in the rural areas.

In Punjab, Assam and Bihar, the total debt of the farmer households from institutional agencies was less than from non-institutional agencies (Situation Assessment Survey of Farmers, 2003). It is worth looking at why the professional moneylenders are a popular choice of the Indian farmers for raising credit. Besides the problem of exclusion, the lack of flexibility, insistence on collaterals and non availability of appropriate products are the real problems of institutional credit. Further, more often than not, the real cost of loans from banks works out more or less equal to the rate of interest charged by the informal agencies. Table 8.15 reveals very clearly what ails rural credit in India. The formal institutions offer poor quality service through inadequately manned branches under a mandatory rural branch posting policy with a short-term stay, which gives little time to the staff to develop knowledge about the area and the people. Where the loan approval takes an average 33 weeks, the farmers cannot rely on this source of financing.

Table 8.15: Aspects of Formal Borrowings and its Costs.

Criteria	*Banks*	*Regional Rural Banks*	*Co-operatives*
Interest rate (in %)	12.5	11	11
Loan amount received as percentage of amount applied	91.8	88.2	83.5
Percentage of household reporting bribes	26.8	27.0	09.7
Bribe as percentage of amount approved	10.1	18.2	19.9
Time taken to process loan application (week)	33.0	28.5	24.0

Source: Balasaheb Vikhe Patel, Agricultural Indebtedness, *E.P.W.*, February 2, 2008.

Characteristics of rural credit in Bihar may be summarized as:

(i) Nearly one third of the cultivated area is reported to be under oral share cropping and, therefore, cannot avail crop loan.
(ii) Almost 40 per cent of cropped area is flood prone and 41 per cent drought prone which also restricts the flow of credit.
(iii) About 80 per cent of the land owning households possess less than 0.30 hectare of land.
(iv) Only 33 per cent farmer households have access to credit as against the national average of 48.6 per cent; and
(v) Only 23 per cent of borrowing farm households get credit from institutional sources as against the national average of 56 per cent i.e. only 8 per cent of the total farm households have access to institutional credit (all-India—27 per cent).

Minikit Programme

It is an accepted fact that majority of small/marginal farmers are not able to take advantage of institutional credit for various reasons as discussed here. Appreciating this problem, Report of Dr. Sen Committee on Agriculture Production in Eastern India 1987, made a specific recommendation for a Minikit programme for marginal/small farmers. The idea is that the small/marginal farmers are given free, a small kit, which contains the required quantity of seed, fertiliser and even pesticide for specific farm size along with instructions for use in local language. The programme was taken up earnestly by West Bengal for summer rice during late eighties and early nineties with very successful results. Under present Bihar conditions, introduction of this programme both for rice and wheat is strongly recommended.

Realising the value of this Programme, the Haryana State Agriculture Department, has decided to distribute 8,050 minikits of hybrid varieties of seeds of pulses and oilseeds to farmers under the Integrated Scheme of Oil Seeds. Pulses, Oil Palm and Maize (ISOPOM) during Kharif-2008 to encourage them to cultivate these crops. This will go a long way to improve productivity. It is difficult for the small farmer to procure quality seeds and fertilizers, etc., in small quantities. This programme meets his most important need. It is high time for the Government of Bihar to benefit from the experience of other States in this respect so as to improve productivity levels of different crops.

FARM MECHANISATION

It needs no reiteration that farm mechanization helps to increase agricultural productivity, reduce the cost of cultivation as well as alleviation of drudgery associated with various farm operations. In fact, productivity of the farm depends considerably on use of farm power derived from efficient farm implements and their judicious utilization. Though farm mechanization is generally construed as use of tractors and basic tractor dawn implements, all other types of machinery, implements and equipments, used in agriculture are also covered under this sub-sector. Farm mechanisation is thus one of the important sub-sectors of agriculture and forms an integral component of the agricultural development strategies.

Productivity of the farm depends considerably on the use of farm power derived from efficient farm implements and their judicious utilisation. Technological advances in other areas of agriculture also necessitated use of mechanised equipment, instruments and other tools for maximising benefits from the technological advances and for optimizing costs. Farm machinery ensures timeliness of various operations which is an important factor affecting crop productivity, particularly wheat mechanisation of various farming operations increases the production and productivity of land, reduces the drudgery associated with farm operations and helps in reducing socio-economic disparity among farmers. It encourages better management of farm enterprise, reduces overall cost, provides employment opportunity and encourages entrepreneurship. It must, however, be appreciated that mechanisation is not an end in itself, but a means of development.

Present Status

Farmers in Bihar have been given incentives to go in for farm mechanization. The state government gives subsidy on identified agricultural machineries, in particular on small farm implements such as power tiller, thresher, willower, zero-tillage reaper, zero-seed drill, and specialized power driven equipments. This is being done under Farm Mechanisation Plan with an outlay of Rs. 18-22 crore for the financial year 2007-08. This programme is included in the Macromode Plan, sponsored by the central government, to encourage farm mechanization in the country. To encourage selected agricultural equipments, the state government would provide a subsidy of 25 per cent in addition to the provisions made under the Macromode Plan. Thus there would be a subsidy of 50 per cent on purchase of the selected farm equipment. However, tractors would continue to have 25 per cent subsidy because small farm implements have been prioritized in the mechanisation programme.

After the repeal of BAPMC (Bihar Agricultural Producers Marketing Co-operative Act), agricultural market in the state are functioning without any formal institutional structure. Following disbanding of BSAMB, the fate of 1500 rural hats under its purview is also hanging in balance. The Board had developed infrastructure during 1978-79 with the World Bank assistance, but at most places these hats are in a very poor shape because of poor maintenance. In the meantime, absence of a fully functional market body has certainly created a significant glitch of non-availability of market intelligence, as there is no information on market arrivals, its prices and destination of dispatch etc.

Compared to other states of India, the adoption of farm mechanisation is lagging in Bihar. The availability of power at the farm level in India has increased considerably to 1.231 kw/ha in 2001 from about only 0.25 kw/ha in 1951. However, the farm power use is still lower in Bihar (0.35 kw/ha). At present, number of tractors in the state are 111.7 thousand. Number of threshers, stonery engine and sprayer are 300, 250 and 29 thousand respectively. The tractor density in the state is also low at 4.93 tractors/1000 ha as against 56.20 tractors/1000 ha in Punjab and 12.2 tractors/1000 ha in Uttar Pradesh. Presently only 11.24 per cent of the total power requirement for farms is estimated to be met from the available tractors in the State. Therefore, considerable potential exists for other farm equipments like power tillers etc; since 80 per cent of the total land holdings belong to small and medium farmers who cannot afford tractors. Considering the present status of mechanised farming activities potential for other agricultural implements is very high. In order to improve the labour productivity of the farmer and timely planting of wheat in particular, introduction of Zero Till machines etc., is important. Haryana and Maharashtra models will come quite handy for the small/marginal farmers (Table 8.16).

Haryana Experiment

One of the major problem which stands in the way of spread of mechanization is the inability of the small/marginal farmer to make an economic use of machinery. To popularize mechanization, the Government of Haryana has introduced a Modified Micro-Management Mode (MMM) scheme of Agricultural Mechanisation. MMM provides for subsidies and other benefits to farmers for farm mechanization. Initially MMM was a failure as farmers viewed it with skepticism due to the cumbersome procedures that were required to be followed to avail benefits for mechanization of farms. Things improved after changes were made in the MMM, keeping in mind the practical difficulties of the farmers in 2006-07. This has led to the tremendous rise in the adoption rate of this scheme among farmers. A study conducted by the Department of Agriculture in Sirsa district shows that just one little change has made Haryana farms more mechanized. The last few months have seen the sale of 217 straw reapers, 110 cotton seed drills and 102 tractor mounted spray pumps, indicating a sudden spurt in buying because of the small change. The study indicates that more and more farmers are now going in for farm mechanization under the MMM scheme. They are willing to adopt new technologies. It has to be realized that the farmer needs to be assured, that he will be purchasing not only a quality product, but also the sole beneficiary of the subsidy that has been offered to him by the Government.[2]

A study of mechanization at the District level shows (Table 8.16) that in districts like Katihar, Madhubani, Saharsa, Sitamarhi and Siwan, even the availability of electric pumpsets per hectare are very small in number. Many a times, farmers use diesel pumpsets to irrigate crops which is an expensive method of irrigation. But in the absence of adequate and stable power supply, they have no other choices.

A New, Cheap Power Tiller

Birla Harit, a power tiller developed by the Uttarakhand based Birla Power Solutions for farmers with small land holdings, is being billed as a substitute for costly and bigger tractors. The power tiller costs Rs. 1.2 lakh, but the company hopes that NABARD will provide subsidy of Rs. 30,000 which will help bring down the cost to less than Rs. 1 lakh. The power tiller will consume less than a litre of diesel in an hour, making it cheaper for the small farmers to till their lands or use the accompanying agricultural appliances. The 11.5 horse power, 12-V machine can be used for tilling, ploughing, weeding, pumping, leveling, puddling, hulling, thrashing and even riding by attaching trailer, Birla Power will produce 250 mini-tractors per month at the Lal Tappar unit near Dehradun. Given the encouraging response it received from farmers at the recent agriculture fair at PUSA Institute in Delhi, the product might in fact help the small farmers in a bigway.

Experiment of a Pune Trader

Here we also quote the example of another simple religious minded person—a trader from Pune. He launched his own 'Prevent Farmer Suicide' Mission. He bought a tractor for Rs. 50,000 from the personal savings of his wife. His mission involves locating poorest of poor farmers and plough their land free of cost. The farmer has to bear only the fuel cost and the driver's wages which come to around Rs. 150 a day. He has set simple rules for lending his tractor. The smaller the farm, the higher the priority. Special concessions are given to poor peasants; those who cannot afford to pay for the driver's fee, are trained and he pays for the driving school and license fees. In a short period of six months, 50-odd small farmers from the area have benefited, resulting in nearly 100 acres of fallow land getting transformed into cultivable land. Add to that another 50 acres elsewhere in Pune district and the result of this initiative becomes apparent.

The problem of the small farmer to maintain a pair of bullocks is no more a practical proposition. It costs him anything between Rs. 50 to Rs. 100 a day for the entire year and their utility period lasts hardly a couple of months in a year. The owners of tractors target only big farmers and they charge up to Rs. 3,000 an acre. Loss in cultivation cost when crop fails is one of the major reasons of farmer suicides in Maharashtra. The effort of this unique mission to cut down the cultivation cost of small farmers, certainly addresses the issue in right perspective *(Indian Express)*. This pattern coupled with micro credit, if replicated across the country with the aid of government and private industry, can possibly go a long way towards improving the fate of marginal/small farmer families. It must be appreciated that the share of this small/marginal farmer is just 13.5 per cent of an average Indian in the national kitty. (Graph 8.5) The present system of credit for machinery has thus hardly any meaning for this '*aam admi*'.

Graph 8.5: Index of Per capita Income by Agriculture Status

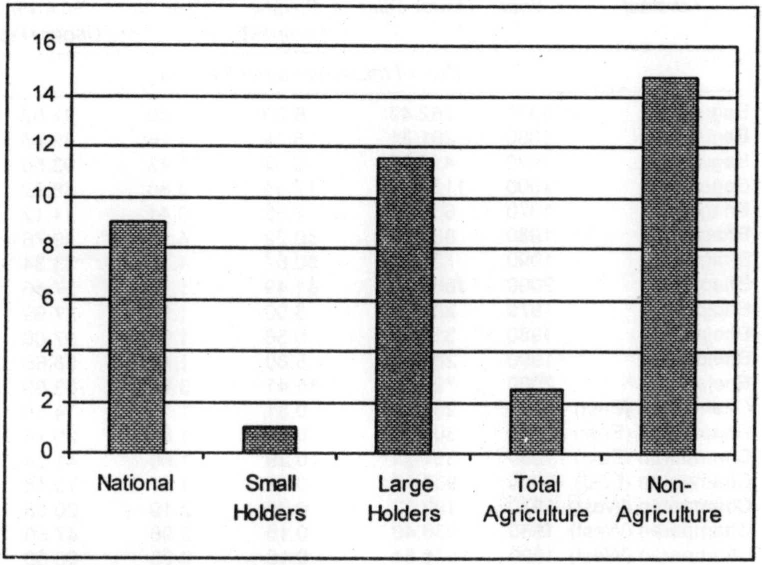

Source: Calculations by the author.

Finally, in order to improve the labour productivity of the farmer in Bihar and timely planting of wheat in particular, introduction of zero till machines, etc., will come quite handy for the small/marginal farmers of the state. As in Punjab, agricultural co-operative societies can also come to the help of poor farmers by renting out various types of agricultural machinery.*

Table 8.16: Changes in Level of Modern Technical Inputs by district

District	*Year*	*Power Tiller*	*Electric Pumpset*	*Tractor*	*% Area Under HYV*
		(No. of Implements per '000 Ha.)			
Aurangabad	1970	486.16	16.15	0.90	8.28
Aurangabad	1980	703.41	20.16	1.12	60.88
Aurangabad	1990	701.42	15.77	0.88	67.52
Aurangabad	2000	1807.74	41.90	2.33	80.47

(*Contd.*)

* For more details refer to Chapter 12.

Table 8.16: (*Contd.*)

District	Year	Power Tiller	Electric Pumpset	Tractor	% Area Under HYV
		(No. of Implements per 'ooo Ha.)			
Begusarai	1970	262.49	6.39	1.40	27.02
Begusarai	1980	291.31	5.75	1.26	39.56
Begusarai	1990	454.91	6.52	1.43	53.50
Begusarai	2000	1172.43	17.34	3.80	88.92
Bhagalpur	1970	631.88	2.68	0.41	11.12
Bhagalpur	1980	859.03	30.22	4.30	29.76
Bhagalpur	1990	725.00	30.67	4.36	33.34
Bhagalpur	2000	1868.50	81.49	11.58	57.46
Bhojpur	1970	289.93	6.00	1.51	27.99
Bhojpur	1980	317.92	6.56	1.65	57.00
Bhojpur	1990	283.15	5.80	1.46	58.65
Bhojpur	2000	729.74	15.41	3.89	63.69
Champaran (East)	1970	258.86	0.51	1.12	14.58
Champaran (East)	1980	306.90	0.27	1.50	29.76
Champaran (East)	1990	351-21	0.29	1.66	41.54
Champaran (East)	2000	905.16	0.78	4.41	73.46
Champaran (West)	1970	192.27	0.29	2.19	20.03
Champaran (West)	1980	930.49	0.15	2.96	47.60
Champaran (West)	1990	1075.81	0.16	3.23	60.00
Champaran (West)	2000	2772.64	0.44	8.58	69.81
Darbhanga	1970	191.50	0.85	0.13	8.46
Darbhanga	1980	88.38	15.53	4.79	26.27
Darbhanga	1990	99.25	17.44	5.37	65.32
Darbhanga	2000	255.78	46.33	14.28	65.82
Gaya	1970	598.44	4.49	0.28	9.63
Gaya	1980	856.97	51.50	0.91	34.80
Gaya	1990	764.35	47.18	0.83	55.60
Gaya	2000	1969.91	125.34	2.22	58.87
Gopalganj	1970	201.70	18.91	1.48	35.79
Gopalganj	1980	226.74	17.78	1.39	57.97
Gopalganj	1990	243.88	16.43	1.29	73.03
Gopalganj	2000	628.53	43.65	3.42	72.60
Katihar	1970	2649.81	1.57	1.06	11.72
Katihar	1980	2546.44	1.38	0.93	22.67
Katihar	1990	2801.35	1.40	0.95	33.23
Katihar	2000	7219.78	3.71	2.52	53.62
Madhubani	1970	314.68	0.27	0.41	11.87
Madhubani	1980	144.17	0.23	0.34	21.88
Madhubani	1990	137.04	0.26	0.39	32.08
Madhubani	2000	353.18	0.69	1.03	62.36
Monghyr	1970	441.47	4.93	0.38	15.54
Monghyr	1980	671.83	8.71	0.99	43.25
Monghyr	1990	684.03	11.14	1.27	42.45
Monghyr	2000	1762.92	29.59	3.36	51.95

(*Contd.*)

Table 8.16: (*Contd.*)

District	Year	Power Tiller	Electric Pumpset	Tractor	% Area Under HYV
		(No. of Implements per 'ooo Ha.)			
Muzaffarpur	1970	279.28	3.20	0.18	6.14
Muzaffarpur	1980	324.81	12.50	10.65	34.74
Muzaffarpur	1990	427.08	11.88	10.12	21.87
Muzaffarpur	2000	1100.69	31.56	26.88	41.61
Nalanda	1970	489.37	68.90	1.27	25.33
Nalanda	1980	437.92	77.30	1.43	56.49
Nalanda	1990	197.98	70.91	1.31	66.61
Nalanda	2000	510.25	188.40	3.48	71.82
Nawada	1970	538.45	25.08	0.32	10.62
Nawada	1980	638.47	39.30	0.50	39.93
Nawada	1990	516.08	33.95	0.43	39.86
Nawada	2000	1330.07	90.20	1.14	69.76
Patna	1970	366.16	25.69	0.60	7.10
Patna	1980	357.88	182.33	17.60	35.28
Patna	1990	244.28	174.61	16.86	97.90
Patna	2000	629.57	463.90	44.79	99.37
Purnia	1970	1949.74	0.17	0.85	8.71
Purnia	1980	1663.73	0.09	1.36	23.12
Purnia	1990	1621.68	0.09	1.36	10.55
Purnia	2000	4179.47	0.24	3.60	9.25
Rohtas	1970	7.08	18.68	2.20	33.18
Rohtas	1980	17.21	20.54	2.42	69.25
Rohtas	1990	15.05	17.97	2.11	70.35
Rohtas	2000	38.79	47.74	5.62	72.34
Saharsa	1970	363.93	0.20	0.65	17.99
Saharsa	1980	308.83	0.38	0.89	38.80
Saharsa	1990	523.49	0.67	1.58	72.59
Saharsa	2000	1349.17	1.78	4.20	72.02
Samastipur	1970	160.64	4.38	0.69	16.55
Samastipur	1980	214.41	4.55	0.72	34.60
Samastipur	1990	351.17	5.18	0.82	41.99
Samastipur	2000	905.04	13.75	2.17	72.13
Saran	1970	160.10	1.18	0.43	24.04
Saran	1980	215.91	5.29	5.24	43.76
Saran	1990	225.05	5.30	5.24	63.82
Saran	2000	580.00	14.07	13.92	70.69
Sitamarhi	1970	329.95	1.13	0.66	14.90
Sitamarhi	1980	427.66	1.07	0.63	24.97
Sitamarhi	1990	829.36	1.35	0.79	10.51
Sitamarhi	2000	2137.47	3.59	2.11	6.18
Siwan	1970	244.71	1.74	1.47	29.91
Siwan	1980	242.99	1.54	1.29	50.80
Siwan	1990	265.30	1.51	1.27	62.78
Siwan	2000	683.73	4.00	3.37	70.17

(*Contd.*)

Table 8.16: (*Contd.*)

District	Year	Power Tiller	Electric Pumpset	Tractor	% Area Under HYV
		(No. of Implements per 'ooo Ha.)			
Vaishali	1970	169.96	11.71	0.58	16.88
Vaishali	1980	231.62	11.48	0.57	38.53
Vaishali	1990	356.09	11.30	0.56	63.76
Vaishali	2000	917.73	30.03	1.50	72.72
Bihar	1970	482.44	9.55	0.88	17.22
Bihar	1980	542.71	21.44	2.73	40.07
Bihar	1990	578.92	20.32	2.73	51.62
Bihar	2000	1492.01	54.00	7.26	63.63

Notes

1. [E.A. Sadique, 1999, Agricultural Instability and Farm Poverty in India, *Indian Journal of Agricultural Economies*, [Vol. 44 (1) 1-16]. Water and Price Policy are discussed elsewhere in this Study.
2. *The Tribune*, July 16, 2007.

9

AGRICULTURE EDUCATION AND RESEARCH

Access to education is one of the critical elements in rural development in general and agriculture in particular. This is a means to upward mobility of the rural population. It enables the persons from rural origin to acquire new skills and get placed in occupations other than agriculture. At the same time education enables them to acquire knowledge and new technology in agriculture. A large number of studies at national and international levels have brought out that educated persons in the rural area get better access to health infrastructure, poverty alleviation programmes and other social security measures of the state. Thus, to make rural population partners in development or what is termed "inclusive growth" access to education and educational attainments is important.

The ever-increasing pace of technological revolution, national emphasis on agro-processing and value addition, increasing farm mechanization, the growing need of treating agriculture as a commercial activity and emerging competitive global economic environment, all call for the vitalization of agricultural education-technology-business relationship. In fact, the collaborative efforts of farm scientists having state-of-the art knowledge and creative industry, including agro-industry, can help in combining biological, engineering, biotechnological, nano-technological, chemical and other related sciences for achieving future breakthroughs in the production of high-quality commercialisable agro-products and agro-economic growth. However, the importance of these linkages has yet to be adequately appreciated.

As of now, the interaction between farm universities and industries is inadequate. At the farm university level, these linkages have mainly remained developmental type and limited to the testing of various kinds of agro-chemicals (fertilizers, pesticides, growth regulators), farm implements, industrial byproducts and pollutants and green as well as glass houses.

Some know-how is also provided to encourage agro-related enterprises such as farm machine industry, oil extraction, textile and sugar industry, fruit and vegetable processing, etc. The industry has so far largely ignored farm education and sponsoring of cutting edge farm research in the country for finding solutions to their problems, upgrading indigenous agro-processing methods or for developing internationally acceptable new agro-products processes, agro-chemicals, farm implements, etc.

The industry also does not prefer to suitably employ farm-trained graduates because agricultural education and training programmes are not adequately designed to satisfy the requirements of industry and business. So effective bridging of gaps in farm education-industry relationships is crucial to resolve these issues. It is time to realize that farm education-industry collaboration will be mutually beneficial through the rapid integration of new knowledge, experiences and technologies. These linkages benefit each other through exchange of information about cost-effective and eco-efficient agro-production and processing, development and efficient management of new hitech farming systems, employment requirements and opportunities international quality standards and bio safety regulations and cross-utilisation of facilities such as laboratories, sophiscated and expensive equipment. This interaction will also contribute in developing new and eco-friendly agro-chemicals, efficient farm implements/machinery, start-of-the art agro-processing infrastructure and waste management other than and non-farm activities.

The emerging trends of technological innovations demand our farm scientists and students to get them to comprehend the competencies, priorities and R&D needs of the industry. School children, rural artisans and farmers should get some industrial and business exposure so that they gain through enhancement of their skills, e.g., repair, fabrication, custom hiring and sale of equipment, transfer of technology, employment opportunities, etc.

Similarly, the involvement of industrial representatives in different education, training and research programmes and periodic reviews of syllabi would ensure incorporation of challenging ideas into the course contents. In fact, industry and business will be the main beneficiaries from these linkages due to continuous availability of newer agro-products, processes, technologies and trained human resources. These will also be of immense competitive advantage through shortening of time between discovery and agro-commercialisation.

Moreover, farm education-industry linkages, will enhance the socio-economic development of the area/region in general mainly due to skill enhancement, entrepreneurship development, and establishment of many companies and businesses around the institutions. The surrounding community will gain through improvements in indigenous agro-production and value-addition systems, development of agro-economic infrastructure, availability of newer products and processes, expanded labour absorption capacity of farm and non-farm sectors and enhancement in overall socio-economic growth.

Our fund-starved farm universities can gain through capital formation from industry supported collaborative adhoc short-term and long tem basic and applied research projects, seminars and workshops, consulting opportunities, demand driven training to industry's human resources and other activities. The students will gain through industry—sponsored scholarships, internships, hands—on experience and exposure to the business world necessarily required for becoming knowledge-workers and self-employed entrepreneurs.

A Central coordinating cell can facilitate industry's access to different education and research programmes and suitable faculty or department for taking up collaborative R&D projects. This cell can also arrange training programmes for small and medium-scale workers, entrepreneurs and industrial representatives and create an employment database, according to need-analysis (skills required) of the industries. It can also offer guidance of faculty members regarding patenting and commercialization of new technologies, products and processes and maintain liaison with other related public and private sector agencies.

In the emerging era of sustainable globally competitive and commercial agriculture of the 21st century, coordinated efforts of farm education and research systems and industry are paramount. However, the policies of both educational institutes and the industry have to be restructured so that collaborators may maintain certain autonomy while remaining interdependent and supportive of professional, entrepreneurial and commercial activities.

Research

While talking about research in India, we need to look at fundamental issues like education and availability of qualified persons. Today, we are about 1.1 billion people against China's 1.3 billion. But the number of graduates, specially technical, coming out of China is much more than that in India. Our IIT graduates may be at par with those from the Ivy League, but there is a problem with focus. We still do not have a degree, even in IITS, exclusively for agriculture. So, if we are strictly talking about research, there is a shortage of specialized talent, and the government is not putting in enough effort to grow the talent.

Even with the researchers, the main problem is that we are not challenging them enough on what they are doing. Research is not a 9-to 5 job. If you have that kind of a mindset, research will not succeed. The way to go about is to create an environment, get the best people, don't worry about the costs, give them a boundary, and challenge them. Unfortunately in India, this is not happening and there are too many cooks. It is time to actively look at applied research in which we have to follow a backward approach to look at the applications first. The development of technology can follow. We need to talk to people, end-users and customers to find out what applications are required today. Research can help develop technologies towards that end. The same applies to agriculture much more where we have to deal with millions of uneducated deprived persons.

There are then no two opinions that application of science and technology in agriculture leads to productivity growth. In a state like Bihar where nearly 77 per cent of the total workforce are dependent on agriculture and where out of 16.4 million land holdings, nearly 92.5 per cent are small and marginal, importance of agricultural research for innovation of commodity-specific and location specific technology and their upgradation along with their extension can hardly be overemphasized. The research-extension—farmers linkage, is also extremely poor in the state and there is a complete lack of focus on areas of relevance, opportunity and comparative advantage.

Studies consistently show high returns to investments in agricultural research in developing countries, averaging over 40 per cent (Table 9.1). A recent paper on comparative returns to public investments in R&D, irrigation, roads, education and electricity in China and India showed that public investments in R&D contributed the most to promoting agricultural growth in both countries.

Table 9.1: Estimated Rates of Return to Investment in Agricultural Research

Region	*Number of Estimates*	*Median Rate of Return (%)*
Africa	188	34
Asia	222	50
Latin America	262	43
Middle East-North Africa	11	36
All Developing Countries	683	43
All Developing Countries	990	46
All	1772	44

Source: Alston J.M., *et al.*, Research Returns.

Graph 9.1: Commodity Focus of Public Expenditure on Agricultural Research in China and India (Triennium Average Ending 1999)

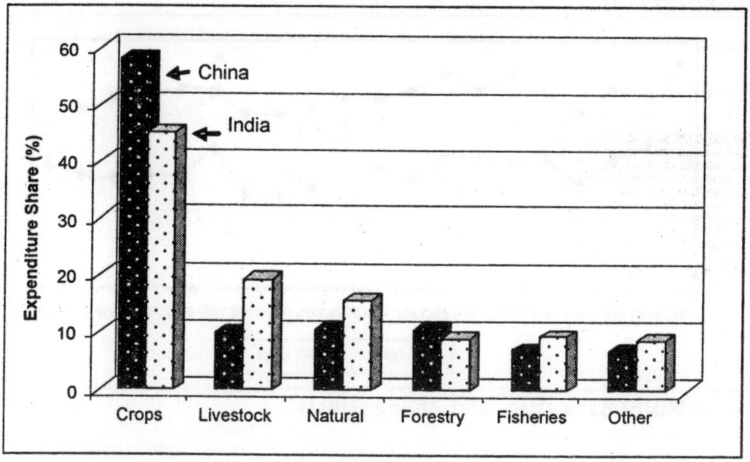

Source: China-Fan *et al..*, 2006 and India: Computed by the author on agricultural bio-technology.

TERI Source: *EPW*, Volume XLIII, No.26-27, 2008.

Importance of agricultural research to create/upgrade agricultural technology and extension of this technology and other support services to the farmers, cannot be over emphasized. In Bihar, there is a huge breach between the technology available in the agricultural universities/institutions and those on the field. Efforts have to be made to transfer the available technology to the farmers in the field. The public sector agricultural research[1] and extension system consists of Rajindra Agriculture University, seven research centres/stations of the Indian Council of Agricultural Research for commodity and location-specific research; and various Departments of the State Government that provide agricultural extension and regulatory services. The agricultural research and extension system has not been fully effective in developing and disseminating appropriate technologies to the farmers in the State. While 0.4 per cent of agricultural GDP is spent on agricultural research and education at the national level, the average for Bihar is only 0.2 per cent. Further, 95 per cent of this spending is used for salaries and 5 per cent for establishment expenses, thereby leaving no funds for operational expenses needed to carry out relevant research.[2] (Graph 9.2)

Graph 9.2: Public Expenditure on Agriculture and Allied Activities (as percentage of Agriculture GSDP)

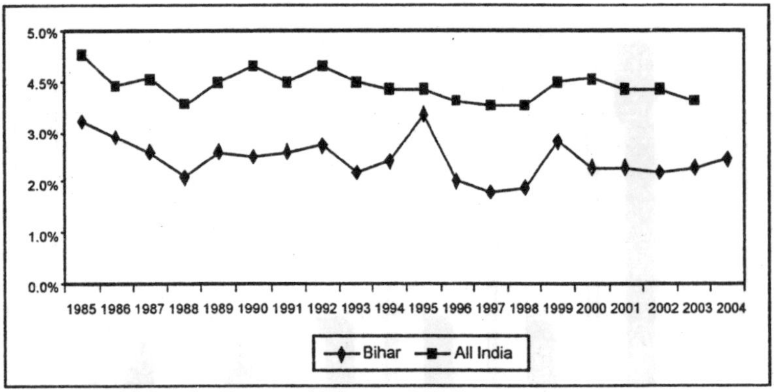

Source: World Bank (Fan, and Zhang; 2002)

Despite all these problems, Rajendra Agriculture University (RAU) has a major contribution towards bringing sustainability in agricultural activity of the state. Research undertaken covers all economically important crops grown in the state with focus on rice, wheat, maize, vegetables, fruits, fisheries, animal husbandry, medicinal and aromatic plants and post harvest technology of different crops. Soil testing based fertilizer equation has been developed and tested for achieving targeted yields of different crops.

This has, however, not been implemented fully. It is important at this stage to establish a close linkage between the RAU and the well spread extension system. RAU must support ATMA, NHM and other central projects through its scientists. Workshops for the different stake holders may be organized. Involvement of KVKs with ATMA which is presently symbolic should be increased at all levels. The line department officers should also be involved in the KVK programmes. Now that the Planning Commission has a proposal to declare RAU as a Central Agricultural University, a much bigger boost to development of agriculture in the State can be expected.

In a national survey conducted by the NSS in 2003, less than 0.5 per cent of farmers in Bihar reported accessing information on modern technology from extension workers, the worst across all states. The human resources base which was developed in 1960s and 1970s had eroded due to retirements, intellectual isolation and lack of in-service training. Other weakness included dilapidated infrastructure, lack of mobility, centralized top down command and control approach, weak research-extension-farmer linkages and a lack of focus on areas of relevance, opportunity and comparative advantage. Consequently Bihar is unable to make any worthwhile use of the 15,000 staff employed in the public sector for addressing the problems faced by the agricultural sector. Details about available research findings in rice, wheat and maize are discussed here.

RICE RESEARCH

Since it is uneconomical to grow rice at the prevailing low yield levels in the State, there is an urgent need to take some remedial measures. Given below in brief are some details about the 'System of Rice Intensification (SRI) and Hybrid Rice. By a proper application of SRI and hybrids in the 8 'rice bowl' Districts in the first few years and spreading the same technology to other irrigated area, we can expect an additional rice production of 1.0 to 1.5 million tonnes at the end of Eleventh Plan.

System of Rice Intensification[3]

The System of Rice Intensification (SRI) is considered to be one of irrigated rice crop management through which production can be substantially increased by effective changes by integration of the management of plants, soil, water and nutrients. SRI practices are likely to make rice cultivation more sustainable, enhancing rice production without heavy dependence on modern high-cost inputs and, reducing cost of production. Being environmental friendly, SRI is also quite accessible by resource poor farmers when increases in yield enhance their food security.

The recent breakthrough in rice cultivation known as System of Rice Intensification (SRI) method is one such case which may be considered as disembodied technology. The disembodied type of technical change is mainly due to improved management methods. Father Henri de Laulanie developed SRI in Madagascar in the early 1980s. In 1990, Association Tefy Saina (ATS) was formed as a Malagassy NGO to promote SRI. Four years later, the Cornell International Institute for Food, Agriculture and Development (CIIFAD) began co-operating with Tefy Saina to introduce SRI around the Ranomafana National Park in eastern Madagascar supported by the US Agency for International Development (USAID). In SRI method, synergic interaction increases land, labour and water use efficiency. SRI method deviates from the traditional method of cultivating irrigated paddy in a number of ways. For instance, five kgs/he. of seeds is sufficient in SRI as against the usual 50 to 60 kgs/ha in traditional method. The transplantation of the seedlings is carried out within two weeks as against four to five weeks with a wider spacing and one seedling per hill. The root development is more and healthier under SRI method; tillering is almost double, the crop does not lodge and the grain weight is more with fewer incidence of pests and diseases in the new method. It involves the application of certain management practices, which together provide better growing conditions for rice plants, particularly in the root zone, than the traditional method. The reduced demand for biodiversity. The un-flooded paddy fields do not produce methane, one of the major use efficient method of rice cultivation is suitable for resource poor farmers and water scarce areas.[4]

Its effect has now been seen in 21 countries of the world including the largest rice producers. Initial trials have shown high potentials for its acceptance as an improved method of rice production. In Cambodia, SRI was unheared of in the year 2000, but by 2003, nearly 10,000 farms had adopted it, a figure that may have touched 90,000 by now. In Andhra Pradesh, results of SRI trials by 50 farmers during rabi season 2003-04 showed an increase of yield from 7.1 tonnes ha. to 9.7 t/ha. with the highest measured yields surpassing 15 tonnes ha. The technology was demonstrated on an extent of 11,300 hectares by extending input subsidy of Rs. 2,000 per hectare and imparting training to 56,600 farmers during 2007-08 by the Tamil Nadu Government. It helped to achieve the per hectare productivity between 10 tonnes and 13 tonnes. Last year, 4.2 lakh hectares were covered. This year's target was 7.5 lakh hectares. Promotion of SRI to maximize rice production would be among the major strategies to achieve the production target.

This method has been shown to increase rice yields tremendously with less amounts of water. Hence, it is a timely step in reducing the water requirement of rice culture. At the present moment, when rice production is quite costly in Bihar, due to non-availability of power, This method coupled with hybrid rice varieties can certainly help increasing productivity levels significantly.

A study about the cost effectiveness of SRI technique in paddy cultivation in the western zone of Tamil Nadu by randomly selecting 15 adopting and 15 non-adopting farmers throws interesting results. The cost incurred for nursery has been found lower (5.47 per cent) in SRI than conventional (14.77 per cent) method. However, from the main field activities up to harvest, the cost incurred has been found higher in SRI (94.53 per cent) than conventional (85.23 per cent method, mainly due to higher manpower requirement in transplanting and harvesting operations. The average yield obtained was 1.35 times higher by SRI adopters than non-adopters in one acre of paddy in the sample farms. The average net profit per acre has been observed higher (Rs. 16,566) by SRI adopters than non-adopters (Rs. 8,848). The major reasons for adoption of SRI technique have been low seed requirement, less area under nursery, less duration of seedlings maturity, less water requirement and higher yield. Higher labour requirement and lack of skill among the farmers have been identified as the major constraints being faced requirement and lack of skill among the farmers have been identified as the major constraints being faced SRI adopters. The study has suggested that training-cum-field visits may be organized for the paddy growers to promote SRI technique on a large scale to increase water-use efficiently in water-scarce areas.[5]

Hybrid Rice

New Hybrid varieties released by Bayer Crop Science gave higher yields than the inbred varieties. Since its introduction, in India, in 2001, Arize 6444 has been cultivated on more than 190,000 ha. A survey of more than 320 sites has shown that Arize 6444 outperforms the local farmer inbred references by +34 per cent in the wet season.

Table 9.2: Performance of Arize 6444 on Farms in India Over 3 Year Period; Survey Conducted Over 328 Sites in Wet Seasons

	Number of sites surveyed	*Average Yield Kg./ha.*		
		Arize® 6444	*Inbred*	*Arize® 6444 vs inbred*
Andhra Pradesh	74	7908	5814	+36%
Karnataka	38	8803	7238	+22%
Orissa	7	5480	4779	+15%
West Bengal	55	5278	3838	+38%
Madhya Pradesh	52	6573	4774	+38%
Uttar Pradesh	54	7509	5294	+42%
Bihar	48	7453	5628	+32%
Total India	328	7175	5348	+34%

Experiments in India were performed both during wet and dry seasons. Experiments conducted in Haryana at the field of the farmers have given much higher yields at 10.5 tonnes/hect. and much higher returns and higher unit prices. The results from 48 experiments carried out in Bihar with Arize 6444 gave a yield of 7453 kg. per hect. which will make rice cultivation in Bihar quite economical.

A NEW FLOOD-RESISTANT RICE VARIETY

In eastern India, 13 million hectares of rice fields are in the low-lying areas, of which 2.2 million hectares are in UP alone and quite a large area in Bihar. This area is often devastated by floods resulting into low and unstable yields. Till the last kharif season (2008), farmers were using Swarna, a rice variety developed by Rice Research Station in Andhra Pradesh. Released in 1982, Swarna is still the most popular rice variety in rain-fed lowlands. The farmers prefer this variety due to its high and stable yield which averages four tonne per hectare, besides the flexibility it offers in terms of planting dates and adaptability to varying nutrient management practices, tolerance to moderate stresses and good grain qualities. Swarna does well up to 30 cm of water depth. However, if the water crosses this mark, it adversely affects the crop.

Swarna was modified and a Sub-I gene was added by a group of Indian Rice Research Institute Scientists. Swarna, in its new avatar, designated as Swarna Sub-I, proved to be effective as indicated by its performance in a series of trials conducted at IRRI. The experiment in UP was conducted jointly under the supervision of scientists of the Narendra Dev University of Agriculture and Technology (ND-UAT). Faizabad and the Nand Educational Foundation for Rural Development (NEFORD).

Initially one farmer, Sitaram got ready for the trials after being reluctant initially. Later, Kisan Goshthis (farmers' conferences) were organized in which farmers from the nearby areas visited Sitaram, who shared his success story with them. In this farm, Sitaram had grown 12 varieties of rice and his field was flooded twice for six and 13 days. Consequently, of the 12 varieties, only four survived, which included Swarna and Swarna Sub-I. However, there was a big difference between these two varieties in terms of survival and yield. The survival rate of Swarna was 8 per cent and yield was 300 kg. per hectare while Swarna Sub-I's survival rate was 70 per cent and the yield, 3,070 kg. per hectare.

WHEAT RESEARCH

Bihar with over 2 million hect. under wheat (8%) can be considered as an important State with regard to this crop. But despite 90 per cent of the area being irrigated, productivity levels are even lower than those of West Bengal, a late comer to produce wheat. Two crucial factors which determine productivity level in this crop are the sowing time in keeping with the weather conditions in a particular area and quality seed—again most suitable for the area. Best results in the region are possible only if the crop is sown early (by November, 15th) with healthy and certified seeds and application of requisite quantity of fertilizers.

Hissar Experiment

The State and farmer bodies in Haryana decided to involve local communities and educate them about the issues like timely sowing, following a set cropping pattern, etc. For this, a campaign was designed to involve rural schools as it was felt that most students in such schools are fully aware of the crops cycles. In order to boost their knowledge about how and what to do to get maximum production, the staff of the Agriculture Department visited and covered all the schools of the district to educate the school children of the need for early sowing (by November 15th) with healthy and certified seeds as well as the application of nitrogen at proper time.

During 2005-06, as much as 203 thousand hectares of land in Hisar district was brought under wheat cultivation. A total of 7.52 lakh metric tonnes of wheat at the rate of 37.04 quintals per hectare was produced. The challenge, therefore, was to increase wheat production so that the farmer rediscovers agriculture as a remunerative occupation and then continues to contribute to the central food pool. Scientists involved with the project pointed out that during the successive years, wheat sowing was getting delayed due to a number of factors resulting in the wheat grains getting affected by the effects of the rise in the day temperature when they were just about maturing. As a result, the grains remained shriveled and weak and even when it appeared that the crop was satisfactory, the end result fell much too short of the expectations. *Nukhar Sabhas* were held all over the district, where panchayats were also involved. For the first time, all the members of the village community were engaged in one way or another, in educating the farmers about—the proper and timely application of technology to get increased yield from the fields. The result of the campaign was extremely encouraging. During the year 2006-07, a record 44.85 quintals of yield per hectare has been obtained. Same experiment is being replicated in other wheat producing districts.[6]

Zero Till Wheat[7]

Rice and wheat are the two prime cereal crops grown in Bihar. They share around 8 per cent of the area each in India under the two crops and account for 70 per cent of the gross cropped area in the State. Wheat sowing is generally delayed in areas where rice crop is taken before wheat. The delay in wheat sowing is mainly due to late harvesting of the preceding crop as well as the need for intensive tillage for wheat. The excessive tillage is needed for seed bed preparation where soil is hard and more crop residues are present. Sometimes heavy soil remains wet for longer period, which also delays the field preparation of sowing. The delay in wheat sowing beyond 25th November leads to yield reduction of about 1.5 to 2 kg./day/

acre under North Indian conditions. This is confirmed by the Hissar experiment quoted above. After harvest of rice, farmers generally undertake 4-5 harrowing and various other operators, accounting for about 8-11 tractor operations. Excessive tillage consumes more time and energy as well as money. Of the total cost of wheat production, about 7-10 per cent is on field preparation and seeding. Therefore, adoption of zero tillage for wheat sowing can help to get rid of excessive tillage operations leading to savings on time, energy and cost of production. Due to these benefits zero Till (ZT) has been proved as an economical technology. It has become popular among the farmers of Haryana and also getting momentum in other parts of the country (Punjab, UP. Bihar and MP.)

Direct sowing of crop using Zero Tillage (ZT) machine after the harvest of previous crop without any field preparation and causing minimum soil disturbance is known as zero tillage. This machine is similar to ordinary seed cum fertilizer drill except that it has inverted T type furrow openers to open narrow slits for simultaneous placement of seed and fertilizers. This technology was mainly developed to reduce the cost of cultivation and also to advance the sowing time in certain situations. Some farmers in District Gaya who have adopted the ZT system found that the tillering of growing wheat was about 10-15 per plant against tradition method of sowing where only 3-4 plants are realized. By using this machine, seed rate was reduced from 100 kg. per ha. to 60 kg. per ha. and fertilizer usage was reduced 60 per cent against recommended doses. This has minimized weed growth for completing nutrient absorption with growing wheat. Similarly, the water requirement is also reduced. For optimum productivity levels, the required package of practices are as in Table 9.3.

Table 9.3: Recommended Wheat Varieties for Various Zones/ Sowing Conditions

Zone	*Cultural Conditions*	*Varieties*	*Time of Sowing*	*Fertiliser dose*
NEPZ	Irrigated Timely sown	HD 2733, PBW 343, PBW 443, K 8804 (K88), DL 78-3, HD 2402 HP 1731, K9107, HP 1761, NW 1012, HUW 468, (HALVIYA 468), HP 1761, K 7410, UP 1102, UP 262, NW	Middle of November	150-60:40 NPK kg./ha. 1/3 N at sowing and 2/3 at first node stage i.e., 35-40 DAS

Advantage of Zero Tillage

1. Less cost of production (a saving of Rs. 1000-1200/acre).
2. Time saving (takes 30 minute/acre for seeding).
3. Less wear and tear of tractor and other field implements.
4. Lower incidence of *minor.*
5. Water saving (sowing can be done on residual soil moisture in standing crop residue which act as mulch and helps in moisture conservation).
6. Less temperature fluctuations.
7. No crop yellowing after first irrigation.
8. Better nutrient use efficiency compared to broadcast.
9. Helps in better management of herbicide resistance problem.
10. Lesser pollution due to lower fuel consumption.
11. Less crop lodging compared to broadcast method.
12. Possibility of increased crop intensity.
13. Lesser Karnal bunt and termite incidence.
14. Less effect of salinity and alkalinity.
15. *In-situ* crop residue decomposition may help in organic carbon build up.

Bed Planting[8]

Rice-wheat cropping system supports more than 150 million people in the world. In Bihar it covers around 5.5 million hect. –70 per cent of the gross cropped area. But the sustainability of the present system is under threat on several fronts. As already discussed, Late planting is a major problem in most rice-wheat areas and to improve productivity, the wheat crop must be planted at the optimal time.

In bed planting system, wheat is planted on top of raised beds that are usually superficially reshaped for sowing the next crop. Residues may be incorporated during reshaping or chopped and left on the soil surface. In this system beds and furrows are prepared using a machine that makes two beds (70 cm width) in each pass. It is possible to plant 2-3 rows of crop on each bed. In this technology the seed bed is prepared conventionally and raised beds and furrows are prepared using a raised bed-planting machine. Sowing of wheat on raised beds is possible with the development of the bed planter, which enables wheat sowing 20 cm apart on a 37.5 cm. wide bed and 30 cm wide furrow between two beds. An adjustment in these parameters is also possible with this planter.

Two of the major constraints on higher yield of wheat in India are weeds and lodging. Both can be reduced in bed planting systems. These practices may raise the input use efficiency, reduce the intensity of machinery use in a sustainable way and reduce production costs. Bed planting is also a crop establishment method of wheat, which is advocated under the rice-wheat system. The following points are also responsible, for advocating the bed planting in wheat crop.

- This simpler and more efficient system improves Management of irrigation water and saves 30 per cent of it.
- It gives higher yield and saves 25 per cent seed and ensures optimum plant population resulting in better tillering, increasing panicle/ear length and bolder grains.
- Fertiliser efficiency is increased because of proper placement, including top dressing applications. Farmers can apply nitrogen and irrigation at grain filling stage to improve protein content without lodging. Reduced lodging gives better yield. Many farmers do not irrigate after heading to avoid lodging. As a result, water can become a limiting factor during grain filling, resulting in lower yields.
- Bed planting facilitates irrigation before sowing and thus provides an opportunity for weed control prior to planting. If pre-sowing irrigation is likely to delay planting, bed planted crops can be irrigated immediately after seeding. Weeds between the beds can also be controlled mechanically, early in the crop cycle.
- Herbicide dependence is reduced and weeding and roughing between rows are easier. The major weed species affecting wheat, Phalaris minor, is less prolific on dry tops of raised beds than on the wetter soil found under conventional method. Raised beds make it easier to apply herbicides as the person spraying can follow a line.
- On raised beds, the plants catch more sunlight, resulting in better growth, and making the soil around the base drier which prevents lodging.
- Compaction of soil due to heavy machinery is limited only to the furrows.

Bed planting holds immense potential for improving irrigated wheat-based cropping system by making them less resource intensive and more sustainable. Even without conservation tillage, bed planting offers new and better weed control, water and fertilizer management opportunities, along with less crop lodging and some reduction in tillage. Farmers report that yields may be improved by at least 10 per cent with the proper variety, production costs can fall by 20-30 per cent and irrigation water requirements can be reduced up to 35 per cent compared to conventional planting.

Some of the major factors responsible for low yield in Bihar are:[9]

- Lack of adoption of recommended practices leading to wider yield gaps between potential and achieved yield. It can be resolved through strengthening extension activities.
- In some of the rice area, field remains wet for a longer time and only single crop is grown. In such area, cropping intensity can be improved through adoption of surface seeding practices.
- Delayed sowing also reduces the yield and reduction can be curtailed through adoption of zero tillage. This technology, besides reducing the cost of cultivation, improves wheat yield. It also reduces the problem of Phalaris minor in wheat under rice-wheat system.

Site-Specific Nutrient Management[10]

The Site-Specific Nutrient Management (SSNM) is basically a systematic approach to apply sound agronomic management to small areas of a field that can be identified as needing special treatment. The components of site specific management may not be new, but now we have the capability with new technology to use them more effectively. The systematic implementation of these practices into site-specific systems is probably our best opportunity to develop a truly sustainable agriculture system. Site-specific management includes practices that have been previously associated with Maximum Economic Yield (MEY) management, Best Management Practices (BMPs), as well as general agronomic principles. The SSNM experiments with rice wheat system were conducted at 10 locations in India.

Data averaged over two years show that grain yields of 15-17 t/ha. were achievable in this system. Average annual grain productivity of the system was 13.3 t/ha., of which the contribution of rice was 60 per cent and that of wheat 40 per cent. At Sbour (Bihar), the yield was 13849 kg. per hect.

Conclusion

These are varied types of tried systems in one or the other part of 2 *Indian Farming*, July 2007.Bihar itself. Since they have shown promising results in different parts of India and abroad, extension services in Bihar have to adapt them to local conditions. Once adopted by the farmers, they will have their own momentum so as to give better yields.

MAIZE RESEARCH

Improved hybrid maize seed varieties were made accessible to farmers by the active presence of maize private seed companies in India in general, and in Bihar in particular. This was triggered following major seed policy reform (1988) which facilitated private sector participation in the maize seed industry. Today a large number of hybrid maize seed distributors in Bihar represent multinational companies which operate mostly out of Andhra Pradesh. The uptake of improved hybrid maize seeds during the winter season is much higher than that in Kharif, thus contributing to much higher maize yields in the winter season.

Discussions with Dr. Jeevan Thakur, M.D. M/s Kanchan Ganga a major maize seed production company have revealed that out of a total of 35-40,000 tonnes in India they are supplying around 20-25 thousand tonnes hybrid maize seed to Bihar. Their recent seed "Polo" has given very promising results in the farmer field (Table 9.4). They could get around 5 tonnes/hect. in the first year.

District level survey data for Bihar carried out by IFAD. Shows that all the 3 survey districts in Bihar are high yields (over 5 tonnes per hect. for rainy season and 6 tonnes for winter.)—both for rainy and winter seasons. Tables 9.5 and 9.6 provide the data along with per hect. cost (C_2) of cultivation.

Table 9.4: Performance of Maize Hybrids for Seed Yield (q/he.) at Various Locations in HP During Kharif 2005

Sl No.	Code	Variety.	Company/ Organisation	Locations					Over-all Mean	% increase over Check	
				Dhaula-kuan	Kangra	Bajaura	Palam-pur	Sunder-nagar		Girija	PMZ 4
1.	M-501	X 6060	Kanchan Seeds	61-9 (9)	57-2 (1)	86.2 (14)	82-9 (8)	55.0 (10)	68.6 (8)	28.8	-9.2
2.	M-503	POLO	-do-	74.1 (1)	53.9 (5)	101.6 (4)	96.4 (1)	52.1 (11)	75.6 (2)	41.9	0

Source: *Courtesy*, M/s Kanchan Ganga Seed Company Pvt. Ltd.

Table 9.5: Yield (t/ha.) of Different Types of Maize Cultivars in Bihar—2001

State	District	Rainy Season			Winter Season		
		Local	Composite	Hybrids	Local	Composite	Hybrids
Bihar	Munger	1.50	2.40	5.00	2.00	2.85	6.00
	Siwan	1.60	2.50	5.50	2.25	3.20	6.50
	Begusarai	2.00	3.00	6.50	3.00	3.50	7.50
	Average	1.70	2.63	5.33	3.00	3.18	6.66

Source: IFAD-CIMMYT-India RRA Surveys, 2001.

Table 9.6: Cost of Cultivation (Rs/ha') of Hybrid Maize Cultivars in Bihar—2001

District	Hybrid Maize Varieties		
	Cost C	Cost B	Cost A
Munger	8540	6003	5215
Siwan	9238	7512	6450
Begusarai	13034	9510	8297
Average	10270.67	7675	6654

Source: IFAD-CIMMYT-India RRA Surveys, 2001.
Note: Cost A, Cost B, and Cost C are same as those of Cost A_2, Cost B_2 and Cost C_2 used by the Directorate of Economics and Statistics.

Cost of production data for maize (Table 9.7) shows that from the major maize producing states in India, but for Karnataka, net returns at around 25 quintals/hect. in the case of Bihar are highest in Bihar.

Table 9.7: Cost of Production of Maize in Important States

(*Rs./he.*)

	Cost A_2 +FL	Yield (Quintals/hec.)	Gross Return	Net Return
Andhra Pradesh	8193	21.15	11421	3228
Bihar	9952	24.88	13435	3483
Himachal Pradesh	6426	12.82	6923	497
Karnataka	8701	35.01	18908	10204
Madhya Pradesh	4032	7.08	4218	209
Uttar Pradesh	8069	16.18	8739	670

Source: Report of the Commission for Agricultural Costs and Prices, 2004.
Note: FL = Imputed value of family labour.

Taking all these factors into consideration from among the cereal crops, maize is one of the most promising one for Bihar which has some comparative advantage over other crops. Driven by the increased availability of improved seed varieties and irrigation which allowed winter and summer cultivation of the crop, both production and productivity have been increasing in recent years. The crop still suffers from a number of constraints like complete absence of training, extension and technology transfer. Farmers are also not getting sufficient training on other improved maize technologies, such as herbicides, pesticides and post-harvest management practices. If necessary remedial steps are taken as discussed, Bihar has the potential to give much higher yields as projected in this study. The State is also capable of producing other high value varieties like pop corn, sweet corn, baby corn and quality protein maize-varieties, particularly in diara areas. (Tables 9.8 and 9.9) India today producers about 15 million tonnes of maize (19 million tonnes during 2007-08), but the demand is on the increase, particularly for poultry. Every unit of chicken or egg needs 1.2 units of grain.

Table 9.8: Quality Protein Maize (QPM) Varieties for Bihar

Hybrid/ Variety	Grain colour	Grain yield Potential (q/ha)		Maturity (days)		Protein content in Kernel%	Tryptophan content in protein (%)
		Kharif	Rabi	Kharif	Rabi		
Shaktiman-1	White	60	80	90	150	9.6	1.01
Shaktiman-2	White	65	85	90	150	9.3	1.04
Shaktiman-3	Orange Yellow	70	90	95	160	9.63	0.73
Shaktiman-4	Orange Yellow	75	95	95	160	9.98	0.93
Shakti-1	Yellow	40	70	85	145	9.5	0.92

Source: Rajendra Agriculture University (2005).

Table 9.9: Baby Corn Maize Varieties for Bihar

Varieties	*Yield Potential (q/ha)*	*Harvesting Period (days)*	*Fodder Production (q/ha)*
Pusa early Hybrid-2	15	45	250
Pusa early Hybrid-3	15	45	250
Shaktiman-1	18	50	350
Shaktiman-2	18	50	300
VL-42	15	50	250

Source: Rajendra Agriculture University (2005).

EXTENSION

"Agricultural extension is not only imparting knowledge and securing adoption of a particular improved practice, but is also aimed at changing the outlook of the farmers to the point where he will be receptive to, and on his own initiative, continuously seek means of improving not only his farm occupation, but also home and family life in totality. Agricultural Extension in India is a state responsibility and has undergone various transformations since independence. Over the years, the role of the State with respect to Extension has undergone transformation from that of a monopolist provider to performing the functions of co-ordinator, facilitator, regulator and arbitrator of conflicts now. Thus, present extension system is one, which includes a mix of both public and private interventions. Extension services which impart knowledge to the farmers in the application of research findings in their fields for crop production improvement are thus no longer the monopoly of government agencies. There is now a mix of both public and private interventions.

The private extension services are primarily of two types. The first is the entirely private, which is directly involved in farming activities through consultants, agri-business, agricultural input industries, etc. The second type consists of farmer's organizations, NGO's, etc., which remain largely dependent on government subsidies/grants. In the last few decades, there is an increasing realization that Public Extension by itself cannot meet the specific needs of various regions and different classes of farmers, particularly—small/marginal farmers and livestock sector. Equipment manufacturers, input suppliers (such as poultry hatcheries), procurement companies like ITC etc. are also involved in extension work on a limited scale with limited focus. Of late, there is also an increasing interest of corporate sector in organised cultivation for retail chain supplies. Accordingly more involvement of private sector in extension, is foreseen.

Bihar Scenario

Department of Agriculture in Bihar is headed by Agriculture Production Commissioner (APC) and is supported by three Directors viz. Directors of Agriculture, Horticulture and Soil Conservation. Director of Agriculture is assisted by Additional Director (Extension), Director (PPM Cell) and six Joint Directors at the headquarters dealing with crop production, plant protection, extension, education, planning, etc. At the Divisional level, there are Joint Directors while at the District level, District Agricultural Officers are incharge. DAOs are assisted by SDAOs at the Sub-Divisional level and Block Agriculture Officers (BAO) at the Block level. At the grass root level, Agriculture Extension Officers (AEOs) and Village Extension Workers are the officials who are responsible for transfer of technology through direct contact with the farmers.

The programmes being implemented under macro-management mode include, Natural Resource Management, Horticulture, Crop Development etc. Natural Resource Management includes programmes like National Watershed Development for Rainfed Agriculture, flood protection schemes and infrastructure for State Land Use Board. Under Horticulture, commercial floriculture, distribution of vegetable seeds, development of fruits, medicinal and aromatic plants are included. Under Agriculture Sector, Intensive Cereal Development Programme, plant protection, special jute development programme and agriculture mechanisation are being implemented. With a view to improving production and productivity, the State Government implements various programmes such as distribution of HYV seeds of cereals, minikit of seeds of pulses and oilseeds, fertiliser distribution, procurement, farmer's training, Agri fair and seed/soil/fertiliser testing etc. The Department of Agriculture implements centrally sponsored as well as State sponsored programmes with the broad objective of improving production/ productivity of crops and increasing income accrual to the farmers.

Agricultural Technology Management Agency

In late 1990s, the Government of India and the World Bank began exploring new approaches to agricultural extension. The result was a new decentralized farmer centered extension approach, which focused more directly on agricultural diversification and increasing farm income and rural employment. The central institutional innovation that emerged to address the problems of the extension system was the Agricultural Technology Management Agency (ATMA) model. It was pilot tested during 1998-2005 in Bihar and eight other states (covering 65 districts) under the World Bank—funded National Agricultural Technology Project (NATP) and the Diversified Agriculture Support Project (DASP). Initially four Districts—Madhubani, Munger, Muzaffarpur and Rural Patna were covered with the basic objective, (i) decentralise decision-making through "bottom-up" planning procedures that would directly involve farmers and the private sector in planning and implementing extension programmes at the block and district level, (ii) integrate extension programmes across the line departments (i.e. more of a farming system approach), and (iii) link research and extension activities within each district. This model was pilot-tested.

Steps have since been taken to reform the above situation in recent years, in particular during 2006. The State Government has universalised Agricultural Technology Management Agencies (ATMAs) to cover all districts of the State. They are designed to decentralize decision-making through 'bottom-up' planning procedures that would directly involve farmers NGOs, other agencies and private sector in planning and implementing extension programmes at the block and district levels. Similarly, linkages with Krishi Vigyan Kendras (KVKs) agri-clinics, established in 36 districts of Bihar is another area of importance. Main objectives of these KVKs and Agri-clinics, established at district level, are to carry standard seeds, fertilizers and agricultural knowledge to the rural masses. However, training programmes need to be strengthened to impart knowledge to farmers. There were 38 KVKs in the State, 29 of them are under State Agriculture University and 9 under non-Government organizations. But linkages with the district level agriculture and allied department are weak. Some of the suggested steps to improve the situation could be selection of demonstrations and beneficiaries jointly, re-orienting training programmes of KVKs as per Strategic Research and Extension Plans (SREPs) and focus on capacity building of Block Technology Teams (BTTs) etc.

In addition, 31 new seed testing laboratories have been sanctioned to give each district its own seed testing laboratory and agri-clinics. Performance of ATMAs in Bihar is as follows:

(1) ATMAs mobilize farming communities and develop public-private partnership.
(2) ATMAs have improved interaction between farmers, extension workers and researchers.
(3) ATMAs have promoted diversification of farming systems and adoption of new technologies.
(4) ATMAs have facilitated development of supply chains in a number of commodities.

Despite all this, the extension machinery at present is weak and inadequately staffed. Most of the available extension workers were attached to panchayats leaving very little scope for advisory services. Performance of public extension services has been seriously affected because of reasons like field vacancies (50 per cent vacant) non up-dating of competence of Village Extension Workers, inadequate programme funds, lack of involvement of beneficiaries in planning and operation, and poor research-extension linkages, etc., The State machinery for supply of seeds, fertilizers, pesticides, oil, spare parts, etc. is crippled. There are reports of rampant sale of spurious seeds/pesticides by some private sales outlets in the villages due to absence of stringent and vigilant quality control mechanism. The entire mechanism for marketing is unorganised and in the hands of private traders. Farmers, having little or no holding power, find themselves at the mercy of traders and many a time sell off their produce on credit and that too at prices lower than the MSP. Interventions by the Govt. agencies in favour of farmers are almost non-existent. (NABARD Report, 2007)

Farmer Club

Number of Farmers' Clubs (FCs) functioning in the state till 31st March, 2006 was 748. Another 120 clubs were launched during 2006-07 till November, 2006. The clubs were organized in all the 38 districts of the state. Of these, 551 FCs (74 per cent) were on RRB's account, 172 FCs (23 per cent) on CBs account and remaining 25 FCs (3 per cent) were in DCCB's Account. The objective is to establish at least one FC in each of the Rural and Semi-Urban branch of the banks. In order to achieve this objective, at least 1000 FCs are to be promoted. Special efforts on the part of all partner agencies resulted in considerable reduction in the number of dormant clubs from 84 as on 31.03.2004 to 28 as on 31.03.2006. The participating banks have been further advised to identify dormant clubs and initiate remedial measures for their revival/closure as per the guideline issued by NABARD.

Table 9.10: District-wise Position of Farmers' Clubs Launched/Sanctioned as on 30.11.2006

Sl. No.	*District*	*No. of FCs as on 31.03.2006*	*Additions during 2006-07 (upto Nov. 2006)*	*Cumulative Farmers' Clubs*
1.	Araria	10		10
2.	Arwal			0
3.	Aurangabad	6		6
4.	Banka	1		1
5.	Begusarai	32	6	38
6.	Bhagalpur	2	4	6
7.	Bhojpur	10	10	20
8.	Buxar	2	6	8
9.	Darbhanga	24		24
10.	East Champaran	70	6	76
11.	Gaya	49	27	76
12.	Gopalganj	32	11	43
13.	Jamui	28		28
14.	Jehanabad	2	2	4
15.	Kaimur			0
16.	Katihar	17		17
17.	Khagaria	29		29
18.	Kishanganj	11		11
19.	Lakhisarai	19	4	23
20.	Madhepura	5		5
21.	Madhubani	26	1	27
22.	Munger	26	2	28
23.	Muzaffarpur	23	5	28
24.	Nalanda	42	12	54
25.	Nawada	2		2
26.	Patna	12	6	18
27.	Purnea	37		37
28.	Rohtas	1		1
29.	Saharsa	5	2	7
30.	Samastipur	66	2	68
31.	Saran	57	5	62
32.	Sheikhpura	14	3	17
33.	Sheohar	1		1
34.	Sitamarhi	4		4
35.	Siwan	20	5	25
36.	Supaul			0
37.	Vaishali	21	1	22
38.	West Champaran	42		42
	TOTAL	**748**	**120**	**868**

Internet Revolution in Bihar[11]

Bihar Government is now working on an ambitious project that plans to take information technology to the grassroots level of this underdeveloped state by setting up common service centres called Vasudha. The plan to surf Internet at their villages into a reality, in the near future under SREI programme, has already been given a "letter of internet" by the Bihar State Electronic Development Corporation, the nodal agency of the Bihar Government for the implementation of e-governance, to develop Vasudha centres in villages. The proposed Vasudha would be named after SREI "Sahaj e-Village" and would be set up in e-kiosks. Besides housing Vasudha, the e-kiosks would reportedly also function as the delivery point for e-governance as well as e-learning centres where the youth could access distant education modules and vocational skills other than normal Internet access.

In the first phase, it has been decided to open 5,540 Vasudha centres across the state. The state Cabinet had also approved funds for the development of the required software by the TATA Consultancy Services. To begin with, the government reportedly planned to link eight departments to Vasudha to provide information and services for mutual purposes like weather forecast, commodity rates, agricultural counseling, health and hygiene. The issuance of birth and death certificates and land records would also be available on the net on the payment of nominal fees. Going by the plan, Vasudha would be owned and operated by selected village entrepreneurs and monitored by the regional control centre. Estimated cost of setting up each of these e-kiosks would be Rs. 1.20 lakh. The selected rural entrepreneurs would either have to invest this amount on their own or by obtaining loan from banks.

Radio Bihar

Once there was All India Radio, beamed down from on high. Then came FM. Now the broadcasting scene seems set to witness a third wave, as five states have expressed their desire to start their own radio stations and FM channels. Madhya Pradesh, Bihar, Delhi, Karnataka and Punjab have put in applications to this effect with the Union information and broadcasting ministry. The State authorities are convinced that the humble radio—with its extensive rural-urban reach is a better disseminator of news and information. Media advisors in Bihar feel it is the best medium to educate farmers in the state. "Our own radio" can give area-specific news to farmers on weather and best farming practices. It can also disseminate information on local *haats* (fairs) and schemes run by the Government far more effectively than any other medium. It is even planned to start teaching English through radio for primary school students. The radio will also be an extension of the CM's weekly durbars. The Bihar Government has worked it out in great detail. The main focus areas have been identified: dissemination of information regarding government schemes; encouraging local participation by the communities and revival of local and folk art forms.

Other Experiments

There are also other experiments of this nature in various other States and Bihar has the advantage to learn from their experience. Fishermen from Kerala, for example have been tapping the power of cell-phones to get the best price for their catch. Now, farmers in Haryana are SMSing a free government helpline to get solutions to their agro-probiems. The Government set up this trouble shooting service in February 2007, the first of its kind in the country. Farmers can contact senior officials of the agriculture department for advice. Even illiterate farmers are taking help of others in the family or neighbourhood to send SMSes. Some farmers take advantage of the services of village teacher/doctor and get satisfactory answers. Because of illiterate peasantry, it will no doubt take time to exploit such technology in the farm sector.

Quick Solutions-Haryana

Department	*No. of Messages Received*
Agriculture	829
Horticulture	273
Fisheries	6
Animal Husbandry	17
Forests	7

> - *How it works*: When they receive the SMS, scientists and officials call up the farmers within 48 hours. The department has also introduced toll free numbers so that farmers can get advice immediately.
> - *Query categories*: Appropriate pesticides and fertilizers, information on seeds, methods to save crops from inclement weather, etc.

All the same, this service will go a long way to educate farmers once they get hooked to it. The department feels that farmers have been deprived of scientific advice as agriculture department officials and professors of agriculture universities have been unable to disseminate information to farmers in remote area. Mobile technology is helping surmount these problems. The service was first available in the Charan Singh Haryana Agricultural University, Hissar, and has now been extended to Regional Research Centres, where the farmers can avail of these services. Currently, farmers can avail of the service every Monday, Wednesday and Friday from Hissar and Bawal centres and on Tuesday and Thursday from the Uchani centre. Soon, it could be available through the week.[12]

Maharashtra Poultry Farmer

Now a poultry farmer sitting in Vidharbha can get a text message on prevailing market rates in Nashik before he sells his produce. The animal husbandry department Government of Maharashtra, has launched a pilot project to set up a market development cell to promote exports and improve market access for farmers. Maharashtra, Andhra Pradesh, Rajasthan, Orissa and Tamil Nadu have been selected for a World Bank-aided Multi State Agricultural Competitiveness Project (MACP) to develop competitive marketing system and empower farmers giving them better access to markets. The World Bank has also approved the proposal to set up a nodal facility to boost animal husbandry products like meat, poultry and eggs for export.

At present, there is no centralized facility in the state to grade the eggs that are required for export. As part of the new project, egg-grading units will be set up at five places in Western Maharashtra, including Pune. There is also a move to start poultry dressing units and an organized slaughter house. Marketing of meat and eggs is still in the unorganized sector and infrastructure for supply of essential inputs, facilities for improved breeding, feeding and management of meat products are too inadequate, The object is to have a more organized network to increase private sector investment in agri-business and promote entrepreneurs who can set up such units.[13]

Edinburgh University Tie-ups

The tie-up will extend credibility to degrees offered by state institutes. Apart from extending support to the state institutes, the tie-up will open up global markets for Bihari students. Meanwhile, the State Government has recently prepared a draft policy to revive the technical institutes of the state. It has declared its intention to re-open three state engineering colleges which have been lying closed for many years and has also decided to open two more technical colleges.

Notes

1. Education, A Meta Analysis of the Returns to Agricultural R&D, IFPRI.
2. Economic Survey, Government of Bihar 2006.
3. Rice Intensification, A New Approach to Rice Cultivation, and Dinesh Kumar, Y.S. Shivay, B.N. Mishra, R.C. Gautam, Indian Agriculture Research Institute, New Delhi, *Kurukshetra*, December, 2005 and *Leisa India*, December, 2006.
4. Technological change in Paddy Production: A comparative analysis of traditional and ARI Methods of Cultivation, H. Basavaraja, S.B. Mahajanashetti and P. Sivanagaraju, *Ind. Jn. Of Agri. Econ.* Vol. 63. No. 4, October-December 2008.
5. M. Anjugam, S. Varadha Raj and S. Padmarani, Cost Benefit Analysis of SRI Technique in Paddy Cultivation, *Agricultural Economics Research Review.*, Volume 21, 2008.
6. *The Tribune* 6/6/07.
7. R.S. Chhokar, R.K. Sharma, M.K. Gathala, A.K. Pundir and Vivak Kumar, Directorate of Wheat Research Karnal. Grow Zero Till wheat for more profit. *Intensive Agriculture*, November-December, 2005.
8. Rajeev Kumar, Sharad Pandey and Apurv Panday, G.B Pant University of Agriculture and Technology. Pant Nagar. Bed Planting: A new Technique for Wheat Sowing, *Intensive Agriculture*, November-December, 2005.
9. Mr. R.S. Chhokar (personal communication), Directorate of Wheat Research, Karnal, Haryana.
10. *Indian Farming*, July 2007.
11. *The Tribune*, December 17, 2007.
12. *The Times of India*, 9/9/07.
13. *The Indian Express*, 2/11/07.

10 INFRASTRUCTURE

According to World Bank, expansion of educational opportunities are critical for maintaining India's growth momentum. The sustainability of India's growth story would critically depend on how the country addresses infrastructure constraints, skill enhancement and education. Infrastructure challenges are bigger and need faster execution as the country is growing at a rapid pace. Successive Human Development Reports have recommended that the rural poor need to be equipped with resources and skills for livelihood options outside the cycle of subsistence agriculture, which needs investment in rural infrastructure. The World Development Report's definition of infrastructure includes public utilities like power, telecom, water, sanitation, drainage and transport like roads, railways, waterways and airports. There is sufficient consensus among economists that high GDP growth rate does not necessarily promote 'Human Development' per-se, by widening opportunities and choices for the marginalized sections. Even political and social freedoms lose their meaning if a citizen has no access to markets for his needs or for his goods and services.

Several village based surveys demonstrate, that rural purchasing power and agricultural productivity are directly linked to transport, irrigation and research infrastructure. A significant part of expenditure on agricultural research, infrastructure and healthcare has to come from government resources. Table 10.1 quantifies how every million rupee spent on roads and agricultural R&D significantly reduces poverty. Healthy environment, clean water and good sanitation affect quality of life and longevity, the two biggest markers of human development. A 2007 field study by Chandigarh-based researchers, concludes on the basis of various health surveys that as high as about 70 per cent of rural mortality can be attributed to poor environmental conditions. Their study of a total sanitation village in Kapurthala district (Punjab) shows that participation of the community, women and self-help groups play a more important role than just money. The findings further underscore the importance of inclusive policies and good governance. NCAER's Rural Infrastructure Report, 2006, recommends a multi-pronged strategy for the cash strapped Central and State Governments. It recommends that the private service providers take care of Telecom and Power Sector, but a system of incentives, limited ownership and user charges must be promoted for sectors like roads, drinking water and sanitation. In the case of the last mile of road connectivity, it recommends provision of building roads by village communities through targeted subsidies. It calls for dismantling legal barriers that prevent local investment in rural areas. The current legal system forbids village or town-level transporters, networks of telephone exchanges or water distributors from operating legally. Some facts about Rural India are:

- 1 out of ten people on the planet is an Indian villager.
- 50 per cent is rural India's contribution to GDP but rural per capita income is 56 per cent less than urban average.
- Rs. 25 lakh crore will be the size of Indian rural market in 20 years; four times the current urban market size.

Table 10.1: Decrease in Poor People Per Rs. one Million Spent in 1990s. by infrastructure category

Road investment	124
Agricultural R & D	85
Education investment	41
Irrigation investment	60
Credit subsidies	42
Power subsidies	27
Fertiliser subsidies	24

Source: Fan, Gulati and Thorat.

Failed Schemes

1. *Accelerated Rural Water Supply Programme—1972-73*: Accelerated the coverage of drinking water, failed due to inadequate funding and lack of technical capacity of Zila Parishads to operate and maintain water supply and sanitation facilities.

2. *Integrated Rural Development Programme—1980*: Provided rural poor, subsidy and bank credit for productive employment. One-time provision of credit without follow-up, sub-critical investments, over-crowding, and lack of market linkages weakened the scheme.
3. *Indira Awaas Yojana—1985-86*: Aimed to give free housing in rural areas, especially to SC/STs and bonded labourers. Failed due to limited coverage, resource constraints and leakages to undeserving candidates.
4. *Employment Assurance*: Provided rural employment during lean agricultural season, unsuccessful due to corruption and uneven distribution.

Bharat Nirman

The Government's plan for developing rural infrastructure over 4 years (2005-2009), it is estimated to cost Rs. 1,74,000 crore.

- *Roads*: Build 1.5 lakh km all weather roads for habitations over 500 people in hilly areas and over 1000 in other areas.
- *Telephone*: Equip 66,822 villages with telephones. Connect 14,183 remote villages by digital satellite phones.
- *Irrigation*: Create 10 million hectare of irrigation potential. Restore 10 lakh hectare potential by renovation and modernization of schemes.
- *Drinking Water*: Cover 55,067 uncovered and 2.8 lakh partially covered habitations. Give potable water to over 2 lakh villages.
- *Housing*: Construct 60 lakh houses for the rural poor.
- *Power*: Electrify 1.25 lakh villages by grid supply.

Proper infrastructure is a basic requirement for development of any activity industrial or agricultural. For a comprehensive and sustainable development to take place in an economy, it is not only economic infrastructure like roads, power, communication network, banking facilities, etc., but there is also the need for social infrastructure facilities in the fields of education, health, etc. These facilities help in strengthening the linkages between agriculture, industry, trade and business, thereby paving the way for development of broad spectrum of economic activities. Even at the cost of repetition, we may quote World Development Report, 1994, which defines infrastructure as:

- *Public Utilities*—Power, telecommunications, piped water supply, sanitation and sewerage, solid waste collection and disposal, and piped gas.
- *Public Works*—Roads and major dam and canal works for irrigation and drainage.
- *Other Transport Sectors*—Urban and interurban railways, urban transport, ports and water ways, and airports.

The quality of Bihar's infrastructure and people's access to it, have been a persistent problem. In most districts, existing infrastructure is inadequate and acts as a major impediment to the development of the economy in general and agricultural sector in particular. Of all the villages in India that are not connected to a road, 13 per cent are located in Bihar. Only half of the state's villages are connected to a road and the condition of the roads is such that year-round service for agricultural inputs and product marketing is not possible. Only 44 per cent of Bihar's roads are surfaced, compared to 58 per cent for the whole country. Average road density per sq. km. of geographical area works out to 77.9 in Bihar as against 128 in Punjab and 159 in Tamil Nadu. Bench marked 100 for India, Infrastructure Development Index for Bihar is 80. Total length of roads in 2000-03 was only 76169.8 kilometres (Table 10.2).

Table 10.2: Some Basic Indicators of Development in Bihar

(*Kilometres*)

Transport and Communication	
Length of road (2002-03)	
1. National highway	3398.30
2. State highway	2382.47
3. Other P.W.D. Road (M.D.R./O.D.R.)	25385.00
4. District Board Road (Urban/Rural) km including unsurfaced road (Urban/Rural)	45004.00
Total	76169.77
Telephone (March 2004)	
1. Total number of telephone centres operating in Bihar	1117
	1205772
2. Capacity of telephone centre	865792
3. Total no. of telephone in operation	200
4. Number of combined telephones	119546
5. No. of Mobile connections	

Source: Directorate of Statistics and Evaluation, Patna.

It is heartening to know that the Bihar Government has attached priority to infrastructure development. Bihar has spent over Rs. 5,000 crore, including substantial financial assistance from the Asian Development Bank, on upgrading over 1,000 km of roads to all-weather highways. It has invested Rs. 5,000 crore to increase power generation and supply, including a joint venture between the Bihar State Electricity Board and the National Thermal Power Corporation. It has signed a Rs. 2,000 crore agreement with the Power Grid Corporation of India to ensure timely, reliable and uninterrupted power supply to domestic and industrial consumers by upgrading transmission lines and securing them against petty theft. Even more important are the efforts being made to shore up social infrastructure and services. In its first two years, the present government recruited over two lakh teachers, a record for Bihar, if not for the country. Primary health services and government hospitals have witnessed large-scale infusion of funds and experienced major improvements in the volume and quality of services delivered. Doctors are back on duty, equipment and medicines are in stock, and dramatically rising number of patients—the relatively well off as well as the poor—are returning to the public health system. Official statistics show that the number of patients visiting government hospitals has jumped from 39 in January 2006 to an average of more than 4,000 per month in the second half of 2007.[1]

Power*

Power is the basic and most essential requirement for economic development of any area. Expansion and creation of industrial capacity even in small industrial sector requires adequate, qualitative and uninterrupted availability of power. Non-availability of power has been mentioned as an input, that is to be given top priority in order to promote industrial activity in the state. Some basic facts about power in Bihar are as in Table 10.3

Power situation in Bihar is grim, with a total installed capacity of 586.1 MW (comprising 540 MW from Thermal and 46.1 units Hydro). The general level of development is low and there is a shortage of basic infrastructure. This has led to a very low level of per capita availability (76 units) in comparison to all-India level (613 units). In addition, there is a lack of awareness about various

* For details on power refer to Appendix at the end of this chapter.

renewable energy resources, available technologies, and systems/ devices/and programme. Most of the power requirement is met by purchases from the Central generating plants: NTPC, Chukha Hydel (PTC) and Ranjit Hydel (NHPC). Bihar needs urgent attempts to expand its power generation capacity and improvement in existing facilities (Barauni and Muzaffarpur). There has been no capacity addition for the last two decades. No work was going on the Kahalgaon Power Project in Bihar for over 9 months. An agreement has now been signed paving the way for adding 500 MW capacity before the year end (2008). The unit, the seventh overall, is the third generator that was to be added in the second phase of the expansion of the plant. Peak demand 2006-07 as estimated at 1400 MW against the actual availability of 1162 MW-leaving a net deficit of 16.9 per cent for the state. It had far reaching implications for economic growth performance of the State.

Table 10.3: Power Position in Bihar

1. *Rural electrification is 72 per cent*: However, it is ineffective in absence of power availability. Power consumption per capita at 147 kWh against 460 kWh for all India is a matter of concern—share of rural areas in total power sales is declining affecting crucial investments on irrigation and high cost of production.
2. The per capita consumption level of electricity in Bihar at 143 kWh is appreciably lower than the national average of 460 kWh.
3. There has been a negative growth rate in power generation in Bihar during 1990-91—2000-01. Purchased power accounts for 80 per cent of the energy consumed in the state.

Though 72 per cent of the villages in the state are electrified as against the national average of 86.50 per cent and 100 per cent villages in Punjab, Haryana, Tamil Nadu and Andhra Pradesh.

Only 5 per cent of the rural households have access to electricity. (Table 10.4). More than 40 per cent villages in Bhagalpur, Katihar, Purnea, do not have electricity connection. In Bihar, power cuts are very frequent, ranging between 6 and 15 hours a day in many places. The lack of assured supply of power acts as a deterrent to private investment, particularly in manufacturing and food processing. Agriculture's share of power consumption in Bihar is around 27 per cent. Farmers, faced with shortage of electricity, shift to costlier diesel pumps for irrigation. According to the NSS estimates, 97 per cent of Bihar farm households irrigate their lands using diesel pumps, the highest proportion for any state in the country.

Table 10.4: Level of Infrastructure Development in Bihar and Other Selected Indian States.

	AP	*Bihar*	*MP*	*Orissa*	*Punjab*	*UP*	*WB*	*All India*
% of Villages connected by Road (1996-97)	86	48	28	49	97	50	49	57
% Rural Households with access to electricity (2001)	60	5	62	19	90	20	20	44

Source: CMIE, Infrastructure, February 2003; M/o Power, Rajiv Gandhi Grameen Vidyutikaran Yojana (RGGVY), April 2005.

Table 10.4A: Power: Peak Demand and Consumption Pattern.

	2006-07	*April-May-2007*
Peak		
Peak Demand (MW)	1399	1415
Peak Met (MW)	1162	1243
Peak Deficit (-) Surplus (+) – MW	-237	-172
Peak Deficit/Surplus (%)	-16.9	-12.2
Energy		
Energy Requirement (MU)	8425	1500
Energy Availability (MU)	7741	1381
Energy Deficit (-) Surplus (+) – MU	-684	-119
Energy Deficit/Surplus (%)	-8.1	-7.9

Source: Report of Special Task Force on Bihar Power Sector, Government of India, 2007.

There is a severe power shortage in Bihar, even under a 'constrained' demand scenario, where the electricity system covers barely 72.2 per cent of the villages and 5 per cent of households. Peak power availability there is about 1162 MW, causing a peak shortfall of 237 MW, which results in widespread shortage in supply to all categories of consumers. (Table 10.4A). The state has only 584 MW of installed capacity, of which 540 MW is thermal and the remaining 44 MW is hydel. However, generation from the thermal power station is very low, and most of the power requirement is met by purchases from the Central generating plants: NTPC, Chukha Hydel (PTC) and Ranjit Hydel (NHPC). Annual per capita consumption in Bihar is also abysmally low, which is 76 units as against a national average of 612 units in 2005. (Table 10.5). The National Electricity Policy aims to increase the average annual per capita consumption to 1,000 units by 2012. The growth is predicated on increasing generation to remove peak electricity shortfalls, as well as implementation of the Rajiv Gandhi Grameen Vidyutikaran Yojana

(RGGVY), which aims to electrify all villages by 2009, under Central Government grants amounting to 90 to 100 per cent.

Table 10.5: Average Generation and Consumption of Power—Bihar and India

	India	*Bihar*
Village electrified (%)	84	41
Household electrification (%)	55.8	10.3
Per capita consumption (%)	612	76
Installed capacity (MW)	1,18,619	584
Actual generation (MW)	71,000-75,000 (approx)	30

Source: Bihar Eleventh Plan.

Next to floods, non-availability of adequate power is the major threat to diversified agricultural development in Bihar. A complete renovation of existing plants at Barauni and Muzaffarpur which have deteriorated over time, would be required. Besides, there is the need to create an adequate transmission system to make full use of all the 700 MW of available power supply from the Eastern Grid. Three prominent coal-and water-rich states—Bihar, Jharkhand and Uttar Pradesh—have decided to set up a power project under the 'joint venture' route. The states have approached the Power Ministry for approval to set up a 1,300 MW coal-fired project with Rs. 6,000 crore investment. The project would use coal from closely-located 4-5 blocks located in the state. Bihar, Jharkhand and Uttar Pradesh are in talks with the Power Ministry for the proposed JV project. The proposal may be given approval soon after Central Electricity Authority (CEA) completes the feasibility study for the project. There are chances of the project being situated in Bihar as it would be easy to source coal from all the three points. The state also has abundant water required for the project. The states may share power on the basis of equity contribution.

The coal blocks that would be used for the project were earlier reserved for tariff-based competitive bidding by power companies. However, the State Governments have favoured a large single project instead of smaller projects in the lesser capacities at different locations. The Power Ministry is also in favour of commissioning large projects as it gives economies of scale and helps in cutting down the cost of power for consumers. The tripartite joint venture is expected to be a win for the states. While Jharkhand has the largest coal reserves, Uttar Pradesh and Bihar could provide water and other infrastructural support for the project. The project's location yet to be finalized. The move is expected to work as a shot in the arm of the Central Government as it is struggling to meet the capacity addition target during the Eleventh and Twelefth Plans. Apart from central utilities, the Government is also expecting substantial contribution from state projects to meet the targeted capacity addition of 78,577 MW during the Eleventh Plan. Forecast for energy requirement for the coming decades is as in Table 10.5A.

Table 10.5A: Energy Requirement: Forecast

Year	*Peak Load (MW)*	*Energy Requirement (MU)*
2006-07	1570	9629
2007-08	1842	11194
2008-09	2177	12874
2009-10	2575	14886
2010-11	3046	17213
2011-12	3607	19905
2016-17	5598	32857
2021-22	9567	58248

Source: Report of the special task force on Bihar's Power Sector; Government of India, 2007.

40 Lakh Biogas Plants under operation, 2nd position globally.
Wind Power Farms producing 7,500 MW Wind Power Capacity, 4th Position globally.
2200 MW Small Hydro Power Capacity.
14 Lakhs Solar Photovoltaic Systems installed.
Solar Water Heating Systems covering 2 million sq.mt. collector area.

Steps should also be taken to encourage renewable sources of energy. The State has highest potential for small hydropower projects, wind energy, bagasse based co-generation power from the existing and proposed sugar factories, jatropha on waste lands, rice husk based biomass gasification, biogas and solar energy. In a situation like this, alternative sources of energy have to be located and every effort made to enhance power availability in the State. Latest position in this respect is as given below.*

* A detailed note on Renewable Energy is given in the Appendix.

Husk Power Systems

The company - Husk Power Systems created by two youngmen in Bihar has begun establishing small power planstrl driven by gases from rice husk, a widely available agricultural waste. There are big plants for the future. They want to have decentralized production. With a small power generation system, the distribution network could be simple and strictly local. This would keep costs down, which was essential for their venture to be financially sustainable.

Wind would not pro duce electricity throughout the year. A few years were spent examining the possibilities of using organic solar cells and biofuels. But neither met their requirements. Biomass was the only option left. In villages, no form of biomass was left unutilized. Rice husk was the one thing that the farming communities did not use. So they decided to use rice husk. The system they engineered does not burn rice husk but heats it up instead. A clean-burning mix of gases is produced that drives an engine. The engine turns a generator that produces electricity.

Gasification is a very well-understood technology. The gasifier could be made in a local workshop. A cheap engine was bought from a company in Agra and suitably modified. In the early hours on Independence Day in 2007, the first such plant began to produce electricity. Husk Power Systems, the company they established, now has 16 plants in place. Each plant generates between 35 kilowatts and 100 kilowatts of electricity. The power is being supplied to about 60 villages at present. Public reception has been "awesome". As soon as a plant was put up, request for connections came from people in the neighbourhood. There is thus no worry about the market or convince anybody about it, it is pretty hard to make economically electricity at a very small level. By using electricity supplied by the company people could cut their cost on alternate forms of energy, such as kerosene, by as much as 50 per cent. Reliability of supply was another factor that attracted customers.

For the most-part, the company was providing electricity to those villages which were not connected to the power grid. The plants operated for only six to 12 hours a day, depending on local demand. The waste left after gasification too can be used. It was good manure and could also be burnt. Besides, it was rich in silica and could be sold to the cement industry. The Foundation was impressed with the company's performance and recently decided to provide a second round to funding for scaling up operations. The Foundation is an independent charity established by the oil and energy giant, Shell Group, and focuses on enterprise based solutions to global poverty and environmental challenges.

Husk Power Systems is using unique technology and processes to tackle the rural energy deficit in India in an environmentally and commercially sustainable way. More than 40 per cent of the country's population, living in approximately 1,25,000 villages, had no access to reliable electricity. Existing energy options in rural communities, such as diesel generator sets and kerosene lanterns, were polluting and prohibitively expensive. Even those villages that did have access to electricity were often subject to frequent power cuts and shortages in power supply. This directly impeded their economic development.

> A Simple and strictly local power generation system has proved that rural Indian communities are willing and able to pay for reliable electricity.

This company was proving that rural Indian communities were willing and able to pay for reliable electricity and that Bihar represented a viable market in which to deliver modern energy services. Each of its plants becomes operationally profitable within six months of starting. Husk Power Systems wants to install 50 to 70 plants next year. It has an ambitious plan to have 2,014 plants up and running by the year 2014. The company would then be able to supply electricity to about one crore customers in over 4,000 villages.

Small Hydropower Projects: For hydropower development, the state has a dedicated agency, viz., Bihar State Hydro Electric Power Corporation (BSHPC), till date, for setting up small hydropower projects, 92 potential sites with an aggregate capacity of about 95 MW have been identified. Out of these, six small hydropower projects of an aggregate capacity of 46.1 MW have been installed in the state.

Biomass Power and Cogeneration: A potential of about 200 MW through setting up of biomass based power projects including co-generation projects exist in Bihar. Rice husk based biomass gasification, combustion technology for industrial application and decentralized power generation are cited as important sources for power generation in Bihar.

Wind Energy: The state's nodal agency is in the process of taking up wind resource assessment programme in association with an agency at Chennai, to identify suitable potential sites to set up wind energy based power projects.

Biogas: It is an important source of renewable energy for meeting the cooking energy needs in villages. The state has a potential of about 9.4 lakhs family size biogas plants to meet the cooking energy needs and 1.25 lakhs biogas plants have already been installed till date.

Solar Energy

India receives solar energy for 300 days, equivalent to 5000 trillion kWh every day and Bihar is equally blessed with sunshine.

E-Charkha

There are a number of innovative systems adopted in India to produce electricity. In Rajasthan villages e-charka has been introduced and bringing light to villages.

New Government Initiatives in Power Sector

Power Plants to be Set Up Through Tariff Based Bidding: An agreement has been signed between Government of Bihar/BSEB and Infrastructure Leasing and Financial Services Limited (IL & FS) for development of additional coal based capacities to meet the growing demand for power in the state. Accordingly, the Joint Venture Company will carry out the power project development activities, initiation of process of getting all key clearances including site selection and land acquisition and selection of private entrepreneur through tariff based bidding for the installation of power projects at various places in Bihar. The plan is to install 3 × 660 MW capacity Thermal Power Plant at three different places.

Power plants to be set up a Nabinagar: NTPC and Government of Bihar/Bihar State Electricity Board (BSEB) have signed a memorandum of understanding (MoU) in October 2007 to promote a joint venture company for setting up and operate a 1980 MW (3 × 660 MW) coal based power station at Nabinagar in the district of Aurangabad in Bihar.

Power plants to be set up through IIP/MPP: BSEB has sanctioned 8 coal based power projects under merchant power plant for which MOU has been signed with Vikas Metal and Power, Kolkata for setting up 2 × 250 MW units at Begusarai. It has signed another MOUN with JAS Infrastructure Capital Pvt. Ltd., Ranchi for setting up 9 × 135 MW units at Banka.

Nuclear Power Projects (4 × 700 MW) at Rajauli: Government of Bihar/BSEB has requested NPCIL for the installation of 4 × 700 MW Nuclear Power Project at an estimated cost of Rs. 22400 crore at Rajauli. A team of experts has already visited Rajauli for site selection.

Road

It has long been argued that rural areas need, connectivity. The cities have been connected to the global economy and have taken off. Do the same for rural areas, and they will take off too. Today, many villages are not even connected by road or telecom to the closest town, let alone the world. Roads are not, of course, the only thing that matter—other rural projects and policies matter a great deal too. But connectivity enhances the value of every other rural investment, since it empowers people through improved mobility and access. People can, buy agricultural inputs and sell their produce more easily. E-chaupal experience of I.T.C. is a big example. Children can go to schools easily, cattle can move easily to get veterinary help and the sick can get to health centres. Remote areas have, by definition, the worst connectivity. They are among the poorest and slowest-growing, but accelerate when given connectivity. A study done by Sukhadeo Thorat *et al.* shows that government expenditure on rural roads pushes up agricultural productivity and increases non-farm employment opportunities. Moreover, it opens up new rural markets, provides villagers access to health care centres, increases school attendance and facilitates social interaction among the rural communities (Annexure 10.1).

Roads can incubate a thousand small business, and can convert villages into towns. Government staff are much more willing to be posted to places with good connectivity, so roads improve administration. Rural productivity cannot be high without roads. This was first demonstrated in India in Punjab and Haryana, which historically had the most dynamic rural economies and lowest poverty rates. The green revolution in Punjab hinged not only on R&D but roads too. Indeed, the returns to road investment were even higher than today. Studies by International Food Policy Research Institute (IFPRI) in China. Vietnam and some African countries point to the same conclusion—rural roads do more for growth and poverty mitigation than virtually anything else. An idea about present level of road development in Bihar is given in Table 10.6. Former President Kalam's theory of Providing Urban Facilities in Rural Areas (PURA) could help as a quick fix on the road to rural development.

Table 10.6: Level of Infrastructure in Bihar

- The Infrastructure Development index is 80 for Bihar bench marked against 100 for All India.
- Bihar only has 77 km of road length per 100 sq km, worse than Orissa which has 169 km. Average teledensity in India is 9 per cent, i.e., nine out of every 100 homes has a telephone; in Bihar it is 2 per cent, the lowest in the country.
- Rural connectivity is about 32 per cent—all weather connectivity is a serious constraint.
- The road density per lakh population at 90.1 kms compares poorly with the national average of 256.7 kms.
- The share of surfaced roads in Bihar is only 37.4 per cent as compared to the all India figure of 56.5 per cent.
- The average growth in road length for India as a whole is 3.5 per cent, the growth rate for Bihar is a meagre 0.6 per cent.
- The public transport system is extremely limited in the state.

A recent IFPRI paper by Fan, Gulati and Thorat estimated the impact of different government programmes on rural growth and poverty reduction in recent decades. For every million rupees spent, roads raised 124 people above the poverty line, and R & D 85. Every million rupees spent on education reduced poverty by 41 people, and on irrigation by 60 people (Table 10.7). It is ironic to note that the lowest returns came from subsidies that are the most popular with politicians—subsidies on credit (42 people), power (27 people), and fertilizers (24 people) (Table 10.1).

Table 10.7: Rural Infrastructure and Poverty

	Returns in rupees (per rupee spending)	*No. of poor reduced (per million rupee spending)*
Roads	5.31	123.8
R & D	13.45	84.5
Irrigation	1.36	60.0
Education	1.39	41.0
Power	0.26	3.8
Soil and Water Conservation	0.96	22.6
Health	0.84	25.5
Anti-poverty programme	1.09	17.8

Source: FAN, Hazel and Thorat.

Exactly the same picture emerged, when the researchers estimated the agricultural growth impact of these factors. Roads and agricultural R&D contributed by far the most to growth. Lower down, came investment in education and irrigation. At the bottom came subsidies for credit power and fertilizers. Let us conclude by recalling what economist Robert Chambers said back in the 1970s. If I had money, I would use it to build roads. If I had more money, I would build more roads. If I had still more money, I would build still more roads. *Old dictum was give me road(s). I will give you development.*

All weather roads in the villages of the country have doubled the income of rural households, raised literacy rate by 10 per cent, and appreciated land prices by up to 80 per cent, says the World Bank. In 2000, about 40 per cent of India's 825,000 villages lacked all-weather roads. With access to roads, Incomes have soared. Household incomes rose by 50 per cent to 100 per cent on average. The multi-lateral lending agency has been supporting India's rural connectivity programme Pradhan Mantri Gram Sadak Yojana (PMGSY). It aids projects in several places like Himachal Pradesh, Rajasthan, Jharkhand and Uttar Pradesh. The year-round connectivity has narrowed gender gap with access to education for girls and raised job opportunities, IDA said adding every Rs.10 lakh spent on rural roads has helped lift 163 people out of poverty. The gender gap has narrowed with easier access to school for girls. There was better availability of school teachers. Access to jobs improved, and new business started up, diversifying the rural economy.[2]

Farmers received better prices for their products by accessing markets directly and cutting out middlemen, and spoilage of perishable produce reduced. Agricultural and animal husbandry practices were modernized. The World Bank has said a second project of $1billion is under preparation. It would aim to improve maintenance, weak capacity, governance and accountability, and would introduced several efficiency measures.

The World Bank estimates that a one per cent increase in the stock of rural infrastructure can lead to one per cent increase in GDP across all countries. But in India between 1993-94 and 2003-04, the share of budgetary expenditure on rural infrastructure and social services has declined from 32 to 25 per cent. The impact of this is that even after six decades of independence, over 40 per cent of India's rural habitants are not connected to all-weather roads; 56 per cent rural households do not have access to electricity and 80 per cent do not have access to sanitation facilities. Really speaking declining investment is not the only constraint for the dismal state of rural infrastructure. Even some well-intended schemes like the integrated Rural Development Programme launched in 1980 failed to deliver, because of serious design deficiencies coupled with corruption, parochialism and power politics.

In planning for rural development, the importance of rural roads cannot be over emphasized. Under the Pradhan Mantri Gram Sadak Yojana (PMGSY) launched in 2004, now a part of Bharat Nirman, the target has been scaled down to connecting all villages above 100 people by 2009. Transparency International estimates that corruption eats up almost a quarter of the total road construction costs. This complaint was echoed recently by the PM when he said that corruption in road construction projects has spread like cancer to every corner of our vast country.

After agriculture, lack of Health infrastructure in the villages is the second most important cause of rural indebtedness. A World Bank Report in 2002 estimated that one episode of hospitalization costs around 58 per cent of the patients annual family income, pushing 2.2 per cent of the rural population below the poverty line. Even more disconcerting is the fact that 40 per cent of those hospitalized had to borrow money or self off assets. Despite numerous promises, rural health service is very unsatisfactory and the national Rural Health Mission presents a bleak picture. The true potential of rural India can only be harnessed through proper infrastructure development with people's participation and by creating employment opportunities using local skills and assets.

The state's precarious health sector has now been administered a major shake up: Primary Health Centres (PHCs) are reporting a spurt at an average of 3,000 patients per month against a dismal average of 39 patients per month earlier, and now stabilizing at an impressive 2,200 patients per month; routine immunization coverage has increased from 11 to 33 per cent in the past 18 months. The PHCs are now open 24 hours under strict Government directive. "The Government is closely monitoring the healthcare centres to know the kind of patients visiting these centres, the drugs that are being administered and whether the PHCs have sufficient supply of drugs. These centres are also being linked with computers to maintain such records. A hoarding in Patna announces. I would not let my daughter-in-law attend nature's call in the open" and the face of the Chief Minister looks out at you. On the ground, in several villages, quiet sanitation revolution is on. A government scheme helps villagers government scheme helps villagers build sanitary toilets. Over 90 per cent of village schools have access to clean drinking water.

Bihar has 81,555 kilometres of roads of which approximately 63,262 kilometres are village roads. This works out to roughly 77.5 per cent. The government has come up with the build, operate, and transfer projects to pave and maintain them. It is expected that through the Bihar Rural Roads Development Agency, some 45,000 kilometres of new roads will be constructed by 2010. Performance of the rural roads had been satisfactory with around 40,000 km. of new link roads having been built and 50,000 km. of existing rural roads being upgraded in the first two years (2005-2007) of Bharat Nirman. Against this target of connecting 66,802 villages with all-weather roads, 13,831 villages have been connected. It was noted that nine states—Assam, Bihar, Chhattisgarh, Jharkhand, Madhya Pradesh, Orissa, Rajasthan, Uttar Pradesh and West Bengal—account for 90 per cent of remaining coverage. Among these, Rajasthan, Bihar and Madhya Pradesh are well on course.

While Bihar has the potential of being the granary of India and a great tourists hub, it suffers from pervasive poverty characterized by the poor level of infrastructure. Improvement in the transport infrastructure and more particularly the rapid development and strengthening of the road network in the State could provide the requisite impetus for its economic development. The existing road network of around 81700 kms in the State is as in Table 10.8

Table 10.8: Existing Road Network

(in kms)

Category	*Pucca*	*Kutcha*	*Total*
NH	3629.00	0.00	3629.00
SH	3232.22	0.00	3232.22
MDR	7714.25	0.00	7714.25
ODR	2828.00	990.00	3818.00
Village	27400.00	35861.63	63261.63
Total	44803.47	36851.63	81655.10

Note: NH—National Highway, SH—State Highway, MDR—Major District Roads, ODR—Other District Roads.

The existing road network in the State is inadequate both in terms of capacity and quality. Even the National Highways are deficient with a large part of them being in poor condition due to flood damage. Development has also been confined along certain corridors and large part of the State is still bereft of good connectivity. The Sector's performance has been constrained due to (i) delay in award of works due to lengthy tendering process; (ii) lack of good contractors and poor response necessitating re-tendering (the contractors of other areas do not want to go to Bihar); (iii) slackness on the part of the contractors for timely completion; (iv) erratic supply of Bitumen (only source is Barauni Indian Oil Corporation); (v) problem in geometric alignment, inadequate lane width, poor drainage, weak pavement and structure; (vi) absence of safety measures. In the case of Road Development Bihar should learn from the experience of Rajasthan where they have increased their capacity four times through small contractors.

Ministry of Road Transport and highways allocates about Rs. 1,500 crore annually on upgradation of state highways and roads to national highways. While states like Goa, Haryana, Karnataka and Rajasthan have spent well over the national average, surprisingly Bihar has also been aggressive in road-building. It is one of the top 10 states, spending nearly 40 per cent of funds, well above the national average under both repair and highway up gradation. This is a very encouraging development in the future of Bihar-particularly agriculture.

Road Sector—New Initiative at a Glance

- It has been decided to construct five additional major bridges, two over the river Ganges between Arrah and Chhapra and between Bakhityarpur and Patroi, one on the Gandak between West Champaran and Gopalganj, one on the Kosi between Saharsa and Darbhanga and one on the Sone between Arwal and Sahar.
- Agreement is in final stage with IDFC and IL& FS for 4-aning of 8 main corridors (approximately 800 kms) under Public Private Partnership (PPP).
- The time limit for the execution of every road project is worked our. Incentive for early completion and penalty for delay are specified. Monitoring process of weekly progress is computerized, for which a computerized Data Centre has been established at the headquarters.
- Rules for registration of contractors have been simplified so that the registration process can be completed within three days.
- An Equipment Bank has been established to provide Plant and Machineries to the contractors on reasonable hire charges.
- Third Party Quality Control has been arranged to ensure quality of work.

- Integrated Improvement-cum-Performance based Maintenance has been introduced for all road projects.
- All work divisions including headquarters have been computerized.

Telephones

Under the communication sector, telephones form an important component of development. As a successful example of private-public partnership helpline experience for the farmers in Barabanki district of Uttar Pradesh, is the best thing to have happened in several years. The helpline has been set up by private telecom player Airtel in partnership with the Indian Farmers Fertilisers Cooperative Ltd., (IFFCO). While a pilot project was launched on April 2, 2006 at the Danyalpur and Mubarakpur IFFCO Societies in Barabanki, the third helpline was set up in Sitapur district in May. Similar helplines will be set up in other parts of eastern UP before the model is replicated across the country. When a farmer in Barabanki calls Lucknow, his query is registered and the call is routed to the IFFCO centre in Delhi. Function of helpline is that experts come up with the solution, which is then explained to the farmer through a customized voice message. The farmer does not have to pay long-distance charges. As part of the scheme, farmers get five free voice messages a day. The messages update farmers on mandi prices, weather forecast and fertiliser availability. (Table 10.9)

Table 10.9: BSNL Details on Tele Capacity

Items/Year	*2000-01*	*2004-05*	*2005-06*	*2006-07*	*2007-08 (upto Dec.,07)*
No. of telephone exchanges	910	1155	1163	1202	1210
Total Capacity (lakh)	7.71	16.71	21.73	13.54	13.40
Total Connections (lakh)	5.88	12.89	17.38	9.86	9.64
Waiting Lists (lakh)	0.65	0.97	0.74	0.62	0.64
Automatic Trunk Exchange Capacity (lakh)	6.75	1.68	1.88	2.03	2.17
No. of Rural Public Phones (lakh)	0.14	0.38	0.38	0.00	0.38
Inernet Connection (lakh)	0.06	0.20	0.19	0.41	0.42
No. of PCOs (lakh)	0.20	0.55	0.63	0.00	0.67
Mobile Capacity (lakh)	—	0.22	6.98	7.74	8.45
Mobile Connections (lakh)	—	2.40	5.71	7.61	9.10
WLL Capacity (lakh)	—	1.44	1.44	2.15	2.32
WLL Connections (lakh)	—	0.98	1.30	1.53	1.58

Bihar itself has now set an example. Many State Governments can now replicate the RTI model of Bihar, in which instead of writing an application for seeking information, one has to dial a helpline number. The Rs. 10 RTI application fee would be added to the telephone call charge.

Along with the ATMAs and Krishi Vigyan Kendras in every District, the Central Government is embarking on a very ambitious programme for Telephone and road connectivity. Though under Bharat Nirman Programme it is envisaged to cover all the villages in the country by telephone link by November, 2007, this may remain a far cry for Bihar as the rural telephone density is still very low and extra effort will be needed to achieve the objective. The position regarding telephones in Bihar as on March 2004 is shown as in (Table 10.9) and Graph (10.1).

Graph 10.1: BSNL Connections

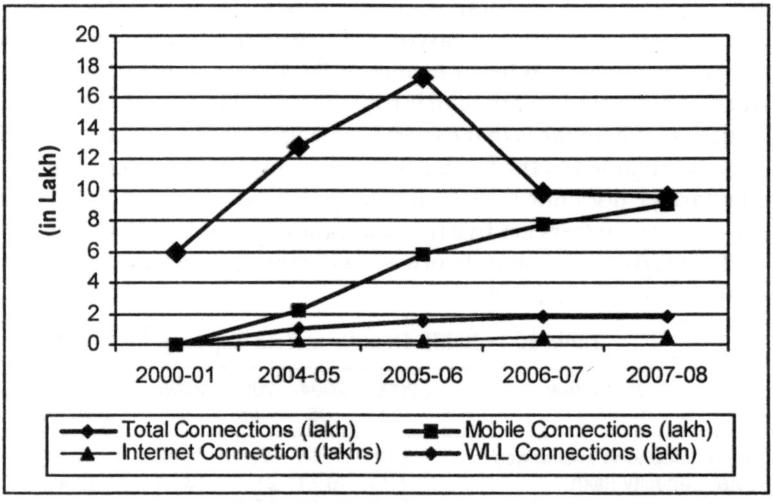

Marketing

Efficient 'marketing' system of agricultural produce is a prerequisite for achieving its faster growth and successful diversification as well as accelerating farmer's income growth. There is also the need for value-addition to production and a portion of which must be channeled to the growers. However, in Bihar, there is lack of proper marketing mechanism. Rural population still has a long way to go, in developing its capacity to become part of globalization. Rural areas of Bihar are ripe with profit potential. They've got eye catching landscapes, which can become beautiful tourists spots. The herbs available here can cure many diseases and ailments. The Handloom work and culinary arts are of world-class quality. People in rural areas have the potential to become major player's in industry. They lag behind due to lack of support and guidance. Besides this, shortage of electricity, marketing problems and poor transport are major drawbacks. Providing such amenities will attract Business houses to invest in rural areas.

Bihar's agricultural sector has shown an impressive growth compared with many other progressive states. However, income growth has not been commensorate with this trend and the challenges ahead are—(i) ensure more value addition, to production, and (ii) that some of that extra value is channelled to farmers. Currently, value tends to be added outside the State, and there is a general lack of marketing mechanisms to respond to farmers' needs. Marketing is crucial to a successful diversification and commercialization of Bihar agriculture. Further income generation from Bihar agriculture and food sector will require addressing product quality, delivery and pricing, despite some technical, institutional and organizational barriers.

Agricultural Marketing

After the repeal of BAPMC (Bihar Agricultural Producers Marketing Cooperative Act), agricultural market in the state is functioning without any formal institutional structure. Following disbanding of BSAMB, the fate of 1500 rural hats under its purview is also banging in balance. The Board had developed infrastructure during 1978-79 with the World Bank assistance, but at most places these hats are in a very poor shape because of poor maintenance. In the meantime, absence of a fully functional market body has certainly created a significant glitch of non-availability of market intelligence, as there is no information on market arrivals its prices and destination of dispatch, etc.

A very small share of local production finds its way to the regulated markets in Bihar. Just as for other Indian states, Bihar has 95 regulated markets (mandis), principal market yards (67), and rural hats (855) that handle some of the marketing of major agricultural commodities in the state. They are supervised and controlled by the State Agriculture Produce Marketing Board, under the provisions of the Bihar State Agriculture Produce Marketing Act (1960). Average area served per wholesale market in Bihar was 392 as against 156 in Punjab (Table 10.10). Market density per lakh hectare of area was less than 2 in most of the districts.

Data on market arrivals indicate that a very small share of agricultural production finds its way to the markets. For instance, between the period from 2001 to 2003, the average arrivals of rice, wheat and maize were only 8.4, 9.7 and 15.9 per cent respectively of their average production. However, the arrival of gram, was as high as 40.9 per cent. (Table 10.11). The reasons behind these poor arrivals in the regulated markets are attributed to poor quality of market infrastructure and support services besides high transportation and informal transaction costs. This is because (a) the amount and quality of market infrastructure and support services is poor, and (b) the transportation costs and informal transaction costs discourage farmers from going to the markets.

Table 10.10: Status of Infrastructure in Bihar *vis-a-vis* Other States

	Road density (per 100 sq.km. area)	*Percentage of villages electrified*	*Average area served per wholesaler market (sq.km)*
Bihar	77.86	72.18	392
Punjab	127.78	100.00	156
Tamil Nadu	158.78	100.00	434
Uttar Pradesh	86.77	79.40	456
West Bengal	85.00	78.10	318
India	74.93	86.50	451

Source: Thorat and Sirohi (2001). Development of Rural Infrastructure in India: Profiles of Districts.

A recent study shows that the quality and quantity of marketing outlets in Bihar is among the worst in the country. Furthermore, a World Bank analysis of some commodity supply chains has shown that up to 40 per cent of the margins along with value chains are lost to wastage and transportation costs. Crop diversification, especially into the relatively more perishable fruits and vegetables, will increase the need for effective marketing linkages supported by modern marketing practices, including introduction of grading, post-harvest management, cold chains, etc.

Standard quality grades for fruits and vegetables are currently not used in Bihar. For milk, fat content and other standards are used, and some processing plants are HACP and ISO certified. The Patna

Table 10.11: Average Estimated Production and Arrivals for Selected Commodities at APMCs in Bihar (2001/03)

	Production MT	*Price (Rs./Kg.)*	*Value of Prod (Rs. lakhs)*	*% Share in VoP*	*Arrival in MT*	*Arrival as % of Production*	*% Share in Arrivals*	*Value of Arrivals (Rs. lakhs)*	*% share in Value of Arrival*
Rice	5130800	813	416878	25	428557	8.4	21.6	34820	30.4
Wheat	4212100	638	268521	16	407669	9.7	20.6	25989	22.7
Maize	1457150	500	72858	4	231254	15.9	11.7	11563	10.1
Gram	69700	1813	12633	1	28508	40.9	1.4	5167	4.5
All oils	115000	1453	16704	1	27175	23.6	1.4	3947	3.5
Potatoes	1436900	425	61068	4	332899	23.2	16.8	14148	12.4
Vegetables	6719100	561	457144	28	403296	6.0	20.3	22605	19.8
Fruits	2957500	1198	354161	21	123514	4.2	6.2	14791	12.9
	—	—	1659966	100	1982871	—	100	133030	100

Source: Bihar Agriculture, Building on Emerging Models of "Success" World Bank.

(Bihar) laboratory of the Directorate of Marketing and Inspection (responsible for quality certification of agricultural products under the "AgMark" scheme), ceased operations at the end of 2006 due to the lack of demand for its services. The Directorate (under the Department of Agriculture and Co-operation—Government of India) has 23 AgMark laboratories that processors can approach for voluntary certification of quality and the right to display the AgMark Stamp. Of the 13000 grading units that exist in the country only 3 are situated in Bihar. The total value of produce that gets graded is negligible, unlike in Maharashtra, Uttar Pradesh and Tamil Nadu, where the share of the value of graded produce is as high as 75 per cent of the national value.

There are significant post-harvest losses. Every year, substantial losses are incurred amongst perishable products. It is estimated that as much as 39 per cent of tomatoes production is wasted. Similarly 39 per cent of mangoes production and 22 per cent of litchis production are lost annually due to the lack of proper post harvest activities and facilities. (World Bank, 2004).

It is recognized that a considerable share of agriculture production in Bihar is being produced for home consumption, thus lowering the marketable surplus. There are also some other important factors which prohibit producers from accessing, the markets. One of the most important issue in this respect is 'non-remunerative price' structure. As discussed elsewhere in this report, cost of production of various crops is much higher than the Minimum Support Price offered. Added to that is the weak marketing infrastructure. In a recent World Bank Study (2004), Bihar ranked third from bottom amongst states in the level and quality of its regulated marketing infrastructure. It also came fourth to the last in the level of farmers satisfaction with existing market conditions. Bihar's density of markets (per million hectares of sown area) is relatively low. There are 7 principal regulated markets for every million ha of area in Bihar compared to 22 in the neighbouring state of Andhra Pradesh.

It is unfortunate that Patna (Bihar) laboratory of the Directorate of Marketing and Inspection, responsible for quality certification under the "AgMark" scheme is going to be closed due to lack of demand. But Bihar's Eleventh Plan has recognized, the necessity of private investment for infrastructure development and promotion of 'contract farming' particularly in perishable products, hence the need for such laboratories. It is encouraging to note that the Bihar's Eleventh Plan recognizes the role of private investment to help in the development of marketing infrastructure. This will help promote contract farming, export of fresh and processed fruits, processing of ethanol, corn oil, in the Feed and Starch industry, export of baby corn, and to make litchi honey.

Priority to Market

There is an even more compelling reason why poverty is destined to remain a spectre at the global level. Poverty will not be eliminated for the very reason that the global developmental paradigm gives priority to the market over government and even the society. Governments everywhere have more or less voluntarily withdrawn from responsibility of distributive justice; and since free markets distribute their rewards according to their own promiscuous and capricious laws, this ensures that wealth flows evently. This means that those who are not beneficiaries of an expanding and increasingly internationalised economy have diminishing redress for their poverty. Of course governments routinely express their desire to create a more equal society, and make provision to alleviate the worst sufferings of the poor. But their capacity to do so is far behind the adroit effects of markets to lavish prizes on those they favour. Indeed Governments have been weakened by a globally integrated economy that permits finance to move so easily, but does all, it can, to prevent the movement of the peoples of the world to where they might command higher wages. Since wealth may rest wherever it can regenerate and reproduce itself most readily, governments throughout the world have had to abandon that modest tribute of wealth that the rich have called Punitive taxation. Even modest demands by government upon the heavy purses of privilege may be evaded by the instant disappearance of billions of dollars into havens, gilded exile and offshore hideaways.

Some Suggestions

The earlier concept of farmers markets, where growers brought their produce to the town, was good. However, inconvenience of going to the town from remote villages deterred many. And some of those who managed to reach were barely able to recover the cost. The need of the hour is to train the farmers in marketing and entrepreneurship. We must popularise the concept of agri-tourism in the villages and create village brands. Combining rural tourism with agri-tourism for the 24 × 7 urban dwellers during the weekends will be a successful venture.

Village level cooperative efforts and promotional campaigns for their village brands will help make them popular. The village Panchayat land should be put to proper use by developing sales outlets on them. Combined with recreational parks, development of heritage houses etc., will ensure the flow of weekend tourists from adjoining urban areas.

Agri-tourism refers to visiting an agricultural farm, agricultural operations for purposes of enjoyment, education and active involvement. As urban work culture has shaped in to 5-day week in these regions and weekend pleasure activities have increased progressively, agri-tourism can be promoted. This will help the urban consumer, desperately looking for outlet for rewinding on the weekends and can combine it with fruit and vegetable shopping. Such direct sales from the farm will benefit both consumers and farmers. Bihar has special agri attractions—Sompur, the site of world's largest cattle fair, is visited by lakhs of Indian farmers and thousands of foreigners every year. Keeping in mind the huge growth potential of tourism in the mineral rich state, which also enjoys a considerable forest cover besides being rich in flora and fauna, the state government has already chalked out a plan to promote tourism in the next two years.

This effort has succeeded in the US and is now taking shape in developed countries through "U-pick farms and orchards". Here, customers themselves harvest the vegetables and fruits. The grower in return gets higher price than what the middleman would pay. There are also some micro-level experiments like making arrangements to rent a fruit tree, where the consumers select their trees in the beginning of the fruit season, pay in advance and pick the harvest.

Storage and Warehousing

The role of storage warehousing is crucial for agricultural growth. It enables farmers realize better returns for their produce as they have the option to store and sell when prices are favourable. The recent policy initiatives regarding issuing of warehouse receipts as an instrument for increasing liquidity, and thus avoiding distress sale of produce, is welcome. On a macro-level, warehousing also reduces post-harvest losses up to 30 per cent by providing efficient cold chain and food security. Increasing commercialization of agriculture, emergence of commodity exchanges and introduction of collateral management agents and accreditation bodies have created a need for quality warehousing. It gives the private sector an opportunity to invest in the construction of warehouses.

The key issues in agri-logistics are non-standard pricing, limited financial capabilities of transporters, opportunistic profiteering, lack of scientific handling of produce and consequent high prices and limited choices for the consumers. Further complications arise from the 'perishable' nature of the commodity and the fact that the requirement of 'cold chain' varies for each commodity, even within the same class. Thus the development of a cold chain network must be produce specific. This would not only reduce wastage but also lead to a greater value in the hands of the farmer. Given the inter-linkages among the individual components of the agri value-chain, there is a need for an integrated and holistic approach with regulatory reforms, infrastructure development, credit availability and innovative business structuring so as to maximize stakeholder benefit and achieve overall development of the agriculture sector.

This will facilitate scale, spruce up cooperatives, encourage farmer associations, define standards and grades for agricultural commodities and develop mechanisms for communication and dissemination of transparent pricing, encourage the system of aggregator producers' cooperatives/bank to pick up the produce of small and marginal farmers, abolish or reduce fees, cess, taxes, duties, etc., on agricultural or horticultural produce procured through any registered contract-farming programme. The number of wholesale markets in the state is also not adequate and for disposal of agricultural produce, the farmers are required to travel a long distance. Nearly 400 sq. km. area in the state is served by each wholesale market as against 156 sq. km. in Punjab and 318 sq. km. in West Bengal. As many of the small and marginal farmers cannot cover such a long distance, they usually dispose off their produce to the village itinerary traders. To the detriment of the interest of the farmer.

Nearly one-third of our horticulture produce, especially fruits and vegetables are wasted mainly on account of poor cold storage facilities. For the same reason the country also experiences fluctuations in prices of horticulture production in Bihar. Capital Investment Subsidy Scheme (CISS) for Construction/Expansion/Modernization of Cold Storage for Horticultural Produce was introduced by the Government of India. The objective is to promote modern cold storage so as to avoid spoilage of perishable horticulture and other items like dairy, meat, fish chicken, etc., provide remunerative price to the farmers and providing business opportunities to cold storage owners throughout the year by establishment of multi-purpose cold storages.

The progress of the scheme has not been very encouraging in Bihar. Considering the annual production of perishable horticulture produce at 152.50 lakh MT in the State during 2002-03, the existing post harvest handling and cold storage capacity is inadequate. Assuming 30 per cent of the marketable surplus produce to be stored, an additional cold storage capacity for approximately 14 lakh MT is required which translates into 281 additional cold storage units of 5000 MT capacity each, and a capital investment of Rs. 562 crore. At present, 27 districts out of 38 are having cold storage facility, which can take care of only about 5 per cent of the total production of fruits and vegetables. Post-harvest handling facilities and cold storage units are required to be established in the remaining 10 districts. Additional facilities may be created in those districts where the number of cold storages and their capacity are inadequate. Rural godowns have also vast potential in the State in view of the fact that majority of the farmers are small and marginal who command little holding capacity. The State Government may identify potential areas and draw up a plan for construction of specialized storage facility for fruits like Litchi, Mango, Banana, etc. and their processed products. This is possible only if an uninterrupted power supply to the Cold Storage units is ensured by the Bihar State Electricity Board.

APPENDIX

RENEWABLE ENERGY

There is no doubt that sufficient energy plays an important role in the continued Indian economic development, more so in Bihar where there is an acute shortage of power. Every sector of the economy—agriculture, industry, transport, commercial, and domestic-needs energy 'inputs'. Energy consumption is rising heavily and dependency on fossil fuels such as coal, oil and gas is seen as a threat for economic development and environment. The share of renewable energy from all world sources today is 14 per cent. (Graph 10.2).

Graph 10.2: Share of Renewable Energy in all World

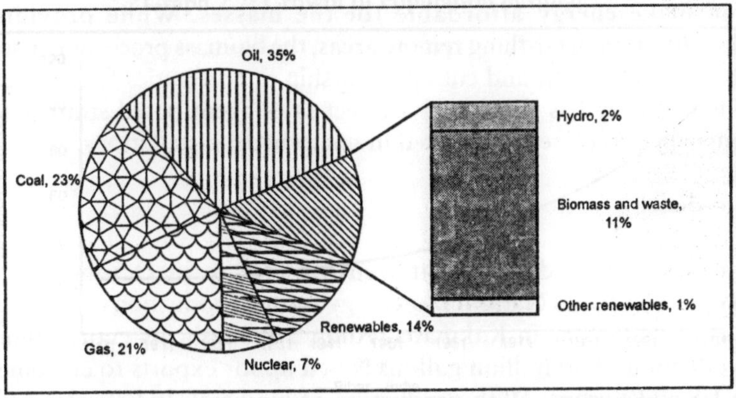

Fortunately India is blessed with a variety of renewable energy sources. The main ones are biomass, biogas, the sun, wind, and small hydro power. Municipal and industrial wastes are also seen as useful sources of energy. The country is implementing one of the world's largest programme in renewable energy. The country ranks second in the world in biogas utilization and fifth photovoltaic production. The total installed capacity of power generation from various non-conventional energy sources reached about 6,800 MW in 2005. Currently, wind energy is the leading renewable energy source in India followed by small hydro-power and biomass. India is the fifth largest wind power producer in the world after Germany, the USA, Denmark and Spain, with a wind power generation capacity of 3,600 MW (March 2005). The Wind Resource Assessment Programme (WRAP) is one of the largest programmes of this kind in the world.* So far, about 11,000 MW grid inter-active power has been installed from renewable energy sources such as wind, hydro, biomass and solar. With this, Ministry for New and Renewable Energy has a share of about 7.5 per cent in the total grid power capacity installed in the country.

Micro Hydel

Tide Technocrats Pvt. Ltd., (TTPL), which specializes in rural energy solutions, is one such company. Its micro-hydel devices provide affordable power in rural areas that lack consistent access to grid power by utilizing locally available natural water flow systems. Its biomass processing systems tap an alternate sources of energy for rural consumption. We address the alternative fuel need for power plants and make energy affordable for the masses. While providing electrification to far-flung remote areas, the biomass processing would create employment and entrepreneurship opportunities for villagers. The company has developed a direct sales agent based strategy for reaching out to areas interested in micro hydel installations.

Bio-diesel

India's first bio-diesel plant went on stream in October 2007. Hyderabad-based Natural Bio-energy started production of the green fuel at its factory in Kakinada, Andhra Pradesh. Its entire annual production of 30 million gallons is tied up for exports to customers in US and Europe. With production of biodiesel, India's place as a source of green energy will get yet another star. Already Pune based Suzlon is one of the leading players in wind energy and Delhi-based Moser Baer is setting a large facility for making solar panels.

Bio-diesel fades in comparison to the performance of gasoline, which western countries are increasingly choosing for fuel. These days, when crude trades over $120 a barrel, bio-diesel is economical. Secondly, states like California in US have already begun increasing use of alternative fuels that are low in carbon emission.

---

* Sitra Reports 63 (2006)—http:www.sitra.fi/kulkaisut/raporti63.pdf.

Bio-diesel, an equivalent to crude derived diesel, is processed from biological sources. Naturol will make its bio-diesel from Jatropha plant with Belgian technology. The plant derived bio-diesel can be used in normal diesel engine vehicles without modifying them. Biodiesel produces between 40-60 per cent lesser carbon dioxide emission but emits more smog forming residues. Vehicle manufactures in Europe, who were initially vary of bio-diesel, are now more willing. European auto makers like Scania now say that their vehicles can run on 100 per cent bio-diesel. Virgin's Richard Branson who is testing the use of bio-diesel in one of his trains, has planned the first commercial flight that will be powered with a 60 per cent bio-fuel-kerosene blend in 2008.

Solar Energy

The potential of solar photovoltaic systems in meeting the growing decentralized energy needs is well recognized. The available solar radiation can be transformed into electricity either directly through the solar photovoltaic route or by converting it into heat through the solar thermal route. Under the scheme a target of 50 MW solar plants has been set for the Eleventh Plan. A maximum of 10 MW capacity in a state will be considered reasonable. It is estimated that about Rs. 1,000 crore will be invested by the private sector. The cost of solar generation is Rs. 15 per unit. It costs Rs. 20 crore per MW but four-five years back, it was Rs. 40 crore per MW. The costs came down because of R&D that went into it. A day will come when solar power will become the cheapest power in the world. Around 11,000 MW grid interactive power has been installed from renewable energy sources like wind, hydro, biomass and solar. This makes our share about 7.5 per cent in the total grid power capacity installed in the country.

The sub-continent also occupies fourth place in the world when it comes to the generation of solar energy, not far behind Japan, Germany and the United States. The sun shines in India for between 2300 and 3200 hours a year, compared to only 1000 to 2000 hours in Germany. IREDA estimates the total potential for renewable energy on the sub-continent at 100 MW. So far only 10 per cent of this total is being used. The Indian government is now increasingly pinning its hopes on wind and sun to satisfy the country's growing hunger for energy and reduce CO_2 emissions.

With a global installed capacity of just about 5,000 MW and relatively at high costs, solar photovoltaic energy has come to be viewed as a futuristic, even some what utopian option. Yet concern is growing over climate-changing carbon emissions and innovative ways are being sought to bring electricity to millions who are not plugged into conventional power grids. Encouragingly, the efficiency of solar photovoltaic cells is improving. In combination with energy-efficient light emitting diodes, solar modules are achieving good results in lighting. The potential of such devices to light up dark homes has been recognized by a Mondialogo engineering award, instituted by Daimler and UNESCO, to a solar-powered LED lantern. As with other simple technology applications, the lantern, developed jointly by teams from the University of Illionis at Urbana-Champaign, United States, and the Jagannath Institute for Technology and Management, Orissa, can potentially revolutionise 'off-grid' lighting. What such projects need is support from Governments to commercialize and achieve commercial scales of manufacture on a fast track. Grid-connected projects seem to be getting fresh attention.

Rajasthan is to set up a big solar farm. However, national capacity to manufacture LEDs (which, incidentally, have been deployed to light up Christmas trees in international cities this year) remains a challenge. Conergy India, through its subsidiary Sun Techniques, has already installed several hundred renewable energy systems and environment friendly solar power projects improving the living standards of over 1 lakh people in 250 villages across the country. About 13 per cent of the 6 lakh villages in the country were in the remote territories, making it very difficult and expensive for the public Grid to supply power there. "Off Grid energy products such as solar powered home and street lighting systems serve as affordable, efficient and eco-friendly alternatives. Masonic Trust Chairman points that of the current 1.44 lakh MW electricity produced in the country, only 8.4 per cent comes form renewable sources.

The issues of solar power are definitely the government's radar. While National Action Plan released on energy recently, describes solar power as having the potential to change the face of the nation, there is goal to increase production of photovoltaics to 1,000 MW/year and generate at least 1,000 MW of solar thermal power. This is a step in the right direction. Thinking small would be tragic, given that the world's most abundant energy resource bestows itself generously over most of our country. Pundits say that dedicating just

0.3 per cent of India's land area for solar power could meet our entire electricity needs. Solar Thermal Electricity Generation (STEG) is a simple technology that consists of curved mirrors that concentrate sunlight onto a receiver tube to heat a working fluid flowing through it. The remaining part of the plant is very similar to a conventional power plant. There are many reasons why STEG is surya's gift to us. STEG is one of the cleanest technologies available today. A 50-MW plant would save around 90-20 million kg of greenhouse gas emissions. Secondly, its energy payback is merely five months compared to a useful life of more than 25 years. Most importantly STEG technology is the only solar technology that provides "firm" power and allows plants to dispatch power when demanded. It can also work in a hybrid mode enabling solar heat to be backed by co-firing with natural gas or coal. Waste heat from the combined generation of heat and power can be used for industrial applications, district heating and cooling and seawater desalination.

The solar industry today is poised for inflexation as cost of traditional energy rises and technology improvements make the once uneconomical solar option the inevitable choice. To put things in perspective, the installed solar photovoltaic (SPV) capacity cross the globe has gone up by about 11 times in the last 10 years. It has shown an annual growth of about 45 per cent during each of the last five years and is expected to be at least 20 times today's installed capacity by 2020. Notwithstanding this projected exponential growth in installed capacity, solar power will constitute only about 3 per cent of the global installed generation capacity and far lesser in terms of generation by 2020. In the Indian context, the installed SPV capacity of about 10 MW (grid plus distributed) as of March 2008 constitutes an almost negligible percentage of the overall installed renewable capacity of over 12,62 MW.

> *India then needs to push for the development of domestic solar industry in an integrated fashion. The sector requires a hub where industry, researchers, government, and NGOs can come together to address the triumvirate of capex costs, efficiency & technology.*

The industry euphoria can be attributed to the Solar Incentive Package Scheme (SIPS) of the Central Government which has so far drawn interest from about 14 players proposing a staggering investment commitment of about Rs. 140,000 crore. The SIPS among other direct and indirect incentives proposes a 20-25 per cent central financial assistance to 3-4 fab units and 10 eco-system units. On the demand side, the MNRE's 10 year generation-based incentive (GBI) scheme has ensured a steady return framework for the solar technologies which has been further expanded by states such as Rajasthan and Punjab. SIPS and GBI together are expected to result in significant capex reduction with long-term revenue assurance that is expected to spur private investment in solar generation. An analysis of the solar value chain shows that the biggest cost component of solar cells is polysilicon, the main raw material for silicon water cells. The solar industry today consumes more of it than the semi-Conductor industry and the subsequent demand supply mismatch has seen spot prices of polysilicon touching $ 200 per tonne. However, our research indicates that the global commitments to new polysilicon capacity addition are likely to outstrip demand by 2010-12 and result in spot prices of silicon falling to about $ 50 per tonne, a significant factor considering that polysilicon alone constitutes about 80 per cent of the solar cell cost.

There are about 1.5 lakh unelectrified villages in India in remote and inaccessible areas where it is not viable for the grid to reach and are suitable for solar lighting. This formidable task can be cost effective in the coming years as mentioned above and we also have a successful model—the "barefoot" solar engineers. Though the elegant and elevating model, pioneered by the Social Work and Research Centre (SWRC). Tilonia in Rajasthan, has achieved remarkable success, it went unnoticed. This model has been established in about 600 villages across the most backward regions of the country. It has been accepted overseas as well.

This model succeeded because it eliminated precisely those aspects of the earlier programmes which led to their failure. The villagers (invariably illiterate, and often women) are themselves trained as solar engineers to deal with all technical problems; maintenance and repair facilities are provided by setting up a fully equipped electronic workshop for every group of villages; and the community is involved both as stakeholder and as participant. Dependence on external expertise and equipment has been completely eliminated. Simple measures, but these have shown remarkable and inspiring results.

This became possible only because the barefoot solar engineers are extremely competent professionals. After a six-month training at the SWRC's Barefoot College, they can assemble, install, operate, manage and maintain the systems. These accomplished solar engineers have replaced the technical experts of the MNCs and large firms who hitherto were supposed to provide these services.

Third Generation Solar Technology: The First Generation cells were of crystalline silicon, fairly high cost, manufactured in relatively small volumes. The Second Generation cells had higher efficiency, whether made of thin wafers or silicon. The Third Generation has more complex, integrated devices that can reduce reflection and capture more of the solar spectrum by using one or a combination of optical methods. They have a wide range. Cost components are photovoltaic (PV) modules, power electronics and installation. One-half of the total cost you pay would be the module cost. The rest is spread over the balance of the system including cabling and connectivity. As with any renewable energy option, solar power is not the cheapest. However, production increases. The costs tend to decreases. The biggest adopter of PV solar energy is Germany. They have an established feed-in tariff system for a number of years now and it costs less because of economies of scale. India's Ministry of Renewable Energy is introducing a similar scheme to provide electricity grid connection to solar energy suppliers. You need to have a mix of renewable energy resources so that you don't depend on any one source of electricity.

As for land areas required to install PV modules, it depends on the location. In Europe, PV modules are fixed on south-facing roofs. The UK Centre for Solar Energy Research has a façade of 1,000 sq. mts. Germany has installed solar panels on roofs of industrial buildings measuring up to 10,000 sq.m. each. China and India are now becoming manufacturing centres. A company in Kolkata, for instance, is working on an innovative technique to make PV modules with thin film made with cheaper compound material to increase conversion efficiency. A UK-based company is to soon market a new type of portable plastic solar cell as low-cost mobile phone charger.

There are two kinds of storage as far as solar energy is concerned: on grid and off-grid. In the UK, it is largely the on-grid that's important. The DC electricity produced by PV modules are converted to AC in the grid. There is two-way energy traffic. The building will supply electricity to the grid during the day an buy from it in the night. This is called feed-in tariff system. In India, there are different variations of the scheme. If you have large tracts of unproductive land that gets sunlight, it's attractive from a seller's viewpoint because you can feed energy to the grid and get paid for that without having to buy back. Off-grid needs battery storage. This is important in India where you have the rural electrification scheme; solar energy can light up India's villages. Solar energy would charge batteries that are then used to power lighting, television sets and refrigerators. Solar energy charged batteries would be for high-efficiency lighting purposes and for other low power domestic requirements.

Solar photovoltaic cell production in the recent past has been at a peak in Japan, Germany, and the United States; global capacity was augmented by 1,727 MW in 2005 alone. Some scientists argue that with sufficient investments on a 'grand solar plan', the United States could meet as much as 69 per cent of its electricity demand and 35 per cent of all its energy demand (including transport by 2020 through solar energy, with provision for demand growth and at prices comparable with those of fossil fuels. For many countries, India in particular, solar energy should be attractive going by a recent solar 'hotspot' map generated by the U.S. space agency, NASA. Australian and the Niger top in harvesting potential, according to the map, but much of India can also reap sun's bounty. Indeed, the Tenth Plan projected an R&D policy for 2025 incorporating a range of alternative energy technologies but the solar option remained a weak element. For a start, an accelerated programme of rural LED lighting and solar thermal technologies for cooking can achieve substantial results. It can contribute to public health by greatly reducing the adverse health effects of indoor smoke from lamps and stoves that use fossil fuels. It is time India got serious about solar energy.

Solar photovoltaic power plants up to 200 kW capacity and those of 1 kW to 200 kW installed for electrification in the unelectrified villages and rural areas have been developed in India. India receives solar energy for more than 300 days, equivalent to 5,000 trillion kWh every year. Daily solar radiations is on an average of 4-7 kW/square metres depending on the locations. Presently, the cost of solar power generation technology is very high. The Ministry will provide financial assistances of Rs. 12/kW hour in case of solar-photovoltaic and Rs. 10/kW hour in case of solar thermal power fed to the electricity grid. If Orissa can take a lead in providing solar light to villages, Bihar should go whole hog in this area.

Wind Energy

Wind energy has been the world's fastest growing renewable energy source growing at a rate of 28 per cent in the last decade and has emerged as an option for grid quality power generation. The costs in respect of wind monitoring stations are shared between the Ministry of Non-conventional Energy sources and state nodal agencies in the ratio of 80:20 (90:10 in case of north-eastern states).

Buoyant Wind Market

India and China are among top five growing markets for wind energy:

- Of the various renewable energy sources, wind energy is making a significant contribution to the installed capacity of power generation and is emerging a competitive option.
- It is used in more than 70 countries and contributes around 1.3 per cent of the global electricity consumption. In some countries, wind energy contributes upto 40 per cent or more of their power needs.
- Germany, the US, Spain, India and China are the top five countries with growing wind markets. At the end of 2007, worldwide capacity of wind-powered generators was 93.849 MW.
- The market for new wind turbines increased nearly tenfold in the last ten years. The overall size of wind turbines market in 2007 was 19,696 MW, up from 15,120 MW in 2006 (30 per cent growth).
- Based on this increase, experts predict that by 2010, the world will generate about 170,000 MW of wind power.
- After a decrease in the average world growth rate between 1999 and 2004, the wind market expanded steadily.
- In 2005, the market grew by 23.8 per cent, in 2006 it grew 25.6 per cent and in 2007 by 26.6 per cent.
- Experts are, however, worried about the uneven distribution of the market growth across the countries.
- Though 2007 was a spectacular year for the wind industry worldwide, the increase in the average growth rate was primarily due to booming wind markets in the top five wind countries.
- Growth of the wind industry, in 20 out of the top 40

countries, slowed down in 2007. Europe, which generates the maximum wind power, decreased its share in installed capacity from 65.5 per cent in 2006 to 61 per cent in 2007.

- Most developing countries in Latin America and Africa, are yet to invest in wind energy in any significant manner. By 2007, these countries fell back in terms of new installations and counted for only 0.6 per cent and 0.4 per cent of the global wind power generation.
- In 2007, Asian countries, however, showed a significant growth in promoting wind power. In 2007 alone, China added 3,313 MW of wind power and India added 1,580 MW of wind to their already installed capacity.
- India and China are already among top five wind markets, but both these countries need to invest more on renewable wind energy to meet their growing need for power in a sustainable manner.*

Wind Energy—India Fourth in the World

The Indian wind energy sector has an installed capacity of 8757.2 MW (as on March 31, 2008). In terms of wind power installed capacity, India is ranked 4th in the world. Today it is a major player in the global wind energy market. The potential is far from exhausted. Indian Wind Energy Association has estimated that with the current level of technology, the 'on-shore' potential for utilization of wind energy for electricity generation is of the order of 65,000 MW. The unexploited resource availability has the potential to sustain the growth of wind energy sector in India for years to come.

The cost of wind power generation varies between Rs. 4-5 crore per MW, with a pay-back period of five to eight years. The machines can be maintained at a cost of Rs. 0.25 to Rs. 0.60 per kWh (kilo watt hour). India ranks first in Asia and fourth in the world among wind power producing countries. It added 1,700 MW to the installed capacity in 2006-07 to aggregate an installed capacity of more than 7,200 MW. Most of the capacity addition has been achieved through commercial projects by private investors. However, India's wind energy Gross Power Potential (GPP) has been assessed at 45,000 MW per year while the Technical Power Potential (TPP) is a dismal 15,000 MW, assuming 22

* *Down to Earth*, May 16-31, 2008.

per cent grid penetration. At present, it is not technically feasible to exploit the full wind power potential in view of total installed power generating capacities from conventional power generating methods.

Graph 10.3: Worldwide Wind Energy

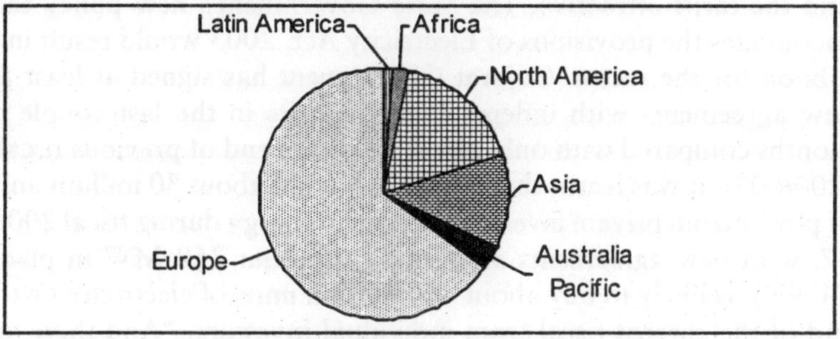

Graph 10.4: Growth in Top Wind Markets

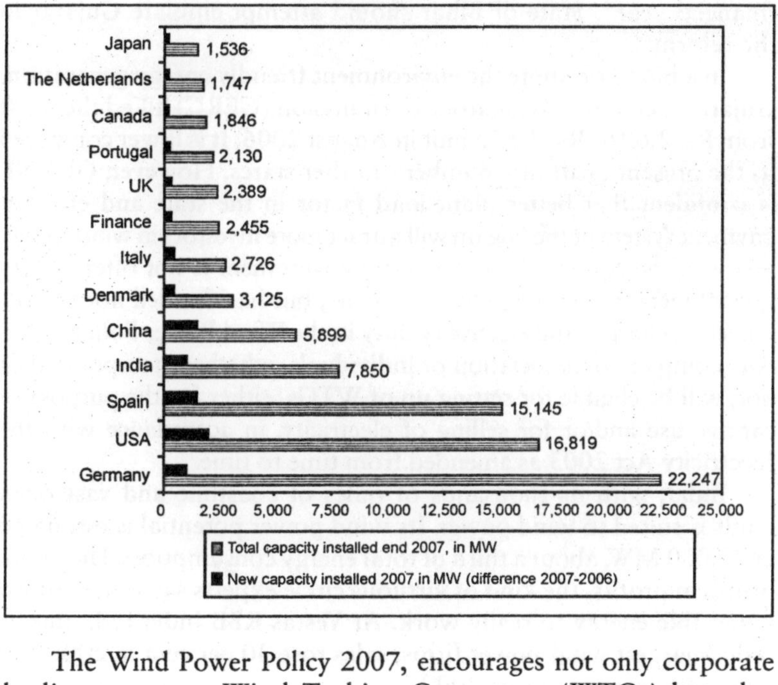

The Wind Power Policy 2007, encourages not only corporate bodies to set up Wind Turbine Generators (WTGs) but also individuals. Thanks to the entitlement for 100 per cent depreciation of WTGs and sales tax benefits, now HNIs aiming for tax savings can also look forward to investing into wind energy. According to the recent reports, celebrities like Aishwarya Rai, Sunil Shetty and Sachin Tendulkar have already made investments in wind mills in Rajasthan for tax-saving purposes. Sources in the energy sector believe that such individuals can now look forward to Gujarat too if they find the tariff attractive. The State Government's new policy that encourages the provisions of Electricity Act, 2003 would result into a boon for the sector. Gujarat Government has signed at least 80 new agreements with independent investors in the last couple of months compared with only about 20 by the end of previous fiscal" (2006-07). It was learnt that GUVNL bought about 30 million units of power from private investors into wind energy during fiscal 2006-07. with new agreements amounting to about 250 MW in place, GUVNL is likely to buy about 400 million units of electricity by the end of the current fiscal from individual investors. "And there are many more agreements that are expected be signed during the current financial year". State of Bihar should attempt emulate Gujarat in the reforms.

In a bid to promote the environment friendly energy generation, Gujarat Electricity Regulatory Commission (GERC) hiked the tariff from Rs. 2.60 to Rs. 3.37 a unit in August 2006. It is lower compared to the present tariff in a number of other states. However, GUVNL is confident that better plant load factor in the state and efficient payment system of the Nigam will attract more investors in wind energy sector in the coming days. The state government is not offering any special incentive to individual investors, but has decided not to hike wheeling charges and electricity duty in the Wind Power Policy 2007. Any company or association or individuals, whether incorporated or not, will be eligible for setting up of WTGs, either for the purpose of captive use and/or for selling of electricity, in accordance with the Electricity Act 2003 as amended from time to time.

India, with its thousands of miles of coastline and vast open lands is suited to wind power. Its wind power potential is estimated at 45,000 MW, about a third of total energy consumption. The boom brings in profits, the kind of virtuous circle experts say is needed for renewable energy to really work. At Vestas RBB India Ltd., one of India's largest wind-power firms, sales rose 30 per cent in 2006 and about 40 per cent during 2007.

Wind power has been helped by some states setting targets that 10 per cent of their power should come from renewable energy. High capital costs and the fact that wind is intermittent-plants often run at a quarter of their capacity compared with 80 per cent capacity for unclear power. It is estimated that wind power costs around Rs. 3.5 a kilowatt hour, compared with Rs. 2.5-3 for imported coal. It can only gain in importance because of concerns about climate change.

According to the latest *World Energy Outlook*, India needed to invest $ 1.25 trillion in energy infrastructure by 2030 and that more than three quarters of this investment should be in power infrastructure. Attracting investment in a timely manner will be essential if economic growth is to be sustained. To date some 412 million people have no access to electricity in India. By 2030, that should fall to nearly 60 million as the electrification rate rises to 96 per cent from 62 per cent in 2005, according to the IEA's reference scenario, or most likely energy forecast. Only in its high growth scenario would all households have access to electricity. According to its reference scenario.

If things move in the right direction, Rajasthan is likely to attract an investment of Rs. 8,000 crore in the renewable energy sector. The state government has signed MoUs with 14 power companies to set up power plants in the field of wind energy, biomass and solar energy. There would be five wind energy projects in Jaisalmer, Barmer and Jodhpur with a generation capacity of 1,600 MW, while eight biomass projects would be set up in Sirohi, Baran, Tonk, Alwar, Sawai Madhopur, Kota, Nagaur, Hanumangarh and Jalore districts.

Currently, Rajasthan has an installed capacity of 490 MW in wind energy and 46.3 MW in biomass. With the installation of the proposed projects, the installation capacity would increase to 2,150 MW in renewable energy sources. Suzlon Gujarat Wind Park, with a proposed investment of Rs. 4,910 crore, leads the table followed by Enercon (India) and Wish Wind Infrastructure, with investments of Rs. 1,072 crore and Rs. 888 crore, respectively. These initiatives in wind electricity are the result of the Wind Power Programme (WPP), which was started in the Sixth Plan in 1983-84. It aims at survey and assessment of wind resources, setting up demonstration projects and provisions of incentives to make wind electricity competitive.

Another case is that of Daman, a city with some 160,000 inhabitants which is the showcase of a new energy era in India. Two of the most important wind tubine builders in the world, the Indian Suzlon Energy and the Indian subsidiary of the German Enercon company, are producing 80 per cent of the sub-continent's wind power stations there. The seventh-biggest country in the world with over a billion inhabitants is a gigantic market for the production and sale of renewable energies. Wind power installations with a total output of 1840 MW were commissioned there during 2006 alone. India has now moved up to fourth place in the ranking of the world's wind energy producers, significantly ahead of China.

Gujarat Experiment—JGY

The Gujarat Government has embarked on an innovative scheme—Jyoti Gram Yojana (JGY). Bihar can learn from the Gujarat experience. Gujarat Government took the next best option. They launched the feeder separation plan that bifurcates industrial, domestic and agri consumers and provides electricity accordingly. This was followed by an innovative scheme, the Jyoti Gram Yojana. The scheme supplies round-the-clock electricity to villages by making them partially contribute towards infrastructure. It was a huge hit. In 2003-04, around 2,516 villages were covered under the scheme. The number rose to 6,203 by March 2005 and in another two years, shot up to 17,826. "This lessens the burden on the state utility and electricity demand too".

Another major initiative came in the form of the Industrial Extension Bureau under the industry commissioner. INDEXTb, an ISO 9001:2000 organization of the Gujarat Government, is a single-point contact for all investment related activities in the state. There is one officer assigned to every investment proposal that gets filed. He constantly tracks it and sees that it materializes. In the process this wings keeps the Chief Minister informed about its progress or lack of it. This makes the job earlier. Provides real time information about every proposal to the seniors.

Notes

1. *Times of India*, January 5, 2008.
2. Teri Source:- *The Economic Times*, November 30, 2009.

ANNEXURE 10.1

Table : Changes in the Level of Rural Infrastructure by District

District	*Year*	*Road Density (Per lakh ha. area)*	*No. of Market (Per lakh ha. Area)*	*% Village Electrified*	*% Literacy*	*Primary School*	*High School*	*Hospital*	*Credit*
Aurangabad	1970	247.77	0.000	20.86	19.58	7.07	1.02	0.10	166.00
Aurangabad	1980	247.77	0.000	50.50	26.14	6.38	1.08	0.12	318.00
Aurangabad	1990	247.77	0.000	91.15	25.64	5.26	1.02	0.10	415.00
Aurangabad	2000	330.55	0.000	91.26	55.50	3.81	0.19	0.08	578.00
Begusarai	1970	478.65	0.000	22.01	10.96	5.62	1.65	0.16	339.00
Begusarai	1980	478.65	0.000	41.84	13.55	4.47	1.71	0.15	464.00
Begusarai	1990	478.65	0.000	70.16	20.83	3.89	1.55	0.19	799.00
Begusarai	2000	638.58	0.000	86.55	46.50	1.99	0.18	0.08	113.00
Bhagalpur	1970	335.04	0.530	6.42	18.54	6.40	0.33	0.16	207.00
Bhagalpur	1980	335.04	0.707	19.01	24.24	5.02	0.87	0.13	312.00
Bhagalpur	1990	335.04	0.707	50.83	28.73	3.96	0.86	0.11	520.00
Bhagalpur	2000	446.99	0.707	52.79	44.40	3.05	0.18	0.10	725.00
Bhojpur	1970	802.27	0.483	20.75	32.39	5.30	1.16	0.12	302.00
Bhojpur	1980	802.27	0.967	30.12	28.45	4.23	0.95	0.12	434.00
Bhojpur	1990	802.27	0.967	61.72	35.73	3.83	0.85	0.12	664.00
Bhojpur	2000	1070.33	0.967	62.61	56.80	3.07	0.18	0.10	925.00
Champaran (East)	1970	910.53	0.252	12.64	12.70	2.41	0.26	0.05	219.00
Champaran (East)	1980	910.53	1.008	30.24	17.16	1.85	0.30	0.05	318.00

(Contd.)

(*Contd.*)

District	Year	Road Density (Per lakh ha. area)	No. of Market (Per lakh ha. Area)	% Village Electrified	% Literacy	Primary School	High School	Hospital	Credit
Champaran (East)	1990	910.53	1.008	77.65	20.73	1.51	0.26	0.05	527.00
Champaran (East)	2000	1214.77	1.008	87.25	35.40	1.58	0.17	0.03	734.00
Champaran (West)	1970	376.05	0.383	12.36	12.70	5.01	0.98	0.06	161.00
Champaran (West)	1980	376.05	0.574	23.12	17.16	4.04	0.79	0.05	259.00
Champaran (West)	1990	376.05	0.574	63.10	20.73	3.39	0.75	0.04	376.00
Champaran (West)	2000	501.70	0.574	63.36	36.00	1.85	0.12	0.03	524.00
Darbhanga	1970	4830.78	0.438	7.09	16.89	3.47	0.97	0.11	128.00
Darbhanga	1980	4830.78	0.438	22.53	21.98	3.30	2.19	0.25	677.00
Darbhanga	1990	4830.78	0.438	81.71	26.65	3.21	0.69	0.09	1160.00
Darbhanga	2000	6444.88	0.438	81.71	41.50	2.31	0.11	0.10	1616.00
Gaya	1970	725.70	0.000	26.78	19.58	2.08	0.54	0.08	84.00
Gaya	1980	725.70	0.000	61.94	26.14	2.66	0.94	0.13	271.00
Gaya	1990	725.70	0.000	85.26	25.64	3.46	0.62	0.09	389.00
Gaya	2000	968.18	0.000	85.73	46.20	2.93	0.12	0.10	542.00
Gopalganj	1970	640.61	0.000	8.60	18.22	5.41	0.96	0.07	251.00
Gopalganj	1980	640.61	0.000	22.44	14.33	4.15	0.73	0.10	374.00
Gopalganj	1990	640.61	0.000	63.32	18.44	3.43	0.65	0.05	532.00
Gopalganj	2000	854.66	0.000	64.94	46.50	2.70	0.18	0.08	783.00
Katihar	1970	47.45	1.963	2.20	8.89	4.93	0.95	0.06	182.00
Katihar	1980	47.45	2.618	9.98	17.25	3.69	0.71	0.09	264.00
Katihar	1990	47.45	2.618	47.24	18.77	3.10	0.59	0.07	445.00
Katihar	2000	63.30	2.618	48.54	31.10	2.39	0.18	0.08	620.00
Madhubani	1970	513.85	0.000	11.41	16.89	7.32	1.98	0.28	302.00
Madhubani	1980	513.85	0.000	29.65	21.98	7.54	3.53	0.54	387.00

(*Contd.*)

(*Contd.*)

District	Year	Road Density (Per lakh ha. area)	No. of Market (Per lakh ha. Area)	% Village Electrified	% Literacy	Primary School	High School	Hospital	Credit
Madhubani	1990	513.85	0.000	93.17	26.65	7.96	3.72	0.39	668.00
Madhubani	2000	685.55	0.000	94.34	41.20	8.43	1.11	0.55	931.00
Monghyr	1970	592.46	0.506	11.85	10.96	7.05	1.75	0.13	149.00
Monghyr	1980	592.46	0.632	33.44	13.55	6.55	1.75	0.16	305.00
Monghyr	1990	592.46	0.632	58.65	20.83	5.84	1.68	0.23	599.00
Monghyr	2000	790.42	0.632	59.28	52.80	8.00	0.47	0.18	835.00
Muzaffarpur	1970	1092.31	0.285	9.83	15.84	1.82	0.32	0.07	174.00
Muzaffarpur	1980	1092.31	1.140	39.38	22.44	6.42	0.84	0.08	656.00
Muzaffarpur	1990	1092.31	1.140	79.00	28.48	1.88	0.51	0.05	1020.00
Muzaffarpur	2000	1457.28	1.140	80.27	44.70	1.45	0.12	0.03	1421.00
Nalanda	1970	373.67	0.000	44.09	24.42	5.77	0.83	0.12	301.00
Nalanda	1980	373.67	0.000	84.70	21.34	4.69	1.12	0.10	32.00
Nalanda	1990	373.67	0.000	98.96	24.46	3.83	1.14	0.09	649.00
Nalanda	2000	498.52	0.000	100.00	50.40	3.14	0.17	0.10	905.00
Nawada	1970	196.15	0.000	24.56	19.58	8.77	1.33	0.13	152.00
Nawada	1980	196.15	0.000	61.22	26.14	6.76	1.02	0.18	337.00
Nawada	1990	196.15	0.000	92.39	25.64	5.64	0.96	0.15	451.00
Nawada	2000	261.69	0.000	93.40	44.70	4.09	0.22	0.14	628.00
Patna	1970	1147.25	2.265	45.91	24.42	3.72	0.62	0.10	222.00
Patna	1980	1147.25	2.589	78.36	21.34	4.29	0.93	0.12	653.00
Patna	1990	1147.25	2.589	97.83	24.46	1.79	0.45	0.05	949.00
Patna	2000	1530.58	2.589	97.97	51.40	1.54	0.14	0.05	1322.00
Purnia	1970	850.84	0.754	1.79	8.89	1.93	0.64	0.08	79.00
Purnia	1980	850.84	1.005	10.49	17.25	3.26	1.07	0.15	157.00
Purnia	1990	850.84	1.005	51.80	18.77	5.40	1.16	0.15	244.00

(*Contd.*)

(*Contd.*)

District	Year	Road Density (Per lakh ha. area)	No. of Market (Per lakh ha. Area)	% Village Electrified	% Literacy	Primary School	High School	Hospital	Credit
Purnia	2000	1135.13	1.005	52.56	31.40	4.79	0.16	0.17	339.00
Rohtas	1970	300.25	0.278	24.57	33.36	6.24	1.04	0.13	223.00
Rohtas	1980	300.25	0.557	42.68	28.50	4.88	1.03	0.12	318.00
Rohtas	1990	300.25	0.557	65.48	34.66	4.02	0.88	0.10	484.00
Rohtas	2000	800.57	0.557	66.33	59.20	3.05	0.17	0.06	674.00
Saharsa	1970	405.44	0.680	5.16	14.44	14.44	2.17	0.25	93.00
Saharsa	1980	405.44	1.529	24.46	18.78	11.97	1.69	0.23	140.00
Saharsa	1990	405.44	1.529	81.39	23.04	13.90	2.12	0.24	355.00
Saharsa	2000	540.91	1.529	81.87	36.10	12.52	0.29	0.11	494.00
Samastipur	1970	731.10	0.000	17.73	15.84	1.92	0.43	0.03	425.00
Samastipur	1980	731.10	0.000	34.18	22.44	5.72	0.17	0.04	594.00
Samastipur	1990	731 10	0.000	93.33	28.48	2.32	0.62	0.06	1018.00
Samastipur	2000	975.38	0.000	93.41	43.90	1.69	0.40	0.03	1418.00
Saran	1970	2990.00	0.371	10.64	18.22	3.50	0.53	0.11	155.00
Saran	1980	2990.00	0.741	33.39	14.33	6.59	1.19	0.11	603.00
Saran	1990	2990.00	0.741	75.45	18.44	4.50	0.33	0.03	965.00
Saran	2000	3989.04	0.741	76.85	50.30	1.38	0.15	0.06	1345.00
Sitamarhi	1970	666.67	1.138	13.16	21.59	3.60	0.85	0.06	282.00
Sitamarhi	1980	666.67	1.896	32.66	18.26	3.07	0.77	0.07	402.00
Sitamarhi	1990	666.67	1.896	76.46	21.48	2.68	0.69	0.07	700.00
Sitamarhi	2000	889.42	1.896	77.90	36.70	2.06	0.14	0.06	975.00
Siwan	1970	591.44	0.000	10.20	18.22	4.20	0.34	0.05	385.00
Siwan	1980	591.44	0.000	22.69	14.33	3.32	0.49	0.06	538.00

(*Contd.*)

(*Contd.*)

District	Year	Road Density (Per lakh ha. area)	No. of Market (Per lakh ha. Area)	% Village Electrified	% Literacy	Primary School	High School	Hospital	Credit
Siwan	1990	591.44	0.000	64.10	18.44	2.63	0.63	0.06	807.00
Siwan	2000	789.06	0.000	68.59	50.50	2.13	0.18	0.05	1124.00
Vaishali	1970	666.07	0.000	10.00	15.84	1.46	0.48	0.05	469.00
Vaishali	1980	666.07	0.000	36.90	22.44	2.71	0.73	0.06	761.00
Vaishali	1990	666.07	0.000	87.71	28.48	2.32	0.68	0.06	1184.00
Vaishali	2000	888.63	0.000	90.00	49.30	1.90	0.14	0.09	1649.00
Bihar	1970	854.68	0.43	15.86	17.87	4.98	0.92	0.11	227.08
Bihar	1980	854.68	0.68	36.50	20.40	4.90	1.11	0.13	398.92
Bihar	1990	854.68	0.68	74.95	24.34	4.16	0.98	0.11	663.33
Bihar	2000	1140.25	0.68	77.02	45.10	3.41	0.23	0.10	884.17

11

POLICY FRAMEWORK

After discussing in brief various policy statements and newly set up commissions/committees, this chapter covers in detail all the major policies adopted, in the agriculture sector.

New Policies

- Sugar Incentive Policy: (to revive the sugar industry and incentives for the new sugar units with power generation and distillery production).
- Supply of diesel at Rs. 10 a litre to drought-hit farmers.
- Mukhya Mantri Rozgar Guarantee Yojana in 15 districts which were denied central scheme.
- Linking of villages with 500 to 999 population with road under Mukhya Mantri Gram Sadak Yojana.
- Integrated education development programme under Mukhya Mantri Smagra Vidyalaya Vikas Yojana.
- Formation of Common School System Commission.
- Decision to Open Nalanda University to revive glorious history.
- Opening of Chanakya Law University of Five-year degree course.
- Assistance from CM Medical Relief Fund to the patients having an annual income of less than Rs. 1 lakh.

Newly-Formed Commissions/Committees

For quick implementation of the government decisions and to bring reform in governance, some Commissions/Committees have been formed.

- Land Reforms Commission: (to take into account all land related issues, renewal modernization and solution of land disputes).
- State Extremely Backward Class Commission: (to provide welfare schemes, identification of their problems and its solutions).
- State Information Commission: (under Right to Information Act, 2005, to bring transparency in official working).
- Farmers Commission: (to look into integrated problem of farmers and growth in agriculture products).
- Bihar Forest and Wildlife Protection Committee: (for Valmiki Tiger Project, Gangetic Dolphin Protection Scheme, forest development, forest product based industries, tourism, opening of zoos, environment protection center and development of villages, close to forest areas).

New Laws and Amendments Effective from 2006

- Bihar Single—Window Clearance Act.
- Bihar Regional Area Development (Amendment) Act.
- Bihar Land Reforms (Maximum Ceiling and Land Acquisition) Act.
- Bihar Water Tax Management Act.
- Bihar Agro-Product Marketing (Abolition) Act.

LAND POLICY

Land Acquisition is coming in the way of Bihar's economic revival plan, particularly in the rural sector. A number of projects already cleared by the State Cabinet have not been implemented as yet because there is no land available. The worst hit is the sugar industry and it is particularly unfortunate for the State, since this industry has the potential to turn around its rural economy. So far, the Government has not been able to arrange land for any of the proposed 14 green field sugar projects and 19 ethanol generation units, cleared by the Cabinet. We have also to take note of existing act regarding ceilings on Land Holdings (Table 11.1).

Table 11.1: Ceilings on Land Holdings in Bihar State

(In Acres)

1. Irrigated land with two crops	15 to 18
2. Irrigated land with one crop	25
3. Dry Land	30 to 45

Source: Agriculture at A Glance, Ministry of Agriculture, 2005.

Farmers are coming up with stiff opposition wherever the Government proceeds to acquire land. Agencies concerned are quietly stepping back as the government has made it clear that there will be no forcible acquisition. The Government certainly is cautious after the experience of Singur and Nandigram in neighbouring West Bengal. After the initial failure, the Government framed an attractive land acquisition policy in February 2007 offering 50 per cent over and above Government rates plus an additional compensation between 30 and 60 per cent. But despite that the farmers are not willing to give up land. A New Rehabilitation and Resettlement Policy has been announced now. This should be able to meet all types of problems being faced today.

There is a proposal by the Central Government to promote 'land-share' companies. According to this concept, farmers can lease their land to a company, say an agro-processing unit, and become shareholders in proportion to the size of their holdings. The land will be considered as the farmer's equity in the company and he will receive a share of the company's profit. The farmers can also take land on lease, including his own land, from the company and cultivate for a fixed rent. The farmer thus stands to gain both from farming as well as his share in the agro-processing unit. The idea of land-share companies is part of a new policy note prepared by the Ministry of Agriculture circulated to all states. At present, some land-share companies exist, though there is no legal framework. The ITC group grows tobacco in Andhra Pradesh and Pepsico recently started growing potatoes, tomatoes, chilles and rice in Punjab, Maharashtra, Karnataka and West Bengal. The Mittals, in Collaboration with UK Bank, Rothschild, have started cultivating several crops in a big way.

There are a few other successful experiments in some States. In 2002, the Tamil Nadu government leased out wasteland to corporates for 30 years to grow crops like cotton, flowers, fruits, vegetables and spices. The Gujarat government has also leased out over

2,000 acres of wasteland. The government hopes that the concept of 'land-share' companies, a middle path between contract farming and direct corporate farming—can ensure faster growth in the farm sector and increase farmer's income. But the fear is that the farmer could be alienated from his land if the firm heads for liquidation. To ensure that this doesn't happen, the government plans to put in some safeguards. One of the several proposals is that majority of the shares, in the land share company would be held by the farmers/land owner's and up to 25 per cent of the company's paid up capital can be subscribed to in cash by others, including an agro-processing unit or a trading company.

Another version is that the land share company can operate on a commercial basis and the management should be vested in an elected board of directors. Voting rights could be on the basis of cooperative principles—one member, one vote irrespective of the value of share of each member. Descendants of farmers can inherit shares in the company. A farmer can sell his share to other farmers, but the shares cannot be traded through public issue as there may be a risk of takeover by corporate houses or other entities. In case of liquidation, the shareholders would get back their land rights.

This type of Land Policy would possibly be easily accepted particularly by the small farmers in the drought affected districts where productivity levels are low and agricultural activity is not quite remunerative. Along with land sharing, the Policy can be further modified that first preference for employment in the processing industry is given to the farmer and his family members. The government should, in fact, organize, to train these farmer members through the District Training institutes in specific trades needed for the concerned industry. This exercise will help to rehabilitate the farmer and his family to make reasonable earning outside the agricultural activity without moving from his village.

Problem Still Unsolved

Despite all these proposals, these problems are getting complicated. The issue reached a flash point late in 2007 when a massive human wave descended upon the power centre of Delhi. It was a wave of thousands of landless villagers who wanted to be heard, a wave of people in pursuit of land reforms and an end to their state of poverty and hunger. The villagers, from 15 states, started their march to Delhi from Gwalior on Gandhi Jayanti. This is the third war of Independence (the Second being the movement of Jaiparkash in Bihar

> *We are here to wake up the government from its slumber. We cannot allow our basic right to life to be compromised.*

organized by Ekta Parishad as they call. Emaciated old men and women with sticks in their hands, young couples and teenagers walked in disciplined rows, holding green-and-white flags. They chanted slogans, but not against the government. They only want their land back, their right to livelihood protected. These non-violent campaigners nearly 25,000 persons, displaced farmers—Dalits and Adivasies—wanted the government to establish the National Land Commission to formulate land policies, provide direction to state governments and monitor progress on land distribution and entitlement by state governments. What is our future they ask? We do not have food, land or water. We do not have a roof over our heads. We have come to Delhi to get all these. It is an initiative to bring land reforms to a centre-stage in rural development policies. We want a land policy which takes into account the needs of the poor. The rally included protesters from several states, including Madhya Pradesh, Orissa, Chhattisgarh, Jharkhand, Tamil Nadu and Kerala. Also among the campaigners were hundreds of women. Around 200 foreigners from 30 countries also took taking part in the rally, lending support to the cause of tribals, adivasis and landless workers.

> *This is equivalent to the third war of independence. This is a non-violent direct people's action to press the government to finally act in favour of the masses.*

The protesters were in Delhi after a three-year long preparation to press for their demands on water, forest and land and for establishing fast track courts to resolve land disputes. Carrying forward the protest, after 60 years of independence, they want the government to give them food and honourable living.

The Government has now decided to constitute a Committee of 'Experts' to look into the ongoing agitation by landless villagers and tribals on the issue of land reforms. Agitators demand included the setting up of National Land Authority, fast-track courts and single window system to deal with land and livelihood disputes. The Committee would carry required field surveys and studies, collate necessary data and make recommendations to the National Land Reforms Council headed by the Prime Minister. The Committee will look into all related issues, including land reforms and make specific recommendations on policies related to land, speedier disposal of land-related court cases and judicial forums.

Broad terms of reference of the Committee include detailed consideration of the problems and making recommendations on issues relating to ceiling on land holdings, distribution of land to eligible persons, including landless and homestead landless and for ensuring their possession, speedier disposal of land-related court cases and necessary mechanisms, including setting up of fast track courts at various levels.

World Experience

The vexed question of converting agricultural land to industrial use is hardly unique to India. Almost all countries big or small, rich or poor, have a legal framework under which land can be acquired by the government for public purpose. However, what the public purpose will be is spelt out in different ways. Each country has different laws for when and how property is assessed, so there could be a discrepancy between what is considered market value and the compensation offered. In China which is closest to India in terms of population and the fact that a huge number of people are dependent on agriculture, the government can take over land for economic, cultural, national defense or public works projects, under the Land Administration Law of 1988.

- *China*: Land Administration Law, Article 21 (1988) Economic, cultural, national defence, public works projects.
- *Brazil*: The Constitution, Article 5,182 and 184 (1988) Public use, social interest, or for agrarian reform of rural property not performing its social function.
- *Mexico*: The Expropriations Law (1936) Infrastructure development, conservation of history/culture, national security, public benefit, equitable distribution of wealth, preservation ecological balance and natural resources.
- *South Africa*: Expropriation Act, No.63 Definitions Article 2

(1975) Public purpose and certain other purposes if connected with provisions of any law by an organ of State.

- *US*: The Constitution, 5th Amendment, existing case law including Kelo V City of New London, 2005 Private property can be taken for public use: include property development.
- *UK*: Town and Country Planning Act (1990) if required for development, redevelopment or improvement or in the interest of proper planning.

Source: PRS Legislative Research.

In all cases of land acquisition by the Government, assurance of jobs to the displaced is provided. Since there is massive investment in both agriculture and industry, and earnings in industrial employment are better, there is not much resistance to conversion of land to industrial use. Other developing countries like Brazil, Mexico and South Africa, too have provisions for takeover of land by the Government for public purpose. This includes conservation of history and culture in Mexico interestingly, in Brazil, there is provision for Government takeover for the purpose of agrarian reform, if the land is not performing its social function. Brazil's Supreme Court is mulling over a petition filed by rice farmers challenging the Government's creation of a reservation for indigenous people spread over 1.7 million hectares in Roraima province. In the US, the Fifth Amendment to the Constitution ensures the Government's right to takeover private land.

In 2005, the Supreme Court interpreted this to include takeover for real estate development also. In the U.K., too, Government can take over land for development or even improvement purposes. In the Western countries there are strong provisions for involvement of local communities in deciding what the land will be used for. In all the cases, the general principle of giving compensation equivalent to the prevailing market value is followed. In China, ownership is not private, but the farmer has the right to its produce. So earning is compensated, on the basis of up to 20 times the average annual income from the produce of the land. China also has provision for resettlement subsidies, which can be up to 10 times the average annual value of the produce.

LAND REFORMS

There is a growing argument that the crisis in agriculture can only be battled if it is recognized that no economy can grow with 60 per cent of its people trapped in land. On the one hand, the vast majority are ensuared on their unproductive patches of land; on the other hand land ceiling laws, land conversion laws and the absence of clear title deeds and land records mean that there is no free buying and selling of land. So a farmer cannot maximize his holdings, cannot increase the size of his assets of farm a productively; plus a battery of other ridiculous laws bear down on the farmer's mobility. If a farmer builds his own ponds and check dams he could be held guilty of violating the Irrigation and Drainage Act.

Land tenancy is widespread with field studies indicating around 25 per cent of cultivated area under short-term oral tenancy—this, together with a moribund organized rural credit system, and small and fragmented farms (average farm size in Bihar of about 0.7 hectare is half the national average), acts as a severe disincentive to private investments on land. Land Reforms have practically remained an unfinished agenda in the state. This when the author of Land Reforms—Wolf Ladejinsky brought out his first famous report on the abuse of '*Bataidari*' system in north Bihar shook the conscience of those then in power.[1] The major objectives of Land Reforms have been the re-ordering of agrarian relations in order to achieve an egalitarian social structure, elimination of exploitation in land relations, realizing the age-old goal of land to the tiller enlarging the land base of the rural poor, increasing agricultural productivity and infusing an element of equality in local institutions. Thus, Land Reforms have the following dimensions.

- (i) Tenancy Reforms—Conferment of ownership rights to tenants;
- (ii) Distribution of surplus land to land less farm houses;
- (iii) Consolidation of holdings; and
- (iv) Updating and Computerisation of Land Records.

Present Position

Looking at the land situation in Bihar, we find that the political economy of Bihar is beset with one paradox after another. It was the first state to pass land reform legislation, including abolition of zamindari, tenancy reforms and anti-fragmentation laws. Yet, after decades, land reforms remain an unfulfilled dream. Bihar had one of the highest levels of political awakening in India, starting with the Kisan Sabha movement, even before Mahatma Gandhi came on the scene. Even so, political leaders have been, largely, unable to mobilize the democratic zeal of the people, especially as it bears on economic and social welfare and improvement. Bihar, in the early 1950s, was heralded as the shining star among the state of India. Even the introduction and adoption of high-yielding varieties and associated technologies could not keep pace with the increasing population.

Bihar, once recognized as a granary of North India had been reduced to pauper status. In the mid-1970s, production enhancing technologies were made available and adopted by many cultivators. However, their impact was never as profound as in other areas, for a variety of reasons. Important ones among these were the existing social, political and economic structure that curtailed continued investment in agriculture. This reversed, in some instances, the process of production-increasing technologies. There is hardly any doubt that the population that must be supported by Bihar's cultivators is vastly larger today. The failure of cultivators to invest in improved technologies is both a response to the existing socio-economic milieu, or realities, as well as an environmental issue. Now, Bihar has become a macro-indicator for the measurement of backwardness as also a reference point for institutional collapse. It led the nation through the Jayaprakash Narayan (JP) movement to fight corruption and misrule, but itself fell prey to corruption and misrule of family fiefdom.[2]

In the absence of effective and implementable land reform legislation, little has changed for the tillers of the soil. Although, *the zamindars* of Bihar were legislated 'out-of-existence' in the 1950s and 1960s, their legacy has never been eliminated. Their influence is still all-pervasive in a negative sense. They have no justifiable reason to invest in production increasing technologies. They continue to dominate rural Bihar and live as parasites.[3] Regressive agrarian structure is considered to be a major impediment to the higher agricultural growth in the State. There is yet another serious problem of a large percentage of households who do not own any land—a group that has risen rapidly during the recent decades.[4] (Table 11.2) The distribution of land is extremely skewed with predominance of marginal holdings (< 1 ha.). This is compounded by the existence of fragmentation of holdings. Though the legislation with regard to consolidation of holdings was enacted in 1956, its implementation started only in seventies and has not made much headway. Till March 2003, only about 39 lakh hectares of area, which is just 41 per cent of the net sown area, is reported to have been consolidated.

It has to be realized that consolidation of land is not purely a technical issue. In a state where land distribution, in most parts, is extremely inequitable and clear land title a rarity, it is well nigh impossible to effect a large-scale land consolidation programme in a democratic manner. Clearly, such consolidation must be preceded by a comprehensive land reform programme. Such an approach is, however, a long-term project and would yield enhanced agricultural productivity only over the next couple of generations. Efforts must, therefore, be made simultaneously to figure out ways by which small farms can increase their productivity in the short term. Simple and low cost technological innovations such as gravity-fed drip-irrigation kits, and human-operated treadle pumps can reportedly increase the earnings—obviously following a concomitant rise in productivity—of quarter—acre farms by as much as Rs. 2,200 a year.[5]

Though legally banned, the problem of 'Tenancy' is still an issue in Bihar and no official figure $ about tenancy and conferment of tenancy right is thus available. Tenancy in the State overwhelmingly consists of leasing in of their land by small and marginal farmers. Distribution of ceiling surplus land among the landless and marginal land holding farmers is, however, being implemented in the state. Till March 31, 2003, nearly 17 lakh hectares of area was declared surplus, out of which nearly 16 lakh hectares was taken possession and 12.4 lakh hectares was distributed among almost 3.8 lakh individual beneficiaries. Progress in the state in this respect may thus be considered satisfactory.

Finally, regular updating of land records is essential for effective and meaningful land reforms. The performance of the State on this count is very much wanting. Computerisation of Land Records, a Centrally Sponsored Scheme, is under implementation in the state. Under this Scheme, the Centre will fully fund computerization of land records including data entry/reentry, data conversion, digitization of maps and integration of textual and spatial data, tehsil/sub-division/

district data centres, state/union territory-level data centres, inter-connectivity among the revenue offices, training and capacity building, including strengthening of the revenue and survey training institutions, village index maps and core GIS, legal changes and programme management.

Table 11.2: Proportion of Households that do not Own Land (2003-04)

(Percentage)

State	*Official Estimates (Households that Do Not Own Any Land Including Homestead)*	*Households that Do Not Own Any Land Other than Homestead*	*Households that Neither Own Any Land Other than Homestead Nor Cultivate on Owned Homestead Land*
Jammu and Kashmir	3.29	10.97	7.29
Himachal Pradesh	15	22.68	21.87
Punjab	4.57	56.89	29.51
Uttarakhand	10.64	26.4	21.15
Haryana	9.21	49.49	25.96
Uttar Pradesh	3.82	26.2	16.31
Rajasthan	5.65	19.95	12.73
Chhattisgarh	12.09	27.31	20.8
Madhya Pradesh	12.05	31.81	22.76
Gujarat	13.6	44.11	35.37
Maharashtra	17.66	44.78	38.27
Andhra Pradesh	14.33	53.19	48.75
Karnataka	14.09	40.47	30.76
Kerala	4.8	68.36	36.74
Tamil Nadu	16.55	64.52	55.43
Bihar	7.6	38.8	31.01
West Bengal	6.15	46.52	34.69
Jharkhand	4.8	39.25	18.43
Orissa	9.56	38.48	31.07
Sikkim	30.67	44.4	37.96
Arunachal Pradesh	21.59	23.5	22.59
Nagaland	8.02	15.45	10.85
Manipur	2.68	30.3	13.53
Mizoram	2.34	14.1	6.67
Tripura	8.69	59.52	33.22
Meghalaya	6.7	29.01	19.93
Assam	8.05	40.3	23
India	10.04	41.63	31.12

Source: Vikas Rawal, Ownership Holdings of Land in Rural India, *EPW*, March 2008.

The Centre and the States will equally share the cost of survey/re-survey and updating of survey and settlement records, including ground control network and ground truthing and modern record rooms, land records management centres at the tehsil taluka circle and block levels. The Centre will give 25 per cent funds and the rest will be contributed by the States for computerization of registration, scanning, and preservation of legacy mutation records, and providing connectivity to the sub-registrar offices with revenue offices.

Land Records in Bihar are outdated. In the absence of computerization, it is difficult to get land possession certificates (LPC). The scheme aims at removing the problems inherent in the manual system of maintenance and updating of land records and meeting the requirement of groups of users. Implementation of the Scheme, however, is reported to be not very satisfactory. In the above backdrop, it may be mentioned that the State Government has constituted the Land Reforms Commission in 2006 to look into all the issues relating to land reforms and recommend suitable measures accordingly.

Bihar needs to act swiftly and concertedly to enhance its farm productivity in order to keep the frightening spectre of global food shortages at bay, and ensure the country comfortably food secure. The stagnation in global agricultural production ever since the mid-1990 has ominous portents. Farmers would have to feed three billion more mouths by 2050. The Russian Government enforced a cap on food prices from October 15 to December 2007. In China, too, food inflation is higher—at roughly 8 per cent—than ever before, and considerably more than its general rate of inflation (a little below 3%). Our policymakers, too, would do well to realize that the current rate of inflation has touched an all time high of nearly 12 per cent. World prices have touched new highs.* (Graphs 11.1 and 11.2). Clearly, our largely rain-fed agriculture has to be freed from the vagaries of weather. And that can be accomplished only through rapid industrialization of the farm sector with special emphasis on developing a scientific and efficient irrigation system. But, such industrialization entails consolidation of land. India's agrarian order—which is primarily subsistence based and where average farm size is four acres—is hardly conductive to such industrialization. Technology *per se* may be scale-neutral, but high levels of capital intensity require economies of scale.

---

* International prices having touched highest peaks in the first quarter of 2008, have now levelled of 2007 and reached normal.

Graph 11.1: Food Commodity Price Indices

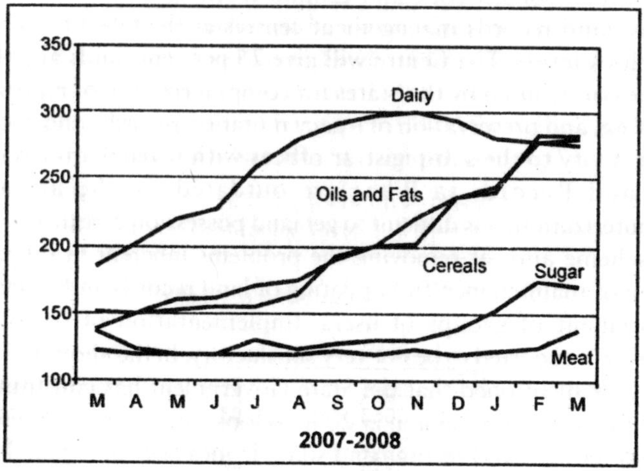

Graph 11.2: FAO Food Price Index

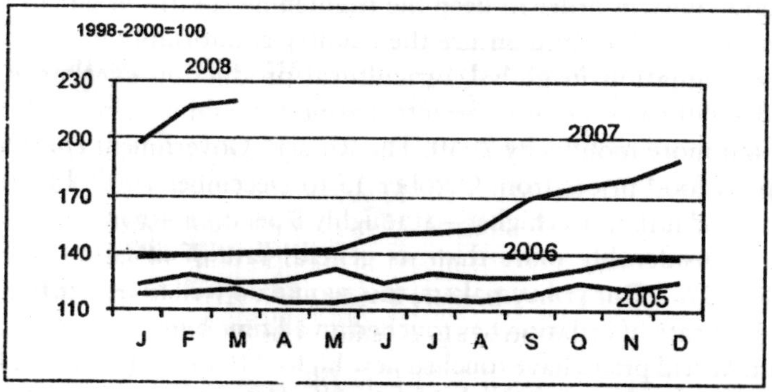

REHABILITATION AND RESETTLEMENT POLICY

The Union Cabinet approved the National Policy on Rehabilitation and Resettlement, 2007 which not only provides substantial benefits to farmers, including a continuous source of income, for land acquired for industrial projects but also makes it simpler for corporates to acquire land with state help. The policy provides for land acquisition up to 30 per cent by state to give a crucial contiguity to developers, be those of Special Economic Zones or industrial projects. The Cabinet also decided to amend the Land Acquisition Act of 1984 to suitably define 'public purpose' for industrial purposes. Given the row over protests against land acquisition in Singur and Nandigram, the policy has followed the principle of rehabilitation before displacement. It provides for an 'order of magnitude' jump in compensation for land acquired by the Government, fixed at the average rate of 50 per cent highest land sale deeds in the last three years or the market rate decided by the State Governments, whichever may be higher. In fact, the State Governments will have to take into account the intended use of land acquired before fixing the market value of the land.

Project-Affected Families (PAFs) will also get 60 per cent compensation, up from the earlier 30 per cent. Under the policy, PAFs will be offered land-for land, to the extent government land is available in resettlement areas. Subject to availability of vacancies and suitability of the affected person, preference will be given for employment in the project to at least one person from each nuclear family. Training will be given for capacity building for suitable jobs, self-employment and scholarships for education will be offered to eligible persons from affected families. Preference will be given to groups of cooperatives of affected persons in the allotment of contracts and other economic opportunities in and around the project site; wage employment to affected persons in project construction work and housing benefits including houses provided to the landless affected families in both rural and urban areas. The policy also proposes that landowners will have an option of either opting for one time monetary compensation or a mix of compensation and equity. In case of PAFs opting for equity, the policy allows a maximum of 50 per cent of the compensation as equity in the project and the remaining as cash.

To check any speculative land transaction, the amendment proposes that the land acquired by the government will have to revert to the Government in case the proposed project does not take off in five years after the acquisition. In case of sale after the project has taken off, the amendment proposes that 80 per cent of the net profit earned during the sale should go to the original landowner. The policy makes it mandatory to conduct a Social Impact Assessment (SIA) in case of displacement of more than 400 families in plain areas and 200 in case of tribal hilly and other scheduled areas. It is mandatory to consult gram sabhas.

To assist quick disposal of cases involving compensation disputes, the policy has decided to introduce a Land Acquisition Compensation Settlement Authority, removed from the normal civil courts, at the local level for speedy disposal of cases to facilitate projects. But the aggrieved can appeal to a High Court and above, against the settlement decided by this authority. Under the policy, there will be a standing relief and rehabilitation committee at the district level and an ombudsman at the state level to monitor rehabilitation under any project. For effective monitoring of the implementation of resettlement plans, the Government will constitute a National Monitoring Committee Cell with which State Governments will have to share information.

The Way Out

All said and done, old land reform approaches, such as blind adherence to land, providing for secure land reform, need reconsideration. Time has come for policymakers to take inventory of India's land reform experience, distill lessons learnt, and use land laws, policies and programmes more innovatively and with the focused objective of broadening land access and strengthening land rights for the poorest and least empowered. The "land route" is not the only route out of poverty, but in a country where landlessness is the best predictor of poverty, it is a route that cannot be ignored.

Along with the traditional type of regulations, there is the need to adopt some new innovative approaches which have given positive results elsewhere. One such scheme is to allocate one fifteenth to one-tenth acre homestead plots which may be the most practicable method of providing meaningful land rights to India's 17 million rural families that are completely landless. Since land in most village habitation areas is scarce and expensive, State Governments should consider the approaches now being used in Karnataka and West Bengal that involve purchasing multi-acre land parcels within one kilometer of a village, dividing the parcel into house-and garden plots of approximately 0.10 acre or more, providing some basic infrastructure to this new colony (e.g. a road, drinking water, electricity, basic housing) and distributing the plots to landless labourers.

The amount of land needed is not substantial. Allocating such plots to each of the 17 million landless families in India would require less than 0.5 per cent of India's agricultural land. Moreover, the land costs per family are affordable. If un-irrigated agricultural land is targeted, typical costs of such land range from Rs. 40,000 to Rs. 1,00,000 per acre or Rs. 4,000 to Rs. 10,000 to benefit each family with a one-tenth acre plot. With land purchase as a feasible option, the Government need not be limited by insufficient existing Government land or face the political and administrative difficulties of expropriating land.

Allocating homestead plots can be accomplished through new government schemes or can be incorporated into existing rural housing, integrated village development, or rural poverty alleviation schemes. Existing government rural housing schemes currently face two limitations in this regard. First, they typically limit the size of qualifying 'house sites' to 1,200 square feet or less, which is not large enough to provide sufficient space for a garden and livestock. Government planners should aim to provide at least 3,000 square feet to each family. Second, housing schemes rarely include resources for land purchase. Thus, those receiving benefits must already have their own land, and the poorest often do not qualify because they do not own a house site. Rural housing programmes could address these limitations by devoting some significant portion of existing resources for obtaining land and increasing the size of the house sites.

The legislative foundations of land law and policy reform in India (tenancy reform, land ceilings, land allocations) were designed to increase the poor's access to rural land. To date, the effectiveness of the legislation has been mixed and progress over the last few years has slowed. But the link between rural poverty and landlessness remains, and a well-considered plan for rethinking and reforming policies and laws in a manner that advances the interests of the poor should lead India's agenda. The plan can be both pro-poor and market friendly, and the required costs are not unthinkable.

Bihar Land Reforms Commission

> *Rural Bihar is being suffocated by the control exercised by the landed gentry who own a disproportionate share of cultivable land and yet have little interest in increasing agricultural production. Those who can and do - the marginal and small peasants—work the land as sharecroppers and have few rights to it. By rejecting the recommendations of the Bihar Land Reforms Commission, the State in Bihar has lost yet another opportunity to reorder production relations through legal means.*

The Commission, headed by D. Bandyopadhyay,[6] who played a key role in land reforms in West Bengal, looked into the entire sweep of issues related to land, from evaluating the success of Bhoodan movement, mutation *bataidari* (share cropping), ceiling, contract farming and rearrangement of land administration. Stating that though Bihar was one of the first states in India to have Land Reforms Act in 1950, over the years, it has been amended and made weaker.

To make land ceiling effective, the Commission asked Bihar Government to allot between one acre and 0.66 acre of ceiling surplus land to the lowest strata of landless agricultural workers consisting of 16.68 lakh households. It also suggested assignment of atleast 10 decimals of land each to 5.84 lakh homeless non-farm rural workers. As a long-term measure, the Commission recommended that distinction between agricultural and non-agricultural land should be abolished. Land should be defined in a simple dictionary manner so that no one has the opportunity or option to wriggle out of the ceiling provisions by showing some as agricultural and some in other ways. Also, the general exemption given for plantations, orchards, mango/litchi groves, fisheries and other categories of land use should be done away with.

The Commission also reiterated that as already provided in the ceiling act, the ceiling for a family of five or more should also be 15 acres. Religious establishments existing since 1950, should also be allowed one unit of 15 acres. Concession of 100 acres given to sugar mills should not only be done away with but Government should also take over land from these mills leaving only 15 acres for them. Arguing that cutoff date for land ceiling had been misused, the Commission said July 9, 1949, should be the accepted date since on this day, Land Reforms Committee of Congress, headed by J.C. Kumarappa, gave its report that was widely accepted by other states. In case of absentee landlords, the Commission suggested that they be given the option whether they would like to utilize this land through personal cultivation or would like the Government to do it. As for *bataidari* system (share cropping), 15-20 per cent of cultivators in the state were sharecroppers. It recommended that there should be a standalone Bataidar Act.

The following approaches, which involve selected revision to existing legislation and adoption of new methods of increasing land access, will help achieve the original equitable objectives of India's land reform legislation: Revitalise tenancy reform by:

(1) Solidifying the gains of past tenancy reform by converting protected tenants into owners; and selectively liberalizing excessive tenancy regulation and restrictions;

(2) Assist beneficiaries of ceiling legislation in realising the benefits by removing obstacles to land distribution and relaxing moratoriums on the transfer of ceiling—surplus land;
(3) Consider adopting the decentralized IKP project approach in AP as a model for solidifying the gains of past wasteland allocation and otherwise providing secure land rights to the poor; and
(4) Explore using land purchase and existing Government land to create new colonies of one-tenth acre house-and-garden plots distribution to landless labourers.[7]

Certain figures regarding Bihar's rural economy are heart-rending. The percentage of rural labour engaged in agriculture and related activities was 89.3 per cent in 1991 which came down to 81.1 per cent in 2001. One component of this labour force was the "cultivator". This component was 43.4 per cent in 1991, which came down to 34.9 per cent in 2001, a fall of nearly 10 percentage points. Such a drastic change in the occupational pattern could not take place over such a short period. Unless there was a definitional change, the matter would require scholarly scrutiny. Even with such a drop, the figure remains high at 81.1 per cent. This labour force produces, on average, 33.4 per cent of the state GDP (the share of the primary sector in the total). The fact that 80 per cent of labour force produces only one-third of Bihar's GDP, clearly indicates the intensity of poverty in the primary sector. On the other hand, only 19 per cent of labour force produces the remaining 66 per cent of the state GDP. This shows how acutely iniquitous is the sharing of income among and between different sectors of the economy.

The truncated Bihar has hardly any possibility of large-scale industrialisation for shifting of surplus labour from the primary to the secondary sector. Even assuming for the sake of argument that some degree of industrialisation would take place through the intervention of the State, the nature of the modern industries being labour shedding rather than labour absorbing, there would hardly be any prospect of mass scale movement of labour from the primary to the secondary sector. Another major stumbling block against labour absorption in the secondary sector is the lack of education.

The Bihar Economic Survey 2008-09 shows that against the overall literacy rate in India in 2001 of 65.4 per cent, Bihar cut a sorry figure with only 47 per cent, with the female literacy rate a paltry 33.6 per cent. Bihar also has an appallinglyhigh rate of school dropout of 83 per cent at the secondarylevel. What a horrendous wastage of manpower and resources! Thus, knowledge-based modern industry and occupation in the secondary and tertiary sectors is already out of bounds to 53 per cent of the overall population and 76.4 per cent of females in Bihar. Therefore, the position of Bihar as the second most poor state in the country should not cause any surprise. Any state intervention for reducing poverty in the primary sector has to be done by restructuring production relations in that sector. That is the logic of the objective realities in the state of Bihar.

The landownership pattern in Bihar remains acutely skewed even now. The National Sample Survey Organisation's (NSSO) Survey Report No. 491, 2003, on landholding patterns reveals a disquieting picture. Marginal and small farmers who constituted 96.5 per cent of the total landowning community owned 66 per cent of land. Medium and large farmers who constituted only 3.5 per cent of the total landowning community owned 33 per cent of the land. Of the latter, the large owners (constituting only 0.1 per cent of the total) own 4.63 per cent of total land. In absolute terms, this 0.1% of the large owners owned a little over 8 lakh hectares or 19.76 lakh acres of land – a colossal amount by Indian standards. Such a high concentration of landholding in the hands of a minuscule number of owners and with archaic production relations is a major roadblock against increased production and productivity in agriculture in Bihar.

Bihar is endowed with nature's bounty. The land is fertile – in fact the best in the Gangetic plains. It gets good rainfall of about 1,098 mm on average per year. But with such abundant natural endowment, the average productivity of rice (including three varieties) is around only 1,287 kg/ha. This figure compares unfavourably even with the low production figure of 2,509 kg/ha in West Bengal. Punjab is way ahead with 4.6 tonnes/ha. And in China, production varies between 6 and 7 tonnes per hectare.

With such a high average annual rainfall, there should not have been much difficulty in providing irrigation to ensure stability of agricultural production. "But unfortunately only about 60 per cent of cultivated area is endowed with some irrigation facility". Even this 60 per cent is not a bad figure. It compares favourably with that in West Bengal. Yet how could one explain the wide difference in average rate of productivity of rice of 2,509 kg/ha in West Bengal with 1,287 kg/ha in Bihar. Does the solution to this riddle lie in technology only or in social engineering of altered production relations? One thought one would get an answer to this question in the "Road Map for Agriculture" given in the *Economic Survey 2008-09* already referred to. Unfortunately, it repeats the usual technical solutions for augmenting production and productivity in agriculture. These technologies are all well known. The question is: Why are not these technical solutions being applied? Here comes the main issue of production relations.

Though there is no comprehensive data regarding the *bataidari* (sharecropping) system, a conservative estimate is that about 35 per cent of cultivable land in Bihar is under the *batai* system. The extortionist arrangement of crop-sharing gives negative impulses to the use of technology. A bataidar has no incentive to apply costly technology. In the first place, he cannot afford it. Second, even if he did so, a significant portion of the incremental production would be siphoned off as rent to the landowner. Thus, in about 35 per cent of cultivable land, agronomy is practised in an indifferent manner thereby retarding growth in agriculture. In another one-third of cultivable land owned by only 3.5 per cent of landowners belonging to medium and large categories, it is the feudal syndrome of rent seeking which gives them both hassle-free income and social status. Land to them is not a factor of production. It is a symbol of power, influence and status. They cannot think of themselves as cultivating peasant farmer by applying the capitalist mode of production. Like sterile gold, it lends lustre to their position in the society. It is only the other one-third of the cultivable land which is being cultivated by the peasant-farmers where there are some signs of buoyancy. But where two-thirds of the cultivable land drags down production, only one-third of the cultivable land cannot make up the loss.

In the first half of the 1990s agriculture in Bihar had a negative growth rate of (–) 2 per cent per year. Later, up to the early part of the first decade of the present century it grew at an imperceptible rate of (+) 0.8 per cent per year. With all the technologies available in the market why is it that the actual tillers are not accessing them. "It is evident that there is a structural bottleneck in Bihar agriculture due to very queer pattern of land ownership and very extortionate system of tenancy-at-will which are causing great impediment to accelerated rate of agricultural growth" (*Report of the Bihar Land Reforms Commission*).

West Bengal experienced a stagnant growth rate in agriculture for a century between 1891 and 1981 when it varied between 0 and (+) 1. But from 1982-83 it witnessed a remarkable rate of growth rate of 5 per cent to 7 per cent p.a. for little over a decade continuing up to the mid-1990s. How did this miracle happen? Operation Barga recorded the names of 1.6 million share-croppers, which gave them security of tenure and heritable right of cultivation. Distribution of one million acres of good culling surplus agricultural lands among a little over two million landless and land poor families resulted in (combination of both) cultivation of 30 per cent to 33 per cent of land sub-optimally to the extent that labour could substitute capital.

Land reforms, particularly Operation Barga, brought about peace in the countryside which induced and prompted the middle peasantry to invest in land, particularly in mini-deep tube wells for irrigation of their own land and for selling the surplus water to other farmers in need of irrigation. In addition, the middle peasantry came to political power through the three-tier panchayats, which were activated in the middle of the 1978. The Sixth Five-Year Plan diverted a large volume of resources to rural development, which were routed through panchayats in West Bengal where the newly empowered middle peasantry who were productive agents, used them for building infrastructure in support of agriculture. It was the combination of these favourable factors which created the critical mass which produced an explosion in agricultural growth breaking as under the century-old stagnation. In the light of the ground situation in Bihar as discussed above, it is better if the role of Bhumidars (Land usurper) is reduced as recommended by the Commission. The earlier the case is reconsidered, the better for Bihar agriculture.

ORGANIC FARMING

Organic farming is a production system which is supportive of environment and avoids or largely excludes the use of synthetically manufactured fertilizers, pesticides, growth regulators and livestock feed additives. This system relies on the legumes, green manures, crop rotations, crop residues, animal manures, bio-fertilizers, bio-pesticides, bio-herbicides, etc. It is a holistic approach of agricultural system, which aims at cultivation of the land in such a way that the soil is kept healthy and dynamic with biochemical and soil microbial activities related to biodiversity. For trade of any commodity and to survive in the competitive market, 'low cost with quality production' is possible with wise use of nature, native and neighbour. Nature means the use of natural resources, e.g. soil, water, climate etc., native are the naturally grown plant in a locality and neighbour means surrounding of field and farmer and the organic farming certification through a society, self group to make it cost effective.

Sir Albert Howard the founder of the organic farming movement, developed the concept in India. The institute at Pusa in Bihar, where he first realized the value of the traditional farming practices of neighbouring peasants, was largely destroyed by an earthquake in 1934. Howard was no rural romantic or luddite. He was ready to point out the inefficiencies of Indian farmers and also noted approvingly the fast spread of the bicycle and sewing machine in rural India as example of how people took to technology when it was of use to them. But the point was that these technologies were not imposed, but agreed on as appropriate by the users, and Howard's position of developing theories based on what people actually did and wanted, rather than just telling them, was radical for British agricultural scientists at that time and, sadly remains radical for many Indian scientists today.

The princely, states offered a way of being in India but outside British bureaucracy. He was invited by the Indore Royal Court to set up an institute there from which he could work across the Central India princely states, and also maintain contacts with other parts of India. In gratitude Howard would name his composting process, the technology on which organic farming was based, after Indore.

Presently there is a growing interest in organic farming practices in several parts of India, partly due to an expectation of higher prices for organically produced farm commodities. Some states like Uttaranchal would like to be known as an "*Organic Farming State*". The National Academy of Agricultural Sciences (NAAS) has issued a Policy Paper on Organic Farming, which concludes that while synthetic pesticides can be avoided, complete exclusion of fertilizers may not be advisable under all situations. NAAS recommends that "a holistic approach involving Integrated Nutrient Management (INM), Integrated Pest Management (IPM), enhanced input use efficiency and adoption of region specific promising cropping systems would be the best organic farming strategy for India." To begin with, the practice of organic farming should be for low volume, high value crops like spices, medicinal plants, fruits and vegetables. NAAS has also emphasized the need for intensive research on soil fertility and plant health management and on issues relating to microbial contamination of food arising from the use of farm yard manures.[8]

Bihar which boasts of having an institute of Organic Agriculture, is one of the leading State in Horticulture and Livestock products. Organic agriculture in this area is quite promising for the State. The benefits of organic agriculture are of relevance to both developed [environmental protection, biodiversity enhancement, reduced energy use and CO_2 emission] as well as developing regions. The customer demand is fuelling much interest in organic agriculture which has the potential to be on par with the conventional agriculture with respect to productivity of soils. Organic agriculture has come a long way from it being tied to the traditional low productivity agriculture initially, as the approach was a passive one with respect to not using synthetic chemicals and pesticides. It is now pro-active and knowledge intensive with definite inputs for nutrient and pest management. The certification bodies are playing a vital role in categorizing the inputs for organic farming. Nutrient management with microbial products for efficient composting, N and P supplementation and pest management with soft herbal products and biopesticides is increasingly being sought after. Dry microbial functional seed dressing is the need of the hour as a cost effective way to bring about improvement in root growth resulting in yield increase. New discoveries for understanding the natural silent work-force i.e. microbes and other organisms for nutrient and pest management and effective use of organic methodologies will make this agriculture more and more lucrative for the farmer. After all, the focus in organic agriculture is on the health of the soil, farmer and customer. So in times to come, the present and future of agriculture and the health of the society lies in "organic" methods only, which are sustainable.

World Picture

- Organic agriculture is practiced in more than 120 countries. Over 31 million hectares in 633.891 farms are certified organic.
- Oceania has about 40 per cent of the world's organic land. Europe has 23 per cent, and Latin America 9 per cent. But Oceania has less than one per cent share of the global organic food and drink sales. The US is the leading producer and exporter of organics. Land under organics in North America has grown by 30 per cent.

- A lot of the organic produce comes from forests. Bamboo shoots account for 36 per cent of total organic produce, followed by fruits and berries (21%) and nuts (19%).
- Demand for organics is concentrated in rich economies of Western Europe and North America. Europe and North America have low production and face a supply shortage of organic food. A drop in demand in the developed world could hit organic production and prices globally.
- Organic farmers in poor regions like Africa seli organic products to the North, while they eat food grown with chemicals. Export is the main driver of organics in Latin America. From coffee and bananas of Central America, to sugar in Paraguay and cereals and meat in Argentina, the trade mostly depends on exports.
- Global sales of organic food and drink have increased 43 per cent from US $ 23 billion in 2002 to US $ 33 billion in 2005. Six of the G7 countries account for 84 per cent of global revenues from organic's sale.
- Lack of acceptance and recognition dog certification and accreditation systems in the world. Their multiplicity affects organic agriculture. In Africa, certification problems for organic produce inhibits people from supplying to the export market.
- 40 per cent of the certification Bodies are approved by EU, 32 per cent are international/national organizations and the remaining 28 per cent are accredited in the US.

Certification Bodies

	2003	*2004*	*2005*	*2006*
Africa	7	9	7	8
Asia	83	91	117	93
Europe	130	142	157	160
Latin America and Caribbean	33	33	43	43
North America	101	97	84	80
Oceania	10	11	11	11

- International Federation of Organic Agriculture Movements (IFOAM) published the first global standards for organics in 1980. Over 60 countries have regulations for organic products. About 400 agencies worldwide offer organic certification services.

- To harmonise the working of certification bodies, IFOAM, Food and Agriculture Organisation and United Nations Conference on Trade and Development formed the International Task Force on Harmonisation and Equivalence in Organic Agriculture on February 18, 2003 in Nuremberg.
- IFOAM urges farmers in poor countries to not to rely on exports and to develop domestic market for organic products.

Certification

Organic certification is an important aspect of adding value to the organic system. Certified farms can explore different domestic and export markets for fetching better price. APEDA should be consulted for setting up a base with a newly formed proposed Bihar Institute of Organic Agriculture BIOA. BIOA can be involved for training and demonstration of package of practices for organic agriculture and capacity building of the farmers. The soil testing labs of the state can be the extension arms of BIOA. Several models from production to harvesting, to value chain management, can be created for establishing the Bihar Organic brand.

CONTRACT FARMING

Just when contract farming was beginning to be seen as the answer to the woes of the country's farming community, comes a damper from the Ministry of Agriculture. Instead of espousing a hands-off policy, the Ministry wants to frame a model code of conduct for contract farming for agriculture produces such as vegetables, fruits, flowers, medicinal plants, pulses, oilseeds, sugarcane, cereals and cotton. The Government's logic is that this will ensure that farmers are not placed under any disadvantageous position vis-à-vis companies contracting these farmers.

One of the options, proposed by the Agriculture Ministry, is that of a direct contract between the producer and purchaser, with the Government as third party for intervention in case of legal disputes. The Government is of the view that this is more advantageous to small farmers than indirect contract through intermediary agencies. Direct contracts would ensure assured and remunerative marketing opportunities to farmers.

Currently, the Central and State Governments through organizations like Food Corporation of India (FCI) and National

Agricultural Cooperative Marketing Federation of India (Nafed) ensure that operation of the Minimum Support Price (MSP) announced by the Government. The Government now feels that contract cultivation based on a well-defined code of conduct will be helpful to small producers in Getting good quality input, a fair price and prompt payment for their produce. At the state level, a monitoring committee comprising farmers and officials may be set up to ensure the spread of a non-exploitative pattern of contract cultivation.

But companies engaged in "Contract Farming" are apprehensive that model code of conduct may be the first step towards Government stepping in to a business, which was projected to play a significant part in boosting agricultural exports to $ 20 billion by 2010. As they see it, the basic objective of contract farming is to integrate the supply chain to ensure timely availability of quality and quantity of materials. Another objective is to reduce the procurement cost by eliminating the middlemen.

Contract farming generally involves the following elements: a pre-agreed price between the company and the farmer, along with measures of quality, quantity, acreage to be farmed, or duration of the contract. In this system, the contractor supplies all the inputs required for cultivation, while the farmer supplies land and labour. However, the terms and nature of the contract, vary according to the crops grown, the agencies involved, the farmers themselves, and technologies and the context in which contract is taken up. The farmer's participation is, however, generally limited to production in the fields.

The Government's move comes as a surprise because Food and Agriculture Minister has been actively pushing the idea that contract farming is an important institutional arrangement that promotes co-ordination between production and marketing activities in the country. He has been of the view that the main issue is to upscale contract farming. This would require both public and private sector investments in roads, cold chains, electrification and processing. Time and again, the Minister had aired his concern that the smallholders should not be left out in the process. Agro-business firms have been asked to integrate farmers on their supply chains through institutions like co-operatives, producers associated and contract farming.

Goading the States

Latest theme song of the Government is that contract farming model to be implemented in India, will ensure that land is permanently owned and cultivated only by farmers. States are encouraged to take to contract farming but not to acquire such lands. The amendment of the Agricultural Produce Marketing Committee Act by 14 States, which allows farmers to sell their produce in open markets, has given a further fillip to contract farming. Pepsico and ITC have already got off the block with significant projects. Reliance and other companies are in the process of doping so.

Companies, however, feel that the move for having a code of conduct, being promoted by officials of the Agriculture Ministry, goes against the initial thrust of the Government on contract farming. The idea of contract farming is to eliminate middlemen. Here, the Government itself is trying to assume the role of middlemen. The agribusiness companies feel that if the government wants to help small farmers through contract farming, a better way would be to encourage them to form co-operatives and producers' associations, so that they can be integrated in the supply chains.

The central problem in agriculture is one of scientific management to replace conventional farming and access to new technology and assured marketing. These can be facilitated if agriculture is run 'as a business' like any other industry. And the best way to run that business is to adopt contract farming with corporate form of management. The Government seems to have recognized this and is now in a mood to look at contract farming and corporate management (particularly for horticulture) as an acceptable alternative. Some of the companies are already engaged in contract farming. To make it really pervasive and effective, the necessary legal framework has to be in place. Contract farming can work only when the interests of the farmer and the firm are complementary.

'Contract farming' is a win-win situation. The farmer who is faced with the crippling-scale inefficiency of niggardly land holdings, is assured of a price and market, and gains from the capital employed to improve the quality of his produce. The company at the other end of the contract can ensure regular supplies, realize better prices by cutting out the middleman, and rake in higher margins on value-added products that can feed a common market not available to unprocessed crops. On the ground, contract farming in India suffers from a fundamental debility: the lack of enforceability of these contracts. If the farmer finds a higher price for his produce, there is nothing to stop him from reneging. Likewise a company is unable to lift the entire crop contracted for when prices fall. Just about 10 states have amended their Agriculture Produce Marketing laws to make such contracts binding.

Corporate management can deliver better results. Companies can be set up with farmers as shareholders to the extent they contribute the land valued at market price; non-farmers become shareholders to the extent they contribute equity. Considering the present socio-economic milieu, the success of corporate farming will depend on allowing the corporate system to work. Appropriate legislation will have to be enacted to enable this new entity to function without any friction. Companies will then be able to adopt professional management, introduce the best available technologies, including biotechnology, for improving productivity, secure the best price for the product using commodity exchanges, promote research and, wherever possible, undertake food processing. The combination of contract farming and corporate farm management can make horticulture a flourishing business which will, at the same time, ensure food security and generate surpluses for export.

Contract Farming in Vidarbha

Having failed to arrest farm suicides in Vidarbha, the Maharashtra government is now looking at the corporate sector. It has invited top retailers, including Wal-Mart, for contract farming. The Director of State government's relief mission on farm suicides, has submitted a report to the Food are Agriculture Ministry which recommends that big corporates be approached to build a contractual arrangement with farmers in Vidarbha. Contract farming in commercial crops like cotton and horticulture in Vidarbha can make farming a profitable venture. It would also lead to an improvement of quality of crops since a business venture would also bring in new technologies and practices. The report names ten private companies, including giants like Reliance Fresh and the world's largest retailer, Wal-Mart, which has tied up with Bharti for its foray in the retail sector in India. These ten companies have been purchasing produce directly from farmers in Maharashtra. The State Government should formalize this arrangement by roping them in contract farming. The Government was open to the proposal. Contract farming has been approved by the Government by amending the relevant sections of the agriculture produce marketing act. What all these companies have to do is to detail their plans and seek license from Director—Marketing.

TECHNOLOGY

We have discussed elsewhere that the role of technology in agriculture is very important. One of the basic reason for China going far ahead of India was the scale it adopted in Technology, Research and Development'. Incidentally, Research by IFPRI (Fan, Gulati and Thorat) in this direction shows that the marginal returns of government spending say, an additional dollar on fertilizer is much lower (only 0.53) than spending on agriculture R & D (6.9), or rural roads (3.2), or investment in rural education (1.5) or even irrigation (1.4).[9]

With regard to agricultural research and technology—prime source of development—the picture in Bihar is rather bleak. Rajindra Agriculture University (earlier Pusa Institute) being the first such University in India, was the pioneer in agriculture research. But it is not able to pay even the salary of its truncated staff. In this scenario the condition of research/technology can very well be appreciated. Area specific research is the basic requirement of agricultural development. This weak link has to be first looked into.

Any technology developed in the laboratory is of no use unless it is carried to the farmer, tested on the farmer's field and then adopted by the farmer community, in general. This 'land to lab' is the other link which is missing under the present conditions where there is a complete break down of the institutional system. All this will need high priority.

A book, Agricultural Research, Livelihoods, and Poverty: Studies of Economic and Social Impacts in Six Countries, Edited by Michelle Adato and Ruth Meinzen—Dick, senior research fellows at IFPRI, suggests ways to design future agricultural technologies for poverty reduction. First, such technologies must be tailored to poor people's conditions, for example, their need to minimize financial risk, work with limited labour pools, and address the effects of gender inequality. Second, the methods used to disseminate these technologies must match the social environment in which poor farmers operate; mismatches can prevent farmers from adopting a technology.

This book should give researchers a valuable sense of the larger context, that affects how their work is used and will help inform the design of future agricultural research programmes. Such programmes can ensure that technologies help poor women and men, escape poverty.

India's current GDP is estimated at around $ 1 trillion

(Rs. 40,00,000 crore). According to the new World Bank report "Unleashing India's Innovation", while there were many islands of excellence in India's heterogeneous economy, the country could do much more to reach its full potential, especially by bringing the benefits of innovation to the poor. Research and knowledge needed to be better converted for commercial use. The output of the economy could increase more than five fold if all enterprises achieve 'national—best practices' based on knowledge already in use in India. Despite India having the largest systems of higher learning, it may face a deficit of 0.5 million workers by 2010. It recommended larger fiscal and managerial autonomy of universities and colleges and suggested greater private participation in higher education.

Report Card

- Increase competition to unleash innovation.
- More early stage funding and angel investing.
- Strengthen efforts to commercialise knowledge.
- Private enterprises need to increase R & D spending.
- Foster more inclusive innovation.

Actions were needed to promote commercialization and to strengthen links among agriculture industry, universities and public R&D laboratories. These could include providing support to technology transfer offices, creating a patent management corporation, strengthening innovation infrastructure, promoting angel investing and early stage pre-venture capital financing. India will especially benefit from fostering more inclusive innovation. This could be achieved by promoting more formal R&D efforts for the poor people as well as by improving the ability of informal enterprises to better use existing knowledge. Existing pro-poor initiatives needed to be scaled up, "Government programmes should promote more knowledge absorption in the productive sector and help extend the power of innovation to the common man. The Report, no doubt addresses India as a whole, yet it equally applies to the state of Bihar.

Post Harvest Technology

Wastage of fruit and vegetables persists due to lack of post harvest facilities and it is estimated that this leads to a colossal loss of more than Rs. 50,000 crore per annum—nearly 30 per cent of the country's production. Since the 100 per cent tax deduction provided for cold chain infrastructure has not worked, FICCI has on behalf of the industry—now suggested that a 10 year tax holiday should be offered to encourage investments in this crucial link. The incentive would be justified since it would also improve quality and boost exports, apart from checking losses. The apex chamber has argued that boost to food processing would result in higher incremental economic activity, as compared to other segments and create significant employment.

Farmers would get better prices and the Government could reduce food as well as fertilizer subsidies. Pre-cooling facilities, cold storages, refrigerated trucks and handling facilities should get the benefit so that farm produce reaches processing units and retail chains without waste. The investment needed in cold storages alone is estimated to be Rs. 15,700 crore. The food processing industry is pushing hard for sops, but so far has not met with much success. While policymakers are aware of the situation, bureaucrats are keeping their fingers crossed to see if the Government goes for an integrated approach rather than providing piecemeal incentives to some sections.

INVESTMENT POLICY

Major challenge for agribusiness lies in increasing investment in agriculture per se' and more particularly public investment. Agriculture produce in the country is bought and sold in the 'mandi', which is a Government regulated market. Over the years, with the volume of products going up the Mandi system has not been able to cope with the pressure. To make marketing efficient, large investments are required for the development of post-harvest and cold chain infrastructure nearer the fields. As per Planning Commission estimates, an investment of Rs. 12,230 crore in the next plan is required for this infrastructure alone. A major portion of this is expected from the private sector.

During the past two decades, per hectare private investment at constant prices increased marginally from Rs. 65.19 in the early 1980's to Rs. 88.54 in late 1990, while public investment declined at the rate of 4.8 per cent per year. During 2000-02, per capita development expenditure in Bihar State worked out to Rs. 3206 only, as against Rs. 6748 at all India level. Bihar's per capita spending on road was Rs. 45 which is about 38 per cent of the national average. Similarly, for irrigation and flood control, Bihar spends annually about Rs. 104 on per capita basis, as against the national average of Rs. 109. The state faces recurrent phenomenon of floods almost every year, and therefore, low expenditure on flood control and water management speaks of wrong priorities. Per capita cumulative investment by financial institutions (up to March, 2001) in Bihar was merely Rs. 552 as against the national average of Rs. 4829. There is also a flight of capital from Bihar. The banking statistics show that per capita bank credit (as on March 2006) was only Rs. 680 in Bihar as against the per capita bank deposit of Rs. 3191. The per capita credit deposit ratio at 32 was the lowest in the country at 73 (as on March 31, 2006). The low credit-deposit ratio could be partly due to lack of proper policy support and environment for investment. Public investment in agricultural research and education has been hardly 0.14 per cent of agricultural GDP as against the All India average of about 0.5 per cent in recent years. Even in absolute terms per hectare investment in agricultural research and education on capital account was almost zero during the past one and a half decade.

In the absence of appropriate tenancy reform, land record and land registration, it is unthinkable, how private investment in agro-processing and rural entrepreneurship would grow on adequate scale and help promote diversified agricultural growth in Bihar. Co-operative institutions which are generally state managed, are both inadequate and inefficient. The number of bank branches and genuine cooperatives are inadequate to meet the requirements of the state. Responsible and farmer friendly administration of these institutions is also lacking.

Apart from lower allocation of public investment/expenditure to agriculture, one of the problems is the greater delivery system inspite of good policies. There is a view that public institutions in agriculture including Seed and other input supplies, extension and marketing have collapsed. There is an urgent need for the revival of public institutions for raising agricultural growth and incomes. Another issue is the political economy of agriculture. There is a feeling that the Governments have become less sensitive to the farmers' problems notwithstanding the pro-farmer rhetoric. There are apprehensions that Governments are more interested in pleasing the corporate sector (e.g. the SEZ policy) rather than helping the agriculture sector which bears 80 per cent of the burden. Let Bihar be the first state to have farmers SEZ.

COMMERCIALISING AGRICULTURE

Agriculture, trade, distribution, processing and services associated with agriculture, are currently estimated at around 41 per cent of GDP. Agri-business, besides providing inputs for farming, links farmers to consumers through handling, processing, transportation, marketing and distribution of food and other farm products. Thus, there are strong synergies between agri-business and the performance of agriculture. Dynamic and efficient agri-business spurs agricultural growth and provides business opportunities in the value-added segments of processing, distribution and marketing. A strong link between agri-business and the farmer can reduce rural poverty and provide better value to the end-consumer. The whole process can be win-win game for all stakeholders—the farmer, the consumer, the Government and the industry—and has the potential or economic transformation. However, declining farm productivity, low investment in infrastructure and inadequate rural incomes have resulted in poor growth of the agriculture sector in the last decade.

Concerted efforts are needed to address these fundamental challenges to realize the transformation potential of the agri-business sector. Near-stagnant farm productivity and a consequent threat to national food security is a serious cause of concern. The decline in production has forced the country to resort to large scale imports of wheat and likely imports of rice in the near future. This only underscores the urgency of boosting agricultural growth in the country. Bihar is a state which can show the way by ushering in the Second Green Revolution which is inclusive and equitable. While the National Food Security Mission is expected to help tackle the situation to some extent, it would need to be supplemented with well-coordinated and properly—aligned policies, encompassing all agriculture and allied sectors, especially industries supplying inputs, such as fertilizers and seeds. A strong last-mile delivery system for inputs and knowledge to farmers can address the second-generation problems of Indian agriculture, such as decreasing input-sue efficiency and soil degradation.

The last and the most important challenge is to increase farm income. Problems related to subsidies, infrastructure, inputs and extension systems have caused small farmers to remain marginalized in the economic development. Lack of access to finance and fluctuating output prices have aggravated the matter. Just as a strong GDP growth requires an inclusive growth from all sectors, a strong agriculture growth needs to include all segments of the farming community. Focus needs to shift from farming to the farmer, by multiplying his income sources, and migrating from agriculture to agribusiness. The first phase of agricultural reforms through the Green Revolution focused on building food security and income equity. The next generation of reforms need to focus additionally on efficiency and improvements at every stage of the value chain. This would help realize the full potential of agri-business. Ideally, a strong agri-business economy would have improved inputs, informed farmers, efficient markets, adequate processing and distribution infrastructure, and an enabling policy environment. *What the country especially the state of Bihar should aim at is not to achieve just blind figures/statistics of production but also as to how many people have been made to cross the* BPL-APL divide. That would be the crowning glory of a welfare state.

What we urgently need today is a genuine new, dynamic and competitive commercialization of agriculture and allied rural activities. This requires adoption of holistic knowledge, intensive integrated farming systems approach and integration of agro-productivity with economics. Future technological recommendations to the farmers should be based on extra productivity and monetary returns per unit to financial investment or per unit of applied input (seed, water, fertilizer etc.).

Agri-scientists, and planners/policy-makers need to look at agriculture as a commercial or business proposition and carry out analysis of different agriculture-related technology and policy issues to bridge the gap between farmers, technology (generation and adoption), market and agro-profitability. It is time to increase public-private partnerships and investments for generating agro-production, value—addition/processing industry—driven rural economy. To achieve this a network of public private funded small and medium sized commodity-specific agro-processing enterprise clusters should be developed in the rural areas. This will ensure sustained increase in job-led rural agro-economic growth and decrease in urban-rural economic gap. For Agro-commercialisation, we, must modernize the existing out dated market infrastructure and support services and adopt a total market approach (right from the sites of production till the product reaches the consumer).

PRICE POLICY AND PROCUREMENT

Agricultural production is a function of technical inputs, Research/ Extension, seed, fertilizers, credit and water etc. In addition to all this, remunerative prices are an equally important adjunct in this production process. In fact this is the driving force to produce marketable surplus. It was during 1965 that the Minimum Support Price (MSP) Policy was introduced and is being announced every year for all the major crops taking into account all the relevant factors including cost of production. It did serve as a catalytic agent in the success of Green Revolution. This price is uniform for the country as a whole, putting Bihar in a very peculiar situation. Because of low productivity, farming in Bihar is not a very profitable occupation. The net returns per hectare of paddy cultivation in Bihar (based on cash expenses plus imputed value of family labour) worked out to only Rs. 3215 in 2003-04 as against Rs. 19,289 in Punjab. (CACP, 2006). Based on C_2 cost, net return from paddy cultivation was in fact negative i.e. Rs. 1263. Thus, even if there is potential for raising the yield level in Bihar, low productivity and low profitability would stand in the way of realizing such potential. In fact, the ratio of gross value of output to A_2F_L was only 1.35 in Bihar, as against 1.99 in Punjab. Similarly, in the case of wheat, the net returns per hectare, based on A_2F_L was only Rs. 4,787 in Bihar as against Rs. 15,946 in Haryana and 15,755 in Punjab. Based on C_2 cost, net returns from wheat cultivation was negative in Bihar (Rs. 177). The ratio of gross value of output to A_2F_L was 1.47 in Bihar, 2.17 in Punjab and 2.10 in Haryana.

Unless paddy and wheat cultivation in Bihar is made profitable, farmers may not be induced to invest in land improvement measures for productivity growth and the much desired acreage shift towards paddy and wheat in Bihar at the expense of area loss in Punjab and Haryana may not be forthcoming. Even maize does not seem to be a very promising crop in this regard. The net returns (gross value of output – (minus) $A_2 + F_L$ (cash expense plus imputed value of family labour) was only Rs. 7625 and based on C_2 cost, it was Rs. 4450. The ratio of gross value of output to A_2F_L was 1.80. Even the neighbouring state of Uttar Pradesh has relatively higher net returns from paddy and wheat cultivation. It would be further seen from Table 11.3 that in several districts including Darbhanga, Katihar, Muzaffarpur, Nawada, Purnea and Sitamarhi, there was hardly any change in total factor productivity at constant prices during the past three decades.

Table 11.3: Changes in Total Factor Productivity Over Time

State	*District*	*Year*	*Index of Value of total crop output at 1993-94 prices*	*Index of Input cost*	*Index of Total factor productivity*
Bihar	Aurangabad	1970	100.00	100.00	100.00
Bihar	Aurangabad	2000	645.53	561.31	115.01
Bihar	Begusarai	1970	100.00	100.00	100.00
Bihar	Begusarai	2000	913.67	538.10	169.80
Bihar	Bhagalpur	1970	100.00	100.00	100.00
Bihar	Bhagalpur	2000	986.26	619.27	159.26
Bihar	Bhojpur	1970	100.00	100.00	100.00
Bihar	Bhojpur	2000	520.16	480.35	108.29
Bihar	Champaran (East)	1970	100.00	100.00	100.00
Bihar	Champaran (East)	2000	982.49	434.78	225.98
Bihar	Champaran (West)	1970	100.00	100.00	100.00
Bihar	Champaran (West)	2000	1106.83	418.47	264.49
Bihar	Darbhanga	1970	100.00	100.00	100.00
Bihar	Darbhanga	2000	180.68	153.10	118.01
Bihar	Gaya	1970	100.00	100.00	100.00
Bihar	Gaya	2000	477.93	286.57	166.78
Bihar	Gopalganj	1970	100.00	100.00	100.00
Bihar	Gopalganj	2000	1455.41	846.56	171.92
Bihar	Katihar	1970	100.00	100.00	100.00
Bihar	Katihar	2000	1224.21	1077.52	113.61
Bihar	Madhubani	1970	100.00	100.00	100.00
Bihar	Madhubani	2000	640.85	605.03	105.92
Bihar	Monghyr	1970	100.00	100.00	100.00
Bihar	Monghyr	2000	770.42	436.41	176.54
Bihar	Muzaffarpur	1970	100.00	100.00	100.00
Bihar	Muzaffarpur	2000	423.26	385.88	109.69
Bihar	Nalanda	1970	100.00	100.00	100.00
Bihar	Nalanda	2000	899.29	621.39	144.72
Bihar	Nawada	1970	100.00	100.00	100.00
Bihar	Nawada	2000	432.65	381.18	113.50
Bihar	Patna	1970	100.00	100.00	100.00
Bihar	Patna	2000	525.82	298.99	175.87
Bihar	Purnea	1970	100.00	100.00	100.00
Bihar	Purnea	2000	512.82	430.02	119.26
Bihar	Rohtas	1970	100.00	100.00	100.00
Bihar	Rohtas	2000	1596.96	572.89	278.76
Bihar	Saharsa	1970	100.00	100.00	100.00
Bihar	Saharsa	2000	713.87	561.73	127.09
Bihar	Samastipur	1970	100.00	100.00	100.00
Bihar	Samastipur	2000	912.45	632.45	144.27
Bihar	Saran	1970	100.00	100.00	100.00
Bihar	Saran	2000	468.94	268.21	174.84
Bihar	Sitamarhi	1970	100.00	100.00	100.00
Bihar	Sitamarhi	2000	477.91	610.21	78.32
Bihar	Siwan	1970	100.00	100.00	100.00
Bihar	Siwan	2000	678.65	452.63	149.94
Bihar	Vaishali	1970	100.00	100.00	100.00
Bihar	Vaishali	2000	709.60	588.47	120.58

The frequent occurrence of floods and drought, as already stated, aggravates the problem further. As discussed in the previous section—crop Insurance, the co-efficient of variation in annual rainfall is high. Similarly prices of agricultural commodities fluctuate widely from year to year. Crop farming in Bihar is not only unprofitable, but also unstable and risky. It has been estimated that, since variability in the rainfall and farm harvest prices are high, production process is quite risky. Along with low profitability, erratic rainfall and fluctuating market prices act as disincentives for the farmers to invest in modernization of agriculture.

With around 90 per cent of small/marginal farmers, marketable surplus is not expected to be very high. All the same market arrival of rice and wheat are of the order of 20-25 per cent. But the quantity procured is rather negligible. This serves as one of the biggest disincentive for higher production of major commodities in the state where the plan envisages much higher level of production. In a situation like this, there is the need for improvement in the marketing system, density of markets, connectivity of the markets within the producing areas, dissemination of market intelligence and institutional arrangement for procurement of the produce which come to the market. The implementation of minimum support price policy has always been poor in Bihar. Even though some surplus pockets of grains are emerging, Food Corporation of India (FCI) makes hardly any effective purchases. This has to change and the Corporation has to make effective purchase arrangements.

In the absence of a proper procurement agency in the State, surplus paddy in Districts like East Champaran is sold by the farmers to middle men who smuggle the same to Nepal. The slow rate of procurement in Bihar has meant that smaller farmers have been forced into distress sales in the absence of requisite infrastructure. There are only 712 centres in all when 2,000 should be opened. The FCI did open procurement centres to ensure that the state's farmers, who had surplus to sell, were not forced to sell their paddy and wheat at a lower rate. A farmer, even went to court early in 2004, praying for more purchase centres and storage space in every block of Buxar, Rohtas and other Districts. The FCI claimed the Government was not lifting grain and that a lot of grain being brought to it was not according to the Government of India's specifications. At the end, the FCI assured the court that the Corporation was keen to maximize its purchases in Bihar and plans had already been made for procurement of the rabi crop. FCI has to play a more pro-active role to ensure adequate returns to the farmer/producer.[10] It is encouraging to note that this year (2008) over 2 lakh tonnes wheat has been procured.

CROP INSURANCE

Water, air, sunlight, and soil are the four factors which contribute to successful farming. The first three elements are directly connected with weather. Farming, particularly in a country like India, is influenced by weather conditions right from preparatory tillage to the stocking of the produce. Even after the safe stocking, the produce is at the risk of price fluctuations. It is said that "farming in India is largely a gamble in weather conditions".

In the light of this phenomina, a scheme of insuring the former against crop loss due to material factors was introduced first in Japan in 1939 and in India two decades ago. The present version, the National Agricultural Insurance Scheme (NAIS) is available since 1999-2000. Despite States' support for coverage of the NAIS being tardy, it is a useful device, especially for farmers growing relatively risky crops. The main problem is that NAIS is not really actuarial insurance. Premiums for most important crops are fixed at All-India level irrespective of risk and government pays for the entire excess of claims over premium received. Moreover, being compulsory for all borrowers from banks in States where it is in force, and with relatively few non-loanee farmers involved, it mainly insures banks against default following poor harvest. Further, its popularity with farmers is limited since it is based on crop-cutting experiments, which delays claims payments till well after harvest and the risk covered is only of yield shortfalls at the block level.

Nearly 60 per cent area under cultivation in Bihar is rain-fed. The co-efficient of variation there in annual rainfall during 1980-81 to 2002-03 was 32 per cent. The monsoons play a critical role in sub-continent as it determines whether the harvest will be bountiful, average, or poor in a given year. The co-efficient of variation in the farm harvest prices of paddy was as high as 45.4 per cent during 1980-81 to 2002-03. In the case of wheat, it was 44.9 per cent and in maize, it was 38.9 per cent. The coefficients of variation in the yields of rice, wheat and maize were 23.4, 16.4 and 30.3 per cent respectively. Production risks refer to the high variability of

Table 11.4: Crop Insurance in Bihar

	Number of insured farmers	Insured Amount (Rs. lakhs)	Compen-sation (Rs. lakhs)	Number of Beneficiary Farmers	% of benefi-ciary farmers	Compen-sation as % of insured amount	% increase over previous year in no. of insured farmers	% increase over previous year in insured amount
Kharif 2004	293442	34084.72	21184.71	268532	91.51	62.15	—	—
Rabi 2004-05	118090	13716.17	2802.49	54613	46.25	20.43	—	—
Total 2004-05	411532	47800.89	23987.2	323145	78.52	50.18	—	—
Kharif 2005	221985	33911.27	10088.61	125958	56.74	29.75	–24.35	–0.51
Rabi 2005-06	187961	28439.99	5299.73	102760	54.67	18.63	59.17	107.35
Total 2005-06	409946	62351.26	15388.34	228718	55.79	24.68	–0.38	30.44
Kharif 2006	344686	59133.07	4017.44	75880	22.01	6.79	55.27	74.38
Rabi 2006-07	339749	59439.59	3410.43	65927	19.40	5.74	80.75	109.00
Total 2006-07	684435	118572.7	7427.87	141807	20.72	6.26	66.96	90.17

Source: Department of Cooperative, Government of Bihar.

production outcomes due to the unpredictable nature of weather, seasonal fluctuations, and pest infestation, diseases which result in yield loss and revenue loss for the farmer. Insurance can play a pivotal role in covering production risks which would help reduce the vulnerability of the farmers to economic losses. Developing appropriate products based on the needs of a particular geography to manage risks associated with agricultural production is the need of the hour. Thus, crop farming in Bihar is not only unprofitable at present, but also unstable and risky. About 71 per cent of the variations in output were explained by variations in rainfall, irrigation and farm harvest prices.

Crop insurance is in operation in Bihar for quite some time. A total of 6.84 lakh farmers were insured for their crops in 2006-07 with the insured amount reaching a sum of Rs. 11,857.3 million. This was an increase of almost 67 per cent over the previous year. The total insured amount also increased by about 90 per cent (Table 11.4). While a larger proportion of farmers take insurance for the kharif crops, over the last three years, there has been relatively greater increase in the farmers insured for the rabi season. As a result, the number of farmers insured for rabi season has increased from 1.2 lakh in rabi 2004-05 to 3.4 lakh in rabi 2006-07. However, compensation as percentage of insured amount has fallen significantly over the last three years from 50.18 per cent in 2004-05 to 6.26 per cent in 2006.07. While this may be a reflection of lesser amount of crop damage, it might as well be indicative of bottlenecks in settlement of compensation claims. To make the distribution of compensation related to crop insurance more transparent, certificates regarding compensation payment are being distributed now through camps.

Notes

1. D. Bandyopadhyay, Does Land Still Matter, *EPW*, March 8, 2008.
2. Rohtak, Ashok Kumar Pankaj (2005); Dream Bihar; *Hard News Media com*. December, HTTP:/www. *Hardnews media com*/portal/2005/12/ 235 (accessed August 19, 2007).
3. J. Albert (2007); "A case of Socio-economic Divergence in North India: Darbhanga and Ludhiana".
4. See *Indo-Asian News Service* 2004; Rajalakshmi 2001; Karan nd; Human Rights Watch nd; Rorabacher 2007 for brief discussions of the landless issues Bihar.
5. Editorial, *The Economic Times*, November 6, 2007.

6. Adopted from D. Bandyopadhyay, *EPW*, November, 2009.
7. For more details refer to, *EPW*, March 8, 2008.
8. Organic Farming; Approaches and Possibilities in the Context of Indian Agriculture, Policy Paper 30, NAAS, February 2005.
9. Fan, S., Peter Hazel and S. Thorat (1999), Linkages Between Government Spending, Growth and Poverty in Rural India, Research Report 110, IFPRI Washington.
10. *Indian Express*, March 19, 2004.

12

REPLICABLE MODELS

The very purpose of introducing this Chapter is that there are a large number of varied types of successful experiments being conducted in different parts of the country. Bihar, as keen it is, to make rapid strides in agricultural development, can adapt any of those so as to achieve its objectives and cut short the time required in experimentation. There are a lot of controversies regarding contract farming. We have provided quite a number of versions with complete success on the ground. Similarly, some details about experiments in mechanization, credit, extension, their own SEZ and dairying or even agri-tourism, have also been provided. These replicable models from within the country, will service as a reference for the government of Bihar in their march towards accelerated and inclusive growth.

CONTRACT FARMING—SOME MODELS

It is disturbing that, in the past five years, agricultural productivity stagnated because the Green Revolution reached a plateau. The Cabinet Committee on prices is now reconciled to domestic food shortages and is planning to set up an intelligence service to help import of right quantities of foodgrains at the right time for the right price. The central problem in agriculture is one of scientific management to replace conventional farming and of new technology to replace the hybrid variety of seeds. These can be facilitated if agriculture is run as a business like any other industry. And the best way to run that business is to adopt contract farming and the corporate form of management. The government seems to have recognized this and is now in a mood to look at contract farming and corporate management (particularly for horticulture) as acceptable alternatives. Some of the companies are already engaged in contract farming. To make it really pervasive and effective, the necessary legal framework has to be in place. Contract farming can work only when the interests of the farmer and the firm are complementary to each other. Even then, it has limited possibilities because of the small scale of farm operations.[1]

Such ground-level initiatives could be a solution to the problem of low productivity of Indian agriculture. India's yield per hect. of major crops is substantially low and in most cases below that of the top five nations excluding China. Further, growth in yield per acre over the past 15 years in India has been lower compared to China and other major producers. Population pressure has meant a decline in average farm size in India. Small farmers typically lack access to information, particularly with regard to price and use of fertilizers, pesticides, technology and capital. Farmers directly producing for the market is one way in which this gap could be bridged. That would enable farmers to understand the requirements of the Indian and world markets. Similarly, large-scale procurement by corporates would also help raise yields. Companies would have to do what used to be called extension work so that farmers produce according to their requirements. The policy aspect of this is that all models need to be tried including allowing foreign retailers to procure from farmers.[2]

Contract farming is a win-win situation. The farmer who is faced with the crippling-scale inefficiency of niggardly land holdings, is assured of a price and marketing, and gains from the capital employed to improve the quality of his produce. The company at the other end of the contract can pencil in regular supplies, realize better prices by cutting out the middleman, and rake in higher margins on value-added products that can feed a common market not available to unprocessed crops. On the ground, contract farming in India, suffers from a fundamental debility: the lack of enforceability of these contracts. If the farmer finds a higher price for his produce there is nothing to stop him from reneging. Likewise a company is unable to lift the entire crop contracted for when prices fall. Just about 10 states have amended their agriculture produce marketing laws to make such contracts binding. Others need to get cracking. One way to get around the problem is to make both parties to a contract stakeholders in the enterprise. Here's where corporate farming comes in—a new-age cooperative—where farmers become shareholders along with a corporate body. The farmer invests his land in the venture, the company's holding is limited to 26 per cent and the land stakes are hereditary.

Corporate management can deliver better results. Companies can be set up with farmers as shareholders to the extent they contribute the land valued at market price; non-farmers become shareholders to the extent they contribute equity. Considering the present socio-economic milieu, the success of corporate farming will depend on allowing the corporate system to work. Appropriate legislation will have to be enacted to enable this new entity to function without any friction. Companies will then be able to adopt professional management, introduce the best available technologies, including biotechnology, for improving productivity, secure the best price for the product using commodity exchanges, promote research and, wherever possible, undertake food processing. The combination of contract farming and corporate farm management can make horticulture a flourishing business, which will at the same time ensure food security and generate surpluses for export. Discussed here are some successful models. Any one of them can be picked up as it is or adapted to meet local conditions.

Farmer Initiatives in Retail

A group of Gujarati farmers have got together to supply to corporate-owned retail chains. The intention here is to get better prices by banding together and producing in larger quantities. Farmers in Maharashtra are planning to go all the way up the value chain, by setting up malls and stocking agri-produce. There have been other experiments, particularly in Maharashtra, where farmers have been able to obtain better deals from companies trying to buy land from them.

Farmers Turn Shareholders

The farmer knows that each of his shares in the Ginger Growers Cooperative Marketing Federation (Gin-Fed) has a face value of Rs. 50. And that it has been rising ever since Gin-Fed cut middlemen out. Assam accounts for 41 per cent of India's annual ginger yield of 2.75 lakh tonnes, which is 33 per cent of the global production. Bulk of the ginger in the state is grown in Karbi Anglong (Assam), and all of it is organic. Most farmers were unaware that the low fibre Manja variety of ginger they grow, is one of the world's finest, and that middlemen had been giving them a raw deal. Manja, ironically, is named after the local ginger-trading settlement.

As a test run, Gin-Fed procured 200 metric tonnes of ginger from member-farmers at a minimum of Rs. 8 per kg. The ginger was sold across northern India for Rs 13 per kg. The profit was distributed among the members after an amount of 0.5 per kg. was deducted towards running the office. Ginger farmers are now upbeat about trading some 20,000 metric tonnes annually through Gin-Fed. The number of shares owned by a member is decided by land holding patterns, much like tribal customary laws.

Involvement of Local Government

The Panchayati Raj ministry's plans to set up rural business hubs across the country to provide linkages of the corporate sector with rural India has generated interest among a host of companies, that include Microsoft, Pantaloon, Reliance and Mahindra and Mahindra, Rural business hubs, a joint initiative of the ministry and CII, Who will join hands in several states, including Kerala, Jammu and Kashmir, Karnataka, Punjab, West Bengal and Uttar Pradesh. These would focus on agri-product processing, handicrafts and handlooms sourcing and power generation in rural India through biomass and other alternative energy sources. While companies like Reliance are already working on their own to set up rural business hubs, the government's initiative can give these players a pan-India presence. Microsoft, which plans to set up 50,000 IT kiosks in rural India in the next three years, has shown interest in working with the Ministry in this project. These kiosks can be housed in the panchayat ghars, where the ministry would make data such as land records available through these kiosks. Discussions are on with Pantaloon and M&M for agri-sourcing, while Fab India and Anokhi are in talks with the Ministry for handloom sourcing. National Institute of Fashion Technology has already signed MoU in Karnataka with Gram Panchayats to help rural craftsmen and weavers upgrade their skills. ITC has also signed an MoU with local panchayat representatives of Nainital to brand and market food products grown in the area. The Panchayati Raj Ministry has already signed over 30 MoUs including three in Haryana for jatropha and one in U.P. for embroidery—chikankari with the local self-help groups.

Citrus in Punjab

An other model is that of a foreign investment firm which is all set to turn citrus farming all the more profitable for farmers of Punjab. In what could be the biggest—ever foreign investment in the agriculture sector in the country, this firm is all set to invest Rs. 800-odd crore for development of citrus in the state. This is the second biggest company after soft drink major Pepsico to see such a huge potential in citrus farming in Punjab. Pepsico is already helping for establishing a citrus nursery at Jallowal near Jalandhar. What the company has proposed is simple. It will recover its investment of $ 185 million from the sale of produce from land taken on long lease from farmers in the state. It has not asked for any guarantee from the State Government and plans to recoup its investment from the profits of the project itself. Started in 2006, the project, has a scheme under which it takes land on lease from farmers for a fixed period, develops citrus orchards in a scientific manner and returns the land and the orchard to the farmer after the stipulated period. The farmer gets rent during the lease period.

Railways—Agri-retail Shops

Driving the Retail Express now is the Indian Railways, which has invited expression of interest from retail chains to set up agri-retail outlets in surplus railway lands. Reliance, Tatas, Bharati, Future Group and Adanis are all in talks with the railways, which has become one of the most enterprising organization in the country. Private retailers will ensure steady cargo traffic for the railways. In return, they would be assured of freight rates lower than the existing market rates. The agri-retail chain is a win-win formula for all the three parties—the railways, the retailers and the farmers. Farmers would be provided ideal remunerative price for their produce that at times rots due to lack of proper storage facilities or transportation.

The railways is also considering retailers proposal to create a refrigerated rail link between the agri hubs for transporting perishable agri-commodities. The hubs would be created in non-metro stations, facilitating transportation of agri-products using the rail network and provide for their sale at the local level. For every hub, retailers would enter into a joint venture with the Railways. The Railways would have land as their equity in the joint ventures. India produces nearly 13 per cent of the world's vegetables and 10 per cent of fruit. However, nearly 25 per cent of the produce is wasted due to lack of proper storage. In the past, the railways had attempted to run refrigerated rakes in the eastern region but the project did not have many takers. One possible reason for this failure could be the fact that no study was conducted to figure out the demand for such rakes.

Farm-Fresh Vegetables in Pune

Even as the top corporate houses in the country are hotfooting into the hitherto unglamorous domain of vegetable retailing, farmers in Pune district have decided to join the race. A bold new initiative has been taken where farm fresh sales in their own villages are a reality—this ensures that they are no longer required to choose between the vexing option of travelling long distances to sell their produce and depending on middlemen to help them sell their produce. Called *apani mandis*, 25 such agri-produce marketing yards have been launched over the next six months across the district under the aegis of National Horticulture Mission (NHM), covering all 14 talukas. The rollout of these mandis is done under the supervision of the state agriculture department. The scheme enables the farmers as well as the women members of various self-help groups to directly market their products through these *mandis*. On the other hand, consumers also get access to fresh farm produce as these mandis are located in the villages.

Corporates and Farm Diversity

Farmers in the green revolution states of Punjab and Haryana and others, who are planning to diversify, now have help in hand with the corporate sector keen to impart knowledge and become partners in growth. From soil testing to crop management, all will come under the attention of the corporate sector, which will help the farmers in deciding which kind of crop to go in for to increase his yield and the quantum of fertilizer to use. The CII-Young Indians-Aries Agro-Limited National Farmers Network has been launched for providing specialized agri-technology to the peasants. The network will be engaging 10,000 progressive young farmers across the nation within a period of three years in a wide range of knowledge dissemination activities. Its activities will involve study tours, soil health assessment, demonstrations and trials seminars, conferences and advisory services among others. Rural to urban knowledge sharing in key areas will also be a part of the network's objectives.

The idea is to expose the farmers to success stories and best practices. There are islands of prosperity in Indian agriculture. However, we need to showcase best practices to create world-class agri-productivity levels across all areas. The farmers network will have spreading awareness as its primary objective. We will bridge productivity gulfs using knowledge as a catalyst. An analysis of the farmers' decision-making pattern depended on the crop management. It is toward this end that the corporate sector could infuse knowledge and be partners in growth. Aries Agro, which has registered 40 per cent growth every year, would focus on Punjab in particular as it was the fourth largest market segment of the company. Aries Agro, which pioneered the concept of agricultural chelates would help the farmers in ascertaining the soil quality and help them in the decision-making process. The Punjab farmers had been heavily using zinc sulphate. However, if they switched to chelates of zinc, it would not only reduce the cost on fertilizer, but also result in yield increases. Chelates were the most environmentally sensitive, cost effective and efficient method of applying micronutrients. Nutrient elements like nitrogen (N), phosphorous (P), potassium (K) which are recognized as major nutrients are required in large quantities. While calcium, magnesium and sulphur were considered as secondary nutrients, several other nutrients called micronutrients; though required in smaller quantities, were equally important for the good growth and development of plants. Chelated micronutrients would help correct deficiencies by allowing the essential metal to move to the root zone for rapid and efficient plant adoption.

Alok Industries—Cotton Improvement

The company Alok industries is planning to start contract farming over 60,000 acres of land in Maharashtra's Vidarbh—a region mostly in news for the suicide of its cotton farmers. There is a huge demand for fair trade and organic textile in the developed world. The company wants to capitalize on this opportunity. Fairtrade and organic textile fetche nearly 30 per cent premium over conventional product to the manufacturers.

The textile maker has partnered Raj Eco-Farms, a contract cotton farming firm based in Madhya Pradesh, to spearhead its programme.

Raj Eco Farms has been in the business for the past 7 years and deals with over 10,000 farmers on different contract farming projects. "Farmers will be provided all agricultural inputs and assured of the offtake of their produce. While Raj Eco Farms would look after the ground operation. Alok industries will finance the whole project and oversee marketing, Organic farming, cuts farmer's cost by half compared to conventional farming, but the production peaks only in the fourth year and till that time farmers need financial support. India currently produces 15,000 tonne of organic cotton, a minuscule compared to the projected global retail sales of $ 2.6 billion by 2008 end. Alok industries' farming project may involve over 12,000 farmers. The farms would have a peak production capacity of over 3 lakh tonnes of cotton that would be consumed entirely by the company, which is involved in the entire textile chain from spinning to fermenting and retail.

Nearly one-fourth of the proposed 60,000 acres would be reserved for fair-trade farming. The Fairtrade Certification Mark, awarded to a product by an international body Fairtrade Labelling Organisation (FLO), is considered to be a guarantee that producers in the developing world are getting a better deal. Many big brands, including Nike, Reebok and Marks and Spencer have introduced organic textile in their apparel and want to raise it further. It's all coming from the consumer's acceptability and willingness to pay a premium for the product.[3]

Spices Production in Assam

Village headman of a remote village (Paroli, Karbi Anglong) in Assam, and other villagers had no idea that their traditional knowledge of spices cultivation could free them from poverty and backwardness. Helpful expert advice came along: invest your land in a company, jointly owned by farmers and the Spices Board, for growing commercial crops such as turmeric, ginger and chilli and become a shareholder. Each company has a full time chairman and managing director. While 600 farmers own 51 per cent of Coinonya Farms, 4000 own 51 per cent of Karbi Farms. Each company will initially have a plantation area of 500 hectares that will be cultivated over a five-year period, with 175 hectares being taken up in the first year. Agreements are being signed with private companies for processing and marketing.

Another company, Coinonya Farm Producers Company Ltd., (CFPC), was formed in partnership with the Spices Board. It was set up under section 581, a new provision in the Companies Act of 1956 that marries the economic advantage of a corporate entity with the 'Social Benefits of a Cooperative' CFPC incorporated the traditional rule prevalent in Karbi—that of community land being under the headman's control. This makes Teron and 599 other local farmers—all of them shareholders-owners of 51 per cent equity in the firm. Spices Board owns the remaining 49 per cent. The arrangement is similar to that adopted by Karbi Farm Producers Company Ltd.

Hundreds of plant genes have been validated in various research institutes that hold tremendous potential for crop improvement. For instance, the institute of Himalayan bio resource Technology has found a plant—named potentilla—in alpine region and validated the gene that has high degree tolerance to below minus 40 degree temperature and remained active at more than 50 degree high temperature. Thus even within the plant kingdom, there is tremendous scope for crop improvement through genetic engineering. The company is bringing all the latest research and technology for the benefit of the farmers.

Joint Venture—Madhya Pradesh—for Agricultural Inputs

As more and more farmers become disillusioned with the cooperative system, they are looking for alternatives to address the problem areas—accessing inputs and marketing their produce. In an effort to tackle the issue, a singular initiative has been taken by the farmers in Madhya Pradesh. More than 1,200 farmers in Rajgarh district have floated a company called the Khujner Agriculture Producer Company Ltd. (KAPCL), which will produce certified seeds and sell them to shareholders at a price 10-15 per cent lesser than the local retail shops (even after taking into account a small margin for KAPCL).

It has started by planting soybean in 65 hectares of land, pooled from shareholder small and marginal farmers of the region—and will sell the seeds to about 5,000 farmers. The company has even got a seed retail sales licence number so that the extra seeds produced by the company can be sold to the non-shareholder farmers in the area. The farmers are forced to buy the soyabean seed at Rs. 22 per kg. in the local market, but the certified seeds sold by KAPCL will be priced at Rs. 18 per kg. including Rs. 1 per kg. margin for the producer company.

The KAPCL will get its seed cultivation certified by the state's seed certification agency before planting it and after the harvest to ensure that the farmers get only certified seeds of good quality, a worrisome issue till now because of the possibility of getting spurious seeds from the local retailers. For long-term sustenance, the KAPCL has also taken license for fertilizer retail sales, APMC is providing input and marketing support to its shareholders and other farmers. Presently, it has the dealership and sub-dealership of markfed, MP Agro-Industries Development Corporation, Seed Corporations and other agricultural equipment manufacturers to provide backward and forward linkages to shareholders.

By taking up the fertilizer, pesticide, seed and agricultural equipment dealership, KAPCL will leverage the cumulative demands of its shareholders to get a better bargain for them directly from the input suppliers. The leverage, forces the suppliers of agricultural inputs to supply directly of KAPCL, which is based in the soyabean region itself. This helps the farmers reduce the burden of transport and time. KAPCL intends to cut short the long business chain involving several layers of dealers engaged in supplying material to the farmers, which led to the increased retail prices. In fact, the KAPCL will also manage the agricultural produce by grading and marketing it directly, which will cut out the necessity of middlemen and merchants to deliver the procured produce to the soyabean oil extraction plants in the region—thus giving better returns for the farmers.

A Farmers SEZ—Maharashtra[4]

In Avasari Khurd village of Ambegaon taluka, in Pune district, the villagers have decided to float their own company and develop the area into a SEZ. A resolution has been passed by the Gram Sabha to that effect. There are around 1,500 farmers in the village. The idea is that each one will put in Rs. 1 lakh and become a shareholder in the company. They will raise its capital against the landholdings and the farmers themselves will develop the SEZ.

It needs to be noted that the farmers are not talking of making a co-operative—a movement which has been historically strong in this part of India—rather they want to have a company. The village does have some unique features. Most importantly, it is only 20 km away from the proposed international airport at Rajgurunagar. Being in the midst of many small industrial towns, the village itself is in any case only partially agricultural. The village realized the importance of all this and decided that instead of someone else coming and developing a SEZ in this area, it would be much wiser for the farmers to develop the area themselves. In this way, they will get all the benefits which usually go to the developer.

The land owners immediately saw economic sense in this and agreed unanimously to set up a SEZ company of their own. Once the SEZ is developed, they plan to start their own manufacturing units or give land on lease. Another interesting feature of this proposal is that whole of the village is not going to be converted into a SEZ—in fact, 2,695 acres will be irrigated farming while 2,507 acres will be the SEZ—this area is currently rain fed. The farmers are planning to grow crops for export in the irrigated area. They are very happy because this will allow them to continue with their farming also. This gave them a sense of security about future. They admitted that if the entire village were to be converted into a SEZ they would have been ambivalent about the proposal.

A Pilot Project by USAID[5]

A pilot project conceptualized by the Growth-oriented Micro Enterprise Development (GMED) intends not only to push the organized retail agri business in the country but also provide gainful employment to the village youth. As a part of its four-year programme, the project called the village extension model, aims attaining the youth in crop, soil, water and pest management practices to help them provide technical solutions to farmer's routine problems. Funded by the United States Agency for International Development (USAID) and implemented by ACDI/VOCA—a Washington-based international development consultancy—in Malerkotla (Punjab) and other places in Andhra Pradesh and Maharashtra, it is targeting to have 10,000 small and marginal farmers under its ambit by September 2008. Eyeing the business potential and acknowledging the model's merits, corporate giants like ITC and Tata have also stepped in. While the GMED is recruiting, training and paying the village extension agents for an initial six months, the corporates that are benefiting from these sub-agents will support them thereafter. When these agents will help the farmers produce good quality fruits and vegetables for the corporates, the farmers will get technical support and a fetching price for their produce.

The sub-extension agents are being trained by the GMED staff in technical areas, including soil testing, pH and fertilizer monitoring, basic entomology, pest protection, pesticide safety, crop monitoring, post-harvest techniques, cold chain control, good harvest practices, field packing, drip irrigation theory, construction of low-cost production tools, farm record-keeping and producer relations. This is an initiative where all are in a win-win situation, be it the corporate house, marginal farmers or underemployed youth. The farmers or cluster of farmers will have a person who understands the local dialect at their disposal 24/7. Now GMED is planning to replicate the model in other parts of the country. In fact, last year, (2006) two similar initiatives were taken with ITC and the Nandani Cooperative in Maharashtra.

Retail Revolution—An Approach

Here we look at some of the facts as they are unfolding in the Indian retail scene today. Then we simulate the future, based on our knowledge about what is happening in several developing countries. And finally, we suggest how this race can be made more inclusive in the Indian context so that those who are fearing today can learn to ride these very horses and participate in the race for their own benefit.

Graph 12.1 and Graph 12.2 show what is happening in the Indian food and grocery market today. The overall market size is growing gradually, but the organized retail is growing at a phenomenal pace, albeit from a very minuscule base of less than 1 per cent of the total market size. Within a year, the presence of organized retail went up from 18 to 73 cities, retail space with the organized players has grown at about 50 per cent during the last two years and their sales grew by 46 per cent in 2006. Pantaloon (Futures) group under the banner of Big Bazar (and Food Bazaar) is so far the biggest player with grocery sales of approximately $ 244 million in 2006, very closely followed by Mother Dairy with sales of about $ 220 million and Subhiksha at $ 160 million. Reliance is now fast rolling out its retail network of Reliance Fresh and Bharti is yet to make its entry in the next five years, an investment of approximately $ 20 billion is on the anvil and if the horses are allowed to run unhindered, Indian organised retail in food and grocery segment will be easily growing at anywhere between 30 and 40 per cent per annum. A revolution in Indian retails is on its way.

Graph 12.1: Growth of Leading Food and Grocery Retailers in India

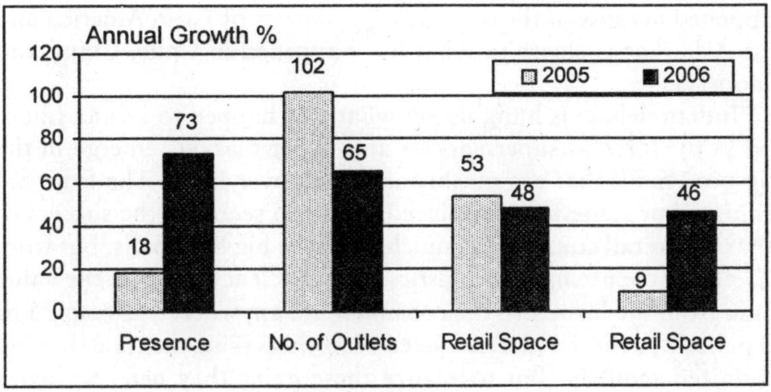

Graph 12.2: Growth of Leading Food and Grocery Retailers in India

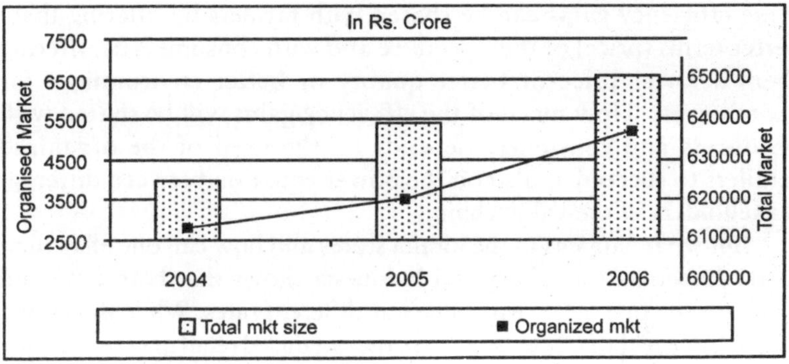

Retail Therapy

- ➢ What remains to be seen is the organized retails willingness to share efficiency gains with producers (farmers) and consumers.
- ➢ If organized retail expands now, small retailers will start feeling the pinch in 10-15 years. During this period, the farmers can co-opt the latter and upgrade them as in China.

An interesting feature of this unfolding retail revolution is that so far it is driven by domestic investments. FDI is banned in multiple brands. If it has played any indirect role through single brands, or through wholesaling (like Metro's cash and carry), it is very minimal.

So this revolution is so far largely *swadeshi*. This is unlike what happened in most of the developing countries of Latin America and east Asia, but is closer to what has happened in Chile, China and Indonesia.

Indian debate is hung up on what will happen to kirana (mom and pop) stores as supermarkets and hyper markets emerge in the 'big box' malls that are mushrooming all over India. The fears are genuine, but somewhat misplaced. The real secret of the success of organized retail come not as much from the big box malls, but from the efficient sourcing and logistics. The efficiency gains in the value chain from the farmer to the consumer are anywhere between 25 to 40 per cent, over the current system. It is this saving that is attracting organized retailers. But to realize these gains they need to invest massively in infrastructure, logistics and institutions to have free access for direct sourcing from farmers or any where else. A part of these efficiency gains can be shared with farmers by offering them better terms (price) or their produce and with consumers by offering them a lower price or better quality or better environment for shopping etc. How much of the efficiency gains will be shared with producers and consumers depends on the need of the organized retailers to succeed as also on the power equation between different stakeholders in the value chain.

But what happens to the kirana stores and how can one allay their fears? Experience in China and Indonesia shows that both exist and grow for the first 5-10 years, albeit at different rates. While the kirana stores grow at about 2 per cent or so, the organized retail may be growing at 20 per cent plus. But the structural changes will start hitting them at some stage, especially when the overall share of organized retail reaches about 25-30 per cent or so. India will take at least 10-15 years to reach that stage. This is the lead time to innovate for greater inclusiveness and train small players to be a part of the race.

Several options are out there that India can experiment with. What we need to remember is that organized retail is not just about big box malls, it is also about the neighbourhood stores (as Subhiksha and Reliance are doing), and even of organized push carts as ITC is rolling out under the banner of choupal Fresh. The organized retailers can co-opt this informal sector of kirana stores and hawkers, upgrade them with a little infusion of capital, design and a bit of training to attend to aspiring customers and organise them under their respective banners through franchise, partnerships, or even employees. It is being done in

China under "200 programme" and even in Japan where big retailers are co-opting convenience stores and upgrading them. The industry associations such as Ficci and CII have to rise to this challenge, duly supported by the government and the civil society, to make it a win-win race for most people of this country. Given its democratic strength and creativity. India can surely innovate several models that will make this inevitable structural change in retailing not only efficient but also sufficiently inclusive for the benefit of its own people, and not bogged down in futile attempts to stop this revolution.[6]

OTHER EXPERIMENTS

Agricultural development is a complex subject, particularly when one is dealing with illiterate small/marginal farmers without any resources. There are varied types of successful experiments carried out in different parts of the country. They have evolved innovative measures to provide quality inputs—seed, fertilizers, machinery and credit etc., one project in Punjab actually teaches the farmers in natural farming. Discussed here are some of the important ones for the benefit of the reader.

Mechanisation—Punjab

For Punjab's farmers plagued by heavy debts and low profits, it's the latest tool for empowerment. Co-operative societies, which rent out agricultural equipment and implements, including tractors and laser land levelers, are beginning to find favour among the state's farmers. The initiative is proving a boon because in the absence of capital investment to purchase costly machinery and tools, the farmers are beginning to optimize production and earn profits. More than 30 farmers at Jeeda village in District Bathinda have sold their tractors after the society bought a tractor and offered to lend it on nominal rent to the farmers. There is no need for the farmer to hire a driver or pay for the escalating cost of diesel, claiming that the cost of cultivation has come down from Rs. 250 per acre to Rs. 100.

A farmer is usually charged Rs. 500 per hour for using the tractor. In an hour, four acres are ploughed. As a majority of farmers have small land holdings, most use tractors for two hours at the most. There is however, no ceiling as far as time is concerned. There are 150 cooperative societies in Punjab, and in various cities in the Malwa belt, nearly 90 tractors are available on rent. The societies in Ludhiana lead the chart, with more than 30 tractors, followed by Moga district societies that boast 25 tractors. Encouraged by the success of this venture, the Punjab Government, in consultation with the Punjab State Farmers' Commission has added another dimension to the programme by deciding to open agro-service centres in all the 143 blocks of the state. Each centre will offer on rent a tractor, a laser land leveler machine, a rotavator and seed sowing machine, besides other implements necessary for farming. The idea to open agro sub-centres, where government will rope in private players by offering 33 per cent subsidy, is the brainchild of eminent agricultural scientist and chairman of the state farmers commission Dr. G.S. Kalkat. The project is likely to get the government's formal approval soon. The agro-service centres will cost Rs. 10 lakh each in all the blocks. It would significantly reduce the capital investment and cost of cultivation, resulting in more profits for the farmers. It would also go a long way in addressing issues like rural indebtedness, economic stress and subsequent suicides by farmers. Notably, Punjab has around 10 lakh farmers and almost every second farmer owns a tractor, making for more than 4.5 lakh tractors in the state. The catch here is that more than 60 per cent of the farmers have less than 10 acres of land and they fail to use the tractor to its optimum capacity due to lesser land holdings.

Agriculture Credit

Rythu Mitra Groups (RMGs) in Andhra Pradesh: NABARD, in partnership with the Andhra Pradesh State Government facilitated a pilot project for promotion and financing of RMGs, during 2004-05. RMGs bring out holistic development in the lives of small, marginal, and tenant farmers through collective action RMGs. Besides accessing loans from banks, they are expected to serve as a conduit for technology transfer, facilitate access to market information and, assist in carrying out various activities for its members. During 2005-06, 4437 RMGs were financed by 18 commercial banks, 9 RRBs and 9 DCCBs involving ground level credit flow of Rs. 28.11 crore. About 62,000 farmers have been assisted under the pilot project.

Credit—Joint Liability Group Approach: Innovations, like SHG-bank linkage programme, have proved to be successful in dealing with providing financial services from the formal banking sector to asset-less or ultra poor. In order to develop effective credit products for mid segment clients, having access to productive assets, NABARD piloted the project during 2004-05 in 8 states of the country through 13 RRBs through the mechanism of joint liability approach. A Joint Liability Group (JLG) is an informal group comprising 4 to 10 individuals coming together for the purpose of availing bank loan either singly or through the group mechanism against a mutual guarantee. The JLG members offer an undertaking to the bank that enables them to avail loans. Unlike SHGs, which are more process oriented, the management of the JLG experiment with select banks have demonstrated that banking system can reach clients like tenant farmers, share croppers, oral lessees and farmers with small holdings without proper land records through the joint liability approach. As on March 2006, the participating banks promoted 870 JLGs and extended finance aggregating to Rs. 10,88 crore to 494 JLGs. Based on the successful experience, NABARD has finalized the guidelines for mainstreaming the programme with commercial banks on an all-India basis.

Menthol in Barabanki (U.P.)[7] Crop Diversification

Defying the farming of traditional crops in the rest of the state, farmers in Begusarai have taken to large scale cultivation of menthol mint, an aromatic commercial crop and a source of menthol oil, which is used in cosmetic and medicinal products. Little wonder then that the district is set to be known as the menthol capital of the state. India is among the top producers of menthol in the world, with Barabanki district in neighbouring Uttar Pradesh leading the country. Given that, till a couple of years ago, Bihar was nowhere on India's menthol map, Begusarai has achieved the feat despite several odds. Realising that mint has remained neglected, despite its export comprising almost 30 per cent of the total export of spices, Minister of State for Commerce announced setting up a mint park in Lucknow. He also announced holding of a mint marketing convention every two years for developing Indian mint brand.

Barabanki's menthol cultivation is spread over 20,000 acres compared to only 8,000 acres in Begusarai. Hopefully, Begusarai will catch up with Barabanki in a few years. State Bank of India extending credit to farmers, has formed cooperatives and is even working to provide buyers for the farmers. There are 200 farmers who have taken to menthol in Bakhri sub-division alone. This is no mean achievement, given the odds. Bihar has neither a market nor a buyer for menthol oil. After extracting the oil in distillation plants set up by them, the farmers carry it 600 km away to Barabanki by booking the containers in trains from Barauni, and sell it for Rs. 500-600 per litre. Barabanki houses an industry that makes menthol crystals and also has a mandi for the oil. Bihar also lacks an institute to educate the farmers on the cultivation of aromatic plants.

It is the profit that is driving the farmers to go for menthol. As per an estimate, farmers earn Rs. 30,000–40,000 a year. The impact of commercial farming on the agrarian economy of the state is already visible in the region. Besides uplifting the economic status of farmers, it has affected the migration of labour in the region decreasing by at least 20 cent. Menthol is a labour intensive crop. Every day 20-25 labourer are needed on a small farm and during harvesting, there is the need for at least 100 of them. The best part is that it is grown during the lean agricultural season which propels the flight of labour. This agricultural revolution came to the government's notice at a recent bankers meet, after which Deputy Chief Minister urged the banks to extend more credit to farmers. (*Indian Express* 14/9/07).

Extension UP Experiment[8]

For the farmers in Barabanki district of Uttar Pradesh, the new helpline is the best thing to have happened in several years. The help line has been set up by private telecom player Airtel in partnership with the Indian Farmers Fertilisers Cooperative Ltd., (IFFCO). While a pilot project was launched on April 2, 2006 at the Danyalpur and Mubarakpur IFFCO Societies in Barabanki, the third helpline was set up in Sitapur district in May. Similar helplines will be set up in other parts of eastern UP before the model is replicated across the country. When a farmer in Barabanki calls Lucknow, his query is registered and the call is routed to the IFFCO center in Delhi. Experts come up with the solution, which is then explained to the farmer through a customized voice message. The farmer does not have to pay long-distance charges. As part of the scheme, farmers get five free voice messages a day. The messages update farmers on mandi prices, weather forecast and fertiliser availability.

Kund Project-Rajasthan[9]

The Rajasthan Government has envisaged a Rs. 22.43 crore Kund horticulture project for 358 Common Interest Groups (CIG) across the state. The project, under the District Poverty Initiative Programme (DPIP), would benefit 3,120 below poverty line farmer families of Churu, Raigarh, Ratangarh, Sardarshahar, Sujangarh, Tara Nagar and Dungarpur districts. DPIP would bear 90 per cent of the cost while the rest would be shared by the GIGS consisting of 10 farmer families each. The project is targeted at poor farmers who have land but lack irrigation facilities. In Churu district itself, they plan to plant 2.5 lakh fruit bearing crops like amla and ber, etc., interspersed with plants needing less irrigation like chana, aradu, moth and gwar in the gaps.

According to estimates, a farmer would be able to earn Rs. 3,000—5,000 per month after one year of implementation of the project and Rs. 15,000—20,000 per month after three years of implementation. Under the programme, each CIG member would get a 30,000 litre water Kund constructed on an acre of land. A plastic tank of 2,000 litre would be mounted 5-6 feet above the ground. The plastic tank would be filled with water through a mechanical hand pump, which would draw water from the kund. As many as 80 fruit-bearing plants and 30 shady plants like neem, ber, sheesham, etc., would be planted at a gap of 6 x 6 metres. These plants would be irrigated through a pipeline connected to the water tank under gravity. A plant can get 10 litres of water in an hour through this scientific technique, which keeps a check on wastage of water and ensures equal distribution among plants. The construction of a tank is expected to cost Rs. 42,000 while the total cost including training and barb wiring of the area works out to be Rs. 70,122 per member. The project has shown good results in Churu district. The payback period is 25 years and it is expected that this technique will not only reduce wastage of water, but also increase the yield by optimizing irrigation facilities.

This so-called drought prone region, with rain water structure in a district like Munger where irrigated area today is low, is the potential source of leading the State in medicinal plants, floriculture, seed production of vegetables and dairy development. Marketing facilities are, however, required to be improved for an easy disposal of all types of perishables like horticulture, floriculture and Dairy products. It is important to examine the possibility of setting up processing industries for these enterprises and also encourage apiculture to provide some additional income and employment to the small and marginal land holders.

Azolla an Alternative Feed[10]

Fodder is an important requirement for cattle. Even if the animals are fed with commercial feeds from the market, fresh green grass or dry fodder availability greatly reduces the expenditure on commercial feeds. The success of a dairy plant depends largely on increasing milk production without escalation in feeding cost Growing fodder grass is a good option. Another is azolla cultivation. Azolla is a floating fern which resembles algae. It is rich in proteins, amino acids, vitamins and minerals. Experiments conducted by the Vivekananda Kendra—Natural Resources Development Project (VK-NARDEP), Kanyakumari district, Tamil Nadu have proved that the quantity and quality of milk yield of cattle went up when they were fed with azolla.

Mixing Ratio: Fresh azolla thus collected can be mixed with commercial feed in the ratio 1:1 or given directly to livestock. It was found that milk production in cattle increased by 10-12 per cent when they were fed with azolla and there was 20-25 per cent saving on buying commercial feeds. Azolla can also be fed to poultry birds. It was observed that the birds grew faster and there was 10-12 per cent increase in their total body weight compared to the birds which were given only normal diet. The egg yolk increased and the egg shell colour became glossy in appearance. In addition the azolla bed acts as a rain harvesting station and live mulch for the fields. About 2 kg. of soil in the azolla bed is almost equal to about 1 kg. of commercial NPK fertiliser after six months

Maize Processing—Karnataka Experience

What Hindustan Unilever, or the erstwhile Hindustan Lever, and Glaxo found unprofitable and dumped along the way, became a pot of gold for a little-known company—Riddhi Siddhi Gluco Boils. In about a decade, a global synergy has helped the small-time starch producer grow into a Rs. 365 crore (March, 2007) firm that manufactures liquid glucose, maize starch, corn gluten and other corn-based ingredients that go into making chocolates and bread, glass and medicines glucose-d (dextrose) and even your clothes. The corn trail runs 12 years into the past when in 1995, glaxo decided to shut its sick unit near Gokak, Karnataka, due to non-availability of raw material. Ridhi Sidhi made a bid and won it against stiff competition form Cargill, British Sugar, Finolex, among others. It was major milestone for the company. The company joined hands with farmers in Karnataka, the largest corn growers in India, for direct supply and cut surplus costs to increase productivity. Soon, the Gokak unit became India's largest corn refining installation. There was no looking back thereafter.

The company emphasized on people—management, steady procurement of raw materials, treatment to value-added products and streaming supply to customers. It also continued the trend of updating product innovations in order to cater to global customers. Today, their client list includes all major brands like Cadbury, Nestle and Perfetti for chocolate making concentrates. In addition, most pharma and paper majors buy starch from Riddhi Siddhi to manufacture their product categories.

The company has surpassed the likes of Sayaji Industries and Anil Starch in terms of revenues. Another big moment for RSGB came when it acquired a functional biopolyment unit at Pondicherry from HLL in 2005 for a consideration of Rs. 6.5 crore. Through the acquisition, the company received an economical share of high end starch advanced R&D facilities and product technologies. By then the company's turnover was Rs. 240 crore. In 2006, Roquette Freres, a leading starch producer, picked up 14.95 per cent stake in Riddhi Siddhi for Rs. 31.7 crore. The deal helped the company get a much required global exposure. The $ 50 billion global starch industry is growing at close to 10 per cent year-on-year. The Indian starch industry pegged at Rs. 15,000 crore is growing at more than 30 per cent a year. Riddhi Siddhi contributes about a quarter of the Indian starch market. The company has now started its new Greenfield project for maize—processing near Uttaranchal. The company has invested Rs. 110 crore to produce 1.65 metric tonnes a year and will increase its exports up to 10 per cent from the current 3.5 per cent a year. Starch is manufactured after various processes of grinding and drying corn. Currently, all its three plants Viramgam and Pondicherry are in the midst of an expansion.[11]

Powering a New Milk Revolution

The White Revolution has offered alternative livelihood to lakhs of farmers in the country. But a section of the farmers in the state of Gujarat are yet to become part of the co-operative movement due to various reasons. Now, with the National Dairy Development Board

(NDDB) adopting six defunct state government dairy plants in Gujarat, the farmers of Suarashtra and Kutch might not have to migrate to other parts of the state. The board has kicked off milk procurement in the region that has not been covered by any milk co-operative yet.

National Dairy Development Board is aggressively working towards establishing a milk procurement network in the rain-deficit areas of north Gujarat. The development programme has helped promote, finance and deliver a variety of supporting services to national co-operative societies, affiliated to 170 district and regional unions and 22 state federations. With the business environment changing, the Board has proposed to put in place new generation methods to replace the traditional model for milk procurement. On the marketing front, NDDB, through one of its subsidiaries, is trying to set up milk marketing networks in major cities in the country. NDDB is also looking beyond milk and milk products. It has set up the Safal Auction Market in Bangalore, a novel idea in collection and marketing of horticultural commodities in India. The auctioning system, if it succeeds, would revolutionize the wholesale trade of agricultural products. The Board is also trying to revive the brand Dhara in the edible oil segment. O late, Dhara has been facing severe competition from private players.

Zero-Budget Natural Farming[12]

For two years, the Sharmas have been sprinkling a concoction of gram flour, jaggery, green chillies, garlic and neem leaves over their crops to keep away pests. That's zero-budget natural farming for you or spiritual farming as Amravati—based environmentalist Subhash Palekar calls it. Over six years, farmers in Punjab—mostly in Faridkot, Bathinda and Ferozpur districts—have been waking up to the benefits of natural farming, which shuns the use of chemicals and does not require buying seeds, fertilizers and pesticides from the market, instead, homemade potions keep the soil pest-free. Practices like mixed cropping and mulching, or covering the soil, help maintain its nutrient balance and bring down water consumption by 40 per cent. Punjab, the biggest user of pesticides and fertilizers in the country, needs that healing touch. Its farms are ecologically degraded, raising the cost of farming and "Farmers' suicides were quite high in the past two decades. Farmers were getting trapped in debt, so we evolved this zero-budget farming. No institutional credit is needed and dependence on hired labour is reduced to a minimum explains Palekar, who has worked on natural farming for 12 years.

> *Unlike organic farming, natural farming avoids costly vermicomposting.*

In the present circumstances the concept of zero-budget is attractive. Today these farms have agro-ecological diversity. Besides cotton, wheat and paddy, they grow fruits, vegetables and sugarcane. Birds and butterflies, absent or scant elsewhere, are in abundance here. Not far away in Bathinda's Raika Kalan village, another farmer brought 15 ha. under natural farming six years ago. His farm boasts cotton, legumes, sugarcane and vegetables. The quality of soil has since improved. Although economists and agricultural scientists argue that natural fertilizers cannot replace chemical fertilizers, it is working for the 1,214 ha. under natural farming in Punjab. Unlike organic farming, natural farming does not use artificially synthesized microbial culture like vermicomposting. Instead it promotes a fermented mix of cow dung and cow urine. Microorganisms developed during fermentation attract earthworms which work on the soil, leaving it puffy, soft and aerated.

Homemade Pesticides

Beejamrutha: A mix of cow dung. Cow urine, water, lime and a handful of soil for protecting the crop from harmful soil-borne and seed-borne pathogens. Seeds and fruits are preserved in this solution. At times, saplings are dipped in the Beejamrutha before being planted. *Ghanjeev Amrit*: is applied to the soil during the sowing season. This is a mix of cow dung, cow urine jaggery gram flour and soil, preferably from forest.

Agni Astra: Kills boilworm, a common pest in cotton fields. Prepared by boiling and fermenting green chillies, garlic, tobacco and neem leaves in cow urine.

Bhrama Astra: Another pesticide prepared by boiling and fermenting leaves of neem, milkweed, dhatura, arjun (*Saraca arjuna*), gilory (*Tinospora cordifolia*), karnaj (*Pongania pinnata*) and guava in cow urine. A mix of neem leaves, cow dung, cow urine and water is yet another pesticide that kills aphids, jassids, mealy bugs and white flies.

Buttermilk is used as an effective fungicide and weedicide.

Yellow Card: A board sprayed with mustard oil and fixed in the fields prevents white flies which get stuck to it.

Cow dung has all the microbes which the costly vermicomposting can provide. Gobar and gomutra of indigenous breeds of cow like sahiwal, tharparkar and rathi give the best results in natural farming. One gram of dung of a desi cow contains 3,000 million bacteria against 5-6 million bacteria in one gram of an American cow's dung. The mix prepared in natural farming attracts the earthworms most suited to Indian soils. These earthworms bring nutrients from the deeper layers of the earth. There is another benefit of using cow dung: repulsed by its smell nilgais and rabbits do not invade crops. Under natural farming, farmers use a lot of other techniques to maintain soil fertility and reduce tilling. One of them mulching with organic residues like leaves and plants reduces tilling, suppresses weed growth, promotes humus formation and enhances the water-holding capacity of the soil, equipping it for drought—like conditions.

Planting leguminous crops helps replenish soil fertility while providing a buffer against total crop failure. Sowing cowpea and other pulses along with companion crops like sorghum and millet (used as cattle fodder) has become a regular practice. Farmers are experimenting with a variety of crops. They had grown ajwain (Bishop's weed), fennel, fenugreek and coriander alongside wheat last year. It is better to grow maize with cotton since 'It attracts birds which will feed on the pests attacking the cotton crop. They had grown vegetables on the same farm in which he grew cotton. The farmers are also using Indian varieties and claim better results. For wheat, most farmers are shifting to the indigenous bansi (Maharashtra), Mundari (Uttaranchal) and C-206 (Rajasthan) varieties. For cotton they use narma seeds. Unlike in chemical farming where seeds last a season, these seeds are preserved for the next season. They are setting up their own seed banks. When natural farming was being promoted in Punjab, many farmers did not buy the idea initially. But negligible input costs and better quality yields are now making them change their opinion. In fact, these became the crucial reason many switched from organic to natural farming (Table 12.1).

Table 12.1: Cost and Yield Comparison Per Acre

	Natural Farming	*Organic Farming*	*Chemical Farming*
Input	Rs. 100.350	Rs. 5,500	Rs. 2,500—Rs. 3,000
Output (of wheat)	1,800—2,000 kg.	1,500 kg.	1,500—1,600 kg*

Note: * Initial yield with chemical fertilizers is 2,000 kg.

Fruit and Vegetable Processing—Punjab Model

A fruit and vegetable processing plant at Hoshiarpur built at a cost of Rs. 38 crore, is the only plant in the world, which can process all types of fruits and vegetables produced in India. The plant has the capability to switch from processing of one fruit to another within four hours. Meanwhile, the government's move to usher in contract farming in the state got a boost with 5,500 farmers opting for the scheme. The government approved a blue print for setting up a network of agri-processing infrastructure on a scale capable of eventually turning the state into the biggest producer and exporter of processed fruit and vegetable products in the world. Under the scheme, farmers who lease out their land for setting up fruit orchids, would get a lease amount of Rs. 8,000—12000 per acre with a 2 per cent hike every year. The owners of the land and the horticulture development council, the organizers of the project—will have a 50-50 profits and risks sharing arrangement. After 12 years the land, with an established orchard, will be handed over to the farmer, if they so like. Otherwise they can continue with the existing arrangement. The orchand will have a life span of 30 years.

Chinese Agriculture

The skeptics of China story often doubt whether there is anything worth while for India to learn as India already has its agriculture in private hands. There are at least three points that one can still learn from China. First China had heavily invested in rural infrastructure, education and health. Although these investments are not strictly counted as investments in agriculture, yet they laid a basic foundation for agricultural growth. In 2004, adult literacy rate in India was still as low as 62 per cent while China's was as high as 92 per cent. Most of villages in China now have access to all weather roads, while India is far, far away from that target. Second China spends almost 1 per cent of its agriculture GDP on agriculture R & D, while India had its expenditure levels below 0.5 per cent for most of the years, which are lately raised to 0.6 per cent last year. Rural roads, and agriculture R & D, in two together can be powerful engines of growth in agriculture and associated rural non-farm activities.

Third, in China, for the last 10 years or so, much of the growth in agriculture is coming from the high-value segment (livestock, fruits and vegetables and fishery which are growing in the range of 5 per cent to 8 per cent per annum while foodgrain segment is growing at less than 1 per cent per annum. China is increasingly importing land demanding commodities like soyabeans (imports exceeding 30 million tonnes) while exporting labour and care intensive commodities (fruits and vegetables). This is all driven by rising incomes, changing consumption patterns, and rational usage of resource endowments. Similar things are likely to happen in India over the next five to ten years. An important lesson to learn here is that high value agriculture needs fast moving infrastructure, and each commodity value chain has its specify. Both public and private sectors need to invest heavily in logistics and infrastructure, processing and modern retailing. This will unlock huge potential in non-farm employment and promote rural urban migration.

It may be noted that in China, during 1978—1997, rural non-farm GDP was growing at more than 24 per cent per annum, and rural non-farm employment at almost 10 per cent per annum. More than 180 million of Chinese farmers migrated to urban sectors since the reform in the late 1970s. Despite rapid overall economic growth, rural urban migration and non-farm sector in India is till embryonic.

China-India Reforms for growth and poverty alleviation. The economic pyramids: Where to start from?

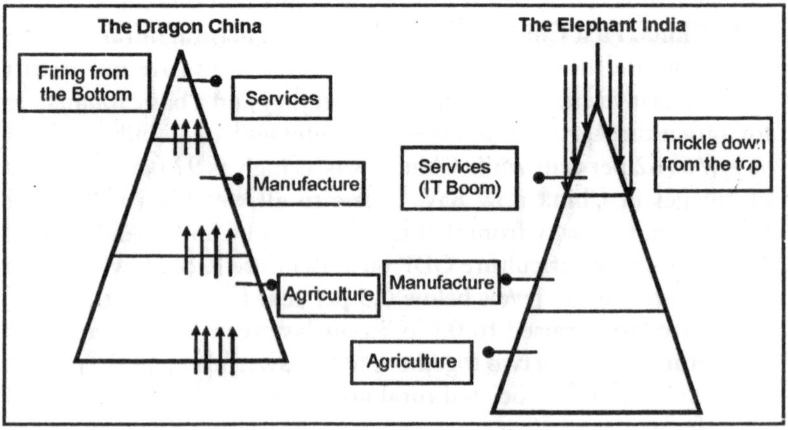

The Malaysian Miracle[13]

What Malaysia has accomplished is impressive—and has much to teach the world, both about economics, and about how to construct a vibrant multiracial, multi-ethnic and multi-cultural society. The numbers themselves say a lot. At independence 50 years back. Malaysia was one of the poorest countries in the world. Its GDP (in purchasing power parity terms) was comparable to that of Haiti, Honduras, and Egypt, and some 5 per cent below that of Ghana. Today, Malaysia's income is 7.8 times that of Ghana, more than five times that of Honduras, and more than 2.5 times that of Egypt. In the global growth league tables, Malaysia is in the top tier, along with China, Taiwan, South Korea, and Thailand. Moreover, the benefits of the growth have been shared. Hard-core poverty is set to be eliminated by 2010, with the overall poverty rate falling to 2.8 per cent. Malaysia has succeeded in markedly reducing the income divides that separated various ethnic groups, not by bringing the top down, but by bringing the bottom up. Part of the country's success in reducing poverty reflects strong job creation. While unemployment is a problem in most of the world, Malaysia has been importing labour. In the 50 years since independence, 7.24 million jobs have been created, an increase of 261 per cent.

There were many reasons not to have expected Malaysia to be a success. Just as Malaysia was gaining its independence, the Nobel Prize winning economist Gunnar Myrdal wrote *Asian Drama*, in which he predicted a bleak future for the region. Malaysia is rich in natural resources. But, with few exceptions, such countries are afflicted with the so-called "natural resource curse", countries with an abundance of resources not only do not do as well as expected, but actually do worse than countries without such benefits.

While natural resources wealth should make it easier to create a more egalitarian society, countries with more resources, on average, are marked by greater inequality. Moreover, Malaysia's multi-racial and multicultural society made it more vulnerable to civil strife, which has occurred in many other resource—rich countries, as one group tried to seize the wealth for itself. In many cases, minorities work hard to garner the fruits of this wealth, at the expense of the majority. Bolivia, one of the many rich countries with poor people, comes to mind. At independence, Malaysia also faced a communist insurgency. The "hearts and minds" of those in the countryside had to be won, and that meant bringing economic benefits and minimizing "collateral" damage to innocent civilians.

Milking Technology

Milking Technology being practiced in Dairies in Maharashtra's village. Unlike in cities where the young and hip use the handheld device as a fashion statement, the farmer in Bhilawadi uses a more complex form of the same machine as a productivity enhancement tool for his dairy and cattle farming business. What's more Radio Frequency Identification (RFID) chips helps him attain efficiencies through proper inventory tracking and management of cows and their yield. Technology—use has got a completely different dimension in this village. "We couldn't have done it on our own. So, we herded up. When an entire village's cattle population is taken collectively, it makes sense for a para veterinarian to obtain information through RFID tags that help record everything from milk yield, health parameters to feed data of this herd. For a chip to work, it can be fixed to an animals ear. The RFID tag can be read only when a pocket PC or PDA is close to it and that prevents any kind of tampering of data. The modern dairy has achieved considerable success with the use of active RFID tags. For its own buffalo farm, the dairy has used Delaval's software DelPro. We use the DeLaval equipment in our farm. But for our suppliers we have used the Bombay, Veterinary Hospital's RFID technology. The technology apparently won the first prize from the World Bank in a competition two years ago. Under the project, an animal is traceable throughout its lifetime. In most cases a farmer brings his animal to a veterinary doctor only if it suffers health problems, while in really 80 per cent animals have subclinical problems. The dairy now has access to all kinds of information and it can plan its business operations on the basis of that. For instance, over-production of milk in any particular season is directly linked to nutrition. "Post monsoon with greenery in abundance, milk production generally rises by 20-30 per cent.

However, the farm has a unique feeding system that feeds the buffaloes correctly. It has installed electronic transponder IDs for animals that ensure that a buffalo can never eat more than a certain programmed portion of daily ration at one time. The biggest advantage that farmers see in this equipment is that it allows detection. Every 21 days delay in impregnating the animal costs the farmer in feed. Ghitale Dairy has a rate of annual calving which is 18-24 months in the field. Additionally, it has installed ultra sound machines that have success rates of 50-55 per cent. If corporates enter this field, investments would be Rs. 3-4 crore and payback would be in six years. Currently, Indian milk production is around 106 million tonnes annually, of which 55 per cent is buffalo milk with a cow population of over 50 million and over 75 million buffaloes.

For milk production to double, the first thing is to improve animal health. While this requires simple things like washing hands before milking and following other basic hygiene practices, a little bit of innovative technology will also help dramatically.

Retail Giants in Rural Business

Typically these hubs are located in a produce-specific area, ensuring consistent yield throughout the year to feed these rural initiatives. For instance, Future Group has set up four such pilot initiatives in rural pockets of Maharashtra. The local produce like fruits, vegetables, grains and other commodities is brought to these collection centres where selling price is determined by sellers. When farmer brings the produce to our hub, he can compare, associate himself with the supply chain and compare produce realization with current system and decide. So far, the pilot initiatives have worked favourably. Its first experiment was set up in 2006 in Vidarbha. In the Akola facility, grains and pulses are procured, sorted, sifted, packaged and finally sent to its Pantaloon retail stores in various locations.

The US varsity is working with these mango farmers on good agricultural practices trying to establish trace ability of produce. If a consumer buys a mango from any of the retail outlets, he will be able to trace the exact farm origins of that mango and the grower. In addition, the produce instead of being handled by various go-between, is sent directly to the retail outlets, ripening naturally in the process.

Farm Tourism

The notion was first embraced by Haryana Tourism, when it convinced 13 farm-owners to open their doors to outsiders in 2004. Three years on, the number of farm resorts in the state has almost doubled, while states like Himachal, Punjab, Rajasthan, Goa and Kerala are also following this route to woo tourists. Owner of the 50 acre Surjivan Farms near Gurgaon was the first to take the plunge in 2004, is so pleased with his farm experiment that two months ago he set up the Association for Promotion of Rural and Agri Tourism (APRAT) to provide free counselling to those keen on taking up farm tourism.

From wheat, gram and vegetables to pulses and oil, he grew everything organically. They make the farm stay, a back-to-the roots experience, by lodging the travellers in mud huts with thatched roofs and preparing the food on *chulhas*. They try to give them slice of rural life while ensuring their comfort, whose kerosene lanterns, often a necessity in the power-starved. Haryana, are a big hit with children at night.

If organic living is the mantra at Surjivan Farms, state-of-the art poultry farming is the strength of Hi-bred farms near Karnal, while camping sites by the Kurali river make the Roots Farm Retreat in Ambala district a big hit with children. The founder of Roots Foundation, says his farm is a great outdoor classroom for city kids. Now wonder, everything at the farm, from the dorm to swimming pool, is tailored for brats. In fact, most farms have a USP. All set to unveil the Citrus Country farm near Hoshiarpur counting on giving people the Aussie homestead experience. "Right from watering and plucking kinnows to grading them, they will be able to do it all on our 80-acre orchard. They will also be able to cycle their way around. Lakhs Farm at Mangar village, equidistant from Gurgaon and Faridabad, promises a quiet weekend with home-cooked food in a farm ringed by the Aravalis.

For most owners, tourism is a part-time business Hi-Bred Farms in Kulwchri village that boasts a 42-year-old poultry collaboration with a US company, took to farm tourism two years ago as it wanted to initiate urban travellers in to the simple joys of village life. A sarpanch for the last 41 years, takes pride in taking his visitors to milk a cow, hitch a bullock-cart ride or lend a helping hand to the local potter. For kids, who think milk comes from Mother Dairy and eggs are made in factories, it's a great learning experience. Farm tourism will help sensitise the city cetizens to the village problems. May be once they come in close touch with the villagers, they will begin to take an interest in their working life.

The Government of India is now stepping in a big way in collaboration with United Nations Development Programme to identify rural destinations across the country. Originally 36 destinations were short-listed but at present 16 are being aggressively promoted. The other 14 rural tourism destinations are :

- Naggar, Kullu district, Himachal Pradesh
- Jyotisar, Kurukshetra district, Haryana.
- Samode, Jaipur, Rajasthan
- Hodka, Kutch Nagar district, Madhya Pradesh
- Chaugan, Madia district, Madhya Pradesh
- Banawasi, Uttar Kannada district, Karnataka.
- Kumbalanghi, Eranakulam district, Kerala.
- Aranmula, Pathanamthitta district, Kerala .
- Pochapalli, Nalgonda district, Andhra Pradesh.
- Bellavpur Danga, Birbhum district, West Bengal
- Nepura, Nalanda district, Bihar.
- Lachen, North district, Sikkim
- Sualkuchi, Kamrup district, Assam.

Land Reforms in China

It is ironic that both India and China are having to grapple, at roughly the same time, with the knotty issue of putting agricultural land to industrial or urban use. India has been beset by a series of protests against the transfer of cultivable land to factory and mining projects. In China, a high-level meeting of the ruling Communist Party discussed sweeping changes last week to allow agricultural land to be used for industrial or non-farming purposes. It is not hard to see the depressing parallels between China and India's predicament—each has a huge rural population, which is largely dependent on land for livelihood. Both countries also desperately need to make land available to support industrialization infrastructure and urban life. But this is only a superficial similarity. The two countries have completely different levels of development, and systems of land ownership. Therefore, they have very different issues pertaining to land acquisition programmes.

In 2007, China's Gross Domestic Product, adjusted to the purchasing power of its currency, was about $ 7 trillion, more than double India's GDP. Industry's contribution to China's GDP was nearly 49 per cent; it was just under 20 per cent of India's GDP India produces 230 million tonnes of foodgrain from 146 million hectares of cultivated land. China produces 400 million tonnes grain from just 100 million hectares. More than 40 per cent of China's population is urban, compared to roughly 30 per cent for India. More to the point, the fundamental difference between the two countries is this: in China, all land, including agricultural, is owned by the state; in India, most agricultural land is privately owned. Therefore, any attempt to use agricultural land for a different purpose raises completely different issues in India, compared to China. In rural China, the village council gives farmers a 30-year land contract under the 1979 household responsibility system. The Chinese farmer is not given the right to use the land for a different purpose. If and when the state plans a different use for the land, such as building a road or a factory, it has to pay the farmer the average price of 20 years worth of produce from that land. The state has also to pay half as much again as resettlement subsidy. There have been several cases of Chinese farmers petitioning the government to allow land to be put to non-agricultural use in order for them to earn good money and move to the city or start up a business. China has also seen protests over alleged corruption in land acquisition.

The situation is exactly the reverse in India. Farmers agitate against the takeover of their land mainly because the price they get for it is low and there is little prospect of viable alternative employment. Land takeover has to been a problem when and where adequate compensation is offered. China is now considering major reform to allow farmers to buy and sell and—use contracts in the hope of boosting the rural economy and the prospects of an estimated 740 million people in the rural hinterland. Time will tell if this will work out but the lesson for India lies elsewhere—in China's massive investment in industry and agriculture over the past decades, making it an enormous power house.[14]

Zhang Xiaosul is the very model of a modern Chinese peasant. Farming a field 10 times larger than any of his neighbour's in this scruffy village in central China, he is on the front lines of a new government drive to transform Chinese agriculture, and with it the lives of 750 million country dwellers.

Under the recently announced land reforms scheme many peasants will turn family pots into a modern farm, and help bang one of the last nails into the coffin of Mao Zedong's collectivist dream. In what the official news agency Xinhua called a "landmark policy document," the Communist Party's Central Committee agreed to allow small farmers to sell their right to till the land. The plan is designed to consolidate land holdings, encourage economic farmers to seek other employment, and boost rural income. The decision did not privatize agricultural land. But "it marks a huge improvement in tenure security for farmers and contains many, many good points".[15]

Some Entrepreneurs[16]

Rahul Gala loves technology. And he loves it so much that he has transformed the methods of agriculture in the arid region of Kutch. Today, he logs on to his computer in the morning, feeds in the data and that's it. The rest is taken care by the system—right from the irrigation to fertigation in his farm Jalbindu Agri-Tech. He grows export—quality dates and mangoes by installing a first—of-its-kind computer aided technology in India. He can feed data for a week's schedule and his system does it for the farm right from irrigation to fertigation.

After returning from Australia to his native village Ratual (near Bhuj), Gala aimed to become an agri-entrepreneur. What helped in installing this technology was his degree in horticulture from Queesland University in Australia. He exploited an untapped opportunity and transformed the dynamics of methods of production. Currently, he grows 'barhi' (a fresh variety of dates) over 12 acres of land and has sown more than 600 dated plants. Each plant is expected to produce 50-70 kg. of dates, which is set for dispatch to Europe and Dubai under his brand Golden Dates. What's more surprising is that fellow farmers near his village are getting accustomed to his technology and are increasingly becoming e-producers. "Agriculture is going to create huge opportunities in India." The need is dynamism and professionalism.

With a hope to make south Gujarat (an NRI-belt) the rose hub of the country, Kumar Patel returned to his village (Kutched 25 kms from Valsad) after working with a few consulting firms in the US. Now Patel, at 34, has just roped in 15 rose producing villagers. Now, they are earning more than 40 per cent profits by growing quality roses. Best roses produces nearly 10 million roses per annum at an average market price of 3.50 per stem, mostly imported from Kenya. It offers a range of roses—bugatti, aloha, aqua, avalance, bordeaux among others. Nisarg Nirman Agro Products, a Mumbai—based firm headed by the Churis grows and sells exotic fruits and vegetables to five-star hotles in India. The company has a business model similar to ITC's choupal, but for the fact that the former only deals in exotic vegetables and fruits. So you have vegetables such as asparagus, artichoke, fennel, curried parsely and zucchini green (and yellow) being supplied to five-star hotels of the likes of Holiday Inn. The

Leela, the Taj Group and also to luxury passenger liners like Libra and Star Cruises. The company first approaches the farmers and check their soild in a laboratory to explore which fruits and vegetables can be best grown. We then give a schedule and also the seeds that we want to source.

Nisarg Nirman Agro Products. The company sources these from the farmers with a 100 per cent buy back guarantee and markets to its five-star customers. Presently around 300 farmers based in Maharashtra, Gujarat, Himachal Pradesh, Tamil Nadu, Punjab and other states supply to Nisarg Nirman, alongside some sourcing from farmers in Thailand. The company grows around 80 varieties of exotic vegetables in India. As you move further towards Jallandhar the potato belt of India, chances are high that the trench fries, potato flakes or frozen green peas that you have been eating might have come from Mandip Singh's firm Satnam Agri Products. His brand Mandy has become popular in many parts of India. Singh says he has invested close to Rs. 100 crore to produce specialized varieties of potatoes for such value-added products which earn premium prices. His daily production has reached 500 tonnes and he claims his brand is set to reach the US, UK and Dubai market very soon.

Ecosan Toilet UP Experiment

Delhi based NGO Foundation of Development Research and Action, The NGO has been campaigning in the area since 2002. They began by talking to the Pradhans of four villages, including Asalatpur (District Ghaziabad) about their methods of farming, fertilizer use and cost. The next step was to introduce the concept of ecosan (short for ecological sanitation) toilets and explain how it would improve sanitation, help them reduce the expenditure on fertilizer, and maintain soil health; human excreta is rich in soil nutrients. Most big farmers saw the point but the majority, used to defecating in the open, were hesitant. The NGO, along with the big farmers, formed a group and organized workshops. It brought delegates from Sweden, from where it gets technical support. It tried various types of toilet pans, like those made of cement and fibreglass, before deciding on using the ceramic ones because they are easy to clean.

Scientific studies in different parts of the world have proved that human excreta particularly urine will become as previous as gold if only mankind knows how to manage it scientifically. Every individual produces on an average 450 to 500 litres of urine and 60 kgs of faeces per annum. Human urine is by far the largest contributor of waste water. About 80 per cent of nitrogen and 50 per cent of phosphorus derived from urine accounts for just one per cent of the volume of wastewater. "If at least 35 per cent of the people in our country having toilets switch over to ecosan toilets, it will enable the country to produce over 65 million tonnes of fertilizer (N.P.K.) which is an environmental friendly form of organic manure."

How it Works

(1) Ecosan toilets do not allow water, urine and excreta to mix. The toilet pan has one hole for faeces and another for urine. The area for washing is separate, water from where goes to fields, faeces collect in a chamber beneath the toilet seat. Once the chamber is full, the toilet is sealed for four months for the excreta to decompose. The gases produced during decomposition escape through a pipe in the chamber. A community toilet room has two toilet seats, which are used alternately.

(2) After defecation the user has to throw a handful of ash down the hole. This saves flushing 10-12 litres of water. Also, ash absorbs moisture in excreta and increases its alkalinity, thereby controlling the growth of pathogens and preventing foul odour.

(3) The urine is collected in 500-litre plastic tanks. Before it can be applied in agriculture fields, the urine is kept in an airtight container—to prevent ammonia loss—for two months to eliminate infections.

(4) Decomposed faeces are used as soil conditioner.

(5) Stored urine is dilute with water and sprayed in fields.

Rain Water Harvesting Distt. Ahmednagar Experiment

Hiware Bazar, a village in Maharashtra's drought prone Ahmednagar district, was sliding into an abyss after degrading its environment. But in less than a decade it turned itself around: into one of the most prosperous village of the country. It used funds from Government schemes, to regenerate its natural resources – forest, watershed and soil – led by strong village body.

In the 1970s, the village famous for her champion Hind Kesari wrestlers lost its fight against ecological degradation. With just 400 mm of annual rainfall (Maharashtra's Marathwada region in which the district lies gets 882 mm), the village needed to project the forests in the surrounding hills – its catchment areas- but didn't. "The naked hills shocked the elders in the village. They were home to magra flowers and fruit trees once." As the hills got denuded, the runoff from the hills ruined the fields. Agriculture became unrewarding. Drought was chronic and acute – a slight drop in rainfall resulting in crop failure. The village faced an acute water crisis, its traditional water storage systems were in ruins. In 1989-90, hardly 12 per cent of the cultivable land could be farmed. The village's well used to have water only during the monsoon. Families began to shift out, first seasonally, then permanently. Today, a fourth of the village's 216 families are millionaires. Hiware Bazar's sarpanch, Popat Rao Pawar, says just over 50 families have an annual income of over Rs. 10 lakh. The per capita income of the village is twice the average of the top 10 per cent in rural areas nationwide (Rs.890 per month). In the past 15 years, average income last has 20 times.

The district was brought under the Joint Forest Management (JFM) programme in 1992. In 1993, the district social forestry department helped regenerate the completely degraded 70 ha of village forest and the catchments of the village wells. With labour donation, the panchayat built 40,000 contour trenches around the hills to conserve rainwater and recharge groundwater. Villagers took up plantation and forest regeneration activities. Immediately after the monsoon, many wells in the village collected enough water to increase the irrigated area from 20 ha to 70 ha in 1993. "The village was just getting a bit of life back". In 1994, the Gram Sabha approached 12 agencies to implement watershed works under EGS. The village prepared its own five – year plan for 1995-2000 for ecological regeneration. The plan was the basis on which EGS was implemented. It ensured that all departments implementing projects in the village had an integrated plan. In 1994, the Maharashtra Government brought Hiware Bazar under the Adarsh Gaon Yojana (AGY). AGY was based on five principles: a ban on liquor, cutting trees and free grazing; and family planning and contributing village labour for development work. The first work it took up was planting trees on forest – land; people were persuaded to stop grazing there. To implement this, the village made another five-year plan. An integrated model of development with water conservation as its core was adopted. An NGO, the Yashwant Agriculture, Village and Watershed Development Trust, was created as the implementing agency for development works under AGY. "Villages and the government should be partners in development; but villager must be in the driver's seat".

The village invested all its funds on water conservation, recharging groundwater and creating surface storage system to collect rainwater. The 70-ha regenerated forest helped in treating the catchments for most wells; 414 ha of contour bunding runoff; and around 660 water-harvesting structures caught rainwater. The state government spent Rs 42 lakh under EGS in the village to treat 1,000 ha of land, at Rs 4,000 a hectare. It was money well spent.

Miracle of Water

Hiware Bazar is now reaping the benefits of its investments. "The little rainfall it receives is trapped and stored into the soil ". The number of wells has increased from 97 to 217. Irrigated land has gone up from 120 ha in 1999 to 260 (see table Intense cropping). Grass production went up from 100 tonnes in 2000 to 6,000 tonnes in 2004. Sakhubai Thange, a 70-years-old villager who has been cutting grass for the last 25 years, recalls the time when overgrazing had made grass scarce. "The efforts put in by the people of the village for soil and water conservation have created a surplus," she says. The grass – cutting season lasts three months, beginning Dussehra. Nearly 80 people go to the forest to collect grass. Rs 100 per sickle has to be deposited with the Village Development Committee.

With more grass available, milch livestock numbers have gone up from 20 in 1998 to 340 in 2003 according to a Government livestock census. Milk production rose from 150 litres per day in the mid-1990s to 4,000 litres now. In 2005-06, income from agriculture was nearly Rs 2.48 crore. Projections are that the 2006-07 figures will be substantially higher, after a good monsoon, with onions alone having been estimated to fetch Rs 1.8 crore.

According to a 1995 survey, 168 families out of 180 were below the poverty line. The number fell to 53 in a 1998 survey. There are now only three such households in Hiware Bazar. "There has been a 73 per cent reduction in poverty, due to profits from dairying and cash crops." The village has developed its own set of BPL indicators: access to two meals a day; capacity to enroll at least two children in a years under these heads, are considered below the poverty line in Hiware Bazar. This is around three times higher than the official poverty line. Nobody asks for work under the National Rural Employment Guarantee Act (NREGA), which has replaced EGS on a nationwide stepped up scale.

Hiware Bazar's strong, participatory institutional set-up has facilitated success. The gram sabha has the power to decide on a range of issues, including identifying sites for water harvesting structures, sharing water and types of crops to be cultivated. The village voluntary body is its implementing arm. The village's biggest innovation is its water budget. The village's second five-year plan (2000-2005) focuses on sustainable uses of the regenerated wealth. "The essence of the experiment in watershed literacy comes from the gram sabha. It is here that decisions are made. The greatest environmental planners are the villagers."

Table 12.2: Water Balance Sheet, 2007 From Scarcity to Surplus

Source of Water	*Quantity of Water in Million Litres*
Runoff from rainfall	3,150
Potential water available	3,077
Water lost	
Water as runoff	88.70
Water lost as water vapour	1,077
Water available	
Stored surface water	153.90
Water that percolates into the earth	307.70
Groundwater from water harvesting structures	526.70
Sub-soil water as moisture or groundwater	923.10
Total water available	1,911.30
Water demand in the village	
Drinking water (humans and animals)	33.90
Irrigation	2,295.50
Additional /peripheral farming activities	38.20
Total demand	2,367.60
Water required (deficit / surplus)	-456.30

Source: *Down to Earth*. January 31, 2008.

Table 12. 3: Intense Cropping Hiware Bazar is getting more out of its land

Land use	*1996-97*	*1998-99*	*2002-03*
Gross cropped area (ha)	821	1,007	1,125
Net cropped area (ha)	723	730	748
Multi-cropped area (ha)	99	276	377
Cropping intensity	1.140	1.380	1.500

Source: *Down to Earth*. January 31, 2008.

Water Pyramid in Rajasthan

Village Roopji Raja Beri near Pachpadra in Barmer district, situated 125 km. from Jodhpur, has a population of 1,000 who normally walk 4 km on an average per day to procure water. Men out of sheer consideration for their women who walk miles to fetch water, normally never drink to their heart's content. The Water Pyramid, towering at a height of nine metres and with a diameter of 30 metres, produces distilled water inside using solar energy while its exterior is used to harvest rainwater during the monsoon. The rainwater is collected separately, purified, and stored in a large ground tank with a capacity of 6 lakh litres. Named "Shiv Jal Dhara"—as it was launched on Mahasivaratri day, the pyramid is only the second of its kind to come up in India. The first one is in the water-scarce Kutch, in Gujarat.

The Water Pyramid, innovated by Marjin Nitzche from The Netherlands, is patented and rewarded by the World bank with the Development Marketplace Award-2006 for small-scale water innovations. It is a uniquely designed inflated foil structure which uses energy from the sun to evaporate brackish source water and condense it to high -quality drinking water. The concept is based on the solar still principle optimized for large areas. Those behind the Good Samaritan act of providing water in the middle of the desert are the Jal Bhagirathi Foundation (JBF) with Gaj Singh, as its chairman, the Aqua-Aero Water Systems BV, The Netherlands, and Acumen Fund of the US. "When JCF was searching for a village where people are ready to experiment on a pilot water pyramid, the community from this village came forward and expressed their willingness.

The plant has a capacity to produce 1,000 litres of safe drinking water per day. The operational cost is minimal since direct sunlight is being used as the energy source. In this particular project the raw water with TDS in the range of 10,000 ppm is purified to ultra-pure distilled water. The State Public Health Engineering Department provides the raw water for the plant. There is provision to manufacture salt also from the pyramid.

E-Charkha

India's Khadi & Village Industries Commission (KVIC) has decided to add 'e' to 'Gandhigiri' in the country by launching 'e-charkha' soon.

The e-charkha developed by RS Hiremath not only produces yarn but also generates electricity using a maintenance free lead acid battery fixed at the bottom, which functions as an inverter. The charkha that generate 6-9 Watts of power can, however, help the weavers as they can generate sufficient power to provide back-up for about six to seven hours of basic lighting LED lighting by spinning the wheel for about two hours. Not to forget that it can also help them play radio. The LED light and the generator (three-phase version) of the e-charkha are of the latest design and both have an extremely long life of at least 35 years. All components have been chosen after a worldwide search so that the end user has a reliable, foolproof, weather-proof and rugged product, which can be enjoyed for many years. With an initial price tag of Rs 3,000, it could be a good option for rural people to achieve back-up power, play radio as well as generate employment and according to KVIC, it plans to induct two lakh pieces of e-charkha at Khadi weaving centres across the country. However, according to Hiremath, e-charkha is just one of the socially relevant products and he has many more such innovations in his sleeve that are being used by poor rural population.

Notes

1. *The Tribune*, December 7, 2006.
2. *The Economic Times*, January 2, 2008.
3. *The Economic Times*, June 29, 2007.
4. *The Economic Times*, October 17, 2007.
5. *The Indian Express*, November 23, 2007.
6. *The Economic Times*, Ashok Gulati and Thomas Reardon.
7. *The Hindustan Times*, June 15, 2007.
8. *The Times of India*, May 10, 2007.
9. *The Indian Express*, April 5, 2007.
10. *The Hindu*, September 20, 2007.
11. *The Economic Times*, January 10, 2007.
12. The S.S. Misra, *Down to Earth*, November 30, 2007.
13. *The Economic Times*, September 21, 2007.
14. *The Times of India*, October 19, 2008.
15. *The Times of India*, October 28, 2008.
16. *The Economic Times*, October 6, 2008.

13

DEVELOPMENT STRATEGY

SWOT Analysis

Development strategy for a concerned sector of the economy of a state or a country must identify its strengths, weaknesses, opportunities and threats (SWOT) before attempting any policy formulation. Today we recognize the fact that agriculture has the potential for generating growth in real incomes, significant multiplier effects into other sectors of the economy and removal of rural poverty and distress. It is in this context that a detailed SWOT analysis has been made.

Strengths

- Abundant natural resources—Fertile soils, plenty of water resources, good rainfall;
- Responsive Government—Keen for satisfactory implementation of development programmes;
- Endowed with good irrigation potential—49 per cent of cropped area is irrigated against national average of 40 per cent. Ultimate irrigation potential more than the geographical area of the state;
- Favourable weather conditions;
- Comparatively developed horticulture—11 per cent of the area in the state accounts for 50 per cent of the total agricultural out put;
- Winter maize, gram and sunflower—A success story; and
- Immense human resources.

Weaknesses

- Large flood prone areas—73 per cent of geographical area

(6.9 million hect.) is affected by floods which is 17.2 per cent of the total flood prone area in the country;

- Poor rural infrastructure-road connectivity, power availability, communication, market access etc;
- Low Education Level—High drop out rates, deficient schools and teachers;
- Ill informed peasantry with very low skill level;
- Absence of proper land policy;
- Low Levels of Farm Profitability with high Instability and Risks;
- Small and fragmented farms and lack of consolidation;
- Non availability of land records easily;
- Skewed distribution of income and land holdings—90 per cent of the rural population sharing some 37-38 per cent of GSDP and; 83 per cent of marginal holder possessing only 41 per cent of land; and
- Poor agricultural research system.

Opportunities

- Central Government Schemes like: National Food Security Mission, NREGS, Sarva Shiksha Abhiyan, Accelerated Irrigation, Benefit Programme (AIBP), Bharat Nirman Programme;
- Prospect of accelerated industrial development based on the interest shown by leading entrepreneurs;
- Irrigation Potentials—100 per cent double cropping possible;
- Developed dairy industry;
- Potential of horticulture, floriculture, makhana and mushroom;
- Organic Basmati and scented rice;
- Processing industry for maize, mango, litchi and other fruits and vegetables;
- A developed model of Decentralised Market Extension and Agricultural Technology Management Agency (ATMA);
- Krishi Vigyan Kendras and Agri clinics at all District Head Quarters;
- Some success in Participatory Irrigation Management (PIM). Some 30 thousand hect. of surface irrigation schemes already transferred to WUAs; and
- Initiatives taken in the social sector—education and health.

Threats

- Ravages of flood damaging extensive cropped areas.
- Naxalite problem.
- High incidence of rural poverty

Policy Framework

The Development Strategy for this study is based on the PAPSL theory (Participatory Assessment and Planning for Sustainable Livelihood) developed by Prof. Chambers, Sussex University. Its starting point is not poverty, but wealth of the poor. Here we first look at the strengths—success stories—and then develop the future prospects, exploiting these strengths and opportunities as a background, but taking due care of problems (weaknesses), failures and difficulties (threats).

Success Stories

Agriculturally Bihar is the only state where ultimate irrigation potential is higher than the geographical area by over 1.5 million hect (Reporting area 9.36 million hect and ultimate irrigation potential 10.9 million hect). Decadal Agricultural growth is one of the best among various states during the period 1980-81 to 1990-91 and 1993-94 to 2005 (Table 13.1). Maize, Gram, and Sunflower yields are higher than national levels and sample farmers maize yields are higher internationally (Table 13.2).

Table 13.1: Decadal Agricultural Growth—1995-96 to 2004-05

State	*1995-96 to 2004-05*	*State*	*1995-96 to 2004-05*
Rajasthan	0.30	Andhra Pradesh	2.69
Madhya Pradesh (old)	–0.23	Uttar Pradesh (new)	1.91
Madhya Pradesh (new)	–0.05	Uttar Pradesh (old)	1.87
Maharashtra	0.10	Tamil Nadu	–1.36
Orissa	0.11	Haryana	1.98
Gujarat	0.48	West Bengal	2.67
Karnataka	0.03	Punjab	2.16
All India	1.85	Kerala	–3.54
Bihar (new)	2.82		

Sources: (1) State Domestic Product (State Series), Central Statistical Organisation, GoI, New Delhi, various issues.

(2) Agricultural Statistics at a Glance, Ministry of Agriculture, GoI, New Delhi.

TERI Source: *EPW*. Vol. XL II No. 26.

Dairy Cooperatives and Horticulture are a success story. The share of Dairy being run under the cooperative system is 10 per cent in the agriculture SGDP. Production of milk during 1999-2000 to 2006-07 has increased by more than 159 per cent. Horticulture with 11 per cent land area accounts for 50 per cent share in agriculture SGDP. Bihar produces over 75 per cent of the Litchi and 90 per cent of Makhana in India. With its share of 10-12 and 6-7 per cent respectively of vegetable and fruit production—the state occupies an important position on the horticultural map of India.

Table 13.2: Bihar Agriculture in the All India Context—2005-06

Area—Million hect.
Production—Million tonnes
Yield—Kg/hect.

Crop	All India			Bihar		
	A	P	Y	A	P	Y
Maize	7.59	14.71	1938	0.65	1.36	2098
Gram	6.93	5.60	808	0.06	0.06	902
Sunflower	2.34	1.44	615	0.02	0.03	1345

Source: Department of Economics and Statistics.

Further, there are very successful farmers in the state who have gone beyond double cropping to mixed cropping and integrated farming. They give priority to conserving natural resources. They are encouraging other farmers to go for organic farming and use vermin compost as well as bio-pesticides to tackle low soil fertility.[1]

Some of the demographic indicators like life expectancy for males (65.66 years), and infant mortality rate (61 per '000 births), are comparable to or better than the national average. Bihar is one of those few states which have shown over 10 per cent improvement from NFHS 2 to NFHS 3 (1998-99 to 2005-06) in various components of immunization coverage. Again Bihar is the only State committed to building a Common School Education System implying a complete restructuring—Right to Education in Bihar. Following this the 'Humer' programme has been started with the National Institute of Open Schooling (NIOS) to provide education specifically to Muslim girls.

As reported in *Economic Times*, higher education had collapsed in the state, but Biharis went to Delhi and elsewhere for higher education. Bihar which has a terrible poor record of law and order, terrible roads and electricity, inefficient and telecom services, has grown at close to 5 per cent per year for over a decade because Biharis could migrate to other states for jobs, send remittances home and bring back new skills. It is absurd to claim that Bihar had little or no development. The situation has changed today, with massive infrastructural development programmes. Bihari labour is no longer interested to go to Punjab/Haryana for work. Sufficient job opportunities have been created within the State itself.

On top of all this, the state has a long history—Buddha, Asoka, Chanakya, the first President of India, Nalanda University and even first Agricultural Research Institute, Pusa—bear testimony to the rich heritage of Bihar.

Failures

- In the social sector a district level study for 2006-07 on per capita expenditure on education and health presented a very disturbing picture. With regard to general education, expenditure in Patna was more than 10 times the State average and while all the other districts had less than this average. With regard to health, Patna again served better, along with some more districts, as compared to the national average. Obviously there is an urgent need to rectify this skewed distribution.
- In the agriculture sector, rice and wheat yields are not only low, but production levels have also stagnated during the last 5-6 years. While there are wide fluctuations in the production of rice, wheat production has declined and even that of maize has not shown any improvement. These low productivity levels along with a skewed distribution of land holdings are responsible for the high poverty ratios (Graph 13.1).
- While irrigation sector in Bihar has attracted an investment of Rs. 11,890 million over the last 5 years, net and gross area sown as well as irrigated has remained stagnant (Graph 13. 2). This sorry state of affairs of the sector can very well be judged from the fact that irrigation revenue collections are hardly 6.7 per cent of the dues and the cost of collecting the same is double the amount collected (Graph 13.3).

- There is a severe shortage of power which is responsible for the inefficient use of the irrigation system.
- Per capita power consumption is only 147 kWh against 460 kWh at the all India level. Instead of improvement, there is a declining trend for rural areas in so far as electricity supply is concerned.

Graph 13.1: Production of Major Crops in Bihar

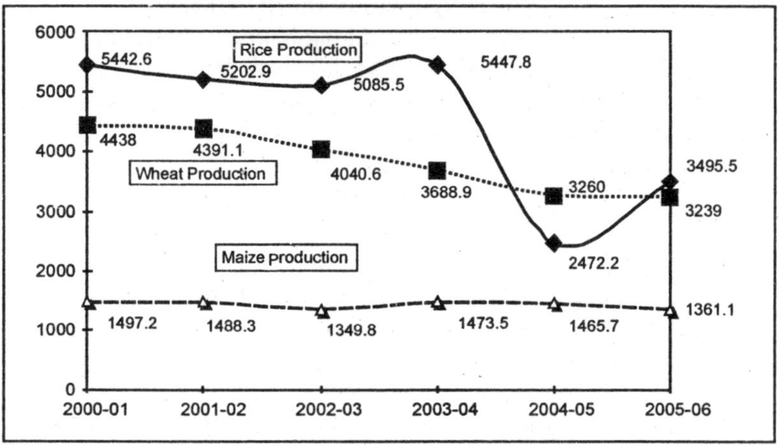

Source: Directorate of Economics and Statistics, Ministry of Agriculture.

Graph 13.2: Land Use Classification and Irrigated Area in Bihar State.

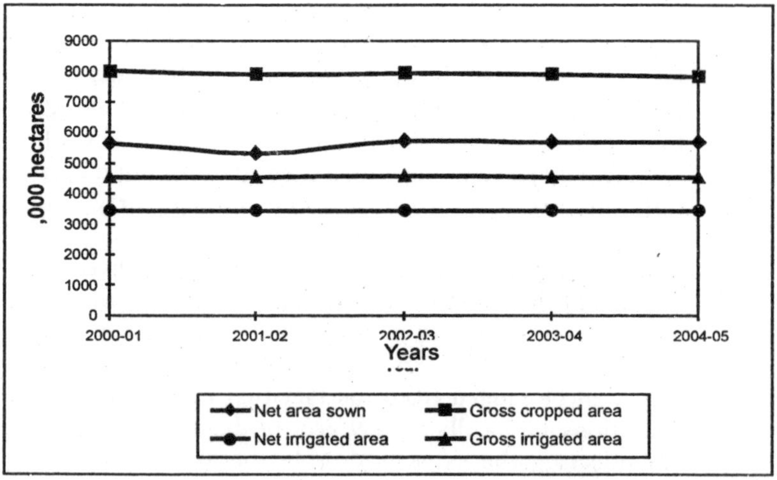

Source: Directorate of Economics and Statistics, Ministry of Agriculture.

Graph 13.3: Falling Irrigation Revenues

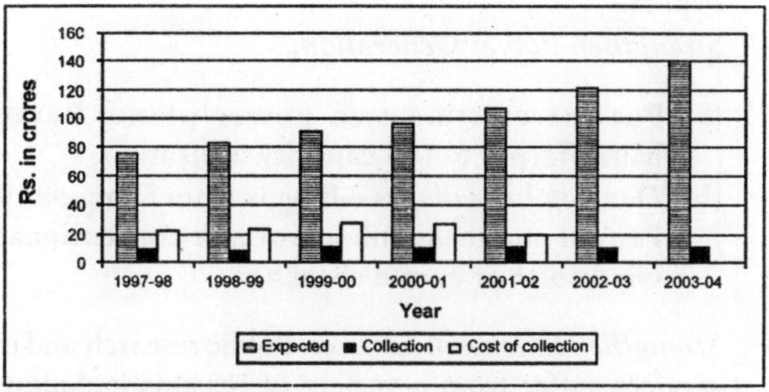

Source: Director (Revenue) Department of Water Resources, Government of Bihar.

Major Constraints

- Lack of quality seeds and Low Seed Replacement Rate.
- Inadequate and Inefficient use of modern inputs.
- Inadequate infrastructure—roads, power, markets, storage, literacy and rural health system.
- High risks and uncertainty due to frequent floods and drought.
- Inadequate policy support—Prices and marketing, land leasing, investment, etc.

The Way Forward

Short Term

- *South Bihar*—Concentrate first on South Bihar which has a low cropping intensity.
- *Strengthen*—A farmer friendly extension system—Involvement of ATMA, Agri clinics and KVKs—(Hissar and U.P. Model).
- *Provide* quality seeds and improve seed replacement rate urgently—Use National Food Security Funds.
- *Ensure*—Balanced use of fertilizers and micro nutrients—Districts like Patna, Khagria, Begusarai consume over 200 kg./hect. fertilizers. Districts Vaishali, West Champaran, Bhojpur and Samastipur consume around 175 kg./hect. Need for introducing organics to preserve soil health—learning from Punjab mistakes.
- *Introduce*--Minikit Programme.

- *Encourage* —Organic horticulture and Basmati rice for export.
- *Strengthen Power Generation:*

  (a) Put the existing two power plants Barauni and Muzaffarpur to full capacity utilization,

  (b) *Develop bio-villages*—Biogas from Dung on the Israeli Pattern and make full use of non conventional energy; wind solar or even e-charkhas.

- *Strengthen Extension Services:* Public research and extension services suffer from a variety of reasons including lack of strategic focus, thin dispersal of available resources, inability to effectively utilize available staff due to limited operating budgets and poor HRD resulting in weak delivery of extension services.
- *Achieve additional 4.3 million hectares of Irrigated Area:*

  (a) One million Shallow Tube Wells give 2 million hect. of irrigated Area.

  (b) Reclamation and new minor irrigation projects—Eleventh Plan target 1.0 million hect.

  (c) Reclamation and new major irrigation schemes—Eleventh Plan target 1.3 million hect.

- *Encourage Cropping Intensification:* With 170 per cent cropping intensity, we will get a total of 10 million hect. gross cropped area from 7.8 million hect. now.
- *Privatize* or take other measures for operationalization of State Tube-wells.
- *Encourage* farmers to avail of the use of subsidized diesel for running tube-wells so that the crops get the required quantity of water.
- *Launch Maegh Pyne Abhiyan (MPA),* a (cloud water campaign) that popularizes a temporary, mobile and low-cost technique for harvesting rainwater to meet drinking water needs of the people, particularly during floods. The project, supported by the Arghyan Trust, has been launched in Khagaria, Saharsa, Supaul and Madhubani—the flood—prone districts.

- *Encourage:* Micro irrigation and Rainwater Harvesting Programmes throughout the State—which has over 1200 mm of rainfall. This will mitigate the problem, if any, of drought prone Districts which really have over 50 per cent of irrigated area Chapter 4, (Table 4.8).
- *Legalize* Land Leasing
  - *Formulate* an appropriate Land Policy, including Rehabilitation Policy for land, agro-industries.
- *Strengthen Credit Support:* Selective Care for Small Farmers—Andhra Model—Rythu Mitra Group and Joint Liability Group. Present system of credit disbursement works against the interests of the small/marginal farmers. The data from Punjab (a developed State) shows that the share of institutional credit increases in direct relation to the size of holding (Graph 13.4). At the same time when we look at the outstanding loans, there is a reverse relation—institutional outstandings increase by size of holding and a steady decline regarding informal outstandings. This means that the poorest the farmer, the least share in institutional credit as well as outstanding. It is the big farmer who took maximum advantage of institutional credit, but refunded the smallest amount-outstanding being the highest: this is true even at the All India level. (Table 13.3A)(Graph 13.5).

Graph 13.4: Share of Different Types of Holders in Total Credit—2002-03.

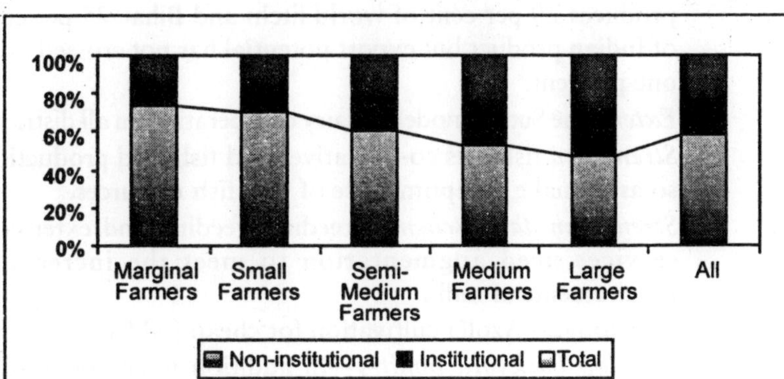

Note: * Household durables and purchase of land.

Source: Sukhpal Singh and M.S. Toor, *Indian Journal of Agricultural Economics*, Conference Number—Volume 60, No. 3.

Graph 13.5: Outstanding Loans-by Size of Holding.

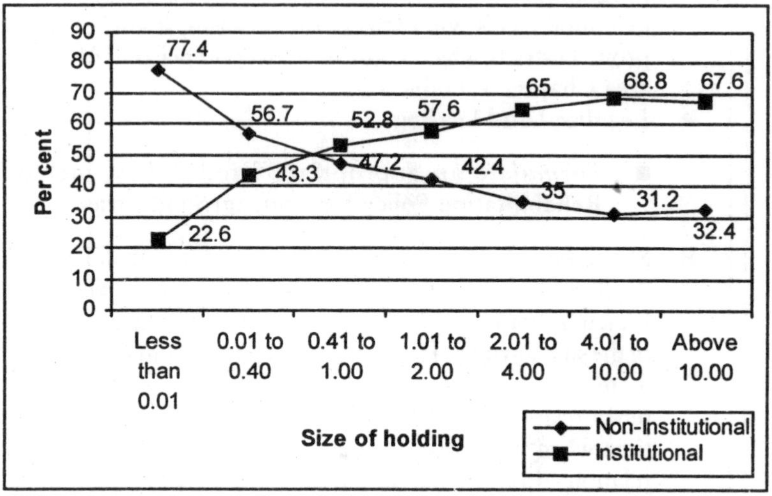

Source: Sukhpal Singh and M.S. Toor, *Indian Journal of Agricultural Economics*, Conference Number—Volume 60, No. 3.

- *Legalize and institutionalize contract farming*, especially in horticulture, vegetable seeds, floriculture, mushroom, medicinal plants, etc., (Punjab Model).
- *Improve and expand good planting material, processing, marketing and storage facilities*—For example, India produces 20 per cent of world litchi and Bihar 75 per cent of Indian product but export potential has not crossed even one per cent.
- *Extend* the Sudha model of dairy co-operatives in all districts.
- *Strengthen* fisheries co-operatives and fish seed production so as to make an optimal use of vast fish resources.
- *Strengthen dairy scheme* breeding, feeding and extension services need augmentation to meet the increasing requirements of milk.
- *Encourage*—Azolla cultivation for cheap fodder.
- *Develop backyard poultry*—In addition to the organized poultry development sector, Bihar should go more for backyard poultry where high yielding birds are now available.

- *Encourage apiculture*—Bihar is the pioneer in apiculture and has special varieties like litchi honey. This is a farmer friendly activity and should be taken up under a mission approach.
- Involve—Local Panchayats in the implementation of various programmes.
- *Introduce an evaluation and monitoring system* by an independent neutral agency and take midterm correction if necessary.

Long Term

Flood Control and Water Management: To achieve flood control and improve water management the following actions are recommended:

- *Pursue* the 2001 proposal with the government of Nepal for construction of dams at upper catchment areas of Kosi, Gandak, Bagmati and Mahananda Rivers;
- *Start* immediate de-silting of the rivers and renovation of traditional community tanks and ponds. There should be a separate department for de-silting water bodies;
- *Introduce* navigation facilities in the rivers; and
- *Interlink* rivers as a step towards integrated water management. This will alleviate problem of frequent floods and droughts and provide additional irrigation facilities.

Power Augmentation

- *Further* programmes for setting up small hydro-power projects, by interlinking of rivers.
- Bihar Government has already taken initiatives to fully exploit big and small Hydro Electric Projects (HEP)—450 MW Interpuri, exploitation of Kaimur region (2570 MW, Sone and Gandak canal projects and Kosi basin as well as Dagmara HEP. Bihar is thus set to become a major player in the national and possibly, at the International level as well. They even believe that Bihar's progress chart will outstrip that of other states in India in the future. Indeed these projects have been designed not just for power generation but for meeting the larger goal of water management. A lot of toil has gone into laying the foundation for the exploitation of

hydro energy resources. This fact, combined with the recent push to expedite matters by the government, will ensure that Bihar becomes a name to reckon with in the hydro sphere in the years to come.

- *Examine the possibility of harnessing*—Wind Energy, Solar Energy, Biogas energy and setting-up of bio villages; bagasse based co-generation power; rice husk based bio-mass gasification; pro-poor, bio-fuel initiatives as developed by ICRISAT.
- Encourage the already tried successful Husk Power System.

Crop Diversification

- *Diversify Crops and Change Land Use Systems*—Indo-gangetic plains presently have predominantly rice-wheat cropping system. In view of changing scenario, marketability and commercialization, there is a need for radical reforms in cropping patterns and land use systems.
- *Install Processing Facilities*: A variety of agricultural produce, namely, milk, makhana, mango, litchi, spices, scented rice, pulses, oilseed, vegetables, medicinal and aromatic plants and flowers have immense potential for commercial agriculture. Setting-up of processing industry will help increase per capita return on agriculture.

  - Honey and ornamental fishes have high potential for commercialization.
  - There is good scope for dal industry in Bihar. Instant dal, roasted grain and dal powder manufacture, green coloured achar, dal sorting' sattu, etc.

- *Expand and Install Storage Facilities*—Huge wastage of horticulture and livestock products. District Aurangabad, for example, is famous for its Maghiya Pan. A private marketing organization is exporting betel leaf to Varanasi. There being no cold storage facilities, the farmers are not able to get good price for their produce.
- *Capitalize on jute and paper industries*—They may prove promising in north-east Bihar region. Kishanganj and Araria districts have a potential for enhancing jute production. National Jute Mission may intervene in developing marketing

and locating specific processing centres. A few Jute marketing yards need to be set up in Bihar immediately under the recently launched National Jute Mission.

- *Install Tea Processing Facilities:* About 20 thousand hect. of area is under tea. It should further be expanded. There are only 2 processing units in the state. There is the need for more such units.
- *Introduce New Researches*: Need for extensive research in major crops to improve productivity. Bihar has just started from a scratch in so far as floriculture, sericulture and apiculture are concerned. The State should take advantage of the available research and other advances made and make a start from a higher plain.

Other Issues

- Computerize land records.
- Consolidate land holdings.
- *Improve Rural Infrastructure*: Such as road connectivity, market, storage, electricity, skill development and expansion of educational opportunities and health facilities, etc. Bihar is one of the top 10 States, spending 40 per cent of the allocated funds on roads, well above the national average. There is a need to integrate this to expand rural connectivity. Under Bharat Nirman Programme—Rajasthan has increased capacity 4 times through small contractors.
- Encourage Research and Development.
- *Improve Human Capital Formation*: Education and health programmes for farmer and poorer sections of the society.
- *Introduce:* Mechanisation on Sirsa, Maharashtra and Punjab Models.
- *Introduce*: Integrated farming system.
- *Strengthen:* Breeding Programmes.
- *Introduce* innovative programmes for additional production of fodder—Azolla and Hydroponics.
- *Develop free range poultry*—This, will help the villagers to produce and eat eggs.
- *Organise* poultry co-operatives in Rural areas.
- *Develop* Fisheries on a Mission Mode approach and establish fisheries cooperatives.

- *Encourage* Honey Production.
- *Create* hygienic conditions for livestock.
- *Create* modern and farmer friendly post-harvest and marketing mechanism.
- *Introduce* well designed agro and other industrial programmes at the village level to provide additional income to the rural poor—a rural diversification policy.
- *Create* adequate non-farm and off-farm employment opportunities—Vaishali, Purnea, Kishanganj and Muzaffarpur as some of potential Districts.
- *Introduce Maize Processing*—Liquid Glucose, starch, and corn glutene, etc.
- *Strengthen Price and Marketing Support*—Open adequate procurement and marketing centres.
- *Revive Sugar Mills Public and Private Karnataka Model*—Shri Renuka Sugar Mills.
- *Examine* possibilities of growing sugar beet.

A Facelift for Bihar

Even as investment proposals have started trickling in the State currently seems to be more interested in preparing the right ground to translate the proposals into reality. As everything has to be started from scratch, the state has been trying to garner help even from outside the box. The innovative approaches, the government claimed, has helped create a climate for private investment and India Inc., while appreciating the moves, has agreed to become Bihar's saviour. While chairman of Aditya Birla Group, Kumaramangalam Birla will look after harnessing the private sector in Bihar, vice-chairman of Mahindra and Mahindra, Anand Mahindra, will be involved in promoting agriculture. Similarly, Sunil Mittal of Bharti Airtel will extend their cooperation in bringing about rapid progress in the rural economy.

These are a few big names from the corporate sector who have agreed to change the face of Bihar through their active involvement and suggestions. This is one of the positive outcome of the first-ever meeting of Bihar Development and Investment Promotion Council (BDIPC). A dozen sectoral groups, known as Abiding Initiatives, each headed by a top CEO, have been formed to help formulate and fine-tune policies. Easing credit flow initiative will be headed by ICICI MD K.V. Kamath, while noted architect Hafeez contractor will draw the blueprint for urban design.

Priya Paul, Chairperson of Apeejay Surendra Group, who showed keen interest in tourism development will head the group on tourism, travel and employment potential and RPG Enterprize V.C. Sanjeev Goenka will over see the energy initiations. Chairman Max India Analjit Singh has been involved in new health paradigm. With a stress on popularizing Brand Bihar—these initiations will facilitate the accelerated development of Bihar.

Threats from Naxalites

The Issues

The Union Home Ministry has provided 33 battalions (over 33,000 personnel) of paramilitary forces to states for deployment in Maoist controlled districts. But even this is not enough. Chattisgarh, for example, has over 13,000 personnel out of the total deployment of central forces, but it has still reported more than 50 per cent of the total casualties (325 out of 601) in 2007. The situation in other states is not very different. While they have all been demanding more forces from the centre, the state governments drag their feet over filling police vacancies. There is also a blame game going on. Central government officials point out that the right to deploy central forces lies with the states. The states, on the other hand, complain that the centre isn't giving them enough man power for taking on the Maoist guerrillas.

The net result is that a major part of the country—some 125—131 districts are badly affected (Tables 13.3 and 13.4). Even Grey Hounds of Andhra Pradesh have achieved only limited success by minimizing casualties. The real solution, however, lies in approaching the issue on humanitarian grounds.

Table 13.3: Red Alert in Heart of India

Highly affected districts	51
Moderately affected	18
Marginally affected	62
Targetted	34
Total affected districts	165
States	13

Table 13.4: Terror-Struck Districts

States	*125 Naxal-hit Districts*
Bihar	25
Jharkhand	20
Andhra Pradesh	19
Orissa	14
Chhattisgarh	10
West Bengal	9
Madhya Pradesh	6
Maharashtra	6
Uttar Pradesh	6
Karnataka	5
Kerala	2
Uttarakhand	2
Tamil Nadu	1

Treating naxalism as a mere law and order problem would be making a grave mistake. It is developing into a revolutionary movement; a revolt by the deprived and the dispossessed, and socially and economically marginalized people. No strategic initiative to counter this grave challenge to national security can succeed without the cooperation of the people. It is unfortunate that a well-meaning initiative in Chhattisgarh of "Salwa judam" has become a controversial subject of partisan politics. The hapless victims of the movement are getting squeezed from all sides, the naxals, the security forces, the unscrupulous politicians and the greedy corporate houses.

The Solution

The solution lies in following a two-pronged strategy:

- *One* restoration, strengthening and extension of the administrative institutions in the affected areas for the supply of essential goods and services like health, education, water, power, and roads; and
- *Two* making the security apparatus more responsive and effective.

There is need to solve the basic problems of poverty, unemployment and social injustice rather than branding them as anti-national and suppressing them ruthlessly. The most-affected districts should receive special attention of both the state and the Central

Governments. Mere allocation of more funds would not do. Unless it is ensured that the allocated funds reach the target groups, these would be total waste and, in fact, only go into the pockets of corrupt politicians and swell the Maoist coffers. The grass-root-level functionaries should be taken care of by committed public servants with a reputation for integrity, and their work should be closely supervised and supported by senior administrators. A study of per capita expenditure on health and education in Bihar shows a skewed distribution. But for Patna which gets the lion's share, the naxal affected districts hardly get any thing. This must be rectified.

The Maoists should be given political space to come over-ground and express their grievances. But the key to this problem lies in accelerating development works, not mega projects but the ones that would give relief to the deprived people in a short period. The people in these areas must be able to see and get the fruits of development. There is need to evolve a broad national consensus about the strategy to deal with such problems.

In a fresh initiative, the Planning Commission has formulated an "arterial road" blueprint to help hot pursuit of Left-wing rebels and to open up the affected districts for trade. The ambitious plan aims to create two major arteries—first, connecting Vijayawada in Andhra Pradesh with Ranchi in Jharkhand, and the second, linking Bastar in Chhattisgarh with the Vijayawada-Ranchi highway or straight with Ranchi. Development Strategy for Naxal Affected Areas has been formulated with four related ministries—rural development, tribal affairs, panchayatiraj and environment and forests.

It has sought creation of a rural electrification 'backbone' under Rajiv Gandhi Grameen Vidhyutikaran Yojana (RGGVY) besides making BSNL, and/or private operator ensure complete mobile phone connectivity in the affected regions. In many of these districts, local administration has been paralysed and a parallel government holds sway. There is a consensus that this is not primarily a security problem but has roots in the feeling of oppression and desperation fuelled by poverty and lack of development, according to the blueprint.

MHA has appended the note to discuss a methodology for more focused action for effective implementation of development schemes in states/districts affected by Naxalite problem. The affected regions are mostly tribal-dominated forested and mineral-rich. "Despite being resource-rich, the local population has remained poor. Forest wealth and minerals have been exploited by "outsiders" without much local value addition. Administration has been thin and weak, connectivity poor and the more difficult areas can only be reached on foot. There is also near absence of education and health facilities.

There was a time, the tribals were driven towards violence by bondage and poverty and the government's insensitivity towards their plight. The tribal belts of Chhattisgarh, Orissa, Jharkhand, and Bihar are still trapped in poverty, but the people there now join the armed rebels for reasons other than their economic condition. In Chhattisgarh, Maoist guerrillas have started forcing young men and women and children to join their ranks. Today, the state is the major training centre for the Maoist recruits coming here from all over the country. The level of poverty in Orissa or Bihar for that matter remains abysmal. As the eastern state fails to check poverty and its many districts are plagued by frequent natural disasters, the state is a perfect breeding ground for extremist ideologies.

In Orissa and Bihar, the Maoists do not follow any formal recruitment system, but try to rope in villagers by cashing in on their personal and social problems. Since, the poor and low-caste people here have limited access to administration and judiciary they often take shelter in the kangaroo courts run by the ultra leftists for easy and quick justice. This allows the Maoist cadres to enter the villages. Instead of carrying out a recruitment drive, Maoist leaders wander through the remote villages help the villagers construct check irrigation canals. They also 'educate' villagers against the "problems caused by the capitalists". This makes an impact. And, through this process, they become a part of the village.

Once inside a village, the extremists offer instant justice for internal problems like theft, cheating, vandalism, and land disputes in the area, drawing villagers closer to them. It is at this stage that the Maoists gain the trust of the villagers who are then ready to protect them from the police. It is only a matter of time before some boys join the red army. Ironically, it is police harassment that drives them into the arms of the red army. Police 'clampdowns' often lead to innocents being rounded up and labeling them as Maoists. To avoid this torture, villagers actually join the extremists, taking shelter under their guns. However, everyone is not forced to traverse this path by police pressure. Some people join these outfits to settle personal scores and family disputes. Others enjoy the VIP status that joining an extremist organization gets them in the village. In some parts of Bihar and Jharkhand, managing the tenders for public works in favour of political leaders is another reason for which some people join the "party" (Maoist organization). And, in return, they get cash and other privileges that improve their living standard considerably. Taking all these factors into consideration, the only solution to the problem lies in special rehabilitation programmes specifically directed towards this class. If 25 districts out of 38 districts are affected by the naxalite activities, what type of development programmes can be expected in the State without the cooperation of these misguided people. Any development plan for its success must, therefore, attach high priority to this issue.

Note

1. *The Hindu*, July 10, 2008. the case of one Mr. Bhagwan Prasad Sharma, Kalayan Gura Village, Samastipur who has only half a hect. of land.

14

AGRICULTURAL PERSPECTIVES

A perspective plan for any state has first to take into consideration the agricultural situation at the national level. The food economy of the country is going through a critical phase. On the one hand, the production of food grains has stagnated over the last 6-7 years (Graph 14.1 and Table 14.1). On the other hand, the Government is committed to meet the food grains requirement of PDS and quite a number of welfare schemes. This means that if Government procurement falls short of its commitment including buffer requirements, it has to resort to imports. The country faced a serious wheat problem during 2005-07 resulting in the import of the commodity to the tune of over 6 million tonnes during 2006-07 (Table 14.2). Despite an all time high estimated production of 231 million tonnes of foodgrains in 2007-08, there is hardly any relief in the prices on the food front. The Government banned the export of non-basmati rice, which it was forced to withdraw under some conditions. Export of maize and pulses has also been banned.

Table 14.1: All India Area, Production and Yield of Rice and Wheat 2000-02 to 2007-08

Year	*Rice*			*Wheat*		
	Area mn.hect.	*Production mn.tn.*	*Yield kg./hect.*	*Area mn.hect.*	*Production mn.tn.*	*Yield kg./hect.*
2001-02	44.90	93.34	2079	26.34	72.77	2762
2002-03	41.18	71.82	1744	25.20	65.76	2610
2003-04	42.59	88.53	2077	26.60	72.16	2713
2004-05	41.91	83.13	1984	26.38	68.64	2602
2005-06	43.66	91.79	2102	26.48	69.35	2619
2006-07*	43.62	92.78	2127	28.14	74.89	2671
2007-08	NA	96	NA	NA	78	NA

Graph 14.1: Production of Foodgrains in India

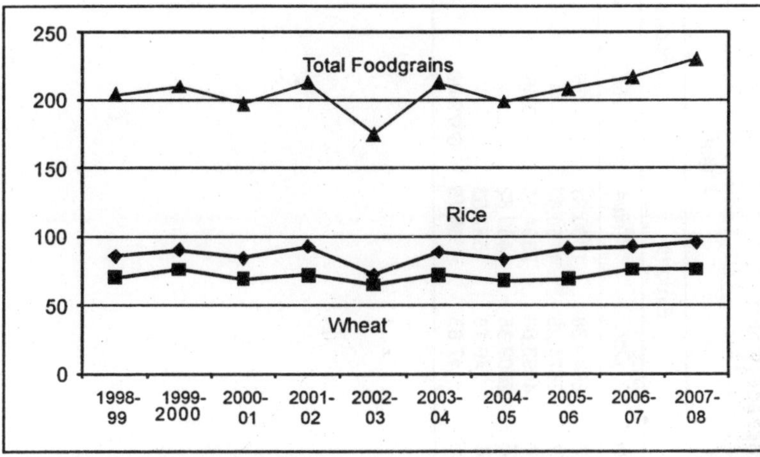

The situation has become quite alarming particularly when the country's rice consumption is increasing by 1 to 1.5 MT every year and the Planning Commission estimates that by 2020 the demand for rice would exceed 120 MT. Concurrently, rapid conversion of cropland to non-farm use, inadequate irrigation facilities and lack of new and healthy seeds have led to a slowdown of rice production. The matter is aggravated further by the fact that in the global market just about 40 MT of rice is traded as compared to over 110 MT of wheat. So, if India's import of 6 million tonnes of wheat could hike the global prices by 30 per cent, we can expect the same for rice, leading to further price rise.

When the country is passing through such a critical situation the role which a state like Bihar has to play becomes all the more, not only important, but also urgent. Our running theme in this study is that a state like Bihar, which has more than sufficient untapped agricultural resources, has to move fast to fully exploit its production potential and occupy its cherished position as the leader of Second Green Revolution.

Plan Synergy

A long-term plan has to be consistent with the realization of several short-term objectives in its overall framework. Thus it is necessary

Table 14.2: Exports and Imports of Rice and Wheat

Year	Rice Exports				Rice Imports		Wheat			
	Basmati		Non-Basmati				Exports		Imports	
	Qty.	Value	Qty.	Value	Qty.	Value	Qty.	Value	Qty.	Value
2001-02	667.07	1842.77	1541.49	1331.37	0.06	0.07	2649.38	1330.21	1.35	0.84
2002-03	708.79	2058.47	4259.08	3772.77	0.87	1.09	3671.25	1759.87	—	—
2003-04	771.49	1993.05	2640.57	2174.94	0.54	0.27	4093.08	2391.15	0.46	0.25
2004-05	1163.00	2823.90	3615.10	3945.02	—	—	2009.35	1459.82	—	—
2005-06	1166.57	3043.10	2921.60	3178.17	0.26	0.34	746.18	557.53	—	—
2006-07	1040.67	2778.31	3704.85	4257.88	0.16	0.41	47.83	36.15	6079.00	N.A

Source: Agricultural Statistics at a Glance.

to identify those steps which can and need to be undertaken within the contours of the overall Plan. For example, output can be increased in the short run through better and more intensive input uses. Hence the Perspective Plan has to be elaborated as a series of short term implementable tasks contributing to the realization of the long term objectives. Similarly, there are tasks which can be attained only over a long term time horizon. These include flood control, power availability, major irrigation, institutional reforms, technology upgradation and management practices, etc.

Programming for the Agricultural Sector cannot be undertaken in isolation. With the current world trend towards less labour intensive agriculture, any development plan may indicate the need to deploy excess agricultural labour in the non-agricultural sectors of Bihar's economy. This will have profound implications and has been articulated in detail. Despite favourable natural endowments, agriculture in Bihar, particularly in crop sector, has almost stagnated, especially in comparison with other States of the Indian Union. The Perspective Plan mainly addresses towards redressing the causes of such stagnation.

Any long term objective for the agricultural sector will principally focus on increased output, employment and efficiency in input-output relations, consequent increases in investment in agriculture and changes required in related infrastructural activities. Given the nature of gestation period involved in obtaining the steady and sustained growth in the agricultural sector in Bihar, the time horizon for the Plan has to be appropriately defined. It is in the light of these general considerations that a comprehensive Perspective Plan has been attempted.

The emerging strategy has to be translated into policies and programmes. Since it is necessary that the Plan is operational and feasible of implementation, the policies and programmes have to be further segmented into targets and micro level action related programmes. In the specific context of Bihar, the outline of a Plan has to be preceded by an assessment of the present situation relating to the agricultural sector. These may be identified in terms of:

- Sector's contribution towards the domestic product in terms of incomes, employment and their inter-relation ships;
- The role and significance of agricultural sector in Bihar's overall economy; and
- Factors inhibiting the growth and development of the Agricultural Sector.

The following is an illustrative list of the various factors that need to be analyzed before embarking on the formulation of a Long Term Perspective Plan. This is actually what has been done in the first 13 chapters.

- Agro-climatic factors;
- Input-output structure and production relation;
- Technology status;
- Capital formation in the sector;
- Institutional factors; labour practices, migration, etc;
- Quality of human resources component;
- State intervention—viz. land reforms etc; and
- Infra-structural factors—past production management and cost-benefit relationships.

The Approach

Traditional Approach

First exercise for any such Plan is the 'busir ess-as-usual' Trend Line approach. This implies no new technology, no special efforts etc., stagnation in crop and irrigation sectors (Graphs 14.2 and 14.3).

Graph 14.2: Production Trends of Foodgrains and Oilseeds in Bihar

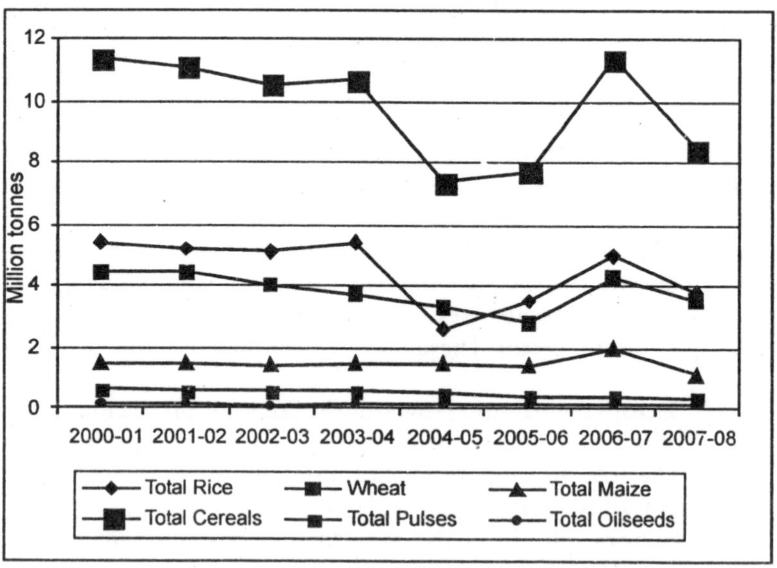

With stagnant agriculture, particularly crop sector, over the recent past, business-as-usual approach will not yield the desired result in terms of achieving accelerated growth. In the light of national requirements and unexploited potential of Bihar, this approach will also not be able to meet the expectations of a Perspective Plan. If progress of 30 years in the Green Revolution States of Punjab and Haryana is to be condensed into say 15 years and Bihar is to be put on the mainstream of India's agricultural development map, concerted efforts will have to be made on war footing. Agriculture cannot wait.

Graph 14.3: Trends of Net and Gross Irrigated Area in Bihar

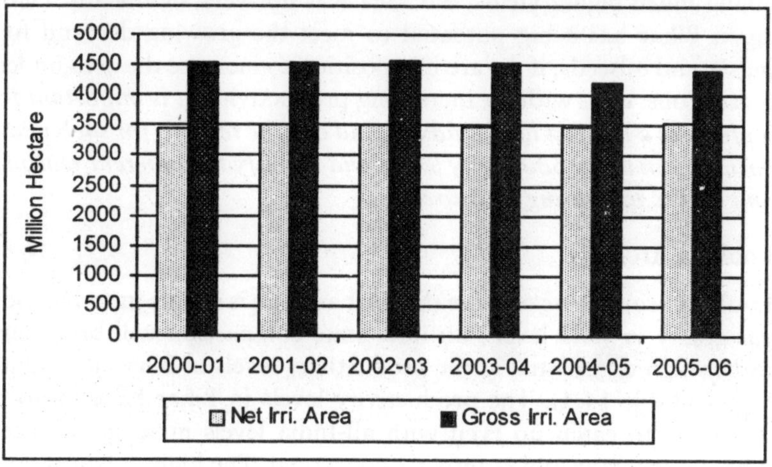

Proactive Approach

The 21st century agriculture, based on technology, will need a new look. The methodology adopted for such a study rests on the philosophy of *'look within, look around and look beyond'*. These are expanded as below.

Looking within

While looking within, a District level study of major crop-yields (Chapter 5)—rice, wheat, maize, gram, sugarcane—showed wide variations from District to District. Winter rice productivity in Bhojpur, Buxar and Rohtas was 3-5 times that of in Gaya, Samastipur and Begusarai. Similarly, Patna, Bhojpur and Rohtas produced 3-5 times wheat per unit area as against some other Districts. The relationship for maize was 1:4 and gram 1:2 (Table 14.3). All the low yielding areas under rice and wheat have 40-50 per cent area irrigated.

The case of maize was all the more interesting. While at the District level Banka, Khagaria and Lakhisarai were the highest producing districts of 4 to 4.5 tonnes/hect., productivity levels of sample farmers at Munger, Siwan and Begusarai were 6-7 tonnes/ hect.—highest levels. Gram where productivity levels in Buxar, Patna and Bhojpur were higher than the all India and international levels, total crop area in the 3 Districts was only 11.7 thousand hect. Rapeseed and Mustard and Sunflower among the oilseeds were also giving much higher yields, but area was not very significant. Here again Bihar has a big potential to meet the growing demand for pulses and oilseeds, if we are in a position to increase the area under these crops, even without increasing productivity. *It is important to undertake a District level study to find out the reasons for such wide variation in the productivity levels and identify the inherent reasons and device mitigating measures.*

Looking Around

Looking around, we observe that just as there is a wide inter-District variability in yield levels, an inter-state comparison will bring out rather high differential in the productivity levels in Bihar and other States (Table 14.4). The productivity levels in Bihar have to go a long way to catch up even with all-India levels in so far as rice, wheat, sugarcane and potato are concerned. The highest wheat and rice producer Districts in Bihar are a shade behind all India productivity levels—leave aside those of Punjab and Haryana. As regards rapeseed and mustard, Bihar productivity level is 83 per cent of all India. The State, however, has superiority over all India in so far as maize, gram, and sunflower are concerned.

Bihar cropping pattern is such that 70 per cent of the area is under rice and wheat and these high productivity level crops account for less than a million hect.—about 12 per cent. Under the food security regime, the country is having a rather higher shortage of pulses and oilseeds. Requirements of maize are rising fast due to the increasing demand from poultry and industry. It has, therefore, to be decided as to how far it is desirable to increase the area under these crops, particularly when it is uneconomic to produce rice and wheat in lower yielding areas. Water requirements for maize, gram and the two oilseeds are also much lower than rice and wheat. *A micro level study at the District level could then be taken up to work out the revised cropping pattern so that Bihar can contribute maximum towards the food requirements of the country.*

Table 14.3: Bihar—Highest/Lowest Productivity Districts-2004-05:

Area'000' hec., Production '000' Tonnes and Yield-kg/hect

	Highest			Lowest	
	Area-hectares	*Yield-kgs./hect.*		*Area-hectares*	*Yield-kgs./hect.*
Rice (Winter)					
Bhojpur	84647	2155	Gaya	52017	336
Buxar	78344	2116	Samastipur	50595	295
Rohtas	195296	2116	Begusarai	18763	242
Rice (Autum)					
Vaishali	10577	1625	Madhubani	33617	608
Supaul	29400	1318	Katihar	11574	735
Madhepura	16933	1385	Samastipur	31324	352
Wheat					
Patna	58755	2102	Sheohar	15504	384
Bhojpur	47752	2591	Katihar	44709	458
Rohtas	115857	2104	Araria	57567	602
Maize (Rabi)					
Banka	2126	4450	Samastipur	24938	1092
Khagaria	29868	4236	Darbhanga	7437	1092
Lakhisarai	541	3862	Sitamarhi	1148	963
Gram					
Buxar	2449	1233	Bhagalpur	2939	663
Patna	11171	1134	Banka	3925	633
Bhojpur	3061	1091	Gaya	3755	611
Sugarcane					
Begusarai	1929	69650	Darbhanga	910	25070
E. Chamaparan	12432	56750	Patna	617	27660
Samastipur	4487	178390	Rohtas	1882	18950
Rapeseed and Mustard					
Khagaria	3111	2026	Nawada	400	453
Lakhisarai	689	1937	Aurangabad	593	593
Begusarai	4840	1937	Nalanda	1183	243
Sunflower					
Patna	110	1474	Katihar	1630	404
Buxar	177	1474	Arwal	2	1000
Khagaria	1044	1469	Rohtas	18	1111

Note: Sample Farmers in Districts Manager, Siwan and Begusarai gave over 6 tonnes with Hybrids.

Table 14.4: Area Production and Yield of Major Crops in Selected States

Year	Bihar			Haryana			Punjab			Andhra Pradesh			All India		
	Area	Produ.	Yield	Area	Produ.	Yield	Area	Produ.	Yield	Area	Produ.	Yield	Area	Produ.	Yield
RICE															
2000-01	3656.3	5442.6	1489	1054.0	2695.0	2557	2611.0	9154.0	3506	4243.0	12458.0	2936	44712.0	84976.6	1901
2001-02	3552.3	5202.9	1465	1028.0	2726.0	2652	2487.0	8816.0	3545	3824.8	11389.8	2978	44904.0	93340.0	2079
2002-03	3584.7	5085.5	1419	906.0	2468.0	2724	2530.0	8880.0	3510	2822.3	7327.0	2596	41176.1	71820.2	1744
2003-04	3578.0	5447.8	1523	1015.0	2790.0	2749	2614.0	9656.0	3694	2975.0	8953.0	3009	42592.5	88526.0	2078
2004-05	3122.6	2472.2	792	1028.0	3023.0	2941	2647.0	10437.0	3943	3086.0	9601.0	3111	41906.7	83131.7	1984
2005-06	3252.4	3495.5	1075	1052.0	3210.0	3051	2642.0	10193.0	3858	3982.0	11704.0	2939	43659.8	91793.4	2102
WHEAT															
	Bihar			Haryana			Punjab			Uttar Pradesh			All India		
2000-01	2067.8	4438.0	2146	2355.0	9669.0	4106	3408.0	15551.0	4563	9239.3	25168.3	2724	25730.6	69680.9	2708
2001-02	2126.3	4391.1	2065	2300.0	9437.0	4103	3420.0	15499.0	4532	9256.0	25498.0	2755	26344.7	72766.3	2762
2002-03	2130.9	4040.6	1896	2267.0	9188.0	4053	3375.0	14175.0	4200	9094.0	23612.4	2596	25195.7	65760.8	2610
2003-04	2076.8	3688.9	1776	2315.0	9114.0	3937	3444.0	14489.0	4207	9150.1	25566.7	2794	26594.7	72156.2	2713
2004-05	2027.6	3263.4	1609	2322.0	9058.0	3901	3482.0	14698.0	4221	8999.8	22513.9	2502	26382.9	68636.9	2602
2005-06	2003.7	3239.0	1617	2304.0	8857.0	3844	3468.0	14493.0	4179	9163.9	24073.8	2627	26483.6	69354.5	2619
MAIZE															
	Bihar			Himachal Pradesh			Karnataka			Uttar Pradesh			All India		
2000-01	620.5	1497.2	2413	298.1	683.6	2293	668.9	2135.7	3193	907.9	1473.0	1622	6611.3	12043.2	1822
2001-02	594.3	1488.3	2504	301.3	768.2	2550	580.1	1451.7	2502	931.0	1516.0	1628	6581.5	13160.2	2000
2002-03	603.6	1349.8	2236	299.8	483.3	1612	649.5	1343.0	2068	755.3	836.4	1107	6635.2	11151.7	1681
2003-04	616.4	1473.5	2390	298.5	729.6	2444	618.1	1209.9	1957	947.2	1318.5	1392	7343.4	14984.3	2041
2004-05	614.4	1465.7	2386	324.0	736.0	2272	850.0	2512.0	2955	876.0	1494.0	1705	7430.4	14172.0	1907
2005-06	648.8	1361.1	2098	295.4	543.1	1839	936.0	2728.0	2915	814.4	1054.3	1295	7588.3	14709.9	1938

A Comparative Picture of Agriculture in Bihar and Punjab

As for inter-state comparisons, Bihar is again far behind one or the other state practically in all the crops. Bihar will then have to travel a long distance, if it has to have its name on the agricultural map of India. When we look at the irrigation potential with good rainfall, along with WUAs practically at all the Districts and institutional success stories like Dairy Cooperatives, one need not loose heart. For this purpose, we are taking up the case of Punjab and Bihar from the pre-Green Revolution period and see how Punjab has come much ahead of Bihar (Table 14.5). Productivity levels of rice and wheat in Punjab and Bihar before the Green Revolution in the sixties showed a low differential in favour of Punjab. The differential, however, widened considerably later. It may be pointed out that agriculture today in Bihar is at a level what it was in Punjab 35 years ago around 1970.

Factors which triggered Green Revolution in Punjab are identified as:

- A high level concentration of quality HYV seeds.
- Higher doses of fertilizers—Not forgetting organics
- An accelerated spread of irrigated area—mostly tubewells with more or less free electricity.
- Practically every village connected with road and provided with electricity.
- With, of course, guaranteed prices serving as a catalytic agent with a network of procuring centres spread all over the state.

While Punjab is applying quality HYV rice and wheat seeds with a replacement ratio of around 30-33 per cent, the ratio in Bihar is only 10 per cent. Fertilizer consumption in Bihar is 99.8 kg./hect., about one-half of 194.6 kg./hect in Punjab. Gross irrigated area in Punjab is 7.8 million hect—one and a half times the total reporting area and 1.8 million hect., over and above the ultimate potential (5.97 million hect.). Punjab in its quest to increase productivity, adopted a number of practices disregarding their harmful effects. Environment-related diseases have spread throughout Malwa region of the State. Unmarried girls in their teens get breast cancer. Degenerative diseases of the bones and joints cripple teenagers; premature graying frosts the hair of children. High levels of pesticides are contaminating water and soil. Therefore the planning in Bihar should take note of these failures and avoid their repetition.

Table 14.5: Historic Comparison of Bihar and Punjab Agriculture

(Area '000 hect.; Production '000 M.T.; Yield kg./hect)

Items		1960-61		1965-66		1970-71		2004-05	
		Bihar	Punjab	Bihar	Punjab	Bihar	Punjab	Bihar	Punjab
Total Reporting Area		9360	5036	9360	5036	9360	5036	9360	5033
Net Sown Area			3757		3796		4053	5712	4243
Gross Sown Area		8200	4734		4889		5678	7802	8069
Net Irrigated Area			2020		2263		2888	3433	4038
Gross Irrigated Area		2000	2556		3136		4242.5	4567	7793
Ultimate Irrigation Potential		10,800	5970	10,800	5970	10800	5970	10,800	5970
Fertiliser Consumption (Kg./hect)		1.5	1	3.2	5.0	9.11	40.3	99.78	194.56
Cropping Intensity			126		128.8	5275	140	136.6	190.2
Rice	A	5240	228	5309	287	4154	389.9	3122.6	2647.0
	P	4545	200	4919	351	787	688.0	2472.2	10437.0
	Y	867	1035	927	1223	1316	1765	792*	3943
Wheat	A	651	1394	636	1563	1259	2299.2	2027.6	3482.0
	P	443	1700	418	2260	957	5145.0	3263.4	14698.0
	Y	680	1235	657	1446	991	2238	1609	4221
Maize	A	805	326	744	383	1113	554.6	614.4	154.0
	P	584	371	584	488	1223	861.0	1465.7	422.0
	Y	725	1135	785	1274	244	1552.0	2386	2740
Gram	A	542	838	466	744	174	357.9	72.1	5.1
	P	282	681	242	666	713	284.0	60.2	4.4
	Y	520	813	519	895		794.0	835	863.
Fruits and Vegetables	A						41.04	1092.7	205.7
	P							3856.9	1611.8
	Y							3530.0	7768.0
Milk Production (thousand tonnes)							2100.0	4743.0	8554.0
Egg Production (lakh)								7894.0	36800.0

Note: Bihar data for years 1960-61, 1965-66 and 1970-71 regarding rice, wheat, maize and gram relates to undivided Bihar (including Jharkhand). Actual figures for Bihar alone should be higher than those given in the Table above. Gross irrigated area in Bihar was also less than 25 percent against 54 percent in Punjab during 1960-61.

Source: Directors of Agriculture, Bihar and Punjab.

This comparison with Punjab gives us both positive and negative lessons. On the positive side, *if Bihar can have 100 per cent quality seed with required quantities of balanced fertilizers along with organics and micro nutrients and, of course, a rationalized use of water, there is every reason for the state to claim its rightful place as the fore-runner of Second Green Revolution in the country.* On the negative side, Bihar has to take care right from now about the health of its soil and water resources for which Punjab is paying heavily today. Fortunately for Bihar, the reporting area in the State is 54 per cent more than that in Punjab. As compared to Punjab, irrigation potential in Bihar (10.8 million hect.) exceeds, more than a million and a half hect., of the reporting area—highest as compared to all the States. Learning from the bad experiences of Punjab where rice/ wheat mono culture has led to a serious water problem and micro nutrients deficiency, Bihar must use all its natural resources judiciously keeping them in pink health for posterity.

Because of significant drop in water levels, higher levels of arsenic concentration have already become a health hazard in Bihar. A government study on ground water in the state has come up with an alarming statistics—12 districts including the capital Patna are in the grip of an arsenic contamination crisis, thanks to the dipping levels of ground water. The study, conducted by the Public Health Engineering Department (PHED) in 2006-07, found that the average arsenic content in the ground water in the districts in question is 500 parts per billion (ppb). World Health Organisation guidelines specify 10 ppb of arsenic as the permissible limit in potable water. In India, the permissible limit for arsenic in potable water has been put at 50 ppb by the health authorities, in view of the climatic conditions and general resistance levels of the people. One of the most worrying aspects is that over the last five years, the problem has spread to several districts, with Darbhanga being the latest. The other districts are Patna, Bhojpur, Vaishali, Bhagalpur, Samastiput, Buxar, Khagaria, Begusarai, Katihar, Chapra and Munger.[1]

Looking Beyond

International comparisons (Table 14.6) show that Bihar has got a lot of potential to increase the yield of major food grains, namely, rice, wheat and maize.

Table 14.6: International Comparisons of Yield, Selected Commodities, 2004-05

Tonnes/hect.

Rice		*Wheat*		*Maize*	
Egypt	9.8	China	4.25	U.S.A.	9.15
India	2.0	France	7.58	France	7.56
Bihar	2.0	Bihar	1.6	Bihar	2.4
Japan	6.4	India	2.71	India	1.18
Myanmar	2.4	Iran	2.06	Germany	6.69
Korea	6.7	Pakistan	2.37	Philippines	2.1
Thailand	2.6	U.K.	7.77	China	4.9
USA	7.8				
World	3.9	World	2.87	World	3.38

Source: F.A.O. Production Year Book.

Agriculture: Where India and China Stand

Both China and India have traditionally been agrarian economies and well over half of their billion—plus people continue to depend on land for their livelihood. Given their large populations and histories of famine, India and China also share similar concerns on issues such as food security. In order to develop an appropriate strategy and development model for Bihar, we have examined the growth strategy of China, India and Bihar. The idea is to understand what has gone wrong with India and Bihar to have failed to keep pace with the development in the two corresponding economies. There is a stark difference between agricultural growth in China and India.

Agricultural development in China has been much faster than India. Reasons for China having outperformed India in agriculture are threefold: *technological improvements accruing from research and development, investment in rural infrastructure and an increasingly liberalized agricultural policy.* Incidentally Research by IFPRI (Fan, Gulati and Thorat, 2007) in this direction shows that the marginal returns of government spending say, an additional dollar on fertilizer is much lower (only 0.53) than spending on agriculture R & D (6.9), or rural roads (3.2), or investment in rural education (1.5) or even irrigation (1.4).

During 2006 China invested $ 1.5 billion as against $ 0.5 billion in 1995 on research in Agriculture and have 1000 developed R&D Centres now. A comparative picture about public investment in agricultural research in China and India is as in Table 14.7. Two third of cotton is BT and 100 per cent rice is of the latest modern varieties. India has the largest number (30,000) of agricultural scientists with a dismal track record. The Commission on Agriculture Costs and Prices says "Even in the States of Punjab and Haryana, which are considered the granary of the nation, the productivity of wheat in the past few years has been gradually declining. Besides the climate, existing wheat varieties, which were released nearly a decade back appear to be showing signs of fatigue".

No. of crop varieties released by the ICAR during 1996-2001 have fallen by 50 per cent. In Bihar, total funds allocated to the Agriculture University, hardly meet the salary of the staff. In China, Government funds are only for research. As for salaries and all other incidental expenses, Research Centres have to generate their own resources—joint ventures with private sector. As for infrastructure, China spend $ 20 billion for buildings and reconstruction of 32000 km. of rural roads during 2006 alone.

This is an important experience for Bihar to learn. Research and Development and infrastructure rural roads, storage and other marketing facilities should be taken up as high priority areas. While India's agricultural sector is projected to grow by about 2.5 per cent this year (a slide from the 2.7 per cent growth in 2006), China's has been steadily growing at between 4 and 5 per cent over the last 15 years. By 2005, China had in fact emerged as the world's third largest food donor. The two widest agriculture—related discrepancies between India and China lie in the diverging productivity levels of various crops and in the differential mix of crop and non-crop segments in the overall composition of the farm sector. According to the Food and Agricultural Organisation (FAO), the average yield of paddy in India between 2003 and 2005 was 3,034 kilograms per ha. In contrast the comparative figure for China was more than double at 6,233 kg./ha. For wheat the corresponding figures were 2,688 kg./ha for India compared to 4,155 kg/.ha for China, while for rape and mustard India averaged 909 kg./ha to China's 1,778 kg./ha.

The data on the trend rise in yields in the 15 years leading up to 2005 are also telling. For rice the trend rise over this period in India was only 1 per cent, less than half of China's 2.1 per cent. According to statistics from the international Rice Research Institute, India produced 124 million tonnes of rice compared to China's 186 million tonnes in 2004, despite having almost double the area under paddy cultivation (42 million hectares vs. 28 million hectares). Regarding

Table 14.7: Public Investment in Agricultural Research in China and India

	1980-85	*1991-94*	*1955-2000*	*2002*
China				
Agricultural research expenditure (annual, million 1999 yuan)	2,232	3,808	4,667	7,837
Agricultural research expenditure (annual, million 1999 international dollars)	1,233	2,104	2,578	4,329
Agricultural research expenditure (as percentage of AgGDP	0.42	0.39	0.34	0.49
Number of FTE researchers	5,443	25,917	29,920	31,927
Expenditure per FTE worker (000 1999 international dollars)	226	81	86	135
	1981-83	*1991-93*	*1997-99*	*2004*
India				
Agricultural research expenditure (annual, million 1999 rupees)	5,214	9,513	12,339	15,691
Agricultural research expenditure (annual, million 1999 international dollars)	603	1,100	1,426	1,813
Agricultural research expenditure (as percentage of AgGDP	0.22	0.28	0.29	0.33
Agricultural research and education expenditure as percentage of AgGDP	0.36	0.44	0.42	0.53
Number of FTE researchers	10,315	12,742	13,092	8884
Expenditure per FTE worker (000 1999 international dollars)	58	86	109	204

Notes: 1. Agriculture includes crops (including horticultural crops), livestock, fisheries and forestry. The investment estimates are for the NARS do not cover the institutes conducting research for agriculture but located outside of the NARS.

2. Number of researchers includes active researchers only (excluding researchers associated with commercial activities retirees and technicians).

Source: India—Pal and Byeriee, (China-Qian *et al.* (2003) and Fan *et al.* (2006).

TERI Source: *EPW*, Volume XLIII, 26-27, 2008.

rape and mustard, the trend rise in China marked an even larger stride 3 per cent compared to the India's mincing step forward of 0.6 per cent. Other crops such as wheat and groundnut reveal similar trends, with China well in the lead.

The widest divergence between India and China, however, is in the profitable horticultural sector with the production of fruits and vegetables in China leaping up from 60 million tonnes in 1980, roughly comparable to India's 55 million tonnes at the time, to 450 million tonnes in 2003, way ahead of India's corresponding 135 million tonnes. China's added advantage lied in the more diversified composition of its agricultural sector, with animal husbandry and fisheries accounting for close to 45 per cent of the total in 2005, compared to less than 30 per cent in India. China has thus clearly developed a more diversified set of instruments than its southern neighbour for increasing net farm incomes.

The key question that arises is why and how China has managed to outstrip India in agriculture when 25 years ago, the two countries were or less on a par on most parameters. The latest report by the Economies Advisory Council to India's Prime Minister, *Economic Outlook* for 2007-08 has a special section on agriculture and points out that the traditional excuses for India's substandard performance in the farm sector are not tired but inadequate. Written by a team headed by C. Rangarajan, the report points out that Indian agriculture is placed favourably when compared to China in terms of quantity of arable land (9161 million hectares *vs.* 130 million hectares), irrigated land (55.8 million hectares *vs.* 54-5 million hectares), average farm size (1.4 hectares *vs.* 7 tractors per 1000 hectares). 15.7 to 7 tractors per '000 hect. Thus most of the usual excuses for India's poor agricultural performance do not hold up when it is compared to China providing yield. There is ample scope for that. The yield of wheat per hectare in India is less than two-thirds that in China, of rice about one half, of groundnut one-third and of cotton one-fifth. Unless we catch up in productivity food security will remain a pipe dream.

Research Gaps

Fifty-two per cent of Krishi Vigyan Kendras (KVKs) and on-farm testing service of the Indian Council of Agriculture Research (ICAR) are still demonstrating old crop varieties released between 1948 and

1997 in frontline demonstrations on oilseeds and pulses. As a result, the latest crop varieties and technologies did not reach the farmer's fields, the Comptroller Auditor General has observed in the report for the year ending March, 2007. The performance audit of the extension activities of the ICAR during the tenth plan revealed that there was significant shortfall in conducting technical activities like training frontline demonstrations and non-farm testing for farmers.

The results of the frontline demonstrations being conducted by the KVKs showed that the demonstrated practices were not cost-effective for adoption by farmers. The total expenditure of the ICAR for its extension activities during the Tenth Five-Year Plan was Rs. 834.77 crore. Further, the report pointed out that only 0.34 per cent of the total rural youth trained were able to gain-self employment. "Inadequate training courses were being conducted for practicing farmers, rural youth and extension functionaries," the report said in scathing criticism of the ICAR. It said that monitoring evaluation and overall management of KVKs were inadequate and needed to be strengthened. The Agriculture Technology Information Centres (ATICs) were not fully successful in providing a single—window supporting delivery system for the availability of technology products, diagnostic service and technology information to the farmers and other end users.

The basic eligibility criteria such as proper site selection minimum requirement of cultivable land (for demonstrations), mortgage of land were not followed in the establishment of KVKs. Further, there were delays in development, no-utilisation and misutilisation of infrastructural facilities in KVKs. The National Research Centre for Women in Agriculture (NRCWA), which is also an agency for frontline demonstrations for women, did not undertake 70 per cent of its activities during the Tenth Plan. It did not create a database on gender specific information, identify hazards in agricultural operations and develop need-based technologies for women. The report has categorically stated that the extension activities of the ICAR needed to be overhauled to ensure better transfer and adoption level of farm technologies and for enhancement of productivity.

Data from the national demonstrations conducted by the Indian Council of Agricultural Research and agriculture universities reveal that if we initiate concerted efforts to bridge the current gap between potential and actual yields, ranging from 1 to 2 tonnes per hectare in Bihar, Rajasthan, Gujarat, and Maharashtra, our future wheat requirements can be fully met.

For harnessing this untapped potential, the National Commission on Farmers (NCF) in its 3rd report presented to Government in December 2005, proposed concurrent attention to soil health enhancement with particular attention to overcoming the deficiency of micronutrients (such as zinc, boron, and sulphur), water harvesting and its efficient use, credit and insurance, technology and inputs, and above all, assured and remunerative marketing through a Minimum Support Price (MSP). Bihar is well endowed with water resources; the major problem, however, is water management not availability. Groundwater has been used only marginally, and with perennial rivers and aquifer recharge during the Southwest Monsoon, there is scope for conjunctive use of surface and ground water. The Ganga, which can help to make India a leading agricultural power, is crying for attention.

The NCF had also proposed the establishment of "Gyan chaupals" or village knowledge centres with internet—radio connectivity to provide both generic and dynamic information relevant to eliminating the know-how-do-how-gap. In the case of rabi and boro rice, high yielding hybrids with good grain quality are available. If grown with efficient agronomic management, an average yield of 5 tonnes per hectare can be obtained. What is essential is an integrated approach and not isolated huge expenditure in the form of drip irrigation and other technologies involving costly market purchased inputs. To be sustainable, agriculture has to become knowledge intensive.

Research results at the farmer field (Table 14.8) which seem to be very conservative and international comparisons (Table 14.6) do indicate further potentialities. Taking all these factors into consideration, an attempt is made about the possible production potential of food grains and oilseeds in Bihar by 1920-21. It has to be understood that horticulture which contributes 50 per cent of Agriculture SGDP from 10 per cent area is one of the most progressive sector, with a high potential. Besides litchi, mango, and banana etc., Bihar has monopoly in water body products like makhana and sanghara. Sericulture is the other area which has been neglected so far. Some area will also be allocated to this sector. Bhagalpur area has potential for tussar silk. There will, however, be a need for technological marketing and credit support.

Table 14.8: Yield Gaps for Various Crops in Bihar (kg./ha.)

Crop	*State Average Yield*	*Potential Yield ***	*Yield Crops Average*	*Yield as % of Potential Yield*	*All-India Yield*
Rice	1,218	3,026	1,818	40	1,840
Wheat	1,816	3,052	1,236	60	2,703
Maize	1,844	4,056	2,212	45	1,810
Gram	1,017	1,380	363	74	792
Arhar	1,379	1,830	451	75	727
Rape-Mustard	1,360	1,850	190	74	1,002
Jute	1,409	2,010	601	70	1,823
Sugarcane	48,856	62,780	13,924	78	70,578
Potato	9,060	23,500	14,440	39	18,161
Mango*	1,000	1,660	660	60	—
Litchi*	1,000	1,850	850	54	—
Banana*	5,000	8,000	3,000	63	—

Note: * Yield numbers per tree, ** Yield achieved in national demonstration and other demonstrations. Yield gap is computed from these figures.

Source: Yadav *et al.* in Prasad 2000.

Recent Developments in Agriculture Social Sector and Infrastructure

Agriculture, Cooperatives, Fisheries and Animal Husbandry

Eager to bring agriculture back to track, the Bihar Government has distributed high quality seeds of wheat, pulses and gram among farmers. The Rainbow Revolution aims to double the yield in farming, horticulture, fisheries, dairy and animal husbandry by 2015.

- Agriculture Road Map' prepared with an outlay of Rs. 6000 crores for integrated agriculture development, 2008 declared as "Agriculture Year".
- Chief Minister's accelerated seed expansion scheme launched.
- Elections to the cooperative bodies to be conducted by an independent authority.
- The membership fee and share capital in respect of SC/ST, women, OBC members of Primary Agriculture Credit Committees (PACs) borne by the State Government.
- Under national agriculture development scheme Rs. 406.91 crores crop compensation paid to 3.32 lac farmers in the year 2007.

- Free drug distribution in all state government veterinary hospitals and free pathological lab facility in 100 sub-divisional veterinary hospitals.
- Special cattle health protection fortnight at Panchayat level for immunization and veterinary care for cattle.
- 500 artificial insemination centers revived, 100 cattle feed banks established.
- Scheme for establishment of 382 Mini Dairy units and 40 cattle breeding units.
- In order to fulfil the demand for milk, production would have to be doubled with the help of village level milk unions. For the plan, a consortium has been formed, where National Bank for Agriculture and Rural Development will raise money and NDDB will provide knowledge and technical—based assistance. With Bihar Dairy of Co-operatives being already strong, they will be able to enjoy maximum benefit under the programme.
- Milk Cooperatives distributed Rs. 1.55 crores dividend for the first time.
- Group accident insurance scheme launched by COMFED.
- Jan Shree Insurance Scheme for fishermen launched.
- To encourage fish seeds production private sector initiatives for fish fingerlings production.
- State government provided additional bonus of Rs. 25 over the support price for wheat procurement (Rs. 1000 per quintal), record procurement of 4.7 lakh M.T. of wheat in 2008.

Health

- Health camps and free spectacles distribution in all Maha Dalit tolas.
- Yoga training facilities in all district and sub-divisional hospitals.
- Telephone based interactive patient problem redressal arrangement-"Samadhan" launched.
- To encourage institutional deliveries and essential nursing care for new borne babies and mothers a new scheme "Mamta" launched.
- In all government hospitals "Rainbow" scheme for ensuring

a new colour coded bed sheet on every bed everyday.

- State's routine immunization coverage increased from 18 to 51 per cent (more than the national average).
- Institutional deliveries increased from 21 per cent to 50 per cent.

Education

- Establishment of new primary schools in 15548 villages/ habitations with populations above 300 having no primary school access.
- Under "Mukhyamantri Samagra Vidhyalaya Vikas Yojana" construction of five rooms in every primary school, ten rooms in every middle school, drinking water sources, toilets, boundary walls and play facilities.
- "Sankalp Abhiyan" in 17 districts to ensure the fundamental right to education for all children aged 6-14 years. The number of out of school children has reduced from 25 lakhs to 10 lakhs.
- "Gyan Jyoti" Training Programme for vocational training of girls in 21 border blocks of the state.
- Appointment of 2.25 lakh teachers in the state, out of which 2.14 lac are primary school teachers and 11676 for middle schools. Of these 50 per cent are women.
- Free text books for all girls enrolled in classes 1 to 8 and to all scheduled caste/tribe, minority Muslim and extremely backward boys.
- One crore 30 lakh children benefited by provision of Mid-day Meal scheme in all schools.
- "Mukhyamantri Balika Cycle Yojana" for all girl students studying in class 9.
- Building construction and teaching commenced at Chanakya Law University.
- Establishment of Nalanda International University; under process, Chandragupt Institute of Management, and Aryabhat Professional University set up.
- Classes commenced at Indian Institute of Technology, Patna and National Institute of Fashion Technology, Patna.

Women's Welfare

- "Mukhyamantri Kanya Vivah Yojana" inaugurated by President of India to provide monetary benefits to women of poor families at the time of marriage.
- "Kanya Suraksha Yojana" to stop female foeticide, and encourage birth of female children.
- "Mukhyamantri Nari Shakti Yojana" for the social, economic and cultural empowerment of women.
- "Laxmibai Samajik Suraksha Yojana" a pension scheme for widows and women's empowerment.

Road and Bridge Construction

In the year 2007-08 following important bridge works completed:

- Khuri river bridge (Nawada district): Rs. 551.84 lakhs.
- Jhauwa bridge (Katihar district): Rs. 216.30 lakhs.
- Dubha Ghat bridge (Sitamarhi district): Rs. 2039.84 lakhs.
- Ram Dhira bridge (Bhojpur district): Rs. 232.16 lakhs.
- Under Chief Minister's Bridge Construction Scheme 130 bridges completed at a cost of 116 crore rupees in 2007-08.
- 1097 kms roads completed from state budget, 465.66 km roads under progress, 354.17 crores rupees utilized.
- Plan for upgradation of all state highways to be two lane highways.
- All district roads to be upgraded to intermediate lane (5.50 meters).

Perspectives

As discussed above, Bihar Government has already embarked on an ambitious progrmme of agriculture and social as well as rural infrastructure development. They would still need to put a higher priority to achieve the desired results. Applying new technology—Zero Till, Bed Planting for wheat, System for Rice Intensification (SRI) Technique for rice, Site Specific Nutrient Management for rice and wheat and hybrids for maize, encouraging results have been achieved within Bihar itself on farmers' fields. To popularize all these new researches among the farmer, Rajindra Agriculture University at Pusa has then to work overtime.

We have discussed in Chapter 4 (Table 4.7), that net and gross irrigated area by 2011-12 and 2020-21 will be around 6000 and

8000 thousand hect., respectively. A projected land use pattern can then be as in Table 14.9

Table 14.9: Basic Changes in Land Use

(Area in'000 hect.)

Year	Net Irrigated	Gross Irrigated	Net Sown	Gross Sown	Cropping Intensity
2004-05	3433	4197	5712	7802	137
2010-12	4500	6500	5712	9000	158
2020-21	6000	8000	5712	11000	192

With regard to irrigated area, it assumes full utilization of the potential already created—7.4 million hect., (Table 4.5) and only 0.6 million hect, of additional area to be brought under irrigation under MSW scheme. This would mean a gross irrigated area of 8 million hect., by 2020-21—an increase of 90 per cent in a period of 16 years. The state had a cropping intensity of 140 when gross irrigated area was only 4300 thousand hect. in recent years. With 11 million hect,, as gross cultivated area, cropping intensity will go up to 190.

Estimated gross cultivated area of 11 million hect., of which nearly 73 per cent will be irrigated, is taken as the base. Future cropping pattern has been then attempted as in Table 14.10. Under the present skewed distribution of cropping pattern in favour of rice and wheat, there will be the need for crop diversification for high productivity crops. It is, therefore, important that a major part of the additional area available is allocated to crops like maize and oilseeds, particularly when it is uneconomic to produce rice and wheat at least in lower yielding areas. Water requirements for maize, gram and the two oilseeds are also much lower than rice and wheat. There is no shortage of water in Bihar, but the cost of drawing water from the ground is very high because of power problems.

Table 14.10: Broad Cropping Pattern 2020-21

(Million hect.)

	2006-07	2011-12	2020-21
Gross Cropped Area	7.9	9.0	11.0
Major Crops, rice, wheat and maize	6.31	6.63	7.0
Balance	1.59	2.37	3.3
Fruits and Vegetables	1.10	1.50	2.2
Miscellaneous Minor Crops	0.49	0.87	1.8

It is noted that there are wide fluctuations in the area and production of rice, wheat and maize—the three major cereal crops in the State. These variations are primarily due to seasonal factors. Looking at the data for the last 7-8 years, we observe that area under wheat has been more or less constant at 2.0 million hect., rice has been declining and we have taken the base as 3.4 million hect. Other minor crops are gram, rapeseed and mustard, sunflower and sugarcane which have been given high priority. The idea is to allocate more area to horticulture floriculture, medicinal plants and crops like maize, gram and oilseeds, etc. where Bihar already has superiority over other States. All these changes involving re-allocation of area to different crops will require comprehensive studies at the district level.

Taking care of the development initiatives already taken, along with other local issues discussed above, the exercise gives us an additional production of 6.7 million tonnes of wheat, 6.7 million tonnes of rice, 7.25 million tonnes of maize, over 5 lakh tonnes of gram and over a million tonnes of oilseeds along with 14 million tonnes of sugarcane (Table 14.11). It is assumed that by 2020-21, Bihar would be able to overcome most of its short/long term problems and even overshoot Punjab in agricultural productivity. With doubling the area under fruits and vegetables and improving productivity by at least 50 per cent, Bihar can take a place of pride among the fruit and vegetable producing States (Table 14.12). The Government of India has a target of 10 million tonnes of rice, 8 million tonnes of wheat and 2 million tonnes of pulses under the Eleventh Plan (National Food Security Mission). The Government of Bihar should take full advantage of the funds allocated under the scheme for the purchase of quality seed, fertilizer and minikit subsidies etc. and can thus become a major partner in this programme.

Table 14.11: Crop Potential in Bihar

Crop	*Area (million hect.)*			*Yield/Kg./Hect.*			*Production (Million tonnes)*		
	2005-06	*2011-12*	*2020-21*	*2005-06*	*2011-12*	*2020-21*	*2005-06*	*2011-12*	*2020-21*
Wheat	2.0	2.0	2.5	1610	2700	4000	3.3	5.4	10.0
Rice	3.4	3.0	3.0	1523	2400	4000	5.3	7.2	12.0
Maize	0.65	1.0	1.5	2390	3750	5500	1.5	3.8	8.25
Gram	0.06	0.2	0.5	1000	1200	1500	0.06	0.24	0.75
Rapeseed & Mustard	0.08	0.2	0.5	926	1100	1300	0.08	0.22	0.65
Sunflower	0.02	0.08	0.4	1345	1500	2000	0.04	0.12	0.80
Sugarcane	0.1	0.2	0.3	41370	45000	6000	4.1	9.0	18.0

Table 14.12: Fruits and Vegetable Perspectives

(*Area thousand hect. production '000 tonnes*)

	Fruits		Vegetables	
	Area	Production	Area	Production
2004-05	300	3400	8.16	13,349.0
2020-21	550	7000	1.65	30,000.0

Graph 14.4: Crop Potential in Bihar

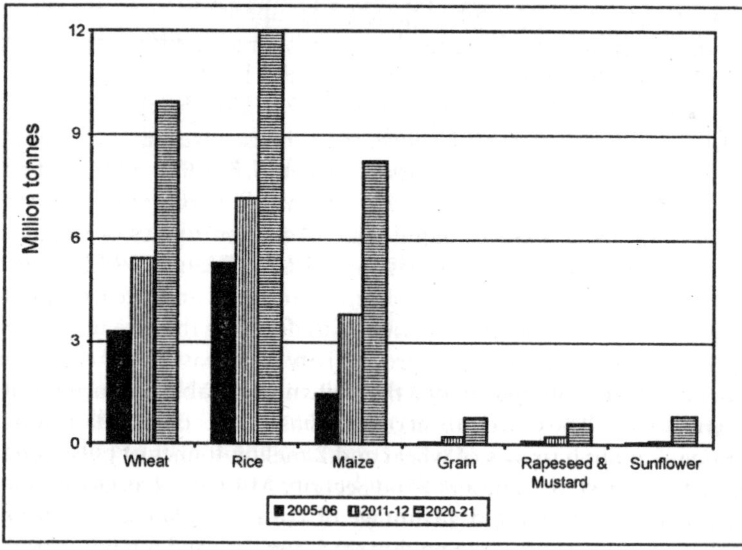

Micro level studies at the District level could then be taken up to work out a revised cropping pattern so that Bihar can contribute maximum towards the food requirements of the county. This approach will call for micro level studies to workout most suitable cropping patterns.

Some of the important districts for such a study are:

Crop	*Districts*
Rice and wheat	Bhojpur and Rohtas
Maize	Banka, Khagaria and Lakhisarai
Sunflower	Khagaria, Patna and Buxar
Rapeseed.	Khagaria and Begusarai
Gram	Bhojpur and Buxar
Sugarcane	Begusarai

Livestock

As for livestock, we have discussed in the livestock chapter that the State Government is alive to the constraints faced by this sector. Veterinary institutional infrastructure in being developed, breeding programmes are being given high priority and growth factors for dairying poultry, fisheries as well as apiculture, have been taken up on high priority basis. Village poultry is awaiting for a complete revolution with the new researches by ICAR and Kegg Farms. All these livestock products have a high growth rate. While dairy and honey projections are more or less based on trend line approach, poultry and fisheries targets are to reach near self sufficiency in the two sub-sectors. (Table 14.13). Future development of the livestock sector must follow the successful experience of dairy cooperatives and extend the system to poultry, psciculture, as well as apiculture.

Feminisation of Agriculture

The perspective Plan for Bihar will not be complete, if it does not take into consideration welfare of the large number of women involved in various agricultural activities. The National Rural Employment Guarantee Programme (NREGP) should, for example, include activities like running crèches, preparing school meals as well as immunization and family planning work, keeping in mind the large number of women in rural jobs. The emerging trend of feminisation of agriculture needed to be addressed through gender-specific needs, such as setting up a Gram Panchayat Mahila Fund as well as strengthening the role of women in the national horticulture mission. According to the 2004-05 census, women constituted 34 per cent of the principal rural workforce and 89 per cent of the subsidiary strength in agriculture. With adverse terms of trade and the extremely low-income in the farm sector, there is the need for integrated off-farm and non-farm employment generation through Government programmes which would establish rural business hubs on the lines of China's Township and Village Enterprises. Bihar, eastern UP, Chhattisgarh, Orissa, West Bengal and Assam have the potential of emerging as the fertile-crescent—if package of technology, services and market opportunities were introduced.

Table 14.13: Livestock Potential—Bihar

	Milk ('000, tonnes)	*Eggs (lacs no.)*	*Meat ('00 tonnes)*	*Fish ('00 tonnes)*	*Honey ('00 tonnes)*
2001-02	2632	7400	1560	2404	
2003	2869	7400	1730	2610	
2004	3175	7800	1730	2665	
2005	4743	7900	1760	2675	
2006	5060	10000	1750	2700	100
2011-12	9000	13500	2000	3500	200
2020-21	17000	21000	4500	6000	300

Note: The Graph presents the growth pattern of each commodity from year to year.

Graph 14.5: Livestock Potential in Bihar

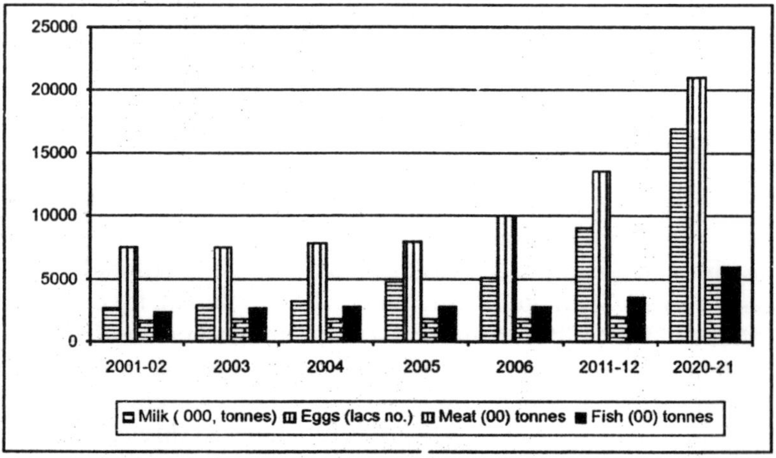

Note: The Graph presents the growth pattern of each commodity from year to year.

Poverty and Agricultural Development

Agricultural development, particularly involving small/marginal farmers and landless labourers and poverty are directly related. In a state like Bihar where poverty is rampant, agricultural perspectives and poverty reduction go hand in hand. The whole strategy of creating surplus value in agriculture rests on synergizing the three crucial resources—people, land and water. This is to be done through a catalyzing combination of research, technology, production inputs, free pricing and marketing options, supplemental non-farming activities and public investment in rural infrastructure. Intensification, diversification and value addition are the key directional goals. Exploiting micro-irrigation potential and liberating the agricultural markets are thus the two leading thrust areas in this exercise. If only about Rs. 5,000 to 10,000 could be added to the annual income of the population hovering around the poverty line, it would substantially reduce the proportion of those living below the poverty line.

The key to reducing inequality is not charity or philanthropy but innovation and entrepreneurship. Poverty and inequity are often worst in rural India. This would call for another Green Revolution that will drive the next wave of agricultural productivity growth and Bihar is possibly one of those states which can accept this challenge. The biggest hope here may be the rural business initiatives of companies like Reliance industries, Bharti Airtel or ITC, investing billions of dollars to create an efficient agricultural supply chain, bypassing scores of middlemen. Or the rural banking initiatives of the likes of SBI, Citibank and ICICI that will deliver affordable credit and insurance, helping them break a historic cycle of poverty.

Graphs 14.6 and 14.7 present the relationship of agricultural growth and poverty in China and India. The sharp decline in Poverty in China from 1981 to 1985 was spurred by agricultural reforms that started in 1978.

Graph 14.6: Urban and Rural Poverty in India

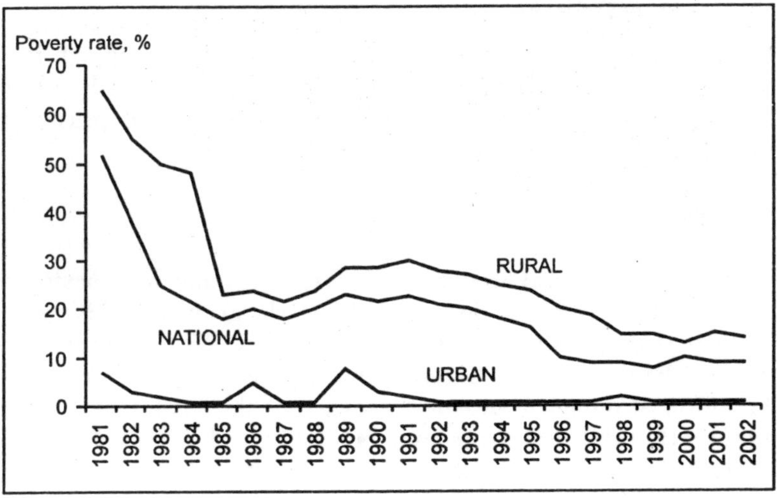

Graph 14.7: Urban and Rural Poverty in India

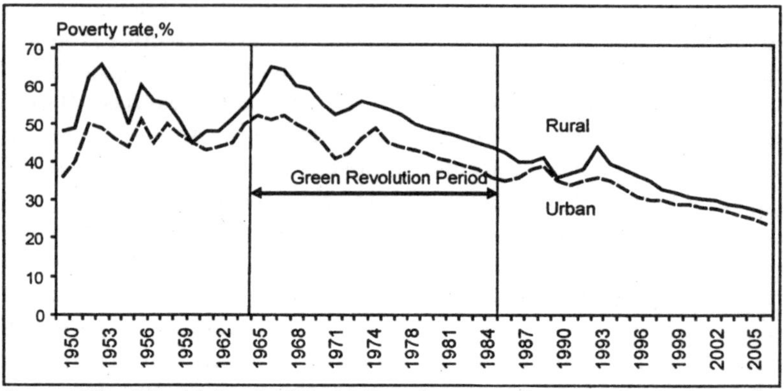

Source: The World Bank, World Development Report, Agriculture for Development.

In the 1960s and 1970s the introduction of semi-dwarf varieties of wheat and rice-in the green revolution—led to dramatic leaps in agricultural production and raised farmers' incomes, especially in north-west India. Rural poverty fell from 64 per cent in 1986.

A great deal of innovation, entrepreneurship and creative public-private partnerships will be required to provide affordable access to essential services to those at the bottom of the pyramid services and schemes. Fortunately, more and more people are waking up to the idea that social commitment is also good business. As such, salable business models that help the poor access markets or essential services represent huge economic opportunities waiting to be tapped. It is these and various other programmes discussed in the study which have to be implemented to raise the living standards of poor peasantry so as to give a boost to agriculture and *vice versa*.

Epilogue

In view of the critical analysis of the strengths and weaknesses in the agriculture sector as discussed above, the growth of this sector in the past at snails pace in Bihar is a paradox. We have therefore, tried to identify the factors responsible for this dismal performance. Adoption of remedial measures for each of the weaknesses in the sector is thus the need of the hour. No doubt, technological factors, in general, are closely associated with the improvement of output per hectare. Cropping intensity, fertilizer use per hectare of net sown area and gross irrigated area as percentage of gross sown area are always positively correlated with output per hectare. But in Bihar, despite creation of huge irrigation potential, a sizable portion has remained unutilized. Irrigation infrastructure which is required for proper usage of water has also not been created in many areas. Further, there is wide disparity in the utilization of various inputs from one district to another and even within a district, from one region to another. Use of fertilizer per unit of area is also one of the poorest in the state.

Proper land reform is a prerequisite for implementation of agricultural development programmes. Fragmented land holdings which we have cited as one of the weaknesses in the state, inhibit adoption of yield raising technologies and usage of labour saving and cost reducing implements. Consolidation of land holding has remained out of the agenda of land reforms. Further, high pressure of population on small and marginal holdings leads to further fragmentation and thereby making the holdings severely uneconomic. This has a cascading effect on the entire economy of the State. This huge surplus agriculture manpower fails to switch over to any alternative vocation and therefore face perpetual distress condition.

In our analysis we have mentioned about immense human resources as one of the strengths of the agricultural economy of the state. At the same time it is also a burden on the national economy as overwhelming majority of the population is not only illiterate but also underfed and diseased. Modernization of agriculture needs a manpower which is capable of adopting latest technologies on their farms. This can be achieved only when they possess the minimum education required for understanding the efficacies of the technologies. Moreover, knowledge empowers them in trading their products in a better way and thereby enhancing their farm income. It is heartening to know about the new initiatives taken by the State Government in the field of infrastructure and education. Here comes the role of the extension staff which is not only inadequate, but also ineffective. The state needs a dedicated extension body for imparting necessary knowledge to the farmers in the state.

One of the major weaknesses that have been identified in the state is low profitability and instability in agriculture. This is mainly due to low productivity of most of the field crops except, of course, maize, gram and sunflower. A number of factors are found to be responsible for low productivity of the crops. In addition to non-availability of various inputs required for achieving high yields, annual floods and droughts in a number of districts damage standing crops extensively. There is also an urgent need for developing location specific high yielding seeds for which more intensified research effort are required. Further increase in farm profitability may be possible by popularization of horticulture, floriculture, herbiculture and apiculture for which there is immense scope.

In fine, agricultural development in the state as well as in the country heavily relies on government finance due to inherent weaknesses in the system which we have already discussed. These discourage investment from private sources. However, in recent years, there has been a decline in public investment due to burgeoning subsidies. It is well known that agricultural research and education attracts very meager public resources. Therefore, public private partnership (PPP) is necessary for development in agriculture in this State which has fallen much behind many other states of India. The State Government, responsive as it is, has called for new initiatives from the private sector in precision farming and contract management of farms, which must be linked to both domestic and external markets through agricultural processing zone, stimulating floriculture and horticulture. Rationalizing public expenditure and shifting investment to key sectors such as rural roads, markets and agricultural research and extension my go a long way in achieving a turnaround in the state's agriculture.

With the initiatives already in hand with regard to land reforms, hydro power education, etc., major weakness will be taken care of. If so, we can say with confidence that Bihar will not be the leader of the second Green Revolution, but also rewrite history to retrieve its ancient glory.

Note

1. *Hindustan Times,* September 6, 2007.

BIBLIOGRAPHY

ATMA (2001) Strategic Research and Extension Plan of Patna District.

ATMA (2002) Strategic Research and Extension Plan of Muzaffarpur District.

ATMA (2003) Strategic Research and Extension Plan of Madhubani District.

ATMA (undated) Makhana – Mithilanchal ki ek Pahchan. ATMA Madhubani.

Bharti, Indu (1993), *Potential for Surplus Land in Bihar and the Ground Realities, Land Reforms in India* (Bihar), Sage Publications, New Delhi.

Bihar Industries Commission (2001). "Industrial Development Imperative for Bihar and Jharkhand – A Report "Thomson Press India.

Bihar Institute of Economic Studies. (1999). "Dynamic of Marketing of Selected Fruits in Bihar". Ministry of Agriculture, Sponsored by the Government of India.

Bihar State Co-operative Milk Producers' Federation Ltd. (2204). "*Resume of Dairy Development Activities in Bihar*" COMFED, India.

Bihar: Towards a Development Strategy", (2005) Washington D.C.: World Bank.

Birthal, P.S., P.K. Joshi and Anjani Kumar (2002), Assessment of Research Priorities for Livestock Sector in India Policy paper 15, National Centre for Agricultural Economics and Policy Research, New Delhi.

Census of India 2001, Series 1, India, Provisional Population Totals, Paper 1 of 2001, Delhi : Controller of Publications.

Census of India 2001, Series 1, India, Provisional Population Totals,

Paper 1 of 2001, Supplement – District Totals. Delhi : Controller of Publications.

Census of India 2001, Series 11, Bihar, Provisional Population Totals, Paper 1 of 2001, Delhi : Controller of Publications.

Census of India 2001, Series 11, Bihar, Provisional Population Totals, Paper 2 of 2001, Rural-Urban Distribution of Population. Delhi: Controller of Publications.

Centre for Monitoring Indian Economy (CMIE). (Various editions 2000-04). India.

Chand, Ramesh (2001), Regional and Sectoral Disparities in Economic Growth in S.S. Acharya and D.P. Chaudhary (eds.) *Indian Agricultural Policy at the Crossroads,* Rawat Publications, Jaipur and New Delhi, pp. 589-602.

Chand, Ramesh, Praduman Kumar and Sapna Sinha (2003), Impact of Agricultural Trade and Related Reforms on Domestic Food Security in India, Preliminary Draft, Institute of Economic Growth (mimeo).

Chandramohan, C., Chitranjan and Sharat Kumar (2001), Dynamics of Foodgrains production in Sharat Kumar and Praveen Jha (eds) *Development of Bihar and Jharkhand – Problems and Prospects.* Pp. 92-118 Shipra Publication, New Delhi.

CMIE (2006) Agriculture-Economic Intelligence Service. March 2006

CMIE (2006) Infrastructure-Economic Intelligence Service, May 2006.

Consultants Report, (2006). "Seeking Success: an examination of selected agri-food chains in Bihar".

"Dairy Developments in Bihar" (1997). *The Bihar Journal of Agricultural Marketing*.Vol.V. No.1 India.

Datta, S.K. and Milindo Chakrabarti (1999), A Perspective on Land Use Planning for Eastern India, in T.Haque (ed) *Land Use Planning in India,* Pp 46-89 National Centre for Agricultural Economics and Policy Research, New Delhi.

Deaton, Angus and Jean Dreze (2002), "Poverty and Inequality in India : A Re-Examination", *Economic and Political Weekly*, Vol. XXX VII, No.36.

Department of Agriculture and Co-peration. (Various editions). "Reports of The Commission For Agricultural Costs and Prices: For the Crops Sown During 2002-2003 Season". New Delhi, India.

Department of Industries (2006) Industrial Policy 2006. Bihar.

Department of Industries (Undated) Investment Opportunities in Agriculture Sector.

Department of Industries (Undated) Investment Opportunities in Industrial Sector.

Directorate of Statistics and Evaluation, Bihar (2004) Bihar at a Glance 2004.

Fan, S., Peter Hazell and S. Thorat (1999), Linkages Between Government Spending, Growth and Poverty in Rural India, Research Report 110, IFPRI, Washington.

Fan, S.,L. Zhang, and X. Zhang, (2002), "Growth, Inequality and Poverty in Rural China : The Role of Public Investments," Research Report 125, International Food Policy Research Institute, Washington, D.C.

"Fish Production in North-Bihar; (2005). An Economic Analysis". *Indian Journal of Agricultural Marketing*. Vol. 14. No.3. December 2003, India.

Giri, A.K. (2006) Cereal Consumption Over Time in the Country Across the States , *Indian Journal of Agricultural Economics*, Vol. 61, No. 3

GOB (1991), Draft Eighth Five Year Plan 1992-97, Planning Department, Government of Bihar, Secretarial Press Bihar, Patna.

Government of Bihar, 1994. The Report of the Second Irrigation Commission, 1994.

Government of Bihar, 1997. The Bihar Irrigation Act, 1997.

Government of Bihar, 2000. Guidelines for Implementing Participatory Irrigation Management. Water Resources Department, Patna.

Government of Bihar, 2003. Bihar Irrigation and Drainage Rules, 2003.

Government of Bihar, 2004. Administrative Report, Year 2003-04, Water Resources Department, Patna.

Government of Bihar, 2005. Status paper on Participatory Irrigation Management: "Bihar Model". Water Resources Department, Patna.

Government of Bihar, Appropriate Eleventh Five Year Plan.

Government of Bihar, Department of Agriculture (Special Programme).

Government of Bihar, Department of Agriculture Census.

Government of Bihar, Department of Agriculture.

Government of Bihar, Department of Animal Husbandry and Fisheries.

Government of Bihar, Department of Co-operative.

Government of Bihar, Department of Disaster Management.

Government of Bihar, Department of Water Resources.

Government of Bihar, Economic Survey.

Government of Bihar, Revenue Department.

Government of Bihar, State Tubewells Minor Irrigation.

Government of Bihar, Status paper on Participatory Irrigation Management: Water Resources Department.

Government of India, 1992. Report of the Committee on Pricing of Irrigation Water, Planning Commission, New Delhi (Chairman : A. Vaidyanathan).

Government of India, Bihar Agriculture Development : Opportunities and Challenges; 2008.

Government of India, Planning Commission.

Government of India. National Sample Survey Organisation. Ministry of Statistics and Programme Implementation. (2005). Indebtedness of Farmer Households : NSS 59th Round (January – December 2003). Report No.498 (59/33/1).

Haque, T (2000), Contractual Arrangements in Land and Labour Markets in Rural India, *Indian Journal of Agricultural Economics*, Vol. 55 (4) 233-252.

"Impact of ATMA Model in *Agricultural Extension System in Bihar (date), – A Case Study of Pilot Project Districts*". Rajendra Agricultural University, Bihar. India.

Indian Horticulture Database – 2002-03 and 2004-05, National Horticulture Board. www.hortibizindia.org.

Indian Institute of Management, Lucknow, (2005), "Impact Assessment Report", National Agricultural Technology Project: Innovations in Technology Dissemination Component.

Institute for Human Development, 2006, Bihar Development Report.

Institute for Human Development. Planning Commission.(2004), Bihar Development Report.

Jha T.N. and K. U. Viswanathan (1999), Problems and Prospects of Agricultural Development in Bihar, Occasional Paper-10, National Bank for Agricultural and Rural Development, Mumbai.

Joshi, P.K., Singh, N.P., Singh, N.N., Gerpacio, R.V., and Pingali, P.L. (2005). "Maize in India Production Systems, Constraints, and Research Priorities". *International Maize and Wheat Improvement* (centre (CIMMYT). India.

Kumar, A (2005) A Presentation of Prospects and Possibilities for Agro-processing and Food Processing Industries in Bihar. Development of Commercial Agriculture and Food Processing Industries in Bihar. Two days Seminar jointly organized by Bihar Industries Association, Patna.

Kumar, A., Haridwar, Kant, K. (2005). "Some Aspects of PoSst-Conflict Face of Poverty: A study of 4 Villages of Musahari Block, Muzaffarpur". Jawaharlal Nehru University, New Delhi, Draft Paper presented at CPRC-IIPA Seminar On 'Chronic Poverty: Emerging Policy Options and Issues".

Kumar, Alok (2001), Agricultural Development of Bihar, in K.P Kaliranjan, G. Mythili and U. Sankar (eds.) *Accelerating Growth Through Globalisation of Indian Agriculture*, Macmillan India Ltd., New Delhi. pp, 167-195.

Kumar, Anjani (2000), Performance of Livestock Sector in India: Implications for Food Security", *Afro-Asian Journal of Rural Development* : 33 (2): 77-9.

Kumar, N.R. and N.K. Pandey, "Indian Institute of Vegetable Research", Varanasi , Uttar Pradesh 'Division of Social Science, Central Potato Research Institute, Shimla, Himachal Pradesh, Constraints in Adoption of Improved Potato Production Technologies in Bihar, *Agriculture Economic Research Review*, Volume. 21, 2008.

Lanjouw, P. and Shariff, A., (2002). "Rural Non-Farm Employment in India: Access, Income and Poverty Impact" National Council of Applied Economic Research (NCAER). Working Paper Series No. 81, India.

M.Anjugam, S.Varadha Raj and S. Padmarani, (2008), Department of Agricultural Economics, Tamil Nadu Agriculture University, Cost Benefit Analysis of SRI Technique in Paddy Cultivation, Agricultural Economics Research Review.

Mathur, Ashok and Mamgain, Rajendra R. (2004) " Human Capital Stocks, their Level of Utilisation and Economic Development in India", the *Indian Journal of Labour Economics*, Vol. 47, No.4.

Mattoo, A., Mishra, D. and Narain, A. (2005). "From Competition at Home to Competing Abroad: A case study of India's Horticulture. " Poverty Reduction and Economic Management Sector Unit, South Asia Region. The World Bank. Washington D.C.

Minor Irrigation Census, Ministry of Water Resources, Government of India.

Mishra, R. K, Jha, Vaidehi and Dehadrai, P.V. (ed.) (Undated) Makhana. ICAR, New Delhi.

Mysore, Sudha (Undated) Socio-Economic Dimensions in Conservation and Use of Native Tropical Fruit Species Biodiversity in Asia – A case of Litchi in India, *Indian Institute of Horticultural Research*, ICAR, Bangalore.

NABARD (National Bank for Agriculture and Rural Development) (2006) State Focus Paper (2006-07).

NABARD (National Bank for Agriculture and Rural Development) (2006) Potential Linked Credit Plan (2006-07).

National Horticulture Board (2006) Indian Horticulture Database 2005.

National Sample Survey Organisation, (2005), "Access to Modern Technology for Farming," NSS 59th Round. Ministry of Statistics and Programme Implementation, Government of India, New Delhi.

National Sample Survey Organisation, Ministry of Statistics and Programme Implementation, (2005). "Some Aspects of Farming: NSS 59th Round" (January-December 2003). Report No. 496 (59/33/3).

National Sample Survey Organisation. Ministry of Statistics and Programme Implementation (2005). "Access to Modern Technology for Farming : NSS 59th Round" (January-December 2003), Report No. 499 (59/33/2).

National Sample Survey Organisation. Ministry of Statistics and Programme Implementation. (2006). Livestock Ownership Across Operational Land Holding Classes in India, 2002-03: NSS 59th round (January – December 2003). Report No.493 (59/18,1/1).

NFHS II (1998-99), National Family Health Survey: Bihar, International Institute of Population Science, Mumbai, India.

Pal, Suresh, S. Pandey and Abedullah (2000), Growth and Variability in agriculture revisited : District-level evidence of rice production in eastern India, in S. Pandey, B.C. Barah, R.A. Villano, and S. Pal (eds.) *Conference Proceedings on Risk Analysis and Management in Rainfed Rice System.*

Parsad, Jagdish (1999). Production and Marketing of Vegetables in Bihar. Study Report. A.N. Sinha Institute of Social Studies, Patna.

Planning Commission (2001), Poverty in India, Press Release, Planning Commission, Government of India, New Delhi.

Prakash, Vijoy, 2001, A note on Financial Management in Universities in (ed.) Bhushan Sudhanshu, Resource Mobilisation, Strategies and Policies in Educational Sector, SIEMAT, Patna.

Rai, P.K (2004) Litchi – production technology & Post harvest handling for Export. Rajendra Agricultural University, Pusa.

Rajendra Agricultural University (2005) Recent Maize production technology for Bihar. Development of Commercial Agriculture and Food Processing Industries in Bihar. Two days Seminar jointly organized by Bihar Industries Association, Patna.

Rajendra Agricultural University (2006 B) 3 Decades of Achievement.

Rajendra Krishi Vishva Vidyalaya (2006 A) Adhunik Kissan Dairy 2006.

Raju, K.V., 2006. *Participatory Irrigation Management* in Bihar—Background Paper.

RBI, Report of the Committee on Agricultural Productivity in Eastern India, Vol. 1, 1984.

Registrar General 1996, SRS Analytical Studies, Report No. 3 of 1996, Below State Level Estimates of Vital Rates, 1987-92, New Delhi: Registrar General, India.

Registrar General 2000, SRS Bulletin, Volume 34, No. I, April 2000.

Registrar General 2001, SRS Bulletin, Volume 35, No. I, April 2001.

Registrar General, India 1998, Vital Rates of India 1971 to 1996 Based on The Sample Registration System (SRS). Delhi: Controller of Publications.

Registrar General, India 1999, SRS Bulletin, Volume 33, No. I, April 1999.

Registrar General, India SRS Bulletins.

"Report on Mango Development in Darbhanga District of Bihar" (date). Draft. Planning Commission. Nambard Consultancy Services (NABCONS). India.

Rodgers, Gerry and Janine Rodgers (2000), "Semi-feudalism meets the Market: A Report from Purnia", working paper 6, Institute for Human Development, New Delhi.

Roy, B.C. (1997), Growth and prospect of fruit and vegetable Processing Industry in India, *Bihar Journal of Agricultural Marketing*, Vol. 5(3) : 356-65.

Sharan, Ramesh 2001, Vocationalisation of Higher Education in Bihar: Issues and Basic constraints in (ed.) Sudhanshu Bhusan.

Shariff, Abusaleh (1999), India Human Development Report : A Profile of Indian States in the 1990s, NCAER, New Delhi.

Sharma, A.N., Karan, A.K. Sarkar, S. 2002, Dynamics of Employment, Poverty and Human Development in Rural Bihar, Institute for Human Development, New Delhi, (mimeo).

Sharma, Alakh N. (2001), Poverty, Employment and Human Development in Bihar paper presented in a Seminar on Bihar's Strategic Direction for Economic Development during 23-24 May, 2001.

Sharma, Alakh N., Anup K. Karan, Sandip Sarkar (2002), Dynamics of Poverty, Employment and Human Development in Bihar, Phase II, (mimeo), IHD, New Delhi.

Sharma, Alakh N., Anup K. Karan, Sandip Sarkar, V. Gayatrhi, and Pushpendra (1999), Dynamics of Poverty, Employment and Human Development in Bihar, Phase I, (mimeo), IHD, New Delhi.

Sharma, V.J. and Gulati, A (2003). Trade Liberalisation, Market Reforms and Competitiveness of Indian Dairy Sector". Markets, Trade and Institutions Division. International Food Policy Research Institute (IFPRI). http://www.ifpri.org..

Sheila, Bhalla (2005) "India's Rural Economy: Issues and Evidence", Working Paper Series No.25, Institute for Human Development, New Delhi, Sept.

Singh, K.M., (2006), "Impact of ATMA Model in Agricultural Extension System in Bihar – A case Study of Pilot Project Districts", Rajendra Agricultural University, Pusa, Bihar.

Singh, K.M., Swanson, B.E. and Singh, J.P. (2005). "Development of Supply Chains for Medicinal Plants: A Case Study Involving the Production of *Vince Rosa* by Small Farmers in the Patna District of Bihar'. India http://ww.globalfoodchainpartnerships.org/chicago/presentations/

Singh, K.M. Medicinal Crops India, Financed by the Government of India and the World Bank.

Singh, Manjit and Anup K. Karan, (2002) Rural Labour Migration from Bihar, Institute for Human Development, New Delhi (mimeo).

Singh, R.K.P. (1996). "Rediscovering Co-operation: Dairy Co-operative: Organisations of the Poor in Bihar". Institute of Rural Management Anand (IRMA). India.

Singh, R.K.P., Prasad, A., and Waris, M.L., (1997). Performance of Dairy Co-operatives in Bihar", Vol.V No.4. *The Bihar Journal of Agricultural Marketing*, India.

Singh, R.K.P., Prasad, K.K., and Chaudhary, A.K. (2000) " Role of Co-operative in Milk Production in Bihar", Co-operative Perspective, Vol. 35, No.3. Department of Agricultural Economics. Rajendra Agricultural University Bihar, India.

Singh, R.K.P., Prasad, K.K., and Chaudhary, A.K. (2000) "Impact of Dairy Co-operatives on Farmers Economy in Bihar", *Indian Journal of Agricultural Marketing* Vol.8(4); 372-382. Department of Agricultural Economics. Rajendra Agricultural University Bihar, India.

Singh, R.K.P., Ranjan, K.P.,and Jha, A. (1996) "*Declining Milk Availability and Nutrition Security in Bihar*", Vol.9 (2). Department of Agricultural Economics. Rajendra Agricultural University Bihar, India.

Singh, R.K.P., Yadav, R.N., and Chaudhary, A.K. (1998) "Farmers' Participation in Management of Dairy Co-operatives in Bihar", *Indian Journal of Agricultural Marketing* (Con.Spl.) 12 (3). Department of Agricultural Economics. Rajendra Agricultural University Bihar, India.

Singh, RKP (2001), Economic Analysis of total factor productivity of irrigated agriculture by district and region, Technical report submitted to National Agricultural Technology Project (NATP).

Singh, RKP and K.K Prasad (2001), Fish production in north Bihar : An Economic Analysis, *Agricultural Economics Research Review* Vol.14 (1) : 65 -73.

Srivastava, L.P. and Jeffrey, D.Brewer, 1994. "Irrigation Management Transfer at Paliganj Canal, Bihar, India". Short Report Series on Locally Managed Irrigation, No.7, IWMI, Colombo.

State Focus Paper on Bihar (2001), National Bank for Agricultural and Rural Development, Mumbai.

"Stocking Density and Species Mix in Composite Fish Culture in North Bihar: (2005). A Techno -Economic Analysis" *Fisheries Economics and Development*. Vol. VI, No. 1 January-June.

World Bank, 2001, World Development Report 2000/2001, Attacking Poverty, Oxford University Press, New York.

INDEX